Advances in Multiphase Flow and Heat Transfer

Vol. 3, 2012

Edited By

Lixin Cheng

School of Engineering
University of Portsmouth
United Kingdom

Dieter Mewes

Institute of Multiphase Process
Leibnia University of Hanover
Germany

CONTENTS

PREFACE

Multiphase flow and heat transfer have been found a wide range of applications in nearly all aspects of engineering and science fields such as mechanical engineering, chemical and petrochemical engineering, nuclear engineering, energy engineering, material engineering, ocean engineering, mineral engineering, electronics and micro-electronics engineering, information technology, space technology, micro- and nano-technologies, bio-medical and life science *etc*. With the rapid development of various relevant technologies, the research of multiphase flow and heat transfer is growing very fast nowadays than ever before. It is highly the time to provide a vehicle to present the state-of-the-art knowledge and research in this very active field.

To facilitate the exchange and dissemination of original research results and state-of-the-art reviews pertaining to multiphase flow and heat transfer efficiently, we have proposed the eBook series entitled ***Advances in Multiphase Flow and Heat Transfer*** to present state-of-the-art reviews/technical research work in all aspects of multiphase flow and heat transfer fields by inviting renowned scientists and researchers to contribute chapters in their respective research interests. The eBook series have now been launched and two volumes have been planned to be published per year since 2009. The eBooks provide a forum specially for publishing these important topics and the relevant interdisciplinary research topics in fundamental and applied research of multiphase flow and heat transfer. The topics include multiphase transport phenomena including gas-liquid, liquid-solid, gas-solid and gas-liquid-solid flows, phase change processes such as flow boiling, pool boiling, and condensation etc, nuclear thermal hydraulics, fluidization, mass transfer, bubble and drop dynamics, particle flow interactions, cavitation phenomena, numerical methods, experimental techniques, multiphase flow equipment such as multiphase pumps, mixers and separators etc, combustion processes, environmental protection and pollution control, phase change materials and their applications, macro-scale and micro-scale transport phenomena, micro- and nano-fluidics, micro-gravity multiphase flow and heat transfer, energy engineering, renewable energy, electronic chips cooling, data-centre cooling, fuel cell, multiphase flow and heat transfer in biological and life engineering and science *etc*. The eBook series do not only present advances in conventional research topics but also in new and interdisciplinary research fields. Thus, frontiers of the interesting research topics in a wide range of engineering and science areas are timely presented to readers.

In volume **3**, there are nine chapters on various relevant topics. Chapter **1** deals with numerical and experimental studies on multiphase flow in stirred tanks. Multiphase flow in a stirred tank was addressed both experimentally and with numerical simulation. The flow field and the distribution of the dispersed phase were regarded in most sections. The Eulerian multi-fluid approach and k-ε turbulence model were employed in numerical simulations. The good agreement found between experimental and predicted results underlines the importance of computational fluid dynamics in future research of multiphase phenomena in stirred tanks. Besides, some novel surface aeration configurations have been provided, which make such operations more reliable and economical. Furthermore, numerical simulation was applied in research of macro- and micro-mixing in stirred tanks. To account for the anisotropy of turbulence in the flow field, the algebraic stress model and large eddy simulation were adopted in the numerical study of hydrodynamics in multiphase stirred tanks as a result of their capacity on retrieving turbulent properties more accurately.

Chapter **2** presents a study on two-phase flow pressure change across sudden contraction and expansion in small channels. The flow of two-phase mixtures across sudden expansions and contractions is relevant in many applications such as chemical reactors, power generation units, oil wells and petrochemical plants. As the two-phase mixture flows through the sudden area changes, the flow might form a separation region at the sharp corner and introduce an irreversible pressure loss. This loss occurs in practical pipeline connections and in the heat exchangers. The small and narrow channels are widely adopted in compact heat exchangers. Also, flow in small rectangular channels is an integral part of CPU cold plate using the liquid cooling with or without phase change. Predictions of these pressure drops had been made using correlations developed for the conventional tubes, extrapolations of these correlations to small diameter tube are questionable. In this study, the authors first give a short overview on the single-phase flow across sudden contraction and expansion, followed by a thorough review of the relevant literature for two-phase flow across sudden contraction and expansion. The applicability of the existing model/correlations for sudden contraction and expansion is then examined with the available data from literature. Comparisons for the pressure change data with the predictions of existing model/correlations indicate that none of them can accurately predict the data.

Chapter **3** presents a topic on coalescence of drops in liquid. Coalescence of drops is an important process for the destabilization of liquid–liquid dispersions. This chapter presents the mechanism of coalescence of drops in dispersion. It also presents the theories of coalescence that are used at present. The film drainage theory and the stochastic theory of coalescence have been discussed explaining their merits and drawbacks. These models are evaluated with numerical examples. The role of van der Waals, electrostatic double layer, steric and hydration forces on the coalescence process has been discussed. The importance of adsorption of surfactant molecules at the liquid–liquid interface on coalescence time has been explained. Possible reasons behind the failure of the film drainage models in predicting the coalescence time in presence of surfactant have been explained. This chapter also presents the calculation of coalescence time in industrial equipment, experimental techniques employed for studying coalescence of a drop at a flat liquid–liquid interface, binary coalescence, and coalescence of two drops in motion. Directions for future research on coalescence have also been presented.

Chapter **4** presents a review on carbon capture and storage with a focus on capillary trapping as a mechanism to store carbon dioxide in geological porous media. Carbon Capture and Storage (CCS) is a feasible short-to-medium term method to dispose carbon dioxide (CO_2) which would otherwise be emitted into the atmosphere and cause potentially massively damaging climate change. In CCS, CO_2 is captured, compressed and injected deep underground into geological formations. There are four main CO_2 trapping mechanisms, namely stratigraphic or structural trapping, dissolution trapping, capillary trapping and mineral trapping. In this text we discuss all these trapping mechanisms with focus on capillary trapping, which has recently been identified as a rapid and reliable CO_2 storage method.

Chapter **5** presents a discrete particle model for dense gas-solid flows. The discrete particle model (DPM) is a mesoscale method used to study the hydrodynamics of dense dispersed flows. In this approach, the particle motion is described in a Lagrangian framework by directly solving the Newtonian kinetic equations of each individual particle while the gas flow is studied in an Eulerian framework. The constitutive relations for the dispersed phase are not required because the particle-particle interactions are modeled through a two-variant collision-handling algorithm. In this chapter a full understanding of the DPM technique is presented through a detailed description of the numerical model and the results of its applications to gas-solid fluidization systems. A multi-component numerical strategy developed by the authors to enhance the DPM efficiency is also presented.

Chapter **6** presents a detailed experimental study involving flow boiling of water in microchannel. The work aims to study the different aspects of the problem such as pressure drop, heat transfer coefficient, pressure instability, and void fraction. Flow visualization has also been performed. Experiments have been conducted in silicon microchannels with trapezoidal or rectangular cross-section of hydraulic diameter 45-140 μm, and a microheater fabricated on the reverse side of the silicon wafer to provide well controlled and metered input power. The experimental data is compared with the annular flow model and various empirical correlations, and complemented with clear discussion of the observed phenomenon. In the two-phase regime, the average pressure drop increases with a decrease in the flow rate and reaches a maximum (with a minimum on either side), while in the dryout regime the pressure drop decreases with flow rate. The results suggest that there are up to four mass flow rate values with same pressure drop penalty and the operating point can be chosen with the maximum heat transfer coefficient. The average pressure drop is found to have a strong dependence on the channel aspect ratio and becomes minimum in rectangular microchannel for width-to-depth ratio of about 1.5. The instability in pressure drop has been quantified and linked to the underlying flow regime. The minimum P_{RMS}/P_{avg} is found to occur when the flow transitions to annular. For the first time, a flow regime map is also obtained for such systems. The flow is found to be predominately annular at high heat flux and flow rate. A breakup of the flow frequency suggests that the flow is bistable in the annular regime. At a fixed location, the flow periodically switches from single phase liquid to annular, and vice-versa. Otherwise, all three regimes: single phase liquid, bubbly and slug are obtained. An image analysis technique has also been developed and utilized to estimate the void fraction as a function of position in the microchannel, heat flux and mass flow rate. The technique has been extended to obtain heat transfer coefficient purely from image analysis. Both void fraction and heat transfer coefficient are found to increase monotonically with position in the microchannel. There are several novel aspects of this study. For example, effect of microchannel aspect ratio on pressure drop is studied for the first time. Some guidelines for choosing the operating point with desired constraints have been proposed. Ways to reduce instability have also been explored. Development of flow regime map and flow visualization technique is not available in the literature currently. Pressure drop data near CHF condition have been presented. The results are interesting from both fundamental and electronic cooling point of view.

Chapter **7** deals with an experimental investigation on two-phase flow boiling flow pattern, pressure drop and heat transfer of R-22, R-134a, R-410A and natural refrigerants of C_3H_8 and CO_2 in horizontal circular small tubes. The experimental data were obtained over a heat flux range of 5 to 40 kW/m^2, mass flux range of 50 to 600 kg/(m^2·s), saturation temperature range of 0 to 15°C, and quality up to 1.0. The test section was made of stainless steel tubes with inner diameters of 0.5 mm, 1.5 mm and 3.0 mm, and lengths of 330, 1500, 2000 and 3000 mm. The effects of mass flux, heat flux, saturation temperature and inner tube diameter on the pressure drop and heat transfer coefficient are reported in this thesis. The experimental pressure drop and heat transfer coefficient were compared with the predictions given by some existing correlations. New correlations of pressure drop and heat transfer coefficient for small tubes that is based on the present experimental data are developed.

Chapter **8** addresses characteristics of flashing-induced density wave oscillations on the basis of the experimental results in a boiling natural circulation system with an adiabatic chimney. Flashing is caused by the sudden increase of vapor generation due to the reduction in hydrostatic head, since saturation enthalpy changes with pressure. Flashing-induced density wave oscillations may, therefore, occur at low pressure. The oscillation period correlates well with the passing time of bubbles in the chimney section regardless of the system pressure, the heat flux, and the inlet subcooling. According to the stability map, the flow became stable below a certain heat flux regardless of the channel inlet subcooling. The stable region enlarged with increasing system pressure. Therefore, the stability margin becomes larger by pressurizing the loop sufficiently before heating.

Chapter **9** deals with nonlinear dynamic characteristics of bubbling fludization. Nonlinear time series analysis techniques were applied to characterize bubbling fluidization. The experiments were carried out in a laboratory scale fluidized bed, operated under ambient conditions and various sizes of particles, settled bed heights, measurement heights and superficial gas velocities. It was found that a minimum in average cycle frequency, wide band energy and entropy with an increase in the velocity corresponds to the transition between macro structures and finer structures of the fluidization system. This minimum was mostly found in the macro structures of the bubbling fluidization system. Hurst exponent of the pressure fluctuations showed that the fluidized bed has a bifractal behaviour. The reciprocal of the break point in the Hurst profile is similar to the domain frequency of the bed. The method of delays was used to reconstruct the state space attractor to carry out analysis in the reconstructed state space. The state space reconstruction parameters, i.e., time delay and embedding dimension, were determined and the results showed that their values are different for various types of methods introduced in literature. Chaotic behaviour of fluidized system was determined by introducing two nonlinear dynamic invariants, correlation dimension and entropy in different ways. The state-space analysis reflected that a low dimension behaviour of the bubbling fluidization system. The nonlinearity test showed that nonlinearity cannot be concluded at all gas velocities.

As the founding editors of the eBook series, we are very happy to see that the eBooks have been available to our readers since 2009 (two volumes per year). All the chapters have been reviewed by experts in the relevant areas. We are very much grateful to the authors who have contributed to the chapters and the reviewers who took their time and energy for their review. It is our great wishes if the eBook series are able to provide useful knowledge for our community and to facilitate the progress of the research in the field of multiphase flow and heat transfer.

We would like to express our gratitude to our families for their great support to our work on the eBook series.

Editor-in-Chief: Dr. Lixin Cheng

School of Engineering
University of Portsmouth
Anglesea Building, Anglesea Road
Portsmouth, PO1 3DJ
United Kingdom

Co-editor: Prof. Dieter Mewes

Institute of Multiphase Process
Leibniz University of Hanover
Callinstraße 36
D-30167 Hannover
Germany

List of Contributors

Amit Agrawal *Indian Institute of Technology Bombay, India*

A. S. Berrouk *Department of Chemical Engineering, The Petroleum Institute, United Arab Emirates*

R.R. Bhide *Indian Institute of Technology Bombay, India*

Jamal Chaouki *Department of Chemical Engineering, Ecole Polytechnique de Montreal, Canada*

Ing Youn Chen *Mechanical Engineering Department, National Yunlin University of Science and Technology, Taiwan*

Jingcai Cheng *Institute of Process Engineering, Chinese Academy of Sciences, China*

Kwang-Il Choi *Graduate School, Chonnam National University, Republic of Korea*

S.P. Duttagupta *Indian Institute of Technology Bombay, India*

Masahiro Furuya *Nuclear Energy System Department, Central Research Institute of Electric Power Industry, Japan*

Pallab Ghosh *Department of Chemical Engineering, Indian Institute of Technology Guwahati, India*

Stefan Iglauer *Department of Petroleum Engineering, Curtin University, Australia*

Xiangyang Li *Institute of Process Engineering, Chinese Academy of Sciences, China*

Zai-Sha Mao *Institute of Process Engineering, Chinese Academy of Sciences, China*

Navid Mostoufi *School of Chemical Engineering, University of Tehran, Iran*

Jong-Taek Oh *Department of Refrigeration and Air Conditioning Engineering, Chonnam National University, Republic of Korea*

A.S. Pamitran *Graduate School, Chonnam National University, Republic of Korea*

S.G. Singh *Indian Institute of Technology, Hyderabad, India*

Rahmat Sotudeh-Gharebagh *School of Chemical Engineering, University of Tehran, Iran*

Arunkumar Sridharan *Indian Institute of Technology Bombay, India*

Chih-Yung Tseng *Energy & Environment Research Laboratories, Industrial Technology Research Institute, Taiwan*

Chi-Chuan Wang *Department of Mechanical Engineering, National Chiao Tung University, Taiwan*

Tao Wang *Institute of Process Engineering, Chinese Academy of Sciences, China*

C. L. Wu *Engineering College, Guangdong Ocean University, China*

Chao Yang *Institute of Process Engineering, Chinese Academy of Sciences, China*

Gengzhi Yu *Institute of Process Engineering, Chinese Academy of Sciences, China*

Reza Zarghami *School of Chemical Engineering, University of Tehran, Iran*

Advances in Multiphase Flow and Heat Transfer, Vol. 3, 2012, 3-54

CHAPTER 1

Numerical and Experimental Studies on Multiphase Flow in Stirred Tanks

Chao Yang[*], Zai-Sha Mao, Tao Wang, Xiangyang Li, Jingcai Cheng and Gengzhi Yu

Institute of Process Engineering, Chinese Academy of Sciences, Beijing 100190, China

Abstract: Multiphase flow in a stirred tank was addressed both experimentally and with numerical simulation. The flow field and the distribution of the dispersed phase were regarded in most sections. The Eulerian multi-fluid approach and k-ε turbulence model were employed in numerical simulations. The good agreement found between experimental and predicted results underlines the importance of computational fluid dynamics in future research of multiphase phenomena in stirred tanks. Besides, some novel surface aeration configurations have been provided, which make such operations more reliable and economical. Furthermore, numerical simulation was applied in research of macro- and micro-mixing in stirred tanks. To account for the anisotropy of turbulence in the flow field, the algebraic stress model and large eddy simulation were adopted in the numerical study of hydrodynamics in multiphase stirred tanks as a result of their capacity on retrieving turbulent properties more accurately.

Keywords: Stirred tank, numerical simulation, multiphase flow, mixing.

INTRODUCTION

Agitation in a stirred tank, one of the most common operations in process industry, can be accomplished in continuous, batch, or fed-batch modes. A good mixing result is important for minimizing investment and operating costs, providing high yields when mass/heat transfer is limiting, and thus enhancing profitability. Over 50% of the world's chemical productions involve these vessels for manufacturing high-added-value products [1]. A stirred tank typically contains an impeller mounted on a shaft, and optionally can contain baffles and other internals, such as spargers, coils, and draft tubes.

Velocity map and distribution of dispersed phases have attracted tremendous attention from chemical engineering community. For experimentation, various flow visualization techniques, including Laser Doppler Velocimetry (LDV), Particle Image Velocimetry (PIV), Electrical Resistance Tomography (ERT) and Magnetic Response Imaging (MRI), have been used to examine the complex nature of the flow fields in stirred vessels, because such visualization has led to an improved understanding of mixing and dispersion of gas, solid or liquid phases within such stirred tanks [2]. The flow field in a stirred vessel has been the subject of investigation by LDV for the last 30 years, the use of other techniques, however, are much more recent, spanning a decade or so. These techniques have not been so popular because of the high cost of instruments. Moreover, it seems that the tested stirred vessels are often of small diameter. Therefore, great efforts have been devoted to numerically resolve the hydrodynamics in stirred tanks and much progress has been achieved with respect to single phase liquid flow.

Multiphase flow and transport in stirred tanks demand more intensive attention, combining numerical and experimental approaches. In most works, the Eulerian multi-fluid approach and the k-ε turbulence model (standard, RNG or realizable k-ε model) are usually employed to simulate multiphase flow. The fluids can be liquid, gas or solid. The Eulerian multiphase model works best for low-volume fraction mixtures (<10%) [1]. The k-ε turbulence model based on isotropy (Boussinesq hypothesis) is not accurate enough in predicting flow characteristics around the impeller region. Assuming that turbulent viscosity is anisotropic, the Reynolds Stress Model computes the stresses individually. For 2D models, this amounts to four

Address coreespondence to Chao Yang: Institute of Process Engineering, Chinese Academy of Sciences, Beijing 100190, China; Tel.: +86-10-62554558; Fax: +86-10-62561822; E-mail: chaoyang@home.ipe.ac.cn

additional transport equations. For 3D models, six additional transport equations are required. As computer capacity and speed have increased rapidly during the past several years, use of the Reynolds stress turbulence model has become more widespread, giving rise to improved accuracy over other RANS-based turbulence models as it has been proved with the closer matching to experimental results for a number of applications. Large eddy simulation (LES) recognizes that turbulent eddies occur on many scales in a flow field. With the LES model, the continuity and momentum equations are filtered prior to being solved in a transient fashion. As a consequence of the investment of finer resolution in space and time domain, the accuracy of numerical simulation of turbulent flow in stirred tanks is raised to much higher levels, while the extension of Reynolds stress turbulence model and LES to multiphase flows in stirred tanks is already under way.

In this chapter, the topic of multiphase flow in a stirred tank is divided into four sections. Firstly, multiphase flows (including two- and three-phase flows) are discussed in detail based on numerical methods using the Eulerian multi-fluid approach and *k-ε* turbulence model. Good agreement between prediction and experiment indicates that computational fluid dynamics has become a powerful tool in the research of stirred vessels. Secondly, some novel surface aeration configurations are introduced for better gas dispersion and high pumping capacity. The hydrodynamic characteristics of multi-impellers and numerical simulation of gas hold-up in surface aerated stirred tanks are also addressed. Then, we present some numerical results in macro- and micro- mixing. Mixing time and fast precipitation process are generally employed to study the both, respectively. In the final section some new advances in numerical simulation are presented. Algebraic stress model (ASM) and large eddy simulation (LES) are recommended for future research of multi-phase flow in a stirred tank. In the following sections, we present extensive experimental and numerical simulation results of recent developments, which are mainly from Chemical Reaction Engineering and Multi-phase Flow Laboratory, Institute of Process Engineering, Chinese Academy of Sciences.

TWO-PHASE FLOW IN STIRRED TANKS

Two-phase flow is a simple but typical case in multi-phase systems. The flow becomes more complex if a second phase is introduced. Tremendous efforts, including experimental and numerical, have been devoted to the understanding the two-phase flow in stirred tanks.

Mathematic Model

In 1982, Harvey *et al.* [3] attempted two-dimensional numerical simulation of a stirred tank. Since then, numerical simulation has developed very quickly. A plenty of literature reports demonstrate that the numerical simulation of single phase flow is generally successful. When a second phase is introduced, the mathematical treatment becomes more complicated. Generally, there are two major approaches for modeling two-phase flow, differing from each other by the treatment of the dispersed phase, the Eulerian-Eulerian and the Eulerian-Lagrangian approaches. In the Eulerian-Lagrangian approach the trajectory of each discrete particle is calculated by its motion equation subject to respective initial conditions and exerted force, while the motion of the continuous phase is followed in a fixed Eulerian grid. On the other hand, in the Eulerian-Eulerian approach the dispersed phase is treated also as a continuum.

In most works, the "two-fluid" approach based on a Eulerian-Eulerian approach [4] and the two-phase *k-ε* turbulence model are usually employed to simulate two-phase flow. In this approach, there are some assumptions: (1) the continuous phase and dispersed phase are considered to be interpenetrating continua with interaction; (2) the pressure field of the system is assumed to be the same for two phases. With such considerations, the time averaged governing equations of momentum balance for phase *k* are

$$\frac{\partial\left(\rho_k\alpha_k\right)}{\partial t}+\frac{\partial\left(\rho_k\alpha_k u_{kj}+\rho_k\overline{\alpha'_k u'_{kj}}\right)}{\partial x_j}=0 \tag{1}$$

$$\rho_k \frac{\partial (\alpha_k u_{ki})}{\partial t} + \rho_k \frac{\partial (\alpha_k u_{ki} u_{kj})}{\partial x_j} = -\alpha_k \frac{\partial P}{\partial x_i} + \frac{\partial (\alpha_k \overline{\tau_{kij}})}{\partial x_j} + F_{ki} + \rho_k \alpha_k g_i -$$

$$\rho k \frac{\partial}{\partial x_j} (\alpha_k \overline{u'_{kj} u'_{ki}} + u_{ki} \overline{\alpha'_k u'_{kj}} + u_{kj} \overline{\alpha'_k u'_{ki}} + \overline{\alpha'_k u'_{kj} u'_{ki}})$$

(2)

$$\sum \alpha_k = 1.0$$

(3)

where α_k is the phase volume fraction (or holdup) and F_{ki} is the interphase momentum exchange term.

For the closure of momentum equations, the turbulent fluctuation correlation terms should be related to the known or calculable quantities *via* either algebraic or differential equations. The triple and higher order correlations are usually omitted and the viscous shear terms are also neglected compared with the turbulent shear terms. With the Boussinesq gradient transport hypothesis, the velocity correlations, named the Reynolds stresses, are modeled following the practice of single-phase flow as

$$\overline{u'_{ki} u'_{kj}} = \frac{2}{3} k \delta_{ij} - v_{k,t} (\frac{\partial u_{kj}}{\partial x_i} + \frac{\partial u_{ki}}{\partial x_j})$$

(4)

The correlation between velocity fluctuations and phase holdup fluctuation $\overline{u'_{ki} \alpha'_k}$, which appears in both continuity and momentum equations due to the presence of a second phase, represents the transport of both mass and momentum respectively by dispersion. However, it has not been included in the mathematical model [5-7]. The simplest way to model $\overline{u'_{ki} \alpha'_k}$ is to assume the same gradient transport, which gives

$$\overline{u'_{ki} \alpha'_k} = -\frac{v_{k,t}}{\sigma_t} \frac{\partial \alpha_k}{\partial x_i}$$

(5)

where σ_t is the turbulent Schmidt number for phase dispersion, whose value is related with the scale of turbulent flow, but few reports are available regarding the systematical experiments and theoretical models on σ_t. It was found that the simulation results were sensitive to σ_t in solid-liquid simulation and the value between 1.0 and 2.0 was suggested [8]. In gas-liquid systems, the value of 1.0 was recommended [9, 10], but Wang and Mao [11] suggested the value of 1.6.

The final closure momentum equation reads

$$\rho_k \left[\frac{\partial (\alpha_k u_{ki})}{\partial t} + \frac{\partial (\alpha_k u_{ki} u_{kj})}{\partial x_j} \right] = -\alpha_k \frac{\partial P}{\partial x_i} + F_{ki} + \rho_k \alpha_k g_i +$$

$$\frac{\partial}{\partial x_j} \left[\alpha_k \mu_{k,\text{eff}} (\frac{\partial u_{ki}}{\partial x_j} + \frac{\partial u_{kj}}{\partial x_i}) \right] + \frac{\partial}{\partial x_j} \left[\frac{\mu_{k,t}}{\sigma_t} (u_{ki} \frac{\partial \alpha_k}{\partial x_j} + u_{kj} \frac{\partial \alpha_k}{\partial x_i}) \right]$$

(6)

Interphase Force

The interphase coupling terms make two-phase flow fundamentally different from single-phase flow. The interphase interaction term, F_{ki}, satisfies the following relationship:

$$F_{c,i} = -F_{d,i}$$

(7)

where subscripts c and d denote continuous and dispersed phases, respectively. For two-phase flow, F_{ki} is modeled as a linear combination of some terms, *i.e.*, interphase drag force, virtual mass force, Basset force and lift force. In most cases, the magnitude of the Basset force and lift force are much smaller than that of the interphase drag force. A report [12] indicated that the effect of virtual mass force is not significant in the bulk region of a stirred tank. Therefore, only the interphase drag force is always considered in two-phase flows. In addition, in a non-inertial reference frame rotating with the impeller axis, the centrifugal force, $\mathbf{F}_{r,k}$, and the Coriolis force, $\mathbf{F}_{c,k}$, must be included:

$$\mathbf{F}_{r,k} = \alpha_k \rho_k (\boldsymbol{\omega} \times \mathbf{r}) \times \boldsymbol{\omega} \tag{8}$$

$$\mathbf{F}_{c,k} = 2\alpha_k \rho_k (\boldsymbol{\omega} \times \mathbf{u}_k) \tag{9}$$

In solid-liquid systems, the expression of drag force is usually as follows:

$$F_{c,\mathrm{drag}} = \frac{3}{4} \alpha_c \alpha_d C_D \frac{|\mathbf{u}_d - \mathbf{u}_c| (\boldsymbol{u}_{d,i} - \boldsymbol{u}_{c,i})}{d_d}$$

$$F_{d,\mathrm{drag}} = -F_{c,\mathrm{drag}} \tag{10}$$

where C_D, the interphase drag coefficient, is a complex function dependent on the dispersed phase holdup and the turbulence :

$$C_D = \frac{24(1 + 0.15Re_d^{0.687})}{Re_d} \quad (\text{for } Re_d < 1000)$$

$$C_D = 0.44 \quad (\text{for } Re_d \geq 1000) \tag{11}$$

where Re_d is the Reynolds number of the particles:

$$Re_d = \frac{d_d |\mathbf{u}_d - \mathbf{u}_c| \rho_c}{\mu_{c,\mathrm{lam}}} \tag{12}$$

In gas-liquid flow, the interphase drag force on a bubble in the control volume is

$$F_n = \frac{1}{2} \rho_c C_D A_n |\mathbf{u}_c - \mathbf{u}_d| (u_c - u_d) \tag{13}$$

where A_n is the projected area of the bubble and C_D is the drag coefficient. The Ishii's expression [13] for C_D, which takes account of bubble-bubble interaction and bubble distortion, can be used here. Assuming that bubbles have the same u_c-u_d in an infinitesimal unit volume, then

$$C_D = \frac{4}{3} r_b \sqrt{\frac{g\Delta\rho}{\sigma}} (1 - \alpha_d)^{-0.5} \tag{14}$$

The bubble shape is assumed to be spherical, and Eq. (13) becomes

$$F_n = \frac{3}{4} \rho_c C_D \frac{V_n}{d_{b,n}} |\mathbf{u}_c - \mathbf{u}_d| (u_c - u_d) \tag{15}$$

Substituting C_D into the above equation, then it reads

$$F_{\mathrm{n}} = \frac{1}{2}\rho_{\mathrm{c}}\sqrt{\frac{g\Delta\rho}{\sigma}}(1-\alpha_{\mathrm{d}})^{-0.5}\left|\mathbf{u}_{\mathrm{c}}-\mathbf{u}_{\mathrm{d}}\right|(u_{\mathrm{c}}-u_{\mathrm{d}})V_{\mathrm{n}} \tag{16}$$

The total interphase force in the control volume becomes

$$F_{\mathrm{d}} = \sum_{n=1}^{N_{\mathrm{b}}}F_{\mathrm{n}} = \frac{1}{2}\rho_{\mathrm{c}}\sqrt{\frac{g\Delta\rho}{\sigma}}(1-\alpha_{\mathrm{d}})^{-0.5}\left|\mathbf{u}_{\mathrm{c}}-\mathbf{u}_{\mathrm{d}}\right|(u_{\mathrm{c}}-u_{\mathrm{d}})\sum_{n=1}^{N_{\mathrm{b}}}V_{\mathrm{n}} \tag{17}$$

where N_{b} is the bubble number in the control volume. Finally, the force component in the *i*th coordinate direction per unit volume reads

$$F_{\mathrm{d},i} = \frac{1}{2}\alpha_{\mathrm{d}}\rho_{\mathrm{c}}\sqrt{\frac{g\Delta\rho}{\sigma}}(1-\alpha_{\mathrm{d}})^{-0.5}\left|\mathbf{u}_{\mathrm{c}}-\mathbf{u}_{\mathrm{d}}\right|(u_{\mathrm{c},i}-u_{\mathrm{d},i}) \tag{18}$$

It should be noted that the interphase drag force becomes independent of bubble size.

Transport Equations for k and ε

To treat the turbulent two-phase flow rigorously, the turbulent model adopted should include interphase turbulence transfer terms accounting for the turbulence promotion or damping due to the presence of the dispersed phase. However, there is no reliable information on such terms, and the proper turbulence model for turbulent multi-phase systems is absent. In two-phase stirred tanks, the turbulence is mainly attributed to velocity fluctuation of the liquid phase because the holdup of the dispersed phase is often quite low in most parts of the tank. The dispersed phase can affect the turbulence of the system through interphase momentum exchange. The liquid phase turbulence effect is modeled using a suitable extension of the standard *k-ε* turbulence model, written in a general form as [9]

$$\frac{\partial}{\partial t}(\rho_{\mathrm{c}}\alpha_{\mathrm{c}}k)+\frac{\partial}{\partial x_i}(\rho_{\mathrm{c}}\alpha_{\mathrm{c}}u_{\mathrm{c}i}k) = \frac{\partial}{\partial x_i}\left(\alpha_{\mathrm{c}}\frac{\mu_{\mathrm{ct}}}{\sigma_k}\frac{\partial k}{\partial x_i}\right)+\frac{\partial}{\partial x_i}\left(k\frac{\mu_{\mathrm{ct}}}{\sigma_k}\frac{\partial\alpha_{\mathrm{c}}}{\partial x_i}\right)+S_k \tag{19}$$

$$\frac{\partial}{\partial t}(\rho_{\mathrm{c}}\alpha_{\mathrm{c}}\varepsilon)+\frac{\partial}{\partial x_i}(\rho_{\mathrm{c}}\alpha_{\mathrm{c}}u_{\mathrm{c}i}\varepsilon) = \frac{\partial}{\partial x_i}\left(\alpha_{\mathrm{c}}\frac{\mu_{\mathrm{ct}}}{\sigma_\varepsilon}\frac{\partial\varepsilon}{\partial x_i}\right)+\frac{\partial}{\partial x_i}\left(\varepsilon\frac{\mu_{\mathrm{ct}}}{\sigma_\varepsilon}\frac{\partial\alpha_{\mathrm{c}}}{\partial x_i}\right)+S_\varepsilon \tag{20}$$

where the values of the Schmidt number are given as σ_k=1.3 and σ_ε=1.0. The source terms in the above equations are

$$S_k = \alpha_{\mathrm{c}}\left[(G+G_{\mathrm{e}})-\rho_{\mathrm{c}}\varepsilon\right] \tag{21}$$

$$S_\varepsilon = \alpha_{\mathrm{c}}\frac{\varepsilon}{k}\left[C_1(G+G_{\mathrm{e}})-C_2\rho_{\mathrm{c}}\varepsilon\right] \tag{22}$$

where *G* is the turbulent generation and G_{e} is the extra dissipation term due to the dispersion phase. Based on the analysis of Kataoka *et al.* [14], G_{e} is mainly dependent on the drag force between the continuous phase and the dispersed phase:

$$G = -\rho_{\mathrm{c}}\alpha_{\mathrm{c}}\overline{u'_{\mathrm{c}i}u'_{\mathrm{c}j}}\frac{\partial u_{\mathrm{c}i}}{\partial x_j} \tag{23}$$

$$G_{\mathrm{e}} = \sum_{\mathrm{d}}C_{\mathrm{b}}|\mathbf{F}|\left(\sum(u_{\mathrm{d}i}-u_{\mathrm{c}i})^2\right)^{0.5} \tag{24}$$

where C_b is an empirical coefficient. When $C_b=0$, the energy induced by the dispersed phase dissipates at the interface and has no influence on the turbulent kinetic energy of the continuous phase. According to the analysis in the literature, the value of C_b has been always set at 0.02 or 0.03.

The reference values of the model constants are the well-accepted ones: $C_\mu=0.09$, $C_1=1.44$ and $C_2=1.92$. Another value of $C_2=1.60$ [15] is also suggested. In consideration of the strong vortex in the discharge zone, C_1 is always modified as follows:

$$C_1 = 1.44 + \frac{0.8R_f\rho_c\varepsilon}{(G+G_e)}$$

$$R_f = \begin{cases} \frac{1}{\varepsilon}\left[\overline{u'_{c,r}u'_{c,\theta}}r\frac{\partial}{\partial r}\left(\frac{u_{c,\theta}}{r}\right)\right], & C-1.5w < z < C+1.5w \\ 0, & z < C-1.5w, z > C+1.5w \end{cases} \tag{25}$$

where C is the clearance between impeller and tank bottom and w represents the axial width of blades.

Source Terms and Diffusion Coefficients

It is known that the flow in stirred tanks is unsteady because of the interaction of the rotating impeller blades with the stationary baffles. However, the flow pattern will become axisymmetrically repeating once it has fully developed. A snapshot of this flow can describe the flow within the impeller blades at this particular instant. Ranade and van den Akker [9] suggested that the time derivative terms in the governing equations can be ignored without much error in most regions of the tank except for the impeller swept volume. The flow field in the impeller swept volume was simulated in a non-inertial reference frame rotating with the impeller. Therefore, the temporal terms in the equations can be omitted. In this way, the resulted model formulation of the mass and momentum conservation equations for phase k in a general form in the cylindrical coordinate system reads:

$$\frac{1}{r}\frac{\partial}{\partial r}\left(\rho_k r\alpha_k u_{kr}\varphi\right) + \frac{1}{r}\frac{\partial}{\partial\theta}\left(\rho_k\alpha_k u_{kq}\varphi\right) + \frac{\partial}{\partial z}\left(\rho_k\alpha_k u_{kz}\varphi\right)$$
$$= \frac{1}{r}\frac{\partial}{\partial r}\left(\alpha_k\Gamma_{\varphi,\,\text{eff}}r\frac{\partial\varphi}{\partial r}\right) + \frac{1}{r}\frac{\partial}{\partial\theta}\left(\frac{\alpha_k\Gamma_{\varphi,\,\text{eff}}}{r}\frac{\partial\varphi}{\partial\theta}\right) + \frac{\partial}{\partial z}\left(\alpha_k\Gamma_{\varphi,\,\text{eff}}\frac{\partial\varphi}{\partial z}\right) + S_\varphi \tag{26}$$

The governing equations can be summarized in Table **1**, with

$$G = 2\mu_{ct}\left[\left(\frac{\partial u_{cr}}{\partial r}\right)^2 + \left(\frac{1}{r}\frac{\partial u_{c\theta}}{\partial\theta} + \frac{u_{cr}}{r}\right)^2 + \left(\frac{\partial u_{cz}}{\partial z}\right)^2\right]$$
$$+ \mu_{ct}\left[r\frac{\partial}{\partial r}\left(\frac{u_{c\theta}}{r}\right) + \frac{1}{r}\frac{\partial u_{cr}}{\partial\theta}\right]^2 + \mu_{ct}\left[\frac{1}{r}\frac{\partial u_{cz}}{\partial\theta} + \frac{\partial u_{c\theta}}{\partial z}\right]^2$$
$$+ \mu_{ct}\left[\frac{\partial u_{cr}}{\partial z} + \frac{\partial u_{cz}}{\partial r}\right]^2 \tag{27}$$

$$\mu_{c,\text{eff}} = \mu_{c,t} + \mu_{c,\text{lam}} \tag{28}$$

$$\mu_{c,t} = \frac{C_\mu k^2\rho_c}{\varepsilon} \tag{29}$$

$$\mu_{d,\text{eff}} = \mu_{d,t} + \mu_{d,\text{lam}} \tag{30}$$

Table 1: The source terms and diffusion coefficients in the equation

Equation	φ	$\Gamma_{\varphi,\text{eff}}$	S_φ
Continuity	1	0	$\dfrac{1}{r}\dfrac{\partial}{\partial r}\left(r\dfrac{\mu_{k,t}}{\sigma_t}\dfrac{\partial \alpha_k}{\partial r}\right)+\dfrac{1}{r}\dfrac{\partial}{\partial \theta}\left(\dfrac{\mu_{k,t}}{\sigma_t}\dfrac{\partial \alpha_k}{r\partial \theta}\right)+\dfrac{\partial}{\partial z}\left(\dfrac{\mu_{k,t}}{\sigma_t}\dfrac{\partial \alpha_k}{\partial z}\right)$
Radial momentum	$u_{k,r}$	$\mu_{k,\text{eff}}$	(see below)
Azimuthal momentum	$u_{k,\theta}$	$\mu_{k,\text{eff}}$	(see below)
Axial momentum	$u_{k,z}$	$\mu_{k,\text{eff}}$	(see below)
Turbulent kinetic energy	k	$\dfrac{\mu_{ct}}{\sigma_k}$	$\alpha_c\left[(G+G_e)-\rho_c\varepsilon\right]$
Turbulent energy dissipation	ε	$\dfrac{\mu_{ct}}{\sigma_\varepsilon}$	$\alpha_c\dfrac{\varepsilon}{k}\left[C_1(G+G_e)-C_2\rho_c\varepsilon\right]$

Radial momentum source S_φ:

$$\frac{1}{r}\frac{\partial}{\partial r}\left(\alpha_k\mu_{k,\text{eff}}r\frac{\partial u_{k,r}}{\partial r}\right)+\frac{1}{r}\frac{\partial}{\partial \theta}\left(\alpha_k\mu_{k,\text{eff}}r\frac{\partial}{\partial r}\left(\frac{u_{k,\theta}}{r}\right)\right)+\frac{\partial}{\partial z}\left(\alpha_k\mu_{k,\text{eff}}\frac{\partial u_{k,z}}{\partial r}\right)$$

$$+\frac{\partial}{\partial z}\left(u_{k,r}\frac{\mu_{k,t}}{\sigma_t}\frac{\partial \alpha_k}{\partial z}\right)+\frac{1}{r}\frac{\partial}{\partial r}\left(ru_{k,r}\frac{\mu_{k,t}}{\sigma_t}\frac{\partial \alpha_k}{\partial r}\right)+\frac{1}{r^2}\frac{\partial}{\partial \theta}\left(u_{k,r}\frac{\mu_{k,t}}{\sigma_t}\frac{\partial \alpha_k}{\partial \theta}\right)$$

$$-\frac{2\alpha_k\mu_{k,\text{eff}}}{r^2}\frac{\partial u_{k,\theta}}{\partial \theta}-\frac{2\alpha_k\mu_{k,\text{eff}}u_{k,r}}{r^2}+\frac{\rho_k\alpha_k u_{k,\theta}^2}{r}$$

$$+\frac{\partial}{\partial z}\left(u_{k,r}\frac{\mu_{k,t}}{\sigma_t}\frac{\partial \alpha_k}{\partial z}\right)+\frac{1}{r}\frac{\partial}{\partial r}\left(ru_{k,r}\frac{\mu_{k,t}}{\sigma_t}\frac{\partial \alpha_k}{\partial r}\right)+\frac{1}{r^2}\frac{\partial}{\partial \theta}\left(u_{k,r}\frac{\mu_{k,t}}{\sigma_t}\frac{\partial \alpha_k}{\partial \theta}\right)$$

$$+\frac{\partial}{\partial z}\left(u_{k,z}\frac{\mu_{k,t}}{\sigma_t}\frac{\partial \alpha_k}{\partial r}\right)+\frac{1}{r}\frac{\partial}{\partial r}\left(ru_{k,r}\frac{\mu_{k,t}}{\sigma_t}\frac{\partial \alpha_k}{\partial r}\right)+\frac{1}{r}\frac{\partial}{\partial \theta}\left(u_{k,\theta}\frac{\mu_{k,t}}{\sigma_t}\frac{\partial \alpha_k}{\partial r}\right)$$

$$-\frac{2}{r^2}\left(u_{k,\theta}\frac{\mu_{k,t}}{\sigma_t}\frac{\partial \alpha_k}{\partial \theta}\right)$$

$$-\alpha_k\frac{\partial p}{\partial r}+F_{k,r}\left\{+\rho_k\alpha_k\left(\omega^2 r+2\omega u_\theta\right)\right\}-\rho_k\frac{2}{3}\frac{\partial(\alpha_k k)}{\partial r}$$

Azimuthal momentum source S_φ:

$$\frac{1}{r}\frac{\partial}{\partial r}\left(\alpha_k\mu_{k,\text{eff}}\frac{\partial u_{k,r}}{\partial \theta}\right)+\frac{1}{r}\frac{\partial}{\partial \theta}\left(\frac{\alpha_k\mu_{k,\text{eff}}}{r}\frac{\partial u_{k,\theta}}{\partial \theta}\right)+\frac{1}{r}\frac{\partial}{\partial \theta}\left(2\alpha_k\mu_{k,\text{eff}}\frac{u_{k,r}}{r}\right)+\frac{\partial}{\partial z}\left(\frac{\alpha_k\mu_{k,\text{eff}}}{r}\frac{\partial u_{k,z}}{\partial \theta}\right)$$

$$+\alpha_k\mu_{k,\text{eff}}\frac{\partial}{\partial r}\left(\frac{u_{k,\theta}}{r}\right)-\frac{1}{r}\frac{\partial}{\partial r}\left(\alpha_k\mu_{k,\text{eff}}u_{k,\theta}\right)-\frac{\rho_k\alpha_k u_{k,r}u_{k,\theta}}{r}+\frac{\alpha_k\mu_{k,\text{eff}}}{r^2}\frac{\partial u_{k,r}}{\partial \theta}$$

$$+\frac{\partial}{\partial z}\left(u_{k,\theta}\frac{\mu_{k,t}}{\sigma_t}\frac{\partial \alpha_k}{\partial z}\right)+\frac{1}{r}\frac{\partial}{\partial r}\left(ru_{k,\theta}\frac{\mu_{k,t}}{\sigma_t}\frac{\partial \alpha_k}{\partial r}\right)+\frac{1}{r^2}\frac{\partial}{\partial \theta}\left(u_{k,\theta}\frac{\mu_{k,t}}{\sigma_t}\frac{\partial \alpha_k}{\partial \theta}\right)$$

$$+\frac{\partial}{\partial z}\left(u_{k,z}\frac{\mu_{k,t}}{\sigma_t}\frac{\partial \alpha_k}{r\partial \theta}\right)+\frac{1}{r}\frac{\partial}{\partial r}\left(ru_{k,r}\frac{\mu_{k,t}}{\sigma_t}\frac{\partial \alpha_k}{r\partial \theta}\right)+\frac{1}{r^2}\frac{\partial}{\partial \theta}\left(u_{k,\theta}\frac{\mu_{k,t}}{\sigma_t}\frac{\partial \alpha_k}{\partial \theta}\right)$$

$$+\frac{u_{k,\theta}}{r}\frac{\mu_{k,t}}{\sigma_t}\frac{\partial \alpha_k}{\partial r}+\frac{u_{k,r}}{r^2}\frac{\mu_{k,t}}{\sigma_t}\frac{\partial \alpha_k}{\partial \theta}$$

$$-\frac{\alpha_k}{r}\frac{\partial p}{\partial \theta}+F_{k,\theta}\left\{+\rho\alpha_k\left(-2\omega u_{k,r}\right)\right\}-\rho_k\frac{2}{3}\frac{1}{r}\frac{\partial(\alpha_k k)}{\partial \theta}$$

Axial momentum source S_φ:

$$\frac{1}{r}\frac{\partial}{\partial r}\left(\alpha_k\mu_{k,\text{eff}}r\frac{\partial u_{k,r}}{\partial z}\right)+\frac{1}{r}\frac{\partial}{\partial \theta}\left(\alpha_k\mu_{k,\text{eff}}\frac{\partial u_{k,\theta}}{\partial z}\right)+\frac{\partial}{\partial z}\left(\alpha_k\mu_{k,\text{eff}}\frac{\partial u_{k,z}}{\partial z}\right)$$

$$+\frac{\partial}{\partial z}\left(u_{k,z}\frac{\mu_{k,t}}{\sigma_t}\frac{\partial \alpha_k}{\partial z}\right)+\frac{1}{r}\frac{\partial}{\partial r}\left(ru_{k,z}\frac{\mu_{k,t}}{\sigma_t}\frac{\partial \alpha_k}{\partial r}\right)+\frac{1}{r^2}\frac{\partial}{\partial \theta}\left(u_{k,z}\frac{\mu_{k,t}}{\sigma_t}\frac{\partial \alpha_k}{\partial \theta}\right)$$

$$+\frac{\partial}{\partial z}\left(u_{k,z}\frac{\mu_{k,t}}{\sigma_t}\frac{\partial \alpha_k}{\partial z}\right)+\frac{1}{r}\frac{\partial}{\partial r}\left(ru_{k,r}\frac{\mu_{k,t}}{\sigma_t}\frac{\partial \alpha_k}{\partial z}\right)+\frac{1}{r}\frac{\partial}{\partial \theta}\left(u_{k,\theta}\frac{\mu_{k,t}}{\sigma_t}\frac{\partial \alpha_k}{\partial z}\right)$$

$$-\alpha_k\frac{\partial p}{\partial z}+F_{k,z}-\rho_k\alpha_k g-\rho_k\frac{2}{3}\frac{\partial(\alpha_k k)}{\partial z}$$

Note: the terms in the curly brackets are present only when a non-inertial reference frame is used.

The formulations for $\mu_{d,t}$ take the following forms in different systems:

(1) In gas-liquid systems, $\mu_{d,t}$ is

$$\mu_{d,t} = \rho_d v_{d,t} \tag{31}$$

where $v_{d,t}$ is given by the Hinze-Tchen formulation:

$$\frac{v_{d,t}}{v_{c,t}} = \frac{1}{1 + \dfrac{\tau_{r,k}}{\tau_t}} \tag{32}$$

with τ_t being the turbulence fluctuation time:

$$\tau_t = 1.5^{0.5} \frac{k}{\varepsilon} C_\mu^{0.75} \tag{33}$$

and $\tau_{r,k}$ the bubble relaxation time:

$$\tau_{r,k} = \frac{\rho_d d_b^2}{18\mu_c} \tag{34}$$

The numerical value of the term $\tau_{r,k}/\tau_t$ is about order 10^{-2} in most parts of a stirred tank, so d_b can be approximately replaced by the Sauter mean diameter d_{32} without introducing much error. Some investigators estimated the ratio between d_{32} and d_{max} for bubbles. Hesketh *et al.* [16] proposed the ratio between d_{32} and d_{max} to be 0.62 for the bubble breakup in horizontal pipe flow. Parthasarthy and Ahmed [17] did experiments on different positions in a gas-liquid stirred tank, and found the value of d_{32}/d_{max} for bubbles to be 0.785. Various d_{32}/d_{max} values have also been reported for droplets in liquid-liquid dispersions. Brown and Pitt [18] found a linear relationship between d_{32} and d_{max} in the kerosene-water system in a stirred tank and found the value of d_{32}/d_{max} to be 0.7. Calabrese *et al.* [19] reported a value of 0.6 for the dispersion of moderate viscosity oils in a stirred vessel. It ought to be noted that despite the diversity of systems, the reported d_{32}/d_{max} values do not show considerable scatter. An average value of 0.68 is resulted for the above mentioned values and is often used. The d_{max} can be obtained by the next equation:

$$d_{max} = 0.725 \left(\frac{\sigma}{\rho_c}\right)^{0.6} \varepsilon^{-0.4} \tag{35}$$

(2) In solid-liquid systems, the turbulent viscosity of the dispersed phase, $\mu_{d,t}$ is expressed in terms of its counterpart of the continuous phase as

$$\mu_{d,t} = K\mu_{c,t} \tag{36}$$

$$K = \frac{\rho_d \overline{u'_{d,i} u'_{d,i}}}{\rho_c \overline{u'_{c,i} u'_{c,i}}} \tag{37}$$

Gosman *et al.* [5] proposed a correlation of u'_d to u'_c derived from a Lagrangian analysis of particle response to eddies which are much larger than the particle diameter:

$$u'_{d,i} = u'_{c,i} \left[1 - \exp\left(-\frac{t_l}{t_p}\right)\right] \tag{38}$$

with $t_l = 0.41 k/\varepsilon$ being the mean eddy lifetime, and t_p, the particle response time, obtained by the Lagrangian integration of the equation of motion of a swarm of particles moving through a fluid eddy of given velocity distribution with the expression:

$$t_p = \frac{4\rho_d d_d}{3\rho_c C_D \alpha_d |\mathbf{u}_d - \mathbf{u}_c|} \tag{39}$$

The laminar viscosity of the solid phase is not directly available from the literature, and the determination of the laminar viscosity coefficient $\mu_{d,\,lam}$ of the dispersion phase is a hot topic, when studying two-phase flow, especially in gas-liquid fluidized beds. In solid-liquid flow simulations, $\mu_{d,\,lam}$ is set as a parameter, varying between 10^{-4} Pa·s and 10^{-2} Pa·s [8]. The influence of $\mu_{d,\,lam}$ on the predicted local solid concentration is very small. Therefore, $\mu_{d,\,lam}$ is always set to be that of the continuous phase laminar viscosity, namely $\mu_{d,\,lam} = \mu_{c,\,lam} = 10^{-3}$ Pa·s.

Numerical Method

Impeller Modeling

There are some difficulties in resolving the complex 3-D turbulent flow in a baffled stirred tank as a result of the rotating impeller, which makes the treatment of the impeller region critical. Gosman *et al.* [5] calculated the flow in solid-liquid stirred tanks treating the Rushton impeller region as a 'black box', hence the experimental data had to be imposed on the surface swept by the impeller blades as boundary conditions. The obvious shortcoming of such an approach is that the experimental data in this region is crucially needed for initiating the simulation and it is not applicable to novel operation conditions without experimental backup. Micale *et al.* [6] and Montante *et al.* [20] applied the inner-outer (IO) iterative procedure developed by Brucceto *et al.* [21] to simulate the flow in solid-liquid stirred vessels. In this approach, the whole vessel volume is subdivided into two partly overlapping zones: an 'inner' domain, containing the impeller, and an 'outer' one, extending from the inner region border to the wall with baffles. The 'inner' and 'outer' simulations are conducted separately under the steady-state assumption in their own reference frames. Since the two frames are different, the information (velocity, turbulence energy and dissipation) is iteratively exchanged on the overlapping region, and is corrected for the relative motion and averaged over the azimuthal direction so as to provide the boundary conditions for the solution of the other domain.

The inner-outer-iterative method does not need experimental data as the impeller region boundary conditions. However, the information on the surfaces of the 'inner' and 'outer' domains is averaged over the azimuthal direction, thus some important features for the flow in the stirred vessel generated by the periodical rotation of the impeller are ignored. Wang and Mao [11] improved the inner-outer iterative procedure by combining it with a snapshot approach proposed by Ranade and van den Akker [9] to keep the unsteady turbulent properties not averaged and applied it to simulate the flow in single-phase and gas-liquid stirred tanks. In this improved approach, keeping the unsteady turbulent properties enables the simulation of the flow parameters with better accuracy.

Since the reference frames in the 'inner' and 'outer' domains are different, the values of flow parameters have to be converted when exchanging between two domains. During the simulation, the parameters are the same as those in the inertial frame, except the tangential velocity in the non-inertial frame. The latter is transformed back onto the inertial frame by adding the product of the angular velocity of the frame and the radial position:

$$u_{k,\theta,in} = u_{k,\theta,out} + \omega r \tag{40}$$

where $u_{k,\theta,in}$ is the tangential velocity in the inertial frame and $u_{k,\theta,out}$ is the tangential velocity in the non-inertial frame.

A crucial feature of the IO approach is the existence of an overlapping region, which requires the iterative match of the two solutions. The width of this region and the exact location of its boundaries are largely arbitrary. On the contrary, in the 'multi frame of reference' (MFR) method by Luo *et al.* [22], the 'inner' and 'outer' steady-state solutions are implicitly matched along a single boundary surface. The choice of this surface is not arbitrary, since it has to be assumed a priori as a surface where flow variables do not change appreciably either with θ or with time.

Solution of Model Equations

In order to solve the governing equations numerically, some approaches have been proposed for discretizing the partial differential equations. The control volume approach using a staggered arrangement of variables with a 'power-law' scheme is popularly adopted in simulation of two-phase flow in stirred tanks. The first step in the control volume method is to divide the domain into discrete control volumes using the staggered grid. The idea is to evaluate scalar variables, such as pressure, density, temperature *etc.*, at ordinary nodal points but to calculate velocity components on staggered grids centered around the cell faces. The staggering of the velocity avoids the unrealistic behavior of the spatially oscillating pressure field. A further advantage of the staggered grid arrangement is that it generates velocities at exactly the locations where they are required for the scalar transport-convection-diffusion-computations. Hence, no interpolation is needed to calculate velocities at the cell faces [23].

The second step is the integration of the governing equations over a control volume to yield a discretized equation at its nodal point P:

$$a_{\mathrm{P}}\phi_{\mathrm{P}} = \sum_{nb} a_{nb}\phi_{nb} + b \tag{41}$$

where b is the source term. Besides, the central, upwind and hybrid differencing schemes are used for discretization. The central differencing method is not suitable for general purpose convection-diffusion problems because it generates large error for solution at large values of the cell Peclet number. The upwind, hybrid or power-law differencing scheme is highly stable, but suffers from false diffusion in multi-dimensional flows if the velocity vector is not parallel to one of the coordinate directions. Higher order schemes, such as QUICK, can minimize false diffusion errors but are computationally less stable. However, if used with care and judgment, the QUICK scheme can give very accurate solutions of convection-diffusion problems.

The final step is the solution of the algebraic equations. The SIMPLE algorithm is often adopted to resolve the coupling between pressure and velocity. The acronym SIMPLE stands for Semi-Implicit Method for Pressure-Linked Equations. The algorithm was originally put forward by Patankar and Spalding [24] and is essentially a guess-and-correct procedure for the calculation of pressure on the staggered grid arrangement. To initiate the SIMPLE calculation process, a pressure field p^* is guessed. Discretized momentum equations are solved using the guessed pressure field to yield the velocity field. Then, the correction p' is defined, as the difference between the correct pressure field p and the guessed pressure field p^*, so that

$$p = p^* + p' \tag{42}$$

Similarly we define velocity corrections u' and v' to relate the correct velocities u and v to the guessed velocities u^* and v^*

$$u = u^* + u' \tag{43}$$

$$v = v^* + v' \tag{44}$$

Substituting the correct pressure field p into the momentum equations yields an equation for deciding the field of correction p'.

The pressure correction equation is susceptible to divergence unless some under-relaxation is used during the iterative process. The new, improved, pressure p^{new} is obtained with

$$p^{\mathrm{new}} = p^* + \alpha_p p' \tag{45}$$

where α_p is the pressure under-relaxation factor. The velocities are also under-relaxed. The iteratively improved velocity components u^{new} are obtained from

$$u^{new} = \alpha_u u + (1 - \alpha_u) u^{(n-1)}$$

(46)

where α_u is the velocity under-relaxation factors with values between 0 and 1.0, u is the corrected velocity component without relaxation in Eq. (43) and $u^{(n-1)}$ represents its value in the previous iteration. A proper value of the under-relaxation factor α is essential for cost-effective simulation. Too large a value of α may lead to oscillatory or even divergent iterative solutions, while too small a value will result in extremely slow convergence. Unfortunately, the optimal value of under-relaxation factor is flow dependent and must be sought on a case-by-case basis [23].

For two phase flow sharing the same pressure field, two continuity equations should be satisfied at the same time. They are combined together so as to obtain the pressure-correction formula [25]:

$$\left[(a_p)_c + (a_p)_s \right] p'_P = \sum_{nb} \left[(a_{nb})_c + (a_{nb})_s \right] p'_{nb} + f_c + f_s + D_c + D_s$$

(47)

where f_k is the turbulent diffusive term due to the asymmetry of the mass diffusive term, and D_k is the mass imbalance term reflecting the phase conservation over a cell due to the pressure field being not compatible with the velocity field. The volume fraction of the continuous phase is obtained by solving its continuity equation, and the volume fraction of the dispersed phase can then be obtained from Eq. (3). The non-linearity in the phase momentum and turbulence equations is handled by standard under-relaxation techniques. The solution is considered converged when the residual in the equations solved become smaller than a prescribed tolerance.

Boundary Conditions

1) Symmetry axis: $u_{c,r}=u_{c,\theta}=u_{d,r}=u_{d,\theta}=0$, and otherwise $\partial\phi / \partial r = 0$.

2) Free surface: the surface is assumed to be flat, then $u_{c,z}=u_{d,z}=0$, and otherwise $\partial\phi / \partial z = 0$.

3) The solid surface: no-slip condition is the appropriate condition for the velocity components at solid walls including the wall, bottom, baffles, shaft and impeller.

4) The wall function is adopted to resolve the flow velocity and turbulent properties at the nodes adjacent to all solid walls when the turbulent flow in stirred tanks is simulated.

Gas-Liquid Flow in Stirred Tanks

Rushton disk turbine is often employed in gas-liquid stirred tanks, because it can provide powerful shear force, which can be used to break up bubbles into smaller ones. Many studies on the numerical simulation of the flow in gas-liquid stirred tanks with a Rushton disk impeller, and the gas holdup distribution in the impeller stream and bulk flow has been reported. Flooding of the tank means that the gas is not well dispersed and rises up in a limited region around the shaft. To show the different gas-liquid flow patterns including flooding, the spatial distribution of gas holdup has been measured in a wide range of stirring speed and gas holdup level in some works. Although the standard k-ε model is the most popular one in use, it often leads to poor results for the flow subjected to complicated distribution of strain rate, *e.g.*, in swirling flow and curved streamline flow.

Improvement of numerical methods and turbulence model is very crucial for obtaining reasonable simulation of stirred tanks. In a stirred tank, the flow around the rotating impeller blades interacts with the stationary baffles and generates a complex, three-dimensional, recirculating turbulent flow. A new swirl number, R_s, is proposed for the gas-liquid flow in a stirred tank, and the k-ε model is modified accordingly by introducing R_s into the energy dissipation equation [26]. The large eddy simulation (LES) model has shown great potential in understanding the fluid flow behaviors in recent years. A Eulerian-Eulerian two-fluid model has also been proposed for gas-liquid flow using LES for both gas and liquid phases in a three-dimensional frame [27].

Flow Field Structure

The typical maps for gas and liquid velocity vectors in stirred tanks are shown in Fig. **1** (Rushton impeller, T=450 mm, ω=27.8 rad/s, Q=1.67×10⁻³ m³·s⁻¹). Two large eddies are formed close to the outer verge of the impeller blades, similar to these in the single phase stirred tank. In the right upper corner, another smaller eddy is formed due to the buoyancy of rising gas. It is also observed that the impeller discharged stream is a little inclined upward by the buoyant action of gas in Fig. **1a**. Gas and liquid behave differently in their motion in stirred tanks. Gas bubbles out of the sparger rise up to the impeller and are dispersed by the impeller to other regions. Compared with the eddies of liquid phase, they are smaller, and located at different positions as in Fig. **1b**.

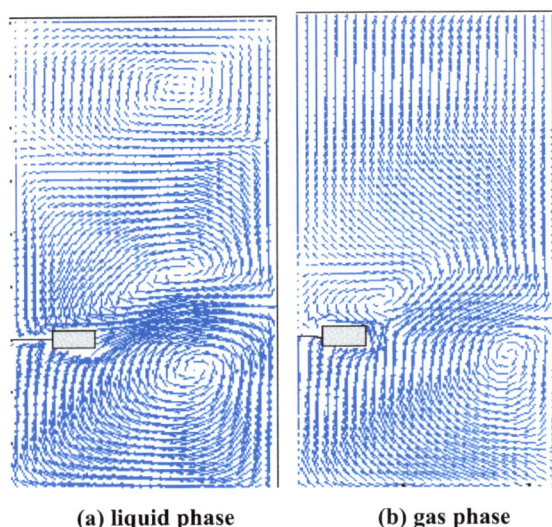

(a) liquid phase   (b) gas phase

Figure 1: Velocity vector maps of gas-liquid flow in the stirred tank [28].

It seems that the swirl modification of the mathematical model is necessary to match better the simulation results with the experimental data. Fig. **2** shows the comparison between the predicted results and experimental data of the resultant velocity u along a vertical line at r=73 mm in a stirred tank with the diameter of 288 mm driven by a single Rushton impeller. It is seen that the swirl modification model provides more reasonable results.

By LES, the instantaneous velocity vector maps of the gas and liquid phases in the r-z plane located midway between two blades are presented in Fig. **3** for a tank with the diameter of 288 mm. It is obvious that the flow pattern in the tank is dynamic and very complex, and such fine structures could not be well predicted by the k-ε model. There are many small vortices in both the gas and the liquid flow fields. Furthermore, the flow in the tank is not symmetrical as most literature presumed it to be.

Fig. **4** shows the profiles of the gas resultant velocities at different radial positions compared with the experimental data. As the gas-liquid turbulent flow in a stirred tank is quite difficult to measure, especially at high gas holdup, few experiments have been reported on gas or liquid velocity field and phase holdup. In comparison to the predictions by the k-ε model, the predictions from the LES are closer to the experimental data, especially in the positions near the impeller tips. This is because the LES simulation is more powerful in capturing the anisotropic nature of turbulence in the impeller region and the impeller stream, much superior to the k-ε model based on the assumption of isotropic turbulent flow.

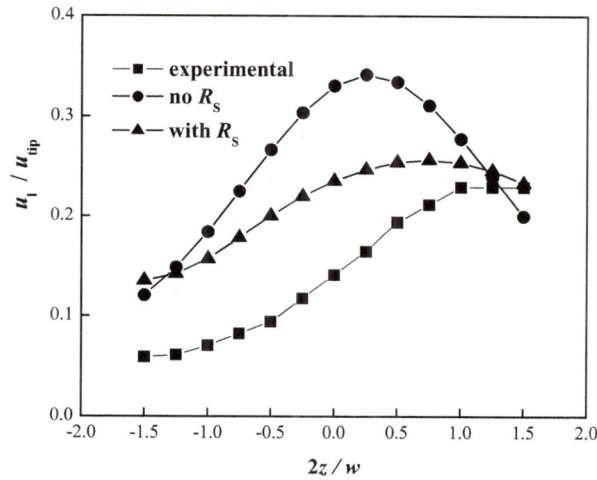

Figure 2: Predictions of mean liquid velocity with swirl modifications compared with the experimental data [26].

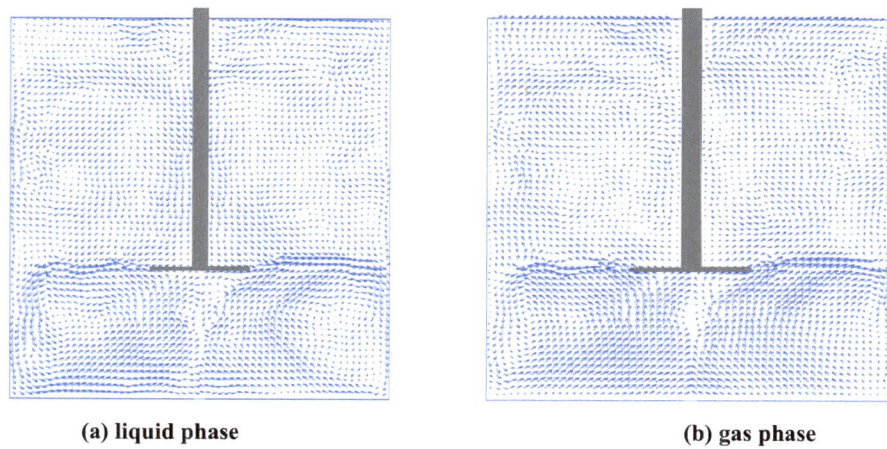

(a) liquid phase   (b) gas phase

Figure 3: Instantaneous velocity fields in *r-z* plane by the LES [27].

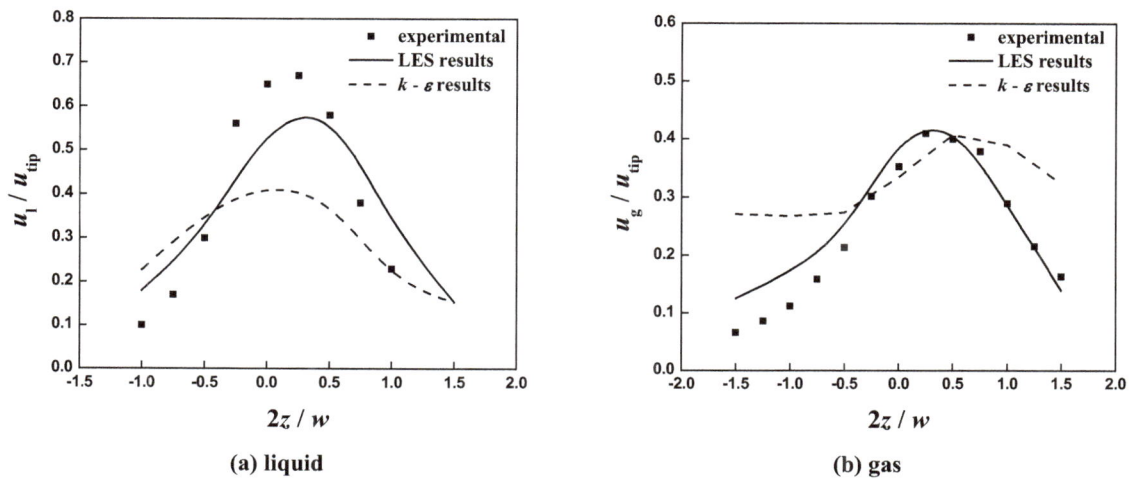

(a) liquid   (b) gas

Figure 4: Predicted velocity profiles by LES [27] compared with the experimental data [29] and the standard *k-ε* model prediction at different vertical positions (ω=62.8 rad·s^{-1}).

Gas Holdup Distribution

Fig. **5** shows the gas holdup contour maps from experiments at the same gas flow rate under different stirring speeds. The gas is dispersed by the impeller and spreads into the upper bulk regions gradually as the stirring speed is increasing. Meanwhile, the recirculation of liquid is also intensified and some gas is entrained into the lower bulk region. Higher stirring speeds lead to more entrained gas in the lower bulk region.

In general, the gas holdup in the impeller stream decreases gradually with the radial positions. When the turbine fails to disperse the gas in Fig. **5a**, the highest gas holdup is around the impeller shaft in the upper bulk flow region. As flooding occurs when gas flow rate is raised above the dispersion capacity of the impeller, the gas holdup presumes a spatial distribution of similar features. On the other hand, the non-flooding normal operations are observed in Fig. **5b-5d**, featured with quite lower gas holdup at the shaft than that in the broad region of the upper bulk region. At even higher stirring speed the topological structure is similar with Fig. **5d**, but the level of gas holdup is further raised significantly. Observations from Fig. **5** suggest that, except for the case of flooding in Fig. **5a** with low stirring speed, there is always an area of high gas holdup at a small distance off the outer edge of impeller blades.

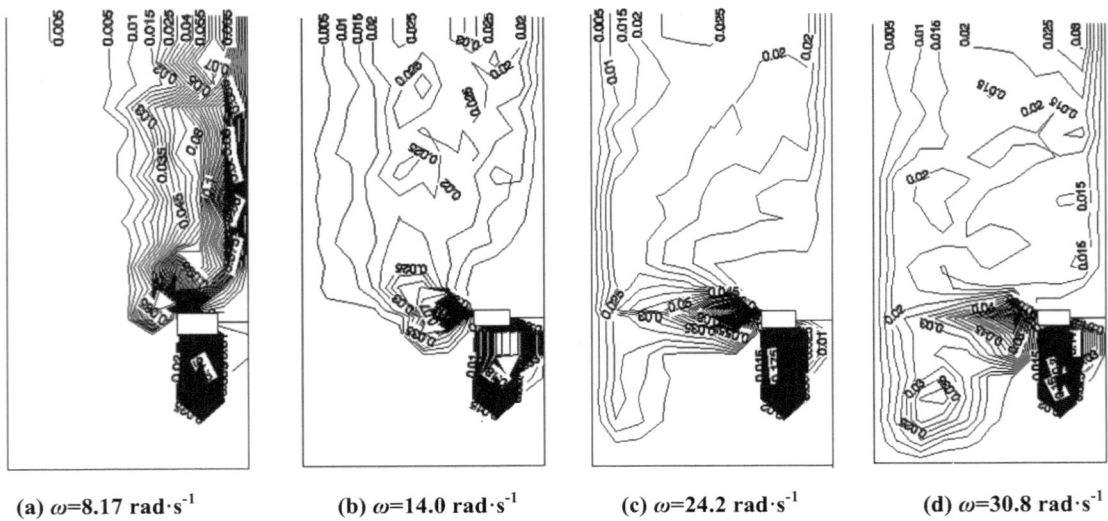

(a) ω=8.17 rad·s^{-1} (b) ω=14.0 rad·s^{-1} (c) ω=24.2 rad·s^{-1} (d) ω=30.8 rad·s^{-1}

Figure 5: Gas holdup contour maps of gas-liquid flow in a stirred tank (Q=2.22×10^{-4} m^3·s, T=H=380 mm).

Simulation of Flooding

Flooding is the least desired operating condition in stirred tanks. From the contour maps in Fig. **5**, it is apparent that Fig. **5a** presents the flooding regime, Figs. **5b** and **5c** are identified as the loaded regime, and Fig. **5d** shows the completely dispersed pattern.

The critical conditions for the transition between flooding and loading flow patterns in a stirred tank have been identified through numerical simulation by Wang *et al.* [30]. It was found that the computation was difficult to converge: the numerical residual error decreased to certain levels and then fluctuated numerically, following the physical flow instability. It was experimentally observed that the transition to flooding happens as a sudden deterioration of liquid pumping capacity and gas holdup uniformity, resulting in the rare appearance of gas bubbles in the lower bulk region. The flooding is characterized by the channeling of gas stream around the impeller shaft instead of the transition from a better dispersion to non-uniform one or the disappearance of gas bubbles in the lower bulk region.

Attempting to correlate the critical condition for the transition between the loaded and flooding regimes, Paglianti *et al.* [31] plotted the Froude number $Fr = N^2D/g$ against the flow number $Fl_g = Q_g/ND^3$ at the flooding/loading transition. The experiment identified two data points on the transition. Point A in Fig. **6** is the transitional state between the flooding (Fig. **5a**) to the loaded regime (Fig. **5b**), and point B is the critical condition for the transition from the complete dispersion (Fig. **5d**) to the flooding regime because of the increase of gas flow rate at the constant stirring speed. The data are in very good agreement with the data from Nienow *et al.* [32] and Paglianti *et al.* [31]. A line based on the correlation is also plotted as a reference:

$$Fr = \frac{1}{30}\left(\frac{T}{D}\right)^{3.5} Fl_g \tag{48}$$

However, the significant correlation between Fr and Fl_g in Fig. **6** seems to demonstrate a typical non-linear relationship other than the linear one.

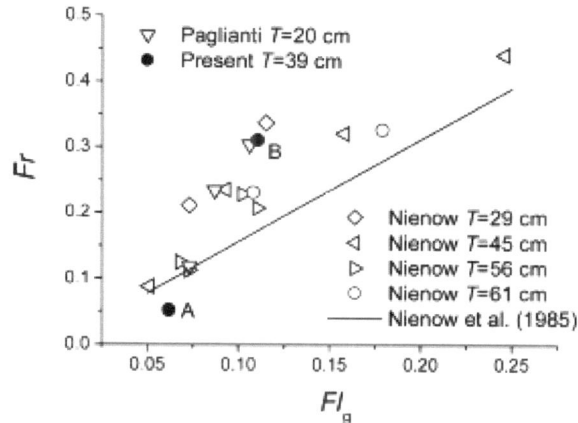

Figure 6: Comparison of the simulation data for the flooding/loading transition with the experimental data and correction [31, 32].

Energy Dissipation Feature

Energy dissipation is an important property in a stirred tank because it provides a direct measure of energy input into the system by the impeller. Fig. **7** presents the distribution of the energy dissipation per mass of fluid in the plane midway between two successive blades throughout the tank. The energy dissipation of the liquid phase is mainly focused on the impeller swept region and the impeller discharged stream. Very high dissipation rate occurs at the blade edges and near the wake of the blades. The region of high dissipation rate is smaller than that in the single phase flow, suggesting that the existence of the bubbles reduces the production of turbulence. It is interesting to note that the region of high energy dissipation of gas is rather large, both in the impeller outflows and in the upper part of the upper bulk circulation zone due to the buoyancy of gas bubbles.

Solid-Liquid Flow in Stirred Tanks

The main purpose of the widely used solid-liquid agitated tanks in process industry is to enhance the heat and mass transfer between two phases. A number of investigations were focused on achieving empirical correlations, mostly on the distribution of solid holdup in stirred tanks and the criteria for off-bottom solid suspension. In recent years, reports about the above topics are available through experimental methods such as LDV, PDA and CT instruments. However, experimental measurements are insufficient to provide the insight of solid-liquid suspension in stirred tanks or for designing and scaling-up purposes. Numerical methods are now more and more popularly adopted to simulate solid-liquid flow in stirred tanks to provide significant details of hydrodynamic information complimentary to experimental methods and empirical correlations. However, there are still some difficulties in simulating multi-phase flows in stirred tanks, for example, in providing an accurate representation of the impeller action and an appropriate model of interphase interaction and flow turbulence.

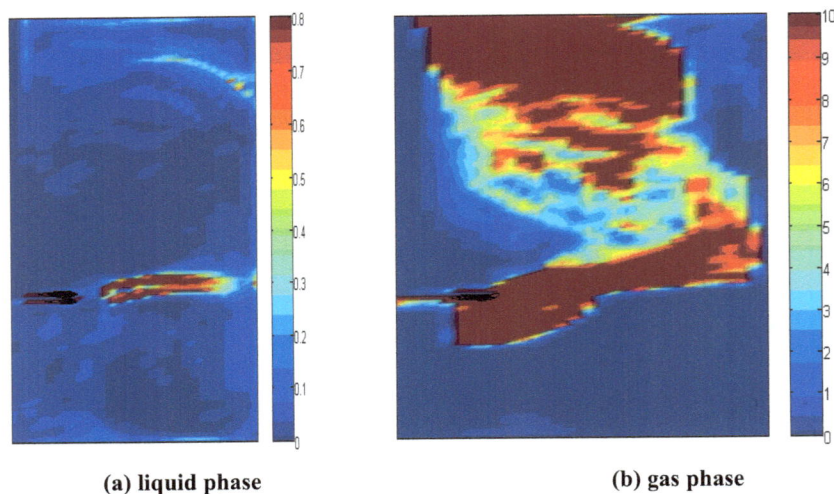

(a) liquid phase   (b) gas phase

Figure 7: Energy dissipation ($m^2 \cdot s^{-3}$) predicted by LES in the *r-z* plane between two impeller blades [27].

Solid-Liquid Flow with a Radial Flow Impeller

Flow Field Information

In Fig. **8**, it is generally observed that the flow pattern of solid phase is similar to that of liquid, revealing that fine solid particles follow the liquid closely. It also shows that the impeller discharged stream is somewhat inclined downward for the settling of solid particles. The circulation patterns are similar for both phases: there are two primary circulation loops in the upper and lower parts of the impeller in the *r-z* plane. In the velocity vector plot of the solid phase, a small recirculation loop appears near the vertical axis of the tank below the impeller which may be the result of the accumulation of the solid particles in this region. This observation is also true for stirred tanks operated under laminar or mild flow condition.

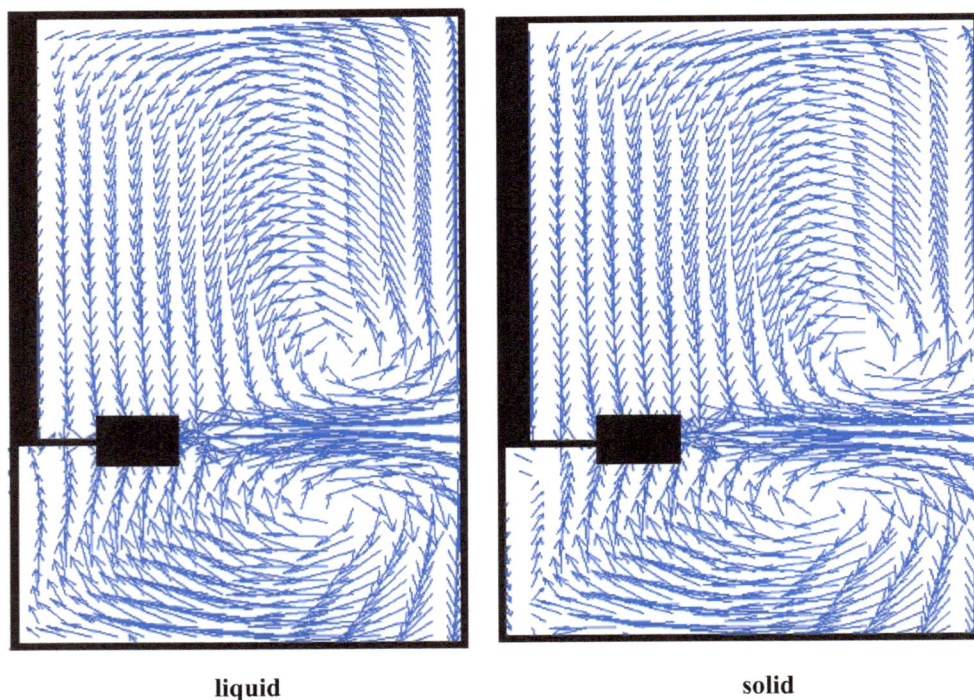

liquid   solid

Figure 8: Velocity vector plots of continuous and dispersed phases (*T=H*=0.294 m, *D=C=T/3,d*$_s$=232.5 μm, *N*=300 rpm) [33].

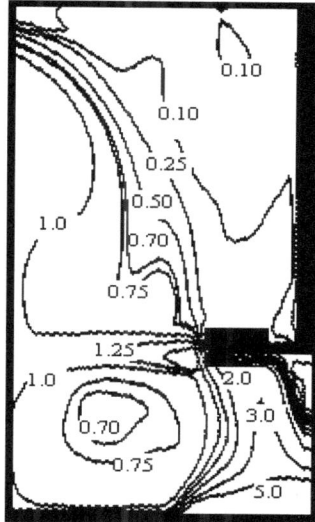

Figure 9: Contour plots of normalized concentration of solid phase ($T=H=0.294$ m, $D=C=T/3$, $d_s=232.5$ μm, $N=300$ rpm) [33].

Distribution of Solid Particles

The distribution of solid particles in the stirred tank with a radial impeller in Fig. **9** suggests that the maximum solid concentration occurs on the center of the tank bottom, and the level gradually decreases from the bottom to the free liquid surface. The solid concentration contour maps show a small circular region with low concentration below the impeller plane. In the region above the impeller plane, there is also a circulation flow but no region with low solid concentration.

Numerical Prediction of Critical Impeller Speed

An important criterion in designing a solid-liquid stirred reactor is the complete suspension of solid particles off the vessel bottom, at which point all of the particles are in motion, and all the solid surface are available for mass and heat transfer and chemical reaction. The impeller speed in accordance with this status is identified as the critical impeller speed N_{js}. Upon further increasing of the impeller speed, the transport rate between two phases will increase only slightly.

The pioneering work conducted by Zwietering [34] in the past decades on the critical impeller speed is considered the most systematic work performed even viewed nowadays. By visual observation of the solid particle motion on the stirred tank floor, the complete suspension state was defined as this in which all particles were in motion and no particle remained on the floor for more than 1 to 2 s. An empirical correlation of N_{js} was proposed as

$$N_{js} = S \frac{v_{c,lam}^{0.1} d_s^{0.2} (g\Delta\rho / \rho_c)^{0.45} X^{0.13}}{D^{0.85}} \tag{49}$$

where S is a dimensionless parameter defined as a function of the tank configuration and the impeller type, *i.e.*,

$$S = f\left(\frac{T}{D}, \frac{T}{C}\right) \tag{50}$$

The work of Armenante *et al.* [35, 36] focused on the effect of impeller clearance off the tank bottom on N_{js}. The correlation of Rushton impeller for $C/D > 1.5$ is

$$N_{js} = 3.06 \left(\frac{T}{D}\right)^{1.33} \exp\left(0.44 \frac{C_b}{T}\right) D^{-0.63} X^{0.13} d_s^{0.20} v_{c,lam}^{0.1} \left(\frac{g\Delta\rho}{\rho_c}\right)^{0.45} \tag{51}$$

Another approach to determine the critical impeller speed is to examine the variation of solid concentration profile with impeller speed. It was found that more solid particles will be suspended and the solid concentration will increase as the impeller speed is increased gradually. When the critical impeller speed is reached, few solid particles can be added to the bulk liquid flow. Further increase of the impeller speed will result in the homogenization of the suspension and the solid concentration will decrease in some locations.

Baldi *et al.* [37] introduced a theoretical model to determine N_{js} based on the assumption that the energy needed to suspend solid particles was proportional to that of turbulent vortices. Another theoretical model was proposed based on the balance between the upward flow velocity and the particle setting velocity [38].

Wang *et al.* [39] employed CFD to analyze the solid-liquid flow in stirred tanks and adopted the following three criteria for predicting N_{js} :

Criterion 1: By observing the solid concentration profiles against the varying impeller speeds at a specific monitor point, N_{js} is defined as the impeller speed corresponding to the peak concentration or that where a sudden change in the slope of the profile occurs.

Criterion 2: N_{js} is determined by inspecting the axial velocity of solid phase in cells closet to the tank bottom at different impeller speeds. When the speed is lower, the velocity in a fairly large region in the vicinity of the tank floor center is negative, meaning that the solid particles settle down. With the increased impeller speed, the velocity at the center of the tank bottom becomes positive, indicating the solid particles are suspended.

Criterion 3: Because the solid particles tend to deposit at the center of tank floor, it is possible to determine N_{js} by inspecting the variation of solid concentration against the impeller speed at such position, experimentally or numerically. The impeller speed to decrease the solid concentration at tank bottom below that of compact sediment is defined as N_{js}.

The predicted values of N_{js} by the above criteria are compared with empirical correlations in Table **2**. The results by Criteria 2 and 3 are both close to the correlations but give somewhat lower predictions. It seems that Criteria 2 and 3 are more consistent and reasonable.

Table 2: Comparison of N_{js} calculated with results from correlations

	Case a	Case b	Case c
Zwietering's Eq. (49)	425	585	717
Armenante's Eq. (51)	472	661	808
Criterion 1	350	485	550
Criterion 2	412	562	637
Criterion 3	412	562	637

Solid-Liquid Flow with an Axial Flow Impeller

Compared with a Rushton turbine, it is more difficult to build the suitable computational grids for an axial flow impeller when dealing with the edge surface of the impeller. It is known that the accuracy of CFD predictions critically relies on the manner in which complex impeller and tank configurations are represented by regular Eulerian computational grids and numerical discretization schemes. In the early 1990s, fully predictive modeling methodologies that were known as the sliding-mesh method [40], the inner-outer approach [41], and the multiple-frames of reference method [22] were developed for the explicit modeling of impeller geometry.

Although, baffled stirred tanks are widely used for better mixing of solid particles and liquid, there are cases in which the use of unbaffled tanks may be desirable. Baffles are usually unnecessary in the case of very viscous fluids ($Re < 20$), because they result in dead zones where vortex formation is inactive by the low stirring speeds and the high friction on the cylindrical wall. Consequently, the mixing performance may actually be worsened. Unbaffled tanks are also used as crystallizers, in which the presence of baffles may promote particle attrition. Besides, higher fluid-particle mass-transfer rates may be obtained in unbaffled tanks for a given power consumption, which may be desirable in many processes. Shan *et al.* [8] conducted both experiments and numerical simulation of an unbaffled stirred tank of 300 mm diameter agitated with a pitched-blade turbine downflow.

Experimental Methods

There are many experimental studies focused on solid-liquid flow in stirred tanks. Also, a lot of methods, for example, optics using an endoscope technique [42] and pressure gauge technique (PGT) based on the measurement for pressure variation on the bottom as a result of the presence of suspended solids [43, 44], have been developed to investigate the local solid hold-up and the critical speed for just-off-bottom suspension. One of the optical methods for local solid concentration was introduced and performed as follows [8]. Firstly, the impeller off-bottom clearance and agitation speed were set and the overall solid holdup was set at a desired level. The measurement of the local solids concentrations, using a PC-6A fiber optic probe (manufactured by Institute of Process Engineering, Chinese Academy of Sciences), was conducted in an obscured environment to prevent daylight from interfering with the optical measurements. The details of this method may be referred to other literature [8, 45, 46].

A Method to Deal with Pitched-Blade Impellers

Both the Rushton turbine and the pitched-blade impeller are widely used in stirred tanks. However, the simulation of the latter is obviously more difficult, because there may be some difficulties in dealing with the moving blade surface with irregular geometry.

The so-called 'vector distance' is introduced to determine whether a node under consideration is in the liquid domain or not [8]. For example, if point A is outside the impeller, as shown in Fig. **10a**, and its distances to the two surfaces of the impeller blade are expressed by vectors $\vec{a}_1$ and $\vec{a}_2$, respectively, their dot product $\vec{a}_1 \cdot \vec{a}_2$ is positive; if point A is inside a blade, as shown in Fig. **10b**, the dot product of two distance vectors $\vec{a}_3 \cdot \vec{a}_4$ would be negative; if A is just on the surface, the dot product is equal to zero. With such a simple geometric rule, all the nodes in the liquid domain can be identified, given that all surfaces of the impeller are already specified. Thus, the smooth blade surface is now approximated by a rough one. This would produce some numerical errors to the simulation results, but the errors are expected to decrease as the grid is refined further.

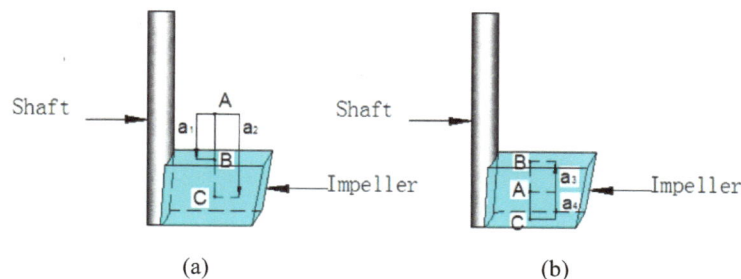

(a)   (b)

Figure 10: Geometric rule for identifying an active node for velocity components and pressure: (a) node outside impeller; (b) node inside impeller.

Simulated Velocity Field

Shan *et al.* [8] provided the two-dimensional vector plots of the continuous phase (water) and dispersion solids (α_{av}=0.005 (volume fraction), ρ_d=1970 kg/m^3, d_s=80 μm). The flow patterns for the two phases seem to be quite similar, naturally because the solid particles have a terminal velocity close to liquid flow. Secondary circulation loops in the flow field are revealed both above and below the impeller, in addition to the main circulation near the blade tips. These circulation regions are the primary cause of the segregation of solid particles in the system, and their positions are dependent on the impeller speed. In the r-θ plane, a vortex is formed behind the impeller in both continuous and dispersion phases, because of relatively low pressure in this region.

Contour Profile of the Solids Concentrations

From the contour profiles of the solids concentration (Fig. **11**), a relatively high concentration region exists below the impeller, which is in accordance with the zone of low pressure in the flow field. The high concentration near the wall in the upper tank region can be attributed to the circumferential flow and centrifugal force. A similar observation was made with a Lagrangian simulation approach [47, 48]. The solids collided with the wall, thus losing momentum, resulting in inability of the liquid to carry them through. Consequently, the particles have a tendency to settle instead of moving with their initial trajectory. The concentration near the shaft is very low, caused by the central vortex. With the increased impeller speed, the concentration below the impeller is also increased, and the concentration of the regions near the free surface and shaft shrank. Furthermore, more and more particles were transformed to the upper tank region. However, further increasing of the impeller speed above the critical suspension speed will not result in higher homogeneity in the tank, which implies that the efficiency that is involved in achieving higher homogeneity simply by increasing the impeller speed above N_{js} is not obvious. Besides, a vortex exists in the lower impeller zone near the bottom of the tank, which can be the result of the high shear stress of the continuous phase.

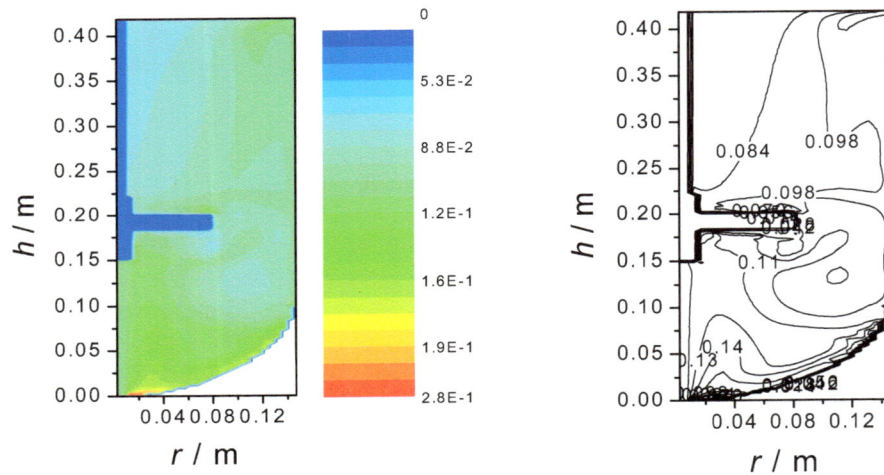

Figure 11: Solid particle concentration distribution (α_{av}=0.1, N=500 rpm) [8].

Liquid-Liquid Flow in Stirred Tanks

Dispersion of two immiscible liquids in mechanically stirred tanks is commonly encountered in chemical and biochemical processes, for instance, in liquid-liquid extraction, in suspension polymerization, *etc.* The purpose of such an operation is to mix two phases and increase the interfacial area by intensifying the dispersion of one liquid into another and to enhance consequently the interphase heat/mass transfer and the chemical reaction.

Investigations previously conducted on liquid-liquid dispersion in stirred tanks were mainly focused on the measurements of the drop size distribution [19, 49], which is a crucial factor for heat and mass transfer. However, the rate of transfer, heavily dependent on the breakage and coalescence of droplets [50], is controlled by the phase dispersion and distribution in the macroscopic flow field [51]. In the past studies, LDA technique was often used to measure the velocity field. On the other hand, more attempts were made to measure the holdup profiles of the dispersed phase by using, for example, the sample withdrawal technique [52-54], a conductivity probe and the ultrasonic method.

CFD method was also employed to predict the hydrodynamic characteristics of liquid-liquid flow in stirred tanks. Compared with other two-phase flows, the difficulties are mainly due to the additional complexity generated by the drop deformation, breakage, coalescence, and occurrence of circulation inside the drops. In the numerical simulations, the Eulerian-Lagrangian approach has been adopted. Zhu and Vigil [55] employed an algebraic slip mixture model to simulate the banded liquid-liquid Taylor-Couette-Poiseiulle flow. Wang and Mao [56] simulated the three- dimensional flow field of stirred tank (Rushton impeller, $T=H=0.154$ m) using the two-fluid approach with incorporation of the phase holdup fluctuation correlations appearing in the Reynolds time-averaged governing equations. The k-ε turbulence model was employed to describe the turbulence in the system.

Mathematical Model

Eulerian-Lagrangian and Eulerian-Eulerian approaches are both used in liquid-liquid two-phase flow simulation. However, the latter shows many computational advantages over the first one in the case of high dispersed-phase concentration. In the Eulerian-Eulerian approach, the droplets are considered as rigid spheres without deformation as far as the interphase friction is concerned, and the fluid flow inside the droplet as well as the processes of breakage and coalescence is neglected.

Interphase Interaction Term

The analysis presented by Joshi [57] suggested that the lift force and the added mass force were important in numerical simulation of multiphase flow in a bubble-column reactor. The drag force and the added mass force proposed by Gosman *et al.* [5], had much influence on gas-liquid and solid-liquid two-phase flow in stirred tanks; Ljungqvist and Rasmuson [7] reported that the added mass force and the lift force had no remarkable influence on the slip velocities in solid-liquid flows in stirred tanks except in the impeller region.

The above three forces were included in the simulation by Wang and Mao [56]. The drag force between the continuous and dispersed phases can be expressed as [9]

$$F_{ci,\text{drag}} = -F_{di,\text{drag}} = \frac{3\rho_c\alpha_c\alpha_d C_D |\mathbf{u}_d - \mathbf{u}_c|(u_{di} - u_{ci})}{4d_d} \tag{52}$$

where C_D is the nondimensional drag coefficient depending on the droplet Reynolds number:

$$Re_d = \frac{\rho_c d_d |\mathbf{u}_d - \mathbf{u}_c|}{\mu_{c,\text{lam}}} \tag{53}$$

and the local value of droplet diameter d_d is suggested by Nagata [58]:

$$d_d = 10^{-2.316+0.672\alpha_d} v_{c,\text{lam}}{}^{0.0722} \varepsilon^{-0.914} (\sigma g / \rho_c)^{0.196} \tag{54}$$

The correlation is adopted for the advantage that the deformation of droplets was taken into account [13]:

$$C_D = \frac{24}{Re_m} \left(1 + 0.1 Re_m{}^{0.75}\right) \tag{55}$$

$$Re_m = \frac{\rho_c d_d |\mathbf{u}_d - \mathbf{u}_c|}{\mu_m} \tag{56}$$

$$\mu_{\mathrm{m}} = \mu_{\mathrm{c,lam}}\left(1-\frac{\alpha_{\mathrm{d}}}{\alpha_{\mathrm{m}}}\right)\exp\left(-2.5\alpha_{\mathrm{m}}\frac{\mu_{\mathrm{d,lam}}+0.4\mu_{\mathrm{c,lam}}}{\mu_{\mathrm{d,lam}}+\mu_{\mathrm{c,lam}}}\right) \tag{57}$$

The added mass force originated from the acceleration of one phase relative to another can be calculated from [59]:

$$F_{ci,am} = -F_{di,am} = -C_{\mathrm{am}}\alpha_{\mathrm{d}}\rho_{\mathrm{c}}\left[\left(\frac{\partial u_{ci}}{\partial t}+u_{cj}\frac{\partial u_{ci}}{\partial x_j}\right)-\left(\frac{\partial u_{di}}{\partial t}+u_{dj}\frac{\partial u_{di}}{\partial x_j}\right)\right] \tag{58}$$

where C_{am} is expressed as [60]

$$C_{\mathrm{am}} = 0.5(1+2.78\alpha_d) \tag{59}$$

For a moving rigid spherical particle, a lift force perpendicular to the average flow direction occurs. The lift force is given by [61]

$$F_{ci,\mathrm{lift}} = -F_{di,\mathrm{lift}} = C_{\mathrm{lift}}\alpha_{\mathrm{d}}\rho_{\mathrm{c}}\varepsilon_{ijk}\varepsilon_{klm}\left(u_{dj}-u_{cj}\right)\frac{\partial u_{cm}}{\partial x_l} \tag{60}$$

where C_{lift} is the lift force coefficient with a value of 0.5.

Mean Flow Field

From the mean velocity field of the continuous and dispersed phases in the form of vector plots in the *r-z* plane, it is seen that two ring vortices exist above and below the impeller plane. In addition, a high-velocity radial impeller stream tilts slightly upward because of the buoyant effect of the lighter dispersed phase. The velocity fields of two phases are very similar to each other in most regions. However, the dispersed phase tends to float upward at the top of the tank close to the impeller shaft, under the effect of buoyancy in this area. From comparative trial simulations it has been found that the added mass and the lift forces are less significant for the model predictions, and that the drag force is the dominating force of interaction.

Turbulence Features

The influence of the dispersed phase on turbulence structure was experimentally investigated and reported in terms of the root-mean-square (rms) of the fluctuating velocities [51]. The rms of the continuous phase with various values of the average dispersed-phase holdup is calculated through Eq. (4) given *i=j*. It has been observed that the rms values of the continuous phase decrease with the increase of the average holdup of the dispersed phase, especially in the impeller region, implying that the presence of a larger amount of the dispersed phase produces a stronger suppression on the turbulence of the continuous phase.

Holdup Profiles of Dispersed Phase

The predicted contour plots of the normalized dispersed phase holdup are shown in Fig. **12** [62]. With a relatively lower impeller speed, the dispersed phase seems to accumulate around the impeller shaft at the top of the tank, in accordance with experimental findings. With an increased impeller speed, the distribution of the dispersed phase becomes gradually homogeneous.

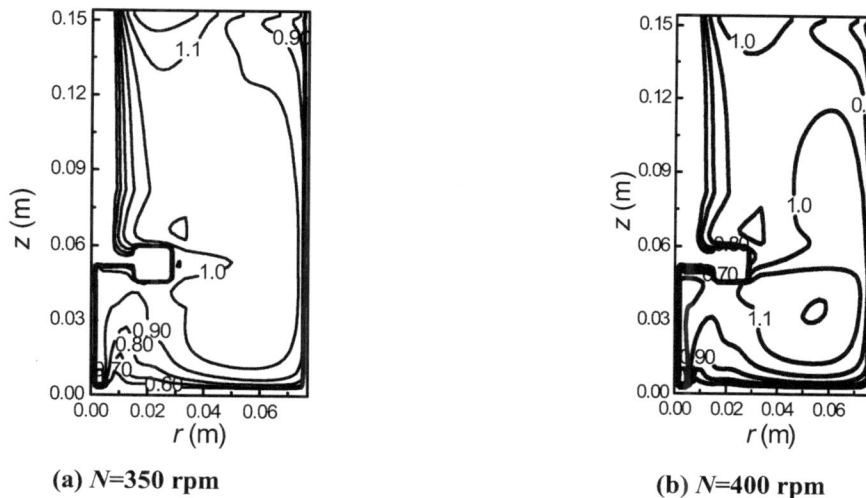

(a) *N*=350 rpm   (b) *N*=400 rpm

Figure 12: Predicted contour plots of normalized holdup of the dispersed phase.

THREE-PHASE FLOW IN STIRRED TANKS

Compared with two-phase flow, three-phase flow is more complex due to the presence of a third phase. In experimental work, the distribution of phase holdup and the rate of heat/mass transfer between phases are hard to be determined. For numerical simulations, the interaction between the two dispersed phases and the contribution of dispersed phases on the turbulence of the continuous phase make the numerical solution of the governing equations even more challenging. Gas-liquid-solid three-phase systems are often reported in the literature, but other three-phase systems are rarely approached numerically.

Liquid-Liquid-Solid Flow

Liquid-liquid-solid three-phase stirred tanks are popular in process industries. Typical applications include reactive flocculation and solid catalyzed liquid-liquid reaction, *etc*. The understanding of the hydrodynamic characteristics, such as suspension of solid particles, dispersion of dispersed liquid phases and their spatial distribution in a stirred tank is critical for the determination of the rates of heat/mass transfer, and therefore of great importance for the reliable design and scale-up of such chemical reactors [63].

Measurement of Phase Holdup

For liquid-liquid-solid three-phase flow in stirred tanks, it is desirable to obtain the information on the state of the dispersion of the dispersed phases. Wang *et al.* [63] provided experimental measurements of axial and radial variations of phase holdups of the two dispersed phases in a lab-scale stirred tank under different operating conditions by the sample withdrawal method. Tap water, *n*-hexane and glass beads were chosen for liquid, liquid and solid phases, respectively.

The measurement results normalized with the respective phase volume fraction are presented in Fig. **13**. It can be seen that the lower local holdup of solids appears close to the tank bottom, and the higher local holdup of oil at the top surface, indicating that both solid and oil phases are not sufficiently dispersed at such impeller speed. Increasing the stirring speed can obviously promote the dispersion of oil phase in continuous phase. It is also observed that the local holdup of the oil phase below the impeller is larger than that at the upper sections, which can be explained by the fact that some droplets adhere to the surface of solid particles as a result of better wettability to oil. This suggests that more ways of phase interaction may appear when a new phase is introduced.

Investigation on Interphase Mass Transfer

Mass transfer between two phases is an important process in multiphase systems. In a liquid-liquid-solid stirred vessel, the introduction of a solid phase is influential to the breaking up and to the coalescence of droplets and circulation in a droplet. Furthermore, the suspension of solid phase also affects the mass transfer between two liquid phases.

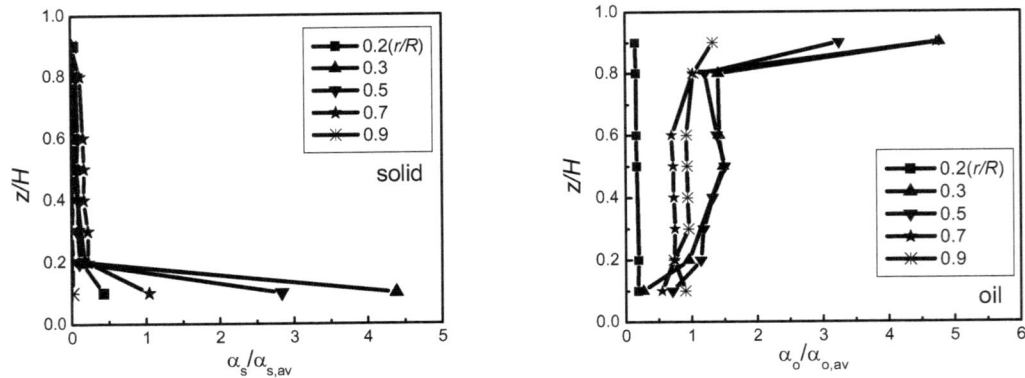

Figure 13: Axial profiles of normalized holdup of solid and oil phases ($\alpha_{s,av}$=0.10 and $\alpha_{o,av}$=0.10). Left: solid; right: oil. (N=300 rpm).

Fang *et al.* [64] examined the typical liquid-liquid extraction system *n*-butanol-deionized water-succinic acid and chose glass beads of various diameters for the solid phase. The effects of operating conditions on the volumetric mass transfer coefficients were studied. At lower agitation speeds, the introduction of a solid phase is harmful to the mass transfer between liquid and liquid phases. But with increasing agitation speed, the presence of solids reinforces turbulence intensity because of the separation of eddies generated by the relative motion between solid particles and liquid droplets, which is beneficial to liquid-liquid mass transfer. Besides, the faster agitating speed is better for uniform solid suspension, resulting in larger mass transfer areas. The mass transfer coefficients also differ by impeller types. Generally speaking, radial impellers perform better than axial ones. However, if the solid particles are not suspended uniformly enough, the enhanced drag force is an obstacle to turbulence, which is also destructive for mass transfer. Furthermore, the influence of particle diameter on $k_L a$ is negative when the diameter exceeds 100 μm [64, 65].

Prediction of Flow Field and Phase Distribution

To date, design and scale-up of multiphase stirred tanks are mainly based on empirical and semi-empirical correlations obtained from experimental data. The strategy of stagewise scale-up is expensive and time-consuming, and satisfactory scale-up of a large-scale reactor is not guaranteed. For the numerical simulation of liquid-liquid-solid three-phase flow, the Eulerian multi-fluid model is preferred. Wang *et al.* [63] treated water, oil and solid phases as different continua, interpenetrating and interacting with each other everywhere in the computation domain under consideration. The oil and solid phases are in the form of spherical dispersed droplets and particles, respectively. The effect of breakup and coalescence of droplets is neglected. The pressure field is shared by all three phases, which are exerted, respectively, by the pressure gradient multiplied by respective volume fraction. Motion of each phase is governed by respective mass and momentum conservation equations.

The RANS version of governing equations for three-phase flow is similar to that of two-phase flows. Reynolds stresses are always modeled by introducing the Boussinesq hypothesis. Although different from two-phase flows, there are still interactions between dispersed droplets and solid particles in liquid-liquid-solid three-phase flow, which have to be modeled. However, this factor was not always included in the reported literature. In the model developed by Padial *et al.* [66] for a gas-liquid-solid three-phase bubble column, the drag between solid particles and gas bubbles was modeled identically to drag between liquid and gas bubbles based on the notion that particles in the vicinity of gas bubble tend to follow the liquid. In the simulation by Michele and Hempel [67], the momentum exchange terms between the dispersed gas and solid phases were expressed as

$$F_{g\text{-}s,\,i} = -F_{s\text{-}g,\,i} = \frac{3\rho_g \alpha_s c_{g\text{-}s} \left| \mathbf{u}_s - \mathbf{u}_g \right| \left(u_{si} - u_{gi} \right)}{4 d_p} \qquad (61)$$

The combination of $c_{g\text{-}s} \left| \mathbf{u}_s - \mathbf{u}_g \right|$ was defined to be a fitting parameter determined by fitting model predictions to measured local solid holdups.

Since the two dispersed phases are presumed to be continua, as mentioned above, it is reasonable to expect the drag between solid particles and droplets behaving in the similar way as the drag force between the continuous and the dispersed phase:

$$F_{o\text{-}s,\,\text{drag},\,i} = -F_{s\text{-}o,\,\text{drag},\,i} = \frac{3\rho_o \alpha_o \alpha_s C_{D,\,o\text{-}s} \left| \mathbf{u}_s - \mathbf{u}_o \right| \left(u_{si} - u_{oi} \right)}{4 d_s} \qquad (62)$$

Numerical simulation of such a liquid-liquid-solid system seems to be successful. In the flow fields of both continuous and dispersed phases, the well-documented flow pattern generated by a disc turbine in a stirred tank is clearly illustrated. Two large ring eddies exist, above and below the impeller plane respectively, and a high-velocity radial impeller stream is also predicted. Overall, the flow fields of the three phases are very similar to each other in most parts of the domain. The velocity field of the dispersed oil phase shows a trend of drifting upwards at lower impeller speeds. This might be due to the fact that the oil phase with lower density accumulates easily in the top section of the tank. The time-averaged flow field of the solid phase reveals a small recirculation zone above the center of the tank bottom, implying that the solid particles tend to settle down in this zone.

The calculated normalized local holdups of the two dispersed phases are presented as contour plots in [63]. It is easy to observe that the distributions of oil and solid phases are both less homogeneous at low impeller speeds. The maximum oil phase holdup occurs in the center of the free surface because of the ring vortices in the upper bulk zone, in qualitative agreement with experimental investigations. The distributions of both dispersed phases become more uniform at higher agitation speeds. The maximum solid concentration is located at the center of the tank bottom due to the density difference and the ring eddy at the bottom, as it is confirmed above.

Dispersion in Gas-Liquid-Liquid Stirred Tanks

Gas-liquid-liquid systems are commonly encountered in catalytic hydrogenation, hydroformylation, carbonylation and liquid-liquid extraction. Yu *et al.* [68] studied the characteristics of three-phase dispersion and gas-liquid mass transfer in an air-water-kerosene system in a self-aspirating stirred tank reactor equipped with a draft tube. The critical speed for gas aspiration and oil dispersion, rate of gas aspiration, gas-liquid mass transfer coefficient and power consumption for some types of impeller were discussed in detail.

The critical speed for oil dispersion (N_o) means that the oil drops appear in the continuous phase (water). It indexes the ability of the impeller to disperse the oil phase. N_o is dependent on impeller diameter and immersion depth. A disk turbine with double rings fixed up and down the blades respectively (labeled as DT) was proved to be more suitable for oil phase dispersion because of its smaller value of N_o. Both rings help to form pressure gradient, which is useful for sucking in the oil phase.

The critical speed for gas inspiration (N_g) is also related to impeller diameter and immersion depth. White and Villers [69] suggested a nondimensional number:

$$C = \frac{N_g^2 D^2}{gS} \qquad (63)$$

From the experimental results, DT and mineral flotation impeller performed better in gas inspiration. In addition, this impeller did a good job in increasing the gas inspiration rate (Q_g).

In summary, DT was confirmed to be adaptive in gas-liquid-liquid stirred tanks. The oil phase is always of great proportions in industrial reactors. It is revealed that the value of N_o is decreased with increasing oil phase fraction, because a larger amount of oil phase makes up the initial interface between oil and water phases closer to the impeller, so that the initial production of oil drops becomes easier. The oil phase fraction will not influence the depth of interface between liquid and gas phase. That is why oil fraction correlates negatively with N_g. However, oil phase fraction has much influence on gas inspiration rate. With increasing oil phase fraction, the system viscosity is also enhanced, which reinforces the stability of gas cavity generated behind the blades. The existence of gas cavity is to reduce the pressure gradient, and results in the decrease of gas inspiration amount.

Dispersion and Phase Separation of Liquid-Liquid-Liquid Systems

A combination of two aqueous phases (mixture of acetone and aqueous solution of ammonium sulfate) and an oil phase (cyclohexane) constitutes a liquid-liquid-liquid system. Yu *et al.* [70] researched the dispersion and phase separation of the triple-liquid phase stirred tank with a CCD camera system. The results showed that different types of impeller had different dispersion abilities. In this specific system, the height of upper phase to the tank bottom was suggested to index the dispersion ability. Generally speaking, radial impellers performed better than axial ones as a result of the great shear generated by vertical blades. For axial impellers, the downward flow type was better than the upflow flow one. Similar to the critical suspension speed in a liquid-solid phase system, Skelland *et al.* [71] defined the minimum agitator speed needed to obtain complete liquid-liquid dispersion, *i.e.*, the rotational speed just sufficient to completely disperse one liquid into the other. The phase volume ratio has a great influence on the dispersion pattern. The increase of middle phase volume ratio has little effect on the critical impeller speed, which has a tendency to increase with the increase of middle to bottom phase volume ratio.

SURFACE AERATED STIRRED TANKS

Surface aeration is an important operation in process industries. In a surface aeration reactor, gas reactant is entrapped into the liquid from the headspace in the upper part of the reactor, avoiding or reducing venting of unreacted gas into the atmosphere or using an external compressor to inject the gas into the reactor. Surface aeration reduces consequently the investment on process equipment and provides savings on operating costs.

Novel Surface Aeration Configurations

Many impellers, *e.g.*, Rushton disk turbine, pitched blade turbine, paddle and propeller, can cause surface aeration when they are placed close to the free surface. Among them, the Rushton disk turbine (RDT) is one of the most commonly used impellers, because of its lower critical impeller speed for gas entrainment, while it provides excellent gas-liquid mixing and higher volumetric mass transfer coefficient [72-74]. In a surface aerator the impeller is often located very near the liquid surface for a higher rate of gas entrainment. However, an impeller being too close to the surface would cause the reduction in efficiency for liquid mixing and even lead to the upper part of the impeller exposed to bulky gas. Furthermore, the gas is dispersed only in a limited region around the impeller, leaving a large bubble-less region in the lower part of the vessel. A second pitched blade turbine can be installed beneath the RDT near the bottom of the vessel to enhance the overall mixing and dispersion [72], but this may cause the decrease of volumetric mass transfer coefficient. For providing better gas-liquid dispersion and mass transfer for surface aeration by a Rushton disk turbine, prior work has reported improved designs of impeller [75-77], use of wall baffles [78], and introduction of a draft tube for enhancing gas entrainment [79]. Roman and Tudose [77] investigated the mass transfer of the modified Rushton impeller with perforated blades in a gas-liquid-solid system, demonstrating 45% higher mass transfer rate than that of the standard RDT. Van' Riet *et al.* [76] examined the power consumption of the modified Rushton turbine with perforated blades in a gas-liquid system and found that the power consumption of the modified Rushton impeller did not increase as expected.

There is much room for innovation of surface aerator with better gas dispersing performances and higher pumping capacity. In this section, two novel surface aeration configurations are introduced.

Self-Rotating and Floating Baffle (SRFB)

A novel surface aeration apparatus called Self-Rotating and Floating Baffle (SRFB) [80] is shown in Fig. **14**, which is characterized by a SRFB and a Rushton disk turbine with a large opening fraction on its disk. Made from a circular light metal sheet, the SRFB has a hole opened at its center to allow it to rotate around the impeller shaft and afloat on the liquid surface. The sheet is radially cut from the edge to 2 mm apart from the brim of the center hole so that a few fan-shaped blades are formed. Each blade is twisted by an angle to the horizontal plane to ensure the baffle floating over the RDT when impinged by liquid stream discharged from the turbine.

Before agitation, the SRFB sits on the spacing ring. When the turbine is set to rotate, the liquid flow discharged by the RDT pushes the twisted fan blades of the SRFB and the baffle begins to rotate and floats gradually to the surface. With the increased impeller speed, the SRFB floats and rotates at a certain position close to the gas-liquid surface and starts to entrain gas into the liquid. Even when the SRFB is raised up the liquid surface, it still works as a lid to prevent the adverse exposure of the RDT to the air. However, this blockage does not prevent gas entrainment because the SRFB has a large opening fraction. Additionally, the SRFB rotates passively at a lower speed than the active RDT, and shear is thus created in favor of gas dispersion and gas-liquid mass transfer. Therefore, the RDT can be immersed to a larger depth to enhance the whole gas-liquid mixing and dispersion, while at the same time discards the possible stagnant zone over the bottom, without deteriorating the performance in gas entrainment. The relatively closed space above the RDT introduced by SRFB helps in producing a radial pressure difference which is in favor of entrainment of gas from the free surface.

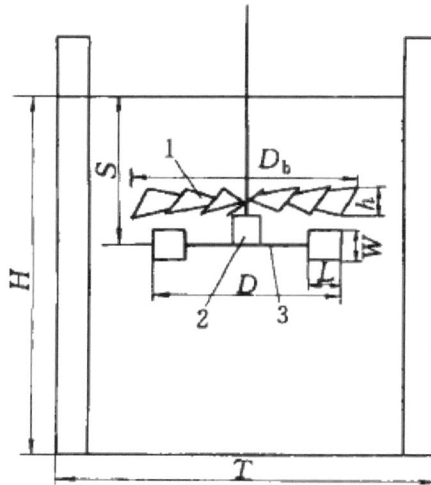

Figure 14: Configuration of the SRFB surface aeration system. 1: self-rotating and floating baffle; 2: spacing ring; 3: Rushton disk turbine.

The general mechanism of surface aeration may be attributed to the wave motion and eddies on the fluctuating liquid surface. Under the condition of low impeller speeds, the liquid surface is approximately smooth. The liquid surface becomes turbulent and starts to suck gas into the tank by eddy or wave motion when stirring speed becomes high enough. The impeller speed for the onset of gas entraining is defined as the critical impeller speed, N_C, and is determined as objectively as possible by visual observation. The experimental measurement of N_C *vs.* immersion depth for the combination of SRFB and the RDT surface aeration has revealed that the critical impeller speed for the SRFB system is lower than when using conventional RDT.

Volumetric mass transfer coefficient $k_L a$ is the result of mixing and contacting between two phases at certain power consumption per volume. The relation between $k_L a$ and impeller speed suggests that the SRFB system provides a higher mass transfer coefficient than the conventional RDT system at the same stirring speed. Besides, it is concluded that the SRFB system is superior to the conventional RDT over the entire range of power input. The values of $k_L a$ of SRFB system are 30%-68% higher than those of conventional RDT systems. It is believed that the combination of breaking up of bubbles by the SRFB fan blades, of the shear generated by the differences of stirring speed between SRFB and the active RDT, and of the enhanced gas entrainment because of the radial pressure gradient along the fan blades, has led to the improved gas-liquid mass transfer.

Self-Rotating Flow Guide (SRFG)

Another novel surface aeration reactor is shown in Fig. **15**, named Self-Rotating Flow Guide (SRFG) [81] for multiple impellers. In this typical configuration, there are two impellers on the shaft and 45°-pitched blade downflow turbines are chosen. The upper one is used for surface aeration, and the one below is for dispersion. The bearing bushing covering on the shaft is a hollow cylinder and the internal diameter is slightly bigger than that of the shaft, allowing the bushing to rotate around the shaft freely. There are a few holes along the circumference on the wall of bearing bushing, which are used to fix the radial bearing arms. Some curved guiding sheets are installed outside of the bearing arms to form an approximately closed flow room around the shaft.

Figure 15: Configuration of self-rotating flow guide.

When the shaft starts to rotate, the SRFG also begins to revolve as being driven by the impeller beneath. Moreover, the flow discharged from the upper impeller is deflected by the curved guiding sheets, and then it enters the impeller zone at a more beneficial angle. Furthermore, the flow inside the approximately closed room is mainly axially downward while outside the space the flow becomes axially upward. Thus an overall circulation is established, which shortens mixing time and improves mixing quality.

From the comparison of results between conventional Rushton disk turbine surface aerators and SRFG systems on mixing time obtained using a conductivity method, it is obvious that the new surface aerator SRFG has a shorter mixing time at the same rotating speeds. Besides, the less stirring power demand and better gas-liquid mass transfer performance at the same power consumption are also achieved.

Hydrodynamic Characteristics of Multi-Impeller in Surface Aerators

As it has been mentioned above, a bubble-less region with very low gas holdup exists in the lower part of the tank. Many industrial vessels have a great aspect ratio which exceeds unity, sometimes with ratio of even larger than two, which discourages use of surface aerators. Thus, multi-impeller configurations have been introduced in surface aeration vessels. The upper impeller placed near the liquid surface is used as aerating impeller for entraining gas. The lower impeller located well below the aerating impeller is employed to carry out other duties like solid suspension, gas dispersion, heat transfer, mixing, *etc*. When the aspect ratio reaches 2.0 or even more, triple impeller configurations should be taken into account [82].

Mass Transfer Coefficient

Li *et al.* [83] investigated the gas-liquid mass transfer coefficients of some dual-impeller configurations at different rotational speeds in a stirred tank with the aspect ratio of 2.3. Compared with the single-impeller system, the overall flow circulation is strengthened in a dual-impeller stirred tank as a result of the use of the lower impeller. Besides, the lower impeller helps breaking up the bubbles into smaller ones, which is beneficial for gas dispersion and interphase mass transfer. It is obvious that k_La increases with increasing stirring speeds. With regard to the lower impeller type, the pitched blade turbine upflow is superior to other types of impellers.

The combination of surface aeration and sparger aeration has provided better gas liquid mass transfer performances. The lower impeller should do a good job in breaking up and dispersing the bubbles produced from the sparger when compared to the configurations without sparger aeration. Thus, a radial flow agitator like Rushton impeller and its modifications are often being adopted.

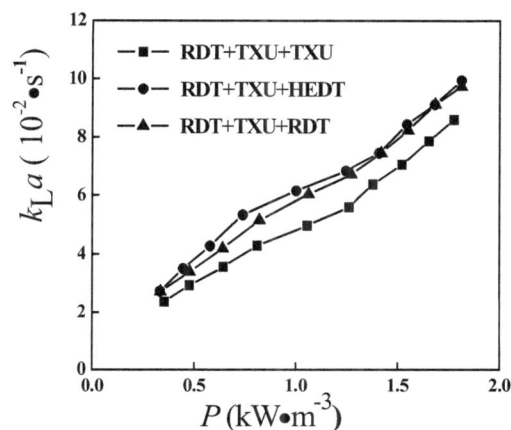

Figure 16: k_La values of three triple-impeller configurations with surface aeration and sparger aeration combined.

Li *et al.* [82] conducted research on a triple-impeller stirred tank. The mass transfer results in the system of combined surface aeration and sparger aeration are showed in Fig. **16**. The Rushton disk turbine (RDT) creates high local shear and is suitable for gas dispersion. It is popular in gas-liquid systems, because its central disk guides the bubbles passing by the blades and makes it effective to avoid flooding. However, the gas cavities theory indicates that the structure of RDT easily results in the formation of large gas cavities behind the impeller blades and consequently leads to low efficiency. Half elliptical blade disk (HEDT) is recommended for the lowest impeller, because HEDT has similar gas dispersion ability as RDT, but its arc shaped blades prevent the forming of such gas cavities in the impeller region. Furthermore, HEDT has a lower power number than RDT. Techmix335 hydrofoil impeller upflow (TXU) is an axial impeller which has a low power number but is not suitable to work as the lower impeller.

Power Consumption

Power consumption per volume (P_V) has often been measured to indicate the ability of power input. The overall power consumption is on the summed contribution from individual ones, but P_V of the double RDT configuration is less than twice that of a single RDT because of the interaction between individual flow patterns.

With sparger aeration, the power input decreases owing to the formation of gas cavities behind the blades, but the degree of reduction differs among various configurations. The power decrease of the configuration with both RDT is the highest among the mentioned configurations because the formation of gas cavities is easier for RDT due to its vertical blades. Furthermore, the gas cavities are always larger because of its powerful shearing. Contrary to RDT, P_V of Techmix 335 (TXU) is not sensitive to aeration also owing to its unique curved blades. It is seen that the power input does not drop when the rotational speed increases up to a certain level.

The power consumption of the triple-impeller configuration RDT+TXU+HEDT is plotted in Fig. **17** [82]. The curve shapes are similar to those in dual-impeller configurations. The main reason for the decrease in energy consumption is the influences of sparged gas on surface aeration can be omitted when compared with its effect on raising gas holdup and the uniformity of bubble distribution throughout the vessel.

Gas Holdup in Surface Aerated Stirred Tanks

The spatial distribution of gas holdup in a surface aerated stirred vessel is an important factor for designing and operating such systems. Gas-liquid mass transfer heavily relies on gas dispersion and is determined not only by the local rate of gas input, but also by the complex interaction between the continuous and dispersed phases. To better understand and quantify the interaction and mass transfer between gas and liquid phases in a surface aerated stirred tank, experimental research on the local hydrodynamic characteristics like interfacial area, local gas holdup and bubble size distribution is necessary. There are many studies on gas holdup for gas-liquid stirred vessels using both experimental and numerical methods. Sun *et al.* [84] determined the gas holdup contour maps with a conductivity probe, while they have also simulated the three-dimensional gas-liquid flow in a surface aerated stirred tank with the standard k-ε turbulence model. Besides, LDA technique [85], ultra sound-Doppler technique [86], PEG [87] and algebraic slip model (ASM) [88], Lagrangian approach [85], multiple reference frame (MRF) methods [89] have also been used in studying gas-liquid stirred tanks.

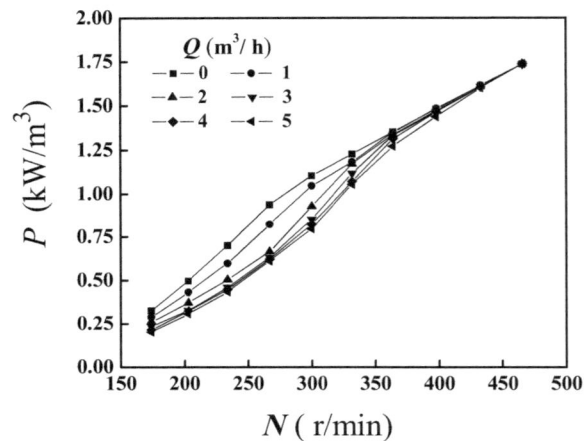

Figure 17: Power consumption versus impeller speed at different gas flow rates stirred by RDT+TXU+HEDT.

Experimental Results of Gas Holdup

Sun *et al.* [90] gave the gas holdup contour maps for a surface aerated stirred tank with a Rushton impeller under different agitating speeds. In Fig. **18**, the surface baffle similar with the novel SRFB mentioned above in section "Self-Rotating and Floating Baffle" is introduced. The very inhomogeneous distribution of gas holdup in the surface aerated tank is observed. The gas bubbles entrained from the air stay near the liquid surface where they are generated. When the impeller rotates at higher speeds, the blades generate many eddies so that the gas is sucked into the impeller region, then discharged radially into other regions in the tank. Because the impeller is placed close to the surface, the total liquid circulation is weak in the bulk, therefore it is difficult for gas to reach the region below the impeller, and the local gas holdup is nearly zero near the bottom.

In the Rushton impeller tank without the surface baffle, there exist two peak gas holdups in its radial profile: one is close to the shaft caused by buoyancy, the other between the tank wall and the impeller blade tip as the result of the hydrodynamic interaction between two streams flowing in opposite directions. One stream is discharged by the impeller, and the other is rebounded off by the wall baffles. As to the vessel with the surface baffle (Fig. **18**), there is one more peak of gas holdup in the radial direction. Two of them are the same as mentioned above, and a third one is generated by the high gas entraining rate at the edge of the surface baffle.

Fig. **18** indicates that for surface aeration, multiple impeller systems should be employed to intensify the total circulation and to achieve good mixing performance and uniform gas distribution below the impeller.

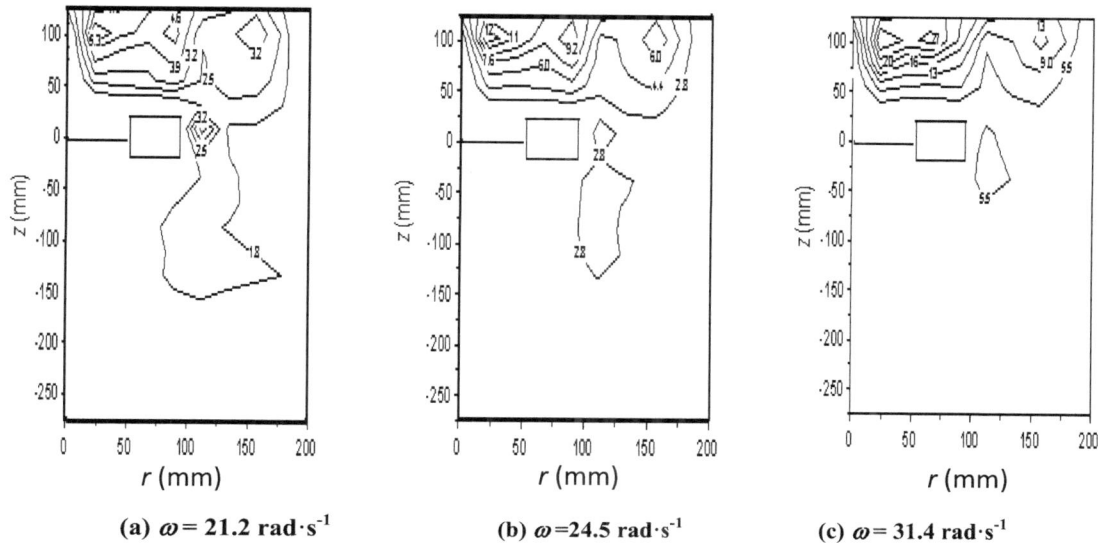

(a) $\omega = 21.2$ **rad·s⁻¹**   **(b)** $\omega = 24.5$ **rad·s⁻¹**   **(c)** $\omega = 31.4$ **rad·s⁻¹**

Figure 18: The gas holdup distribution in the stirred tank with the $T/2$ Rushton impeller with a surface impeller baffle.

Mathematical Modeling

A two-fluid approach combined with the k-ε turbulence model is always used for gas-liquid flow simulation in a stirred tank including the case of surface aeration. For the model details, see section: Two-Phase Flow in Stirred Tanks.

Rate of Surface Aeration

The key factor to characterize the efficiency of surface aeration is the rate of gas entrainment from the headspace above the liquid surface. However, it is difficult to measure this parameter, because the gas is entrapped into the liquid phase from the surface and at the same time bubbles may escape back into the gas phase *via* the same interface. A general method is to measure the gas concentration variation in the liquid phase and derive the rate of surface aeration in combination with mass conservation of several components. Topiwala [91] and Veljkovic *et al.* [92] obtained the surface aerating rates and established empirical correlations. Wang *et al.* [93] used the Na_2SO_3 fed-batch method to measure the aeration rate for a stirred tank with different surface aerating impellers. Matsumura *et al.* [94] provided a useful correlation:

$$V_s = 7.15 \times 10^{-6} N^{1.90} D^{3.95} T^{-2.5} \sigma^{-2.40} \mu^{-0.15} \tag{64}$$

So far few fundamental equations for local surface aeration rates are available in the literature, though they are essential for the numerical simulation of a surface aerating stirred tank. A preliminary effort was made to suggest such a basic equation. The mechanism of surface aeration is related to several hydrodynamic characteristics such as surface shear rate, maximum bubble diameter, turbulent eddy frequency, turbulent length scale, local energy dissipation and so on, namely

$$V_s = f(\tau, d_{\max}, \eta, L, P_v) \qquad (65)$$

From general fluid mechanics, the agitating power is

$$P_V = \varepsilon \rho_1 \qquad (66)$$

The maximum bubble size may be estimated by

$$d_{\max} = 0.725 \left(\frac{\sigma}{\rho_1} \right)^{3/5} \varepsilon^{-2/5} \qquad (67)$$

Uhl and Gray [95] suggested the relation of shear rate with the agitating power and bubble size:

$$\tau \propto \rho_1 \left(P_V \frac{d_{\max}}{\rho_1} \right)^{2/3} \qquad (68)$$

Moreover, the Kolmogorov theory of isotropic turbulence indicates that the pulsating frequency of turbulent eddies and their length scale are given respectively as [96]

$$\eta = \left(\frac{\varepsilon}{v} \right)^{1/2}, \quad L = \left(\frac{v^3}{\varepsilon} \right)^{1/4} \qquad (69)$$

These Eqs. (65)-(69) reveal that the local energy dissipation rate is an important factor related to the parameters mentioned above, causing thus the intuitive assumption that local surface aeration rate is a complex function of local energy dissipation rate. By the Kolmogorov theory, the equilibrium turbulent feature is determined by ε and v only when the Reynolds number is large enough, and the characteristic velocity is

$$V = (v\varepsilon)^{1/4} \qquad (70)$$

As a first approximation to a real aeration rate equation, it is assumed that the gas entrainment velocity is proportional to the turbulent velocity scale:

$$V_s = k(v_1 \varepsilon_1)^{1/4} \qquad (71)$$

with the coefficient k to be determined or estimated from experimental data.

This constitutive equation could be used in numerical simulation of surface aeration as the boundary condition at the free surface. Another constraint is also necessary, that is, that gas is entrained according to Eq. (71) only when the axial velocity component of liquid is negative (downwards). As a comparison, Eq. (64) is also used as the local rate of uniform surface aeration in the simulation.

When the surface baffle is involved in the simulation, the rotating speed measured in the experiment is specified to the surface baffle. The no-slip conditions are applied to the surface underneath of the baffle. The open holes in the baffle are treated as the free surface for surface aeration according to Eq. (71) or (64).

Flow Field and Gas Holdup Distribution

Observations in the gas and liquid velocity vector maps and the gas holdup contour map (Figs. **19** and **20**), reveal that the gas is discharged from the impeller region into the impeller stream, and two vortices are formed close to the blade tip [84]. The gas holdup contours show that there is a high holdup region near the blade tip. Another recirculation zone near the surface is formed from the interaction between the vertical upflow of the gas, and the gas is sucked from the liquid surface.

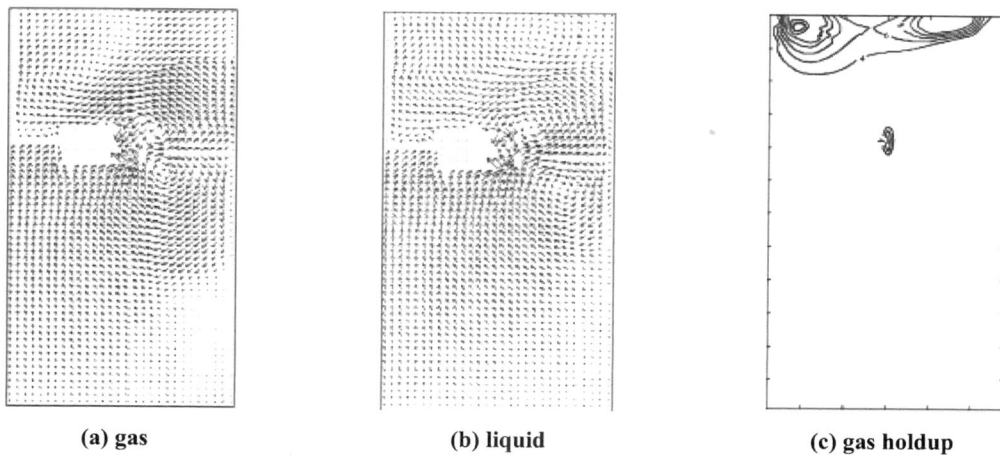

| (a) gas | (b) liquid | (c) gas holdup |

Figure 19: The gas and liquid flow field and the gas holdup distribution in *r-z* plane. (with the *T*/2 Rushton impeller without impeller baffle at ω = 31.4 rad/s).

As for the liquid flow, two vortices are formed in the upper and lower bulk regions respectively, and they are rather symmetrical with respect to the impeller plane. A third liquid vortex is formed close to the impeller shaft at the surface as it has also been observed experimentally. This vortex causes gas bubbles to accumulate there creating a high void region.

The role of the surface baffle becomes obvious when compared with the configuration without surface baffle. The lower gas eddy becomes weaker and closer to the impeller plane, and the gas holdup in the lower bulk region is lower. However, owing to the presence of the surface baffle, the increase of gas holdup in the upper bulk region and the total gas holdup is obvious. More air is entrained from the opening in the surface baffle and the third gas holdup peak occurs at the surface.

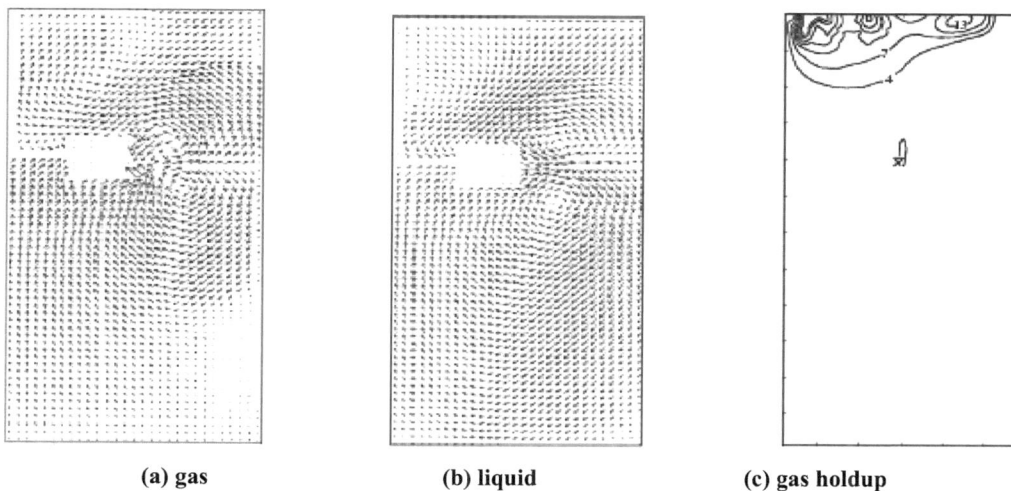

| (a) gas | (b) liquid | (c) gas holdup |

Figure 20: The gas and liquid flow field and the gas holdup distribution in *r-z* plane. (with the *T*/2 Rushton impeller and impeller baffle at *ω* = 31.4 rad/s).

The contour lines are concentrated in the upper bulk region above the impeller. Another region with dense contour lines is at the impeller blade tip, but rather few lines appear in the lower bulk region. This suggests that intensive whole-tank circulation of liquid phase is necessary for uniform distribution of the entrained gas in the whole tank. The turbulence model needs to be upgraded to more sophisticated versions to account for the swirling and anisotropic nature of turbulent two-phase flow in stirred tanks. Furthermore, the gas bubble size distribution is considered to be a major problem to be resolved in the numerical program for greater improvement in the accuracy of the simulation.

MACROMIXING AND MICROMIXING

Macromixing and micromixing are important processes in stirred tanks. Mixing in a stirred reactor occurs as a result of the medium motion at three levels: molecular, eddy and bulk motions. Usually, the bulk motion is superimposed on either molecular or eddy diffusion or both.

Macromixing

Mixing on the vessel scale is called macromixing. Since the mixing of reactants in stirred tanks shows great effect on conversion rate, on selectivity and on productivity, studies on macromixing have intensively been conducted using both tracer experiments and computational fluid dynamic (CFD) simulations based on mathematical models [97].

In a tracer experiment, certain quantity of tracer (usually in the form of a pulse) is injected at some location in a reactor and the tracer concentration as a function of time is monitored to obtain the mixing time, which is often used to characterize the macromixing process. Mixing time is defined as the time required achieving a certain degree of uniformity of tracer concentration. For tracer tracking in the bulk fluid flow one can use chemical species, thermal disturbance or other substances, and the measurement technique is then selected correspondingly, such as visual observation [98], conductivity technique [99-101], laser-induced fluorescence technique [102] and liquid-crystal thermography [103]. Empirical equations correlate experimental data with design and operating parameters [101, 104, 105], and some semi-empirical models such as the dispersion model [98, 106], the circulation model [107] and the network of zones model [108] have been reported in the literature.

Numerical Simulation in Single-Phase Systems

Numerical simulation presents a flexible predicting tool for mixing time under diversified tank configurations and operating conditions for studying macromixing. Ranande *et al.* [108] firstly used CFD for the mixing studies. Commercial CFD software was also popularly employed. Jaworski *et al.* [109] predicted the flow pattern and mixing process in a stirred tank with double Rushton turbines by Fluent. Zhou *et al.* [110] used the commercial code CFX to simulate the mixing process by a single Rushton turbine and revealed that the mixing time was strongly dependent on the flow field, the feeding and detecting positions. Wang *et al.* [97] modeled numerically the mixing time and residence time distribution (RTD) of stirred tanks by solving the tracer transport equation based on the numerically resolved flow field in a baffled tank with a Rushton disk turbine.

Effect of Parameters and Conditions on Mixing Time

The effect of injection points, impeller speed, impeller clearance and draft tube on mixing time has been discussed in detail. It has been found that the mixing time at different injection points were similar. Mixing time, as measured by using different detectors, decreased with increasing impeller speed. Different correlations of dimensionless mixing time with various design and operation parameters have been proposed in the literature. For the stirred tank driven by a disk turbine, the following correlation of the mixing time was proposed by Joshi *et al.* [104]:

$$N\theta_{\mathrm{m}} = 9.43\left(\frac{aH+T}{T}\right)\left(T/D_{\mathrm{d}}\right)^{13/6}\left(W_{\mathrm{d}}/D_{\mathrm{d}}\right) \tag{72}$$

and Shiue *et al.* [105] provided another:

$$N\theta_{\mathrm{m}} = 5.01\left(T/D_{\mathrm{d}}\right)^{2.4} \tag{73}$$

The dimensionless mixing time $N\theta_{\mathrm{m}}$ remains constant if the impeller speed is kept just in the turbulent region while other conditions are not changed. The comparison of the dimensionless mixing time predicted by CFD simulation with correlations [104, 105] and experimental data [111] in the literature is listed in Table **3**. The difference between the predicted results and experimental data may be due to the inaccuracy of flow pattern, which is achieved through solving the RANS equations using the standard k-ε model. Wang *et al.* [97] determined the mixing time at different impeller clearances and the results revealed that the mixing time increased with the decreased impeller clearance. Similar conclusion can be drawn from Brennan and Lehrer's experimental results [98]. To provide gross circulation and improve solid suspension in agitated reactors, a draft tube is often installed especially in crystallizers and precipitators. The influence of a draft tube on mixing time with different geometric characteristics has been predicted numerically and the results show that the effect of the draft tube on mixing time is not evident. However, the presence of a draft tube promotes flow homogeneity in a stirred tank in favor of precipitation operation, because the distribution of circulation time of particles and liquid elements becomes narrower.

Table 3: Comparison of $N\theta_{\mathrm{m}}$ predicted by CFD and the literature correlations

Impeller speed (r/min)	Joshi *et al.* [104]	Shiue and Wong [105]	Experimental results [111]	Numerical prediction [97]
200	47.5	70	52	65
250	47.5	70	—	67.1
300	47.5	70	50	67.5
400	47.5	70	—	66.6
533	47.5	70	51	66.6

Residence Time Distribution (RTD)

Another factor to characterize the macroscopic mixing is the residence time distribution (RTD), *i.e.*, the exit age distribution density function. RTD can also be approached either with experiment or numerical simulation, with the latter requiring solving both the turbulent flow and the mass transport equations. The RTD curves obtained from both experiment and simulation [97] have revealed that the outlet tracer concentration decreases in an exponential manner, similar to the ideal CSTR behavior curves except for the very short residence time. The comparison between the experimental and simulation results at different impeller speeds has indicated that stirring speed hardly affects the mean residence time. Besides, the draft tube seems to exert no remarkable influence on the nature that a stirred tank operated in the turbulent regime is an ideal mixed reactor from a macroscopic view.

Mixing Time in a Gas-Liquid Stirred Tank

Some studies have reported correlations of mixing time from experimental data in gas-liquid stirred tanks [78, 112]. Although these correlations have proved useful, they were not expected to be valid outside the range of conditions reported in the literature.

The conductivity and tracer response technique was usually employed to measure mixing time. The tracer solution is injected onto the liquid surface as a pulse. A monitoring conductivity electrode is located at various positions. As bubbles in the reactor would significantly affect the conductivity, the electrode is surrounded by a mesh to prevent bubble interference. Mixing time is defined as the period of time necessary to achieve the desired level of homogeneity, which can be expressed in terms of the degree of mixing:

$$Y = \left| \frac{c(t) - c_0}{c_\infty - c_0} \right| \tag{74}$$

where $c(t)$ is the tracer concentration at time t, c_0 and c_∞ are the initial and final average concentrations of tracer respectively, and the value of 0.95 is often chosen for Y.

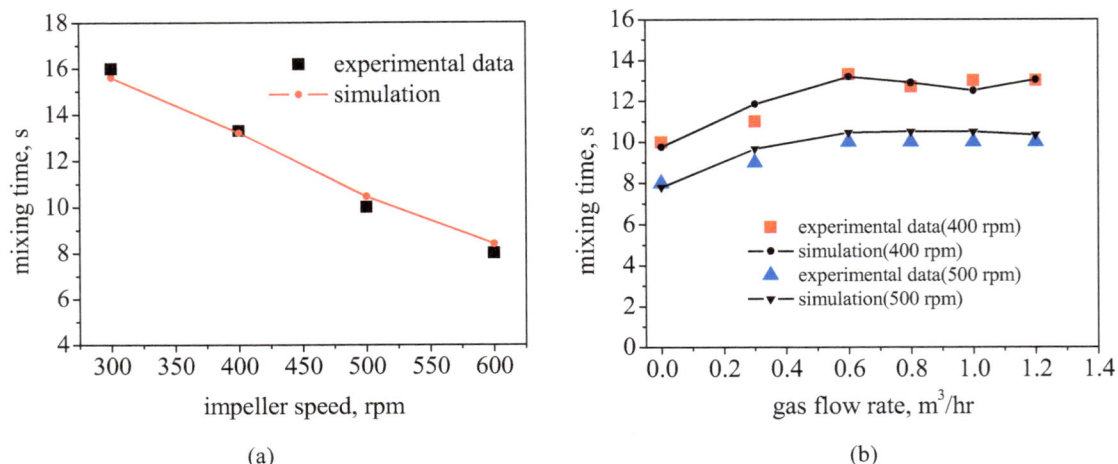

$$\text{(a)} \qquad\qquad\qquad\qquad \text{(b)}$$

Figure 21: Effect of two factors on mixing time (a: stirring speed; b: gas flow rate).

Impeller speed is a key factor influencing the mixing time (Fig. **21a**) [113]. It is obvious that the increased stirring speed leads to the decreasing of mixing time because of promoted liquid circulation and turbulent intensity. It can be seen that the simulated results are also in good agreement with the experimental data. Gas flow rate is another parameter affecting the mixing time in a gas-liquid stirred tank (Fig. **21b**). There exist many contradictions in the literature about the influence of gas flow rate on mixing time. An increase in gas flow rate can increase, decrease, or hardly affect mixing time, as reported by different studies. Bouaifi and Roustan [114] studied the mixing behavior in a stirred tank with various axial and mixed dual-impeller and their results revealed that the mixing time was prolonged when the gas flow rate increased under given configuration and impeller speed. However, Shewale and Pandit [115] reported opposite experimental results. Zhang *et al.* [113] found that the mixing time became longer with increase of gas flow rate within a certain range; once beyond such a range, the value of mixing time leveled off. Pandit and Joshi [116] reported that the mixing time increased with the value of Q_G at any impeller speed larger than N_F (the critical impeller speed for onset of flooding). N_F can be calculated by the following expression [117]:

$$N_F = 0.54 Q_G^{0.3} D^{-1.8} \tag{75}$$

The calculated values of N_F are always lower than the operating impeller speeds. The mixing time of gas-liquid phase is longer than that of single phase, because the presence of gas reduces the impeller energy input and pumping capacities.

Micromixing

Precipitation is a fast reaction process and the mixing in precipitators may be the rate-controlling step, so it is usually employed to study the microscopic mixing. One of the mostly used model reaction is the precipitation of barium sulfate in an aqueous solution. Various kinds of models have been used to study micromixing, e.g., the exchange-with-the-mean (IEM) model, the environment model, the engulfment model, and the segregated feed model. In the IEM model, a typical clump is followed from its birth at the feed inlet. The clump is assumed to act as a batch vessel having a uniform concentration and exchange of matter with an environment where the concentration is taken to be the mean value at any time of all clumps in the vessel. The IEM model has been used to study the effect of micromixing on crystal size in precipitation by Pohorecki and Baldyga [118] and Garside and Tavare [119]. The environment model is another commonly used model for micromixing in the precipitation process, where it is assumed that the vessel volume consists of two or more separate environments having extreme states of micromixing. The reactor is modeled by splitting the feed into completely segregated and maximum mixed parts and combining them at the reactor exit, with no interaction among the parallel environments. The engulfment-deformation-diffusion model has been widely employed to study the mixing in the barium sulfate precipitation process in recent years [120]. In the engulfment model, the vessel volume is divided into two zones, i.e., the mixing-precipitation zone and the environment zone. All reactants take part in reactions in both zones.

Computational fluid dynamics (CFD) is especially helpful for investigating the mixing–controlled reactions. In later years, plenty of numerical simulation efforts have been made to study the precipitation process with CFD in precipitators of simple geometry such as a tubular reactor [121], a rectangular flat reactor [122] and stirred tanks [123-125]. Garside and Wei [123] and Jaworski and Nienow [124] simulated the precipitation process with the mixing at the molecular level (micromixing) completely neglected. However, the key role played by micromixing in the precipitation process, which affects the rate of chemical reaction, nucleation and crystal growth, is generally recognized. Therefore, it is essential to incorporate a suitable micromixing model into a reactive precipitation process to predict final product characteristics. Baldyga and Orciuch [126, 127] coupled CFD with the presumed-beta probability density function (PDF) micromixing model in tubular reactors for the barium sulfate precipitation process. The finite-mode PDF (FM-PDF) model was utilized to describe precipitation processes in tubular and Taylor-Couette reactors.

Wang *et al.* [128] have combined computational fluid dynamics, finite-mode PDF model, population balance and kinetic modeling into a mixing-precipitation model to simulate the barium sulfate precipitation process in a continuous stirred tank agitated by a Rushton disk turbine. The effect of various operating conditions such as impeller speed, feed concentration, feed position and mean residence time on barium sulfate precipitation process has been clearly demonstrated.

Mathematical Model

In setting up the mathematical model, the following assumptions are made [128]:

1) The presence of solid phase with very small diameter (ca. 10 μm) does not affect the flow field;

2) The solid phase is assumed to follow the liquid flow because the crystals are very small and the sedimentation of solids is neglected;

3) Only primary nucleation is considered;

4) The linear crystal growth is modeled to be independent of crystal size; and

5) Aggregation and breakage as well as the dissolution of particles in the region of undersaturation are neglected.

Simulation of the Flow Field

The transports of all dependent variables, such as the velocity components, u, v and w, turbulent kinetic energy, k, and viscous dissipation, ε, under steady state are similar to those used in the section: "Two-Phase Flow in Stirred Tanks".

FM-PDF Model

The finite-mode PDF model was chosen to model the micromixing by Wang *et al.* [128]. Every cell in the computational domain contains N different modes or environments, which correspond to the discretization of the composition PDF in a finite set of delta function [129]:

$$f_\varphi\left(\psi;x,t\right)=\sum_{n=1}^{N}p_n\left(x,t\right)\prod_{\alpha=1}^{m}\delta\left[\psi_\alpha-\left\langle\varphi_\alpha\right\rangle_n\left(x,t\right)\right] \tag{76}$$

where $f_\phi(\psi; x, t)$ is the joint PDF of all scalars, *i.e.*, concentrations, moments *etc.*, appearing in the precipitation model, N is the number of modes, $p_n(x,t)$ is the probability of model n, $\langle \phi_a \rangle_n (x,t)$ is the value of scalar a corresponding to mode n, and m is the total number of scalars. By definition, the probabilities, p_n, sum to unity and the average value of any scalar is defined by integration with respect to ψ. For the system under consideration, the knowledge of mixture fraction and a reaction progress variable suffice to predict the reactant concentration in each environment. Thus, the first scalar is specified to the mixture fraction $\phi_1(x,t) \equiv \zeta(x,t)$. From Eq. (76), the average value of the mixture fraction is given by

$$\langle \xi \rangle = \sum_{n=1}^{N} p_n \langle \xi \rangle_n \tag{77}$$

Previous results [130] showed that three modes were sufficient to work with good accuracy. The 3-mode PDF model is taken as the discretization of the reacting system in three environments, where environments 1 and 2 correspond to unmixed reactant A and B respectively, and environment 3 corresponds to the environment in which reactants A and B are mixed together and react.

At steady state, the scalar transport equations for the probabilities of modes 1 and 2, and for the weighted mixture fraction in environment 3, $s_3 \equiv p_3 \langle \xi \rangle_3$, are

$$\langle u_j \rangle \frac{\partial p_1}{\partial x_j} = \frac{\partial}{\partial x_j} \left(\Gamma_{\text{eff}} \frac{\partial p_1}{\partial x_j} \right) + \gamma_s p_3 - \gamma p_1 (1 - p_1) \tag{78}$$

$$\langle u_j \rangle \frac{\partial p_2}{\partial x_j} = \frac{\partial}{\partial x_j} \left(\Gamma_{\text{eff}} \frac{\partial p_2}{\partial x_j} \right) + \gamma_s p_3 - \gamma p_2 (1 - p_2) \tag{79}$$

$$\langle u_j \rangle \frac{\partial s_3}{\partial x_j} = \frac{\partial}{\partial x_j} \left(\Gamma_{\text{eff}} \frac{\partial s_3}{\partial x_j} \right) - \gamma_s p_3 \left(\langle \xi \rangle_1 + \langle \xi \rangle_2 \right) + \\ \gamma p_1 (1 - p_1) \langle \xi \rangle_1 + \gamma p_2 (1 - p_2) \langle \xi \rangle_2 \tag{80}$$

where $p_3 = 1 - p_1 - p_2$, Γ_{eff} is the turbulent diffusivity, $\langle u_j \rangle$ is the mean velocity in j direction, and γ and γ_s are the micromixing rate and the spurious dissipation rate, respectively:

$$\gamma = C_\phi \frac{\varepsilon}{k} \frac{\langle \xi'^2 \rangle}{\left[p_1 (1 - p_1)(1 - \langle \xi \rangle_3)^2 + p_2 (1 - p_2)(\langle \xi \rangle_3^2) \right]} \tag{81}$$

$$\gamma_s = \frac{2\Gamma_{\text{eff}}}{1 - 2\langle \xi \rangle_3 (1 - \langle \xi \rangle_3)} \frac{\partial \langle \xi \rangle_3}{\partial x_j} \frac{\partial \langle \xi \rangle_3}{\partial x_j} \tag{82}$$

where C_ϕ is a constant of the order of unity and $\langle \xi'^2 \rangle$ is the mixture fraction variance defined by

$$\langle \xi'^2 \rangle = \langle \xi^2 \rangle - \langle \xi \rangle^2 \tag{83}$$

In Eq. (83), $\langle \xi^2 \rangle$ is the second moment of the mixture fraction defined by

$$\langle \xi^2 \rangle = \sum_{n=1}^{N} p_n \langle \xi^2 \rangle_n \tag{84}$$

As environments 1 and 2 contain only reactant A or B respectively, the mixture fractions are $\langle \xi \rangle_1 = 1$ and $\langle \xi \rangle_2 = 0$. Thus, Eq. (76) can be written as

$$\langle \xi \rangle = p_1 + s_3 \tag{85}$$

where the mixture fraction variance can be given by

$$\langle \xi'^2 \rangle = p_1 + \frac{s_3^2}{p_3} - \langle \xi \rangle^2 \tag{86}$$

Moment Method

The CSD can be described by the population balance [131]. To obtain a simpler expression, the population balance can be expressed in terms of the moments of the particle number density function. It should be noted that since nucleation and growth occur only in environment 3, the population balance only needs to be applied to one environment. With this approach, the governing equations for the CSD moments at steady state are

$$\frac{\partial}{\partial x_j} \left(\langle u_j \rangle \overline{m_0} \right) = \frac{\partial}{\partial x_j} \left(\Gamma_{\text{eff}} \frac{\partial \overline{m_0}}{\partial x_j} \right) + B \left(\langle c_A \rangle_3, \langle c_B \rangle_3 \right) p_3 \tag{87}$$

$$\frac{\partial}{\partial x_j} \left(\langle u_j \rangle \overline{m_1} \right) = \frac{\partial}{\partial x_j} \left(\Gamma_{\text{eff}} \frac{\partial \overline{m_1}}{\partial x_j} \right) + G \left(\langle c_A \rangle_3, \langle c_B \rangle_3 \right) \overline{m_0} \tag{88}$$

$$\frac{\partial}{\partial x_j} \left(\langle u_j \rangle \overline{m_2} \right) = \frac{\partial}{\partial x_j} \left(\Gamma_{\text{eff}} \frac{\partial \overline{m_2}}{\partial x_j} \right) + 2G \left(\langle c_A \rangle_3, \langle c_B \rangle_3 \right) \overline{m_1} \tag{89}$$

$$\frac{\partial}{\partial x_j} \left(\langle u_j \rangle \overline{m_3} \right) = \frac{\partial}{\partial x_j} \left(\Gamma_{\text{eff}} \frac{\partial \overline{m_3}}{\partial x_j} \right) + 3G \left(\langle c_A \rangle_3, \langle c_B \rangle_3 \right) \overline{m_2} \tag{90}$$

$$\frac{\partial}{\partial x_j} \left(\langle u_j \rangle \overline{m_4} \right) = \frac{\partial}{\partial x_j} \left(\Gamma_{\text{eff}} \frac{\partial \overline{m_4}}{\partial x_j} \right) + 4G \left(\langle c_A \rangle_3, \langle c_B \rangle_3 \right) \overline{m_3} \tag{91}$$

where B is the nucleation rate, G is the growth rate, $\overline{m_j}$ is the mean value of the jth moment of the particle number density function $\overline{m_j} = p_3 \langle m_j \rangle_3$, and $\langle c_A \rangle_3$ and $\langle c_B \rangle_3$ are the local reactant concentrations in environment 3 that can be calculated by introducing the reaction progress variable Y:

$$\frac{c_A}{c_{A0}} = \xi - \xi_s Y, \quad \frac{c_B}{c_{B0}} = (1 - \xi) - (1 - \xi_s) Y \tag{92}$$

$$\frac{\partial}{\partial x_j} \left(\langle u_j \rangle \langle Y \rangle \right) = \frac{\partial}{\partial x_j} \left(\Gamma_{\text{eff}} \frac{\partial \langle Y \rangle}{\partial x_j} \right) + \frac{\rho_3 k_V G \overline{m_2}}{M \xi_s c_{A0}} \tag{93}$$

where c_{A0} and c_{B0} are the inlet concentrations of the two reactants in their separate feed streams, ρ is the crystal density, k_V is the crystal shape factor, taken as $\pi/6$ by Wang *et al.* [128], M is the crystal molecular weight, and ξ_s is the ratio defined as

$$\xi_s = \frac{c_{B0}}{c_{A0} + c_{B0}} \tag{94}$$

Using the moment method, the mean crystal size of the product precipitate can be calculated from the parameters of the outlet stream by the following expression:

$$L_{43} = \frac{\overline{m_4}}{\overline{m_3}} \tag{95}$$

Another parameter characterizing the crystal size distribution for analysis is the coefficient of variation, *C.V.*, defined as

$$C.V. = \left(\frac{\overline{m_2 m_0}}{\overline{m_1}^2} - 1 \right)^{0.5} \tag{96}$$

Precipitation Kinetics

Given the ion concentrations c_A and c_B, the supersaturation Δc and the supersaturation ratio S_a are defined as

$$\Delta c = \sqrt{c_A c_B} - \sqrt{K_{sp}}, \quad S_a = \sqrt{\frac{c_A c_B}{K_{sp}}} = \frac{\Delta c}{\sqrt{K_{sp}}} + 1 \tag{97}$$

where K_{sp} is the solubility product for BaSO$_4$ (K_{sp}=1.10×10^{-10} kmol$^2 \cdot$m^{-6} at room temperature). The parameters are taken from the literature [121].

The nucleation rate, *B*, as a function of the local supersaturation, Δc, is proposed by Baldyga *et al.* [132] as

$$B = 2.83 \times 10^{10} \left(\Delta c \right)^{1.775} \quad \Delta c < 10 \ (\text{mol} \cdot \text{m}^{-3}) \tag{98}$$

$$B = 2.53 \times 10^{-3} \left(\Delta c \right)^{15.0} \quad \Delta c > 10 \ (\text{mol} \cdot \text{m}^{-3})$$

The crystal growth rate, *G*, is described by a two-step diffusion-adsorption model [133] and the rate expression is

$$G = k_r \left(\sqrt{c_{As} c_{Bs}} - \sqrt{K_{sp}} \right)^\sigma = k_d \left(c_A - c_{As} \right) = k_d \left(c_B - c_{Bs} \right) \tag{99}$$

where c_{is} is the concentration of species *i* near the surface of the crystal at the limit of an adsorption layer and k_d is the mass transfer coefficient that is fixed in the range between 10^{-8} and 10^{-7} m·s^{-1}.

Numerical Procedure

Due to the nature of coupling among model equations, the CFD simulation proceeds in the following steps:

1)   The turbulent flow field is solved first to get the mean velocity, effective viscosity, turbulent kinetic energy and energy dissipation rate in the tank, and these variables are used for the next step of the precipitation simulation;

2)   Then, the iterative procedure starts with the solution of the transport equation for the weighted mixture fraction in each environment;

3)   The reaction progress variable is solved;

4)   The transport equations for the mean value of the 0-4th moments of CSD are solved; and

5)   Go back to step 2 to continue the iteration, if the residual error of any equation is above the convergence criterion and the asymptotic values of the exit stream are not reached. Otherwise, the simulation stops and the statistics are obtained.

Flow Field and Local Supersaturation

An example map of the mean velocity vector and the spatial distribution of the supersaturation ratio in the vertical cross-section of the stirred precipitator are provided by Wang *et al.* [128]. The contour maps of supersaturation are also presented.

Effect of Constant C_ϕ

In the FM-PDF model, C_ϕ is a constant that is ca. 1 or 2 in Eq. (81), which expresses the micromixing rate, γ, derived from associating the turbulent frequency with the scalar mixing rate for a fully-developed scalar spectrum, as detailed by Piton and Fox [130]. The effect of C_ϕ on the simulation results was investigated by Wang *et al.* [128], and the value of 2 was recommended. Besides, Piton and Fox studied the effect of C_ϕ on the barium sulfate precipitation process in a tubular reactor, and found that C_ϕ did not affect the simulation results when it was greater than 4.

Effect of Impeller Speed

Impeller speed is an important parameter affecting mixing performance in a stirred tank. There are major contradictions in the literature about the influence of impeller speed on the mean crystal size. An increase in impeller speed can increase, decrease, produce a minimum or hardly affect the mean crystal size at all. Fig. **22** shows the effect of the impeller speed on the precipitation process. The mean crystal size increased with the faster impeller speed and the *C.V.* was found to be roughly independent on the impeller speed.

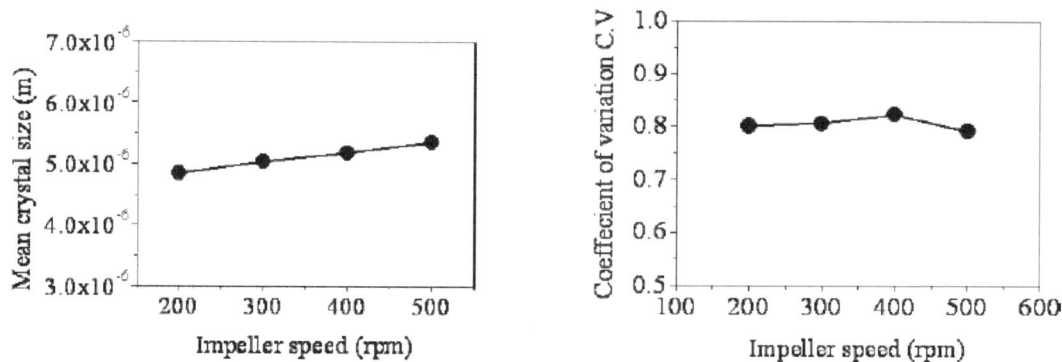

Figure 22: Mean crystal size and *C.V.* at different impeller speeds.

The experimental results of Fitchett and Tarbell [134] and Leeuwen *et al.* [135] showed that increasing the impeller speed resulted in a large mean crystal size of precipitated barium sulfate. The simulation results are also compared with the results predicted by Jaworski and Nienow [124] who neglected the role of micromixing completely. Their prediction values were much greater than the experimental data, indicating the importance of micromixing in continuous stirred tanks.

Effect of Feed Concentration

The mean crystal size is also very sensitive to feed concentration. The FM-PDF model predicted a decrease of the mean crystal size with increasing feed concentration, which was confirmed experimentally by

O'Hern and Rush [136]. The reason is that the supersaturation was greater in magnitude and its distribution became more non-uniform with increasing feed concentration. At lower feed concentration, the rate of crystal growth, $\alpha \Delta c^2$, is a stronger function of feed concentration than that of the heterogeneous nucleation rate, $\alpha \Delta c^{1.775}$. However, while increasing the feed concentration, the homogeneous nucleation rate, $\alpha \Delta c^{15}$, becomes a very strong function of the feed concentration at higher supersaturation values, causing the decrease of the mean crystal size. The discrepancies between model predictions and experimental data are probably due to the crystal agglomeration that is not considered in the model but has a great impact on the product as the feed concentration increases.

Effect of Feed Position

To investigate the dependence of the mean crystal size on the feed position, the precipitation processes at different feed positions were simulated [128]. It can be concluded that placing the feed in the impeller region produced a larger mean crystal size and lower *C.V.* values. This is because the feeding near the impeller tips where turbulent intensity is high would facilitate quick micromixing of the reactants in the solution and thus achieve more uniform supersaturation in the reactor. With the drop of the local level of supersaturation, the faster homogeneous nucleation rate is replaced by the much slower heterogeneous nucleation process. Therefore, a larger mean crystal size was achieved. The decrease of the value of *C.V.* is related to the shorter period of fast nucleation.

Effect of Mean Residence Time

The residence time is actually the time for the crystal growth in the reactor. The precipitation processes with different residence time by changing the feed flow rate were predicted. The effects of mean residence time on the mean crystal mean diameter are shown in Fig. **23**. The simulation results demonstrate that an increase of the mean crystal size and a decrease of the *C.V.* occur while increasing the mean residence time. An explanation of the observed trend is that with the increase of the mean residence time, the growth time of the crystal increases and results in the increase of the mean crystal size.

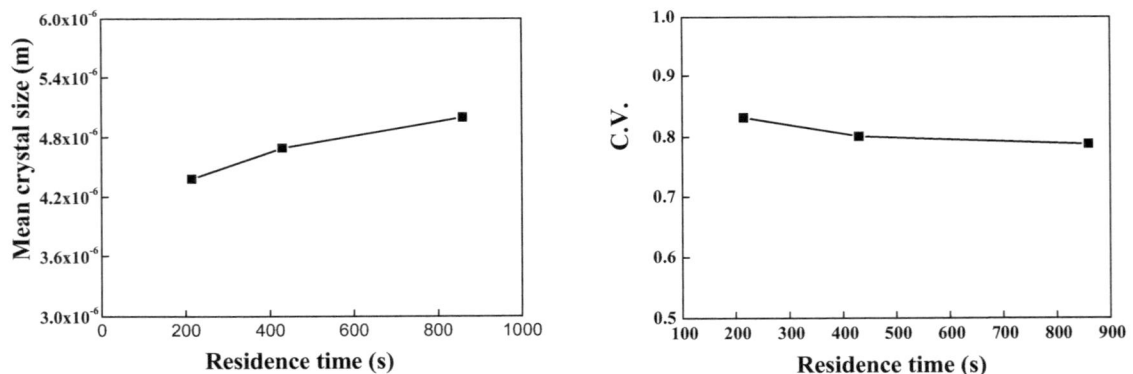

Figure 23: Mean crystal size and *C.V.* at different residence time [128].

SUMMARY AND PERSPECTIVE

The computational fluid dynamics (CFD) method has showed great prospect and attracted extensive attention in recent years owing to its powerful capacity for understanding the physical reality of multiphase flows. The standard k-ε turbulence model based on the assumption of isotropy with Eulerian-Eulerian multi-fluid approach is widely employed in most works in the literature. To account for the anisotropy of turbulence in the flow field, the anisotropic turbulence model is desirable. Anisotropic k-ε turbulence model [137] and algebraic stress models (ASM) [138] have been used for numerical simulation of the flow field in a stirred tank. Moreover, most reports employed Reynolds Averaged Navier-Stokes (RANS) methods to close the momentum equations involved with Reynolds stress. This method is able to predict the mean flow in a rational manner, however, the method proved unsuccessful in gaining the flow information like mean velocity and kinetic energy in the impeller region. Since the large eddy simulation (LES) method is capable of providing detailed and time-dependent flow information, it has been applied more and more extensively with regard to stirred vessels.

Algebraic Stress Model (ASM)

To solve the time-averaged equations of the turbulent flow field in an agitated tank, a turbulent model is needed for the calculation of Reynolds stress terms. The standard k-ε turbulence model, based on the Boussinesq hypothesis of eddy viscosity, is the most widely employed isotropic turbulence model. The Reynolds Stress Model (RSM) is established when the Reynolds stress terms are solved from Reynolds stress transport equations. It requires too much computing resources to solve the Reynolds stress transport equations in engineering applications. In addition, the RSM simulation is heavily dependent on suitable initial conditions. The approximate and compromising form of RSM is the algebraic stress model (ASM), in which the Reynolds stress terms are expressed by a set of implicit algebraic equations. ASM has been applied successfully in many approximately parallel flows, where Reynolds stresses are almost explicitly related to the mean velocity field. In the case of general and complex three-dimensional flow in a stirred vessel, it is very difficult to keep the numerical solution stable, because there is no diffusion or damping in the ASM equation set. Consequently, it is important to employ very small time step or very small under-relaxation factors for achieving a convergent solution. This may increase computational load so that the advantage of simplicity of the ASM model is lost. For a remedy, an explicit form of the stress-strain relation is required. Pope [139] firstly proposed a methodology to obtain an explicit relation for the Reynolds stress tensor from the implicit algebraic stress model that Rodi [140] developed. The methodology producing explicit algebraic stress model is based on the principle of linear algebra. Because of the complexity of the algebra, only a solution of two-dimensional turbulent flows was obtained by Pope. Since then, some research [141-145] extended the results of Pope to three-dimensional turbulence flows.

Mathematic Model

The current form of averaged steady-state Navier-Stokes equation, continuity equation, turbulent kinetic energy and turbulence dissipation in the cylindrical coordinate system is

$$
\frac{1}{r}\frac{\partial}{\partial r}\left(\rho r V_r \varphi\right) + \frac{1}{r}\frac{\partial}{\partial \theta}\left(\rho V_\theta \varphi\right) + \frac{\partial}{\partial z}\left(\rho V_z \varphi\right)
$$
$$
= \frac{1}{r}\frac{\partial}{\partial r}\left(\Gamma_\varphi r \frac{\partial \varphi}{\partial r}\right) + \frac{1}{r}\frac{\partial}{\partial \theta}\left(\frac{\Gamma_\varphi}{r}\frac{\partial \varphi}{\partial \theta}\right) + \frac{\partial}{\partial z}\left(\Gamma_\varphi \frac{\partial \varphi}{\partial z}\right) + S_\varphi
\tag{100}
$$

where φ denotes the general parameter, Γ_φ denotes the diffusion coefficient of φ, and S_φ is the source term in the equation.

The stress tensors of incompressible fluid in the inertial frame are [143]

$$
\overline{v_i v_j} = \frac{2}{3}k\delta_{ij} - 2C_\mu^* \frac{k^2}{\varepsilon}\overline{S}_{ij} - \beta_1 \frac{k^3}{\varepsilon^2}\left(\overline{S}_{im}\overline{W}_{mj} + \overline{S}_{jm}\overline{W}_{mi}\right) +
$$
$$
\beta_2 \frac{k^3}{\varepsilon^2}\left(\overline{S}_{im}\overline{S}_{mj} - \frac{1}{3}\overline{S}_{mn}\overline{S}_{mn}\delta_{ij}\right) + \beta_3 \frac{k^3}{\varepsilon^2}\left(\overline{W}_{im}\overline{W}_{mj} - \frac{1}{3}\overline{W}_{mn}\overline{W}_{mn}\delta_{ij}\right)
\tag{101}
$$

The detailed information about the above equations is provided by Sun *et al.* [138]. It would be difficult to achieve convergence of these transport equations owing to their strong nonlinearity. In fact, the chance of convergence would be minimal if the computation begins with an arbitrarily guessed Reynolds stress field. Thus, a two-step procedure is recommended [138]. First, the general momentum equations are solved along with the conventional standard k-ε turbulence model with

$$\overline{v_r v_r} = \frac{2}{3}k - 2v_t \overline{S}_{rr}, \quad \overline{v_\theta v_\theta} = \frac{2}{3}k - 2v_t \overline{S}_{\theta\theta},$$

$$\overline{v_z v_z} = \frac{2}{3}k - 2v_t \overline{S}_{zz}, \quad \overline{v_r v_\theta} = -2v_t \overline{S}_{r\theta}, \tag{102}$$

$$\overline{v_r v_z} = -2v_t \overline{S}_{rz}, \quad \overline{v_\theta v_z} = -2v_t \overline{S}_{\theta z}$$

After some iterations, the Reynolds stresses based on Eq. (101) are invoked.

Velocity Field Prediction

The comparison of the values of velocity components calculated by standard k-ε turbulence model in a single-phase stirred tank, by the ASM and the experimental data by LDV [146] is shown in Fig. **24**. The velocities are normalized with the impeller tip velocity. It is obvious that the ASM model is superior to the standard k-ε turbulence model. Efforts to extend the application of ASM to simulation of gas-liquid stirred tanks are under way.

Large Eddy Simulation (LES)

It is known that a Direct Numerical Simulation (DNS) of turbulence has to take into account explicitly all scales of motion, from the smallest (the Kolmogorov dissipation scale for instance) to the largest (imposed by the existence of boundaries). For the flow with a high Reynolds number in a stirred tank, there are large and intermediate scale eddies, which are dependent on the tank and impeller geometries. There are also smaller vortices of the Kolmogorov scale, driven by the turbulence cascade and are therefore likely to be independent of the species of the stirred tank. It is difficult in the near future to simulate explicitly all the scales of motion of fluids. However, it is the large scales that are more important and contain the desired information about turbulent transfer of momentum and mass. The motion on this scale is desired to be simulated on the computer. The LES was born in 1970, seeking to predict the motion of eddies which are larger than the user-chosen length scale (mesh size). In LES, not only the mean flow field, but also the dynamics of a range of energetic large scales of motion is directly computed. The effect of the unresolved small-scale turbulence needs to be accounted for through subgrid models [147, 148].

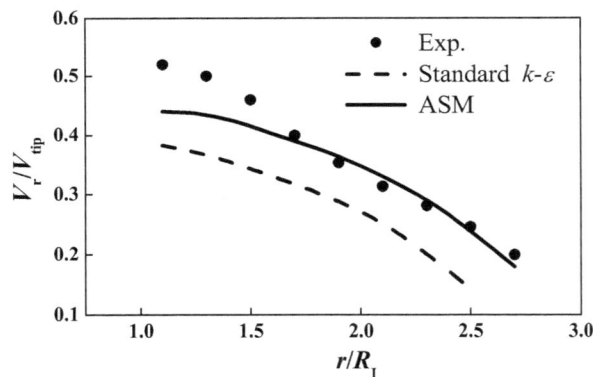

Figure 24: Radial profiles of mean radial velocity in the impeller stream (T=H=270 mm).

The large eddy simulation was successfully applied in the single phase simulation in stirred tanks [149-154]. Some groups [155-157] have attempted to employ the LES in two-phase flow simulation and in most of the reported literature the Lagrangian approach was used to model the dispersed phase. Derksen [48] simulated the liquid-solid flow describing the solid phase by the Lagrangian frame. The Eulerian formulation is preferred in the engineering application. Smith and Milelli [158] and Deen *et al.* [159] used a combination of the large eddy simulation and the Eulerian-Eulerian two phase flow approach to simulate the gas-liquid bubble column. Recently, Zhang *et al.* [27] firstly attempted to apply the large eddy simulation to a gas-sparged stirred tank with two different models for turbulent viscosity (LES model and standard k-ε model).

The most widely used eddy-viscosity model was proposed by Smagorinsky [160], who simulated a two-layer quasi-geostrophic model, representing large (synoptic) scale atmospheric motions. He introduced an eddy viscosity which was supposed to model three-dimensional turbulence that follows approximately a three-dimensional Kolmogorov $k^{-5/3}$ cascade in the subgrid scales. In fact, Smagorinsky's model is too dissipative for quasi two-dimensional turbulence, but it is still very popular for engineering applications.

Zhang *et al.* [27] employed a conventional Smagorinsky sub-grid model to model the turbulence of liquid phase, with a third order QUICK scheme used for convective terms. Some results have been presented in the section "gas-liquid flow". In fact, it is well known that the plain Smagorinsky model is too dissipative close to a wall. This can be shown through expansions of the velocity components in powers of the wall distance. So the model does not work for transition in a boundary layer on a flat plate, starting with a laminar profile to which a small perturbation is added: the flow remains laminar, due to an excessive eddy viscosity coming from the mean shear [147]. A nice alternative to Smagorinsky model is its dynamic version. The dynamic model relies on a LES using a "base" subgrid-scale model such as Smagorinsky's model.

ACKNOWLEDGEMENTS

The authors acknowledge the financial supports from the National Natural Science Foundation of China (Nos. 20990224, 20906090), 973 Program (2010CB630904, 2007CB613507), the National Project of Scientific and Technical Supporting Program (2008BAF33D03) and 863 key project (2007AA060904). Authors wish to thank Prof. Jiayong Chen in our Institute, for his valuable advice and continuous encouragement. We are grateful to many other former Ph.D. and M.S. students of Prof. Mao and Prof. Yang for their contributions to the experimental and numerical studies of stirred tanks. We also thank Prof. Vassilios Kelessidis at Technical University of Crete for valuable discussions.

NOMENCLATURE

C_D	interphase drag coefficient
C_1, C_2	coefficients in k-ε equation
C.V.	coefficient of variation in crystallization
d	diameter of particle
D	diameter of impeller
D_k	mass imbalance term
d_{32}	mean particle size
f_k	turbulent diffusive term
F	interphase force
$F_{r,k}$	centrifugal force
$F_{c,k}$	Coriolis force
Fr	Froude number
Fl	flow number
g	acceleration of gravity
G	dissipation function
G_e	extra dissipation function
H	liquid depth
$k_L a$	mass transfer coefficient
N	impeller speed
N_{js}	impeller speed for critical suspension
$N\theta_m$	dimensionless mixing time
P	pressure
P_V	power consumption per volume

Q	gas flow rate
rpm	revolutions per minute
rad	radian
Re	Reynolds number
t	time
t_l	mean eddy lifetime
t_p	particle response time
T	tank meter
u	velocity
Y	degree of mixing

Greek Symbols

ρ	density
α	phase volume fraction
τ	shear force
τ_t	turbulence fluctuation time
$\tau_{r,k}$	bubble relaxation time
δ	Kronecker delta
v	kinetics viscosity, $m^2 \cdot s^{-1}$
v_t	turbulent kinetics viscosity, $m^2 \cdot s^{-1}$
σ	surface tension
σ_t	turbulent Schmidt number
μ	dynamic viscosity, $Pa \cdot s$
ω	angle velocity, $rad \cdot s^{-1}$
ε	turbulent energy dissipation, $m^2 \cdot s^{-3}$
Γ	turbulent diffusivity, $Pa \cdot s$

Subscripts

am	added mass force
av	averaged
b	bubble
c	continuous phase
d	dispersed phase
drag	drag force
eff	effective
F	flooding
g	gas phase
i, j	principal direction
k	phase
l	liquid phase
lam	laminar
lift	lift force
o	oil phase
r	radial coordinate
s	solid phase
t	turbulent
z	axial coordinate
θ	azimuthal coordinate

REFERENCES

[1]   E. Paul, A. Atiemo-Obeng, and S. Kresta, *Handbook of Industrial Mixing: Science and Practice*. New York: John Wiley and Sons, Inc., 2004.

[2]   R. Stevenson, S. Harrison, M. Mantle, A. Sederman, T. Moraczewski, and M. Johns, "Analysis of partial suspension in stirred mixing cells using both MRI and ERT, " *Chem. Eng. Sci.,* vol. 65, pp. 1385-1393, 2010.

[3]   P. Harvey and M. Greaves, "Turbulent flow in an agitated vessel-Part II: Numerical Solution and Model Predictions," *Chem. Eng. Res. Des.,* vol. 60, pp. 201-210, 1982.

[4]   M. Ishii, *Thermo-Fluid Dynamic Theory of Two-Phase Flow* Paris: Eyrolles, 1975.

[5]   A. Gosman, C. Lekakou, S. Politis, R. Issa, and M. Looney, "Multidimensional modeling of turbulent two-phase flows in stirred vessels," *AIChE J.,* vol. 38, pp. 1946-1956, 1992.

[6]   G. Micale, G. Montante, F. Grisafi, A. Brucato, and J. Godfrey, "CFD simulation of particle distribution in stirred vessels," *Chem. Eng. Res. Des.,* vol. 78, pp. 435-444, 2000.

[7]   M. Ljungqvist and A. Rasmuson, "Numerical simulation of the two-phase flow in an axially stirred vessel," *Chem. Eng. Res. Des.,* vol. 79, pp. 533-546, 2001.

[8]   X. Shan, G. Yu, C. Yang, Z.-S. Mao, and W. Zhang, "Numerical Simulation of Liquid Solid Flow in an Unbaffled Stirred Tank with a Pitched-Blade Turbine Downflow," *Ind. Eng. Chem. Res.,* vol. 47, pp. 2926-2940, 2008.

[9]   V. Ranade and H. Van den Akker, "A computational snapshot of gas-liquid flow in baffled stirred reactors," *Chem. Eng. Sci.,* vol. 49, pp. 5175-5192, 1994.

[10]   W. Lin, Z.-S. Mao, and J. Chen, "Hydrodynamic studies on loop reactors (II) Airlift Loop Reactors," *Chin. J. Chem. Eng.,* vol. 5, pp. 11-22, 1997.

[11]   W. Wang and Z.-S. Mao, "Numerical simulation of gas-liquid flow in a stirred tank with a Rushton impeller," *Chin. J. Chem. Eng.,* vol. 10, pp. 385-395, 2002.

[12]   A. Khopkar, A. Rammohan, V. Ranade, and M. Dudukovic, "Gas–liquid flow generated by a Rushton turbine in stirred vessel: CARPT/CT measurements and CFD simulations," *Chem. Eng. Sci.,* vol. 60, pp. 2215-2229, 2005.

[13]   M. Ishii and N. Zuber, "Drag coefficient and relative velocity in bubbly, droplet or particulate flows," *AIChE J.,* vol. 25, pp. 843-855, 1979.

[14]   I. Kataoka, D. Besnard, and A. Serizawa, "Basic equation of turbulence and modeling of interfacial transfer terms in gas-liquid two-phase flow," *Chem. Eng. Commun.,* vol. 118, pp. 221-236, 1992.

[15]   V. Ranade and J. Joshi, "Flow generated by a disc turbine: Part II Mathematical modelling and comparison with experimental data," *Chem. Eng. Res. Des.,* vol. 68, pp. 34-50, 1990.

[16]   R. Hesketh, T. Russell, and A. Etchells, "Bubble size in horizontal pipelines," *AIChE J.,* vol. 33, pp. 663–667, 1987.

[17]   R. Parthasarathy and N. Ahmed, "Sauter mean and maximum bubble diameters in aerated stirred vessels," *Chem. Eng. Res. Des.,* vol. 72, pp. 565-572, 1994.

[18]   D. Brown and K. Pitt, "Drop size distribution of stirred non-coalescing liquid-liquid system," *Chem. Eng. Sci.,* vol. 27, pp. 577-583, 1972.

[19]   R. Calabrese, T. Chang, and P. Dang, "Drop breakup in turbulent stirred-tank contactors. I: Effect of dispersed-phase viscosity," *AIChE J.,* vol. 32, pp. 657-666, 1986.

[20]   G. Montante, G. Micale, F. Magelli, and A. Brucato, "Experiments and CFD predictions of solid particle distribution in a vessel agitated with four pitched blade turbines," *Chem. Eng. Res. Des.,* vol. 79, pp. 1005-1010, 2001.

[21]   A. Brucato, M. Ciofalo, F. Grisafi, and G. Micale, "Numerical prediction of flow fields in baffled stirred vessels: A comparison of alternative modelling approaches," *Chem. Eng. Sci.,* vol. 53, pp. 3653-3684, 1998.

[22]   J. Luo, R. Issa, and A. Gosman, "Prediction of impeller induced flows in mixing vessels using multiple frames of reference," *IChemE Symp. Ser.,* vol. 136, pp. 549-556, 1994.

[23]   H. Versteeg and W. Malalasekera, *An introduction to computational fluid dynamics: the finite volume method*: Prentice Hall, 1995.

[24]   S. Patankar and D. Spalding, "A calculation procedure for heat, mass and momentum transfer in three-dimensional parabolic flows," *Int. J. Heat Mass Transfer,* vol. 15, pp. 1787–1806, 1972.

[25]   M. Carver and M. Salcudean, "Three dimensional numerical modeling of phase distribution of two-fluid flow in elbaus and return bends," *Numerical Heat Transfer,* vol. 10, pp. 229-251, 1986.

[26]   Y. Zhang, Y. Yong, Z.-S. Mao, C. Yang, H. Sun, and H. Wang, "Numerical simulation of gas-liquid flow in a stirred tank with swirl modification," *Chem. Eng. Technol.,* vol. 32, pp. 1266-1273, 2009.

[27]   Y. Zhang, C. Yang, and Z.-S. Mao, "Large eddy simulation of the gas-liquid flow in a stirred tank," *AIChE J.,* vol. 54, pp. 1963-1974, 2008.

[28]   W. Wang and Z. Mao, "Numerical simulation of gas-liquid flow in a stirred tank with a Rushton impeller," *Chin. J. Chem. Eng.,* vol. 10, pp. 385-395, 2002.

[29]   W. Lu and S. Ju, "Local gas holdrup, mean liquid velocity and turbulence in an aerated stirred tank using hot-film anemometry," *Chem. Eng. J.,* vol. 35, pp. 9-17, 1987.

[30]   W. Wang, Z.-S. Mao, and C. Yang, "Experimental and numerical investigation on gas holdup and flooding in an aerated stirred tank with rushton impeller," *Ind. Eng. Chem. Res.,* vol. 45, pp. 1141-1151, 2006.

[31]   A. Paglianti, S. Pintus, and M. Giona, "Time-series analysis approach for the identification of flooding/loading transition in gas–liquid stirred tank reactors," *Chem. Eng. Sci.,* vol. 55, pp. 5793-5802, 2000.

[32]   A. Nienow, M. Warmoeskerken, J. Smith, and M. Konno, "On the flooding-loading transition and the complete dispersal condition in aerated vessels agitated by a Rushton turbine," in *Proceedings of the 5th European Conference on Mixing,* 1985, pp. 143–154.

[33]   F. Wang, W. Wang, and Z.-S. Mao, "Numerical study of solid-liquid two-phase flow in stirred tanks with Rushton impeller (I) Formulation and simulation of flow field," *Chin. J. Chem. Eng.,* vol. 12, pp. 599-609, 2004.

[34]   T. Zwietering, "Suspending of solid particles in liquid by agitators," *Chem. Eng. Sci.,* vol. 8, pp. 244-253, 1958.

[35]   P. Armenante, E. Nagamine, and J. Susanto, "Determination of correlations to predict the minimum agitation speed for complete solid suspension in agitated vessels," *Can. J. Chem. Eng.,* vol. 76, pp. 413-419, 1998.

[36]   P. Armenante and E. Nagamine, "Effect of low off-bottom impeller clearance on the minimum agitation speed for complete suspension of solids in stirred tanks," *Chem. Eng. Sci.,* vol. 53, pp. 1757-1775, 1998.

[37]   G. Baldi, R. Conti, and E. Alaria, "Complete suspension of particles in mechanically agitated vessels," *Chem. Eng. Sci.,* vol. 33, pp. 1-25, 1978.

[38]   K. Wichterle, "Conditions for suspension of solids in agitated vessels," *Chem. Eng. Sci.,* vol. 43, pp. 467-471, 1988.

[39]   F. Wang, Z.-S. Mao, and X. Shen, "Numerical study of solid-liquid two-phase flow in stirred tanks with Rushton impeller (II) Prediction of critical impeller speed," *Chin. J. Chem. Eng.,* vol. 12, pp. 610-614, 2004.

[40]   J. Luo, A. Gosman, R. Issa, J. Middleton, and M. Fitzgerald, "Full flow field computation of mixing in baffled stirred vessels," *Chem. Eng. Res. Des.,* vol. 71, pp. 342-344, 1993.

[41]   A. Brucato, M. Ciofalo, F. Grisafi, and G. Micale, "Complete numerical simulation of flow fields in baffled stirred vessels: the inner-outer approach," *Can. J. Chem. Eng.,* vol. 71, pp. 269-289, 1994.

[42]   R. Angst and M. Kraume, "Experimental investigations of stirred solid/liquid systems in three different scales: Particle distribution and power consumption," *Chem. Eng. Sci.,* vol. 61, pp. 2864-2870, 2006.

[43]   G. Micale, V. Carrara, F. Grisafi, and A. Brucato, "Solids suspension in three-phase stirred tanks," *Chem. Eng. Res. Des.,* vol. 78, pp. 319-326, 2000.

[44]   G. Micale, F. Grisafi, and A. Brucato, "Assessment of particle suspension conditions in stirred vessels by means of pressure gauge technique," *Chem. Eng. Res. Des.,* vol. 80, pp. 893-902, 2002.

[45]   H. Zhang, P. Johnston, J. Zhu, H. De Lasa, and M. Bergougnou, "A novel calibration procedure for a fiber optic solids concentration probe," *Powder Technol.,* vol. 100, pp. 260-272, 1999.

[46]   Y. Zheng, J. Zhu, N. Marwaha, and A. Bassi, "Radial solids flow structure in a liquid–solids circulating fluidized bed," *Chem. Eng. J.,* vol. 88, pp. 141-150, 2002.

[47]   A. Ochieng and A. Lewis, "Nickel solids concentration distribution in a stirred tank," *Mineral Eng.,* vol. 19, pp. 180-189, 2006.

[48]   J. Derksen, "Numerical simulation of solids suspension in a stirred tank," *AIChE J.,* vol. 49, pp. 2700–2714, 2003.

[49]   G. Zhou and S. Kresta, "Correlation of mean drop size and minimum drop size with the turbulence energy dissipation and the flow in an agitated tank," *Chem. Eng. Sci.,* vol. 53, pp. 2063-2079, 1998.

[50]   K. Wichterle, "Drop breakup by impellers," *Chem. Eng. Sci.,* vol. 50, pp. 3581-3586, 1995.

[51]   F. Svensson and A. Rasmuson, "LDA-measurements in a stirred tank with a liquid-liquid system at high volume percentage dispersed phase," *Chem. Eng. Technol.,* vol. 27, pp. 335-339, 2004.

[52]   A. Skelland and J. Lee, "Agitator speeds in baffled vessels for uniform liquid-liquid dispersions," *Ind. Eng. Chem. Process Des. Develop.,* vol. 17, pp. 473-478, 1978.

[53]   S. Okufi, E. de Ortiz, and H. Sawistowski, "Scale-up of liquid-liquid dispersions in stirred tanks," *Can. J. Chem. Eng.,* vol. 68, pp. 400-406, 1990.

[54]   P. Armenante and Y. Huang, "Experimental determination of the minimum agitation speed for complete liquid-liquid dispersion in mechanically agitated vessels," *Ind. Eng. Chem. Res.,* vol. 31, pp. 1398-1406, 1992.

[55]   X. Zhu and R. Vigil, "Banded liquid–liquid Taylor-Couette-Poiseuille flow," *AIChE J.,* vol. 47, pp. 1932-1940, 2001.

[56]   F. Wang and Z.-S. Mao, "Numerical and experimental investigation of liquid liquid two-phase flow in stirred tanks," *Ind. Eng. Chem. Res,* vol. 44, pp. 5776-5787, 2005.

[57]   J. Joshi, "Computational flow modelling and design of bubble column reactors," *Chem. Eng. Sci.,* vol. 56, pp. 5893–5933, 2001.

[58]   S. Nagata, *Mixing: Principles and Applications*: Halsted Press, 1975.

[59]   T. Anderson and R. Jackson, "A fluid mechanical description of fluidized beds," *Ind. Eng. Chem. Fund.,* vol. 6, pp. 527–538, 1967.

[60]   L. Van Wijngaarden, "Hydrodynamic interaction between gas bubbles in liquid," *J. Fluid Mech.,* vol. 77, pp. 27-44, 1976.

[61]   T. Auton, "The lift force on a spherical rotational flow," *J. Fluid Mech.,* vol. 183, pp. 199-218, 1987.

[62]   F. Wang and Z. Mao, "Numerical and experimental investigation of liquid liquid two-phase flow in stirred tanks," *Ind. Eng. Chem. Res.,* vol. 44, pp. 5776-5787, 2005.

[63]   F. Wang, Z.-S. Mao, Y. Wang, and C. Yang, "Measurement of phase holdups in liquid-liquid-solid three-phase stirred tanks and CFD simulation," *Chem. Eng. Sci.,* vol. 61, pp. 7535-7550, 2006.

[64]   J. Fang, C. Yang, G. Yu, and Z.-S. Mao, "Preliminary investigation on interphase mass transfer in agitated liquid-liquid-solid dispersion," *Chin. J. Process Eng.,* vol. 5, pp. 125-130, 2005.

[65]   S. Asai, Y. Konishi, and Y. Sasaki, "Mass transfer between fine particles and liquids in agitated vessels," *J. Chem. Eng. Japan,* vol. 21, pp. 107-112, 1988.

[66]   N. Padial, W. VanderHeyden, R. Rauenzahn, and S. Yarbro, "Three-dimensional simulation of a three-phase draft-tube bubble column," *Chem. Eng. Sci.,* vol. 55, pp. 3261-3273, 2000.

[67]   V. Michele and D. Hempel, "Liquid flow and phase holdup-measurement and CFD modeling for two-and three-phase bubble columns," *Chem. Eng. Sci.,* vol. 57, pp. 1899-1908, 2002.

[68]   G. Z. Yu, R. Wang, and Z.-S. Mao, "Dispersion characteristics and gas-liquid mass transfer in a gas-liquid-liquid self-aspirating reactor," *Petroleum Processing and Petrochemicals (China),* vol. 31, pp. 54-59, 2000.

[69]   D. White and J. De Villiers, "Rates of induced aeration in agitated vessels," *Chem. Eng. J.,* vol. 14, pp. 113-118, 1977.

[70]   Q. Yu, G. Yu, C. Yang, and Z.-S. Mao, "Dispersion and phase separation characteristics of liquid-liquid-liquid systems," *Chin. J. Process Eng.,* vol. 7, pp. 229-234, 2007.

[71]   A. Skelland and G. Ramsay, "Minimum agitator speeds for complete liquid-liquid dispersion," *Ind. Eng. Chem. Res.,* vol. 26, pp. 77-81, 1987.

[72]   H. Wu, "An issue on applications of a disk turbine for gas-liquid mass transfer," *Chem. Eng. Sci.,* vol. 50, pp. 2801-2811, 1995.

[73]   M. Tanaka and T. Izumi, "Gas entrainment in stirred-tank reactors," *Chem. Eng. Res. Des.,* vol. 65, pp. 195-198, 1987.

[74]   M. Tanaka, S. Noda, and E. O'shima, "Effect of the location of a submerged impeller on the enfoldment of air bubbles from the free surface in a stirred vessel," *Int. Chem. Eng.,* vol. 26, pp. 314-318, 1986.

[75]   J. Rushton, E. Costich, and H. Everett, "Power characteristics of mixing impellers," *Chem. Eng. Prog.,* vol. 46, pp. 395-404, 1950.

[76]   K. Van't Riet, J. Boom, and J. Smith, "Power consumption, impeller coalescence and recirculation in aerated vessels," *Chem. Eng. Res. Des.,* vol. 54, pp. 124-129, 1976.

[77]   R. Roman and R. Tudose, "Studies on transfer processes in mixing vessels: effect of particles on gas-liquid mass transfer using modified Rushton turbine agitators," *Bioprocess and Biosystems Eng.,* vol. 17, pp. 361-365, 1997.

[78]   W. Lu, H. Wu, and M. Ju, "Effects of baffle design on the liquid mixing in an aerated stirred tank with standard Rushton turbine impellers," *Chem. Eng. Sci.,* vol. 52, pp. 3843-3851, 1997.

[79]   Y. Hsu, R. Peng, and C. Huang, "Onset of gas induction, power consumption, gas holdup and mass transfer in a new gas-induced reactor," *Chem. Eng. Sci.,* vol. 52, pp. 3883-3891, 1997.

[80]   G. Yu, Z.-S. Mao, and R. Wang, "A novel surface aeration configuration for improving gas-liquid mass transfer," *Chin. J. Chem. Eng.,* vol. 10, pp. 39-44, 2002.

[81]   G. Z. Yu and Z.-S. Mao, "Performance of a novel surface aerator with self-rotating flow guide," *Petroleum Processing and Petrochemicals (China),* vol. 35, pp. 45-48, 2004.

[82]   X. Y. Li, G. Z. Yu, C. Yang, and Z.-S. Mao, "Experimental study on surface aerators stirred by triple impellers," *Ind. Eng. Chem. Res.,* vol. 48, pp. 8752-8756, 2009.

[83]   X. Y. Li, G. Z. Yu, and Z.-S. Mao, "Optimization of dual-impeller configuration of gas-liquid surface aerator," *Chin. J. Process Eng.,* vol. 5, pp. 601-604, 2005.

[84]   H. Sun, Z.-S. Mao, and G. Yu, "Experimental and numerical study of gas hold-up in surface aerated stirred tanks," *Chem. Eng. Sci.,* vol. 61, pp. 4098-4110, 2006.

[85] G. Patterson, "Measurements and modelling of flow in gas sparged, agitated vessels," in *7th European Conference on Mixing* Brugge, Belgium, 1991, pp. 209-215.

[86] S. Broring, J. Fischer, T. Korte, S. Sollinger, A. Lübbert, and D. Forschungsgemeinschaft, "Flow structure of the dispersed gasphase in real multiphase chemical reactors investigated by a new ultrasound-Doppler technique," *Can. J. Chem. Eng.,* vol. 69, pp. 1247-1256, 1991.

[87] S. Alves, C. Maia, and J. Vasconcelos, "Experimental and modelling study of gas dispersion in a double turbine stirred tank," *Chem. Eng. Sci.,* vol. 57, pp. 487-496, 2002.

[88] K. Pericleous and M. Patel, "The modelling of tangential and axial agitators in chemical reactors," *PhysicoChemical Hydrodynamics,* vol. 8, pp. 105-123, 1987.

[89] G. Lane, M. Schwarz, and G. Evans, "Modelling of the interaction between gas and liquid in stirred vessels," in *10th European Conference on Mixing*, Delft, the Netherlands, 2000, p. 197.

[90] H. Sun, Z. Mao, and G. Yu, "Experimental and numerical study of gas hold-up in surface aerated stirred tanks," *Chem. Eng. Sci.,* vol. 61, pp. 4098-4110, 2006.

[91] H. Topiwala, "Surface aeration in a laboratory fermenter at high power inputs," *J. Ferment. Technol.,* vol. 50, pp. 668-675, 1972.

[92] V. Veljkovi, K. Bicok, and D. Simonovi, "Mechanism, onset and intensity of surface aeration in geometrically-similar, sparged, agitated vessels," *Can. J. Chem. Eng.,* vol. 69, pp. 916-926, 1991.

[93] A. H. Wang, G. Z. Yu, and Z.-S. Mao, "Characteristics of gas-liquid mass transfer of surface aerator with self-rotating floating baffle," *Petroleum Processing and Petrochemicals (China),* vol. 35, pp. 51-55, 2004.

[94] M. Matsumura, S. Hideo, and Y. Tamitoshi, "Gas entrainment in a new entraining fermenter," *J. Ferment. Technol.,* vol. 60, pp. 457-467, 1982.

[95] V. Uhl and J. Gray, *Mixing: Theory and Practice* vol. 2: Academic Press, New York, 1967.

[96] W. Frost and T. Moulden, *Handbook of turbulence. Volume 1-Fundamentals and Applications.* New York, London: Plenum Press, 1977.

[97] Z. Wang, Z.-S. Mao, and X. Shen, "Numerical simulation of macroscopic mixing in a Rushton impeller stirred tank," *Chin. J. Process Eng.,* vol. 6, pp. 857-863, 2006.

[98] D. Brennan and I. Lehrer, "Impeller mixing in vessels experimental studies on the influence of some parameters and formulation of a general mixing time equation," *Chem. Eng. Res. Des.,* vol. 54, pp. 139-152, 1976.

[99] D. Holmes, R. Voncken, and J. Dekker, "Fluid flow in turbine-stirred, baffled tanks-I: Circulation time," *Chem. Eng. Sci.,* vol. 19, pp. 201-208, 1964.

[100] H. Kramers, G. Baars, and W. Knoll, "A comparative study on the rate of mixing in stirred tanks," *Chem. Eng. Sci.,* vol. 2, pp. 35-42, 1953.

[101] R. Biggs, "Mixing rates in stirred tanks," *AIChE J.,* vol. 9, pp. 636-640, 1963.

[102] L. Li and J. Wei, "Three-dimensional image analysis of mixing in stirred vessels," *AIChE J.,* vol. 45, pp. 1855-1865, 1999.

[103] K. Lee and M. Yianneskis, "A liquid crystal thermographic technique for the measurement of mixing characteristics in stirred vessels," *Chem. Eng. Res. Des.,* vol. 75, pp. 746-754, 1997.

[104] J. Joshi, A. Pandit, and M. Sharma, "Mechanically agitated gas-liquid reactors," *Chem. Eng. Sci.,* vol. 37, pp. 813-844, 1982.

[105] S. Shiue and C. Wong, "Studies on homogenization efficiency of various agitators in liquid blending," *Can. J. Chem. Eng.,* vol. 62, pp. 602-609, 1984.

[106] R. Voncken, D. Holmes, and H. Den Hartog, "Fluid flow in turbine-stirred, baffled tanks--III: Dispersion during circulation," *Chem. Eng. Sci.,* vol. 19, pp. 209-213, 1964.

[107] W. McManamey, "A circulation model for batch mixing in agitated, baffled vessels," *Chem. Eng. Res. Des.,* vol. 58, pp. 271-276, 1980.

[108] V. Ranade, J. Bourne, and J. Joshi, "Fluid mechanics and blending in agitated tanks," *Chem. Eng. Sci.,* vol. 46, pp. 1883-1893, 1991.

[109] Z. Jaworski, W. Bujalski, N. Otomo, and A. Nienow, "CFD study of homogenization with dual Rushton turbines—comparison with experimental results Part I: Initial Studies," *Chem. Eng. Res. Des.,* vol. 78, pp. 327-333, 2000.

[110] G. Zhou, Y. Wang, and L. Shi, "CFD study of mixing process in stirred tank," *J. Chem. Ind. Eng. (China),* vol. 54, pp. 886-890, 2003.

[111] M. Lunden, O. Stenberg, and B. Andersson, "Evaluation of a method for measuring mixing time using numerical simulation and experimental data," *Chem. Eng. Commun.,* vol. 139, pp. 115-158, 1995.

[112] E. Fox and V. Gex, "Single-phase blending of liquids," *AIChE J.,* vol. 2, pp. 539-544, 1956.

[113] Q. Zhang, Y. Yong, Z.-S. Mao, C. Yang, and C. Zhao, "Experimental determination and numerical simulation of mixing time in a gas-liquid stirred tank," *Chem. Eng. Sci.,* vol. 64, pp. 2926-2933, 2009.

[114] M. Bouaifi and M. Roustan, "Power consumption, mixing time and homogenisation energy in dual-impeller agitated gas–liquid reactors," *Chem. Eng. Process.,* vol. 40, pp. 87-95, 2001.

[115] S. Shewale and A. Pandit, "Studies in multiple impeller agitated gas–liquid contactors," *Chem. Eng. Sci.,* vol. 61, pp. 489-504, 2006.

[116] A. Pandit and J. Joshi, "Mixing in mechanically agitated gas-liquid contactors, bubble columns and modified bubble columns," *Chem. Eng. Sci.,* vol. 38, pp. 1189-1215, 1983.

[117] W. Lu and H. Chen, "Flooding and critical impeller speed for gas dispersion in aerated turbine-agitated vessels," *Chem. Eng. J.,* vol. 33, pp. 57-62, 1986.

[118] R. Pohorecki and J. Baldyga, in *Proc. of Industrial Crystallization* North-Holland, Amsterdam, 1979.

[119] J. Garside and N. Tavare, "Mixing, reaction and precipitation: limits of micromixing in an MSMPR crystallizer," *Chem. Eng. Sci.,* vol. 40, pp. 1485-1493, 1985.

[120] R. Phillips, S. Rohani, and J. Baldyga, "Micromixing in a single-feed semi-batch precipitation process," *AIChE J.,* vol. 45, pp. 82-92, 1999.

[121] H. Wei and J. Garside, "Application of CFD modelling to precipitation systems," *Chem. Eng. Res. Des.,* vol. 75, pp. 219-227, 1997.

[122] M. Van Leeuwen, O. Bruinsma, and G. Van Rosmalen, "Influence of mixing on the product quality in precipitation," *Chem. Eng. Sci.,* vol. 51, pp. 2595-2600, 1996.

[123] J. Garside and H. Wei, "Pumped, stirred and maybe precipitated: Simulation of precipitation processes using CFD," *Acta Polytech. Scand. Chem. Technol.,* pp. 9-15, 1997.

[124] Z. Jaworski and A. Nienow, "CFD modelling of continuous precipitation of barium sulphate in a stirred tank," *Chem. Eng. J.,* vol. 91, pp. 167-174, 2003.

[125] H. Wei, W. Zhou, and J. Garsides, "Computational fluid dynamics modeling of the precipitation process in a semibatch crystallizer," *Ind. Eng. Chem. Res.,* vol. 40, pp. 5255-5261, 2001.

[126] J. Baldyga and W. Orciuch, "Barium sulphate precipitation in a pipe—an experimental study and CFD modelling," *Chem. Eng. Sci.,* vol. 56, pp. 2435-2444, 2001.

[127] J. Baldyga and W. Orciuch, "Closure problem for precipitation," *Chem. Eng. Res. Des.,* vol. 75, pp. 160-170, 1997.

[128] Z. Wang, Q. Zhang, C. Yang, Z.-S. Mao, and X. Shen, "Simulation of barium sulfate precipitation using CFD and FM-PDF modeling in a continuous stirred tank," *Chem. Eng. Technol.,* vol. 30, pp. 1642-1649, 2007.

[129] R. O. Fox, "On the relationship between Lagrangian micromixing models and computational fluid dynamics," *Chem. Eng. Process.,* vol. 37, pp. 521-535, 1998.

[130] D. Piton, R. Fox, and B. Marcant, "Simulation of fine particle formation by precipitation using computational fluid dynamics," *Can. J. Chem. Eng.,* vol. 78, pp. 983-993, 2000.

[131] A. Randolph and M. Larson, *Theory of Particulate Processes,* 2nd ed. New York: Academic Press, 1988.

[132] J. Baldyga, W. Podgorska, and R. Pohorecki, "Mixing-precipitation model with application to double feed semibatch precipitation," *Chem. Eng. Sci.,* vol. 50, pp. 1281-1300, 1995.

[133] A. E. Nielsen, "Electrolyte crystal growth mechanisms," *J. Crystal Growth,* vol. 67, pp. 289-310, 1984.

[134] D. Fitchett and J. Tarbell, "Effect of mixing on the precipitation of barium sulfate in an MSMPR reactor," *AIChE J.,* vol. 36, pp. 511-522, 1990.

[135] M. Van Leeuwen, O. Bruinsma, and G. Van Rosmalen, "Three-zone approach for precipitation of barium sulphate," *J. Crystal Growth,* vol. 166, pp. 1004-1008, 1996.

[136] H. A. O'Hern and F. E. Rush, "Effect of mixing conditions in barium sulfate precipitation," *Ind. Eng. Chem. Fundam.,* vol. 2, pp. 267-272, 1963.

[137] H. Sun, W. Wang, and Z.-S. Mao, "Numerical simulation of whole three-dimensional flow field in a stirred tank with anisotropic turbulence model," *J. Chem. Ind. Eng. (China),* vol. 53, pp. 1153-1159, 2002.

[138] H. Sun, W. Wang, and Z. S. Mao, "Numerical simulation of the whole three-dimensional flow in a stirred tank with anisotropic algebraic stress model," *Chin. J. Chem. Eng.,* vol. 10, pp. 15-24, 2002.

[139] S. Pope, "A more general effective-viscosity hypothesis," *J. Fluid Mech.,* vol. 72, pp. 331-340, 1975.

[140] W. Rodi, "A new algebraic relation for calculating the Reynolds stresses," *Zeitschrift fuer angewandte Mathematik und Mechanik,* vol. 56, pp. 219-221, 1976.

[141] A. Yoshizawa, "Statistical analysis of the deviation of the Reynolds stress from its eddy-viscosity representation," *Phys. Fluids,* vol. 27, pp. 1377-1387, 1984.

[142] C. Speziale, "On nonlinear *k-l* and *k-ε* models of turbulence," *J. Fluid Mech.,* vol. 178, pp. 459-475, 1987.

[143] T. Gatski and C. Speziale, "On explicit algebraic stress models for complex turbulent flows," *J. Fluid Mech.,* vol. 254, pp. 59-78, 1993.

[144] R. Rubinstein and J. Barton, "Nonlinear Reynolds stress models and the renormalization group," *Phys. Fluids A: Fluid Dynamics,* vol. 2, pp. 1472-1476, 1990.

[145] D. Taulbee, "An improved algebraic Reynolds stress model and corresponding nonlinear stress model," *Phys. Fluids A: Fluid Dynamics,* vol. 4, pp. 2555-2561, 1992.

[146] H. Wu and G. Patterson, "Laser-Doppler measurements of turbulent-flow parameters in a stirred mixer," *Chem. Eng. Sci.,* vol. 44, pp. 2207–2221, 1989.

[147] M. Lesieur, *Turbulence in Fluids* vol. 84. Dordrecht, The Netherlands: Springer, 2008.

[148] H. Yoon, S. Balachandar, M. Ha, and K. Kar, "Large eddy simulation of flow in a stirred tank," *J. Fluids Eng.,* vol. 125, pp. 486-499, 2003.

[149] J. Eggels, "Direct and large-eddy simulation of turbulent fluid flow using the lattice-Boltzmann scheme," *Int. J. Heat Fluid Flow,* vol. 17, pp. 307-323, 1996.

[150] J. Derksen and H. Van den Akker, "Large eddy simulations on the flow driven by a Rushton turbine," *AIChE J.,* vol. 45, pp. 209-221, 1999.

[151] J. Revstedt and L. Fuchs, "Large eddy simulation of flow in stirred tanks," *Chem. Eng. Technol.,* vol. 25, pp. 443-446, 2002.

[152] S. Yeoh, G. Papadakis, K. Lee, and M. Yianneskis, "Large eddy simulation of turbulent flow in a Rushton impeller stirred reactor with sliding-deforming mesh methodology," *Chem. Eng. Technol.,* vol. 27, pp. 257-263, 2004.

[153] R. Alcamo, G. Micale, F. Grisafi, A. Brucato, and M. Ciofalo, "Large-eddy simulation of turbulent flow in an unbaffled stirred tank driven by a Rushton turbine," *Chem. Eng. Sci.,* vol. 60, pp. 2303-2316, 2005.

[154] Y. Zhang, C. Yang, and Z.-S. Mao, "Large eddy simulation of liquid flow in a stirred tank with improved inner-outer iterative algorithm," *Chin. J. Chem. Eng.,* vol. 14, pp. 321-329, 2006.

[155] Q. Wang, K. Squires, and O. Simonin, "Large eddy simulation of turbulent gas-solid flows in a vertical channel and evaluation of second-order models," *Int. J. Heat Fluid Flow,* vol. 19, pp. 505-511, 1998.

[156] S. Apte, M. Gorokhovski, and P. Moin, "LES of atomizing spray with stochastic modeling of secondary breakup," *Int. J. Multiphase Flow,* vol. 29, pp. 1503-1522, 2003.

[157] B. Shotorban and F. Mashayek, "Modeling subgrid-scale effects on particles by approximate deconvolution," *Phys. Fluids,* vol. 17, pp. 1-4, 2005.

[158] B. Smith and M. Milelli, "An investigation of confined bubble plumes," in *Proc. 3rd International Conference on Multiphase Flow,* Lyon, France, 1998, pp. 8-12.

[159] N. Deen, T. Solberg, and B. Hjertager, "Large eddy simulation of the gas–liquid flow in a square cross-sectioned bubble column," *Chem. Eng. Sci.,* vol. 56, pp. 6341-6349, 2001.

[160] J. Smagorinsky, "General circulation experiments with the primitive equations," *Mon. Weath. Rev.,* vol. 91, pp. 99-164, 1963.

Advances in Multiphase Flow and Heat Transfer, Vol. 3, 2012, 55-83 55

CHAPTER 2

Two-Phase Flow Pressure Change Across Sudden Contraction and Expansion in Small Channels

Ing Youn Chen[a,*], Chih-Yung Tseng[b] and Chi-Chuan Wang[c]

[a]*Mechanical Engineering Department, National Yunlin University of Science and Technology, Yunlin, Taiwan 640;* [b]*Energy & Environment Research Laboratories, Industrial Technology Research Institute, Hsinchu, Taiwan 310 and* [c]*Department of Mechanical Engineering, National Chiao Tung University, Hsinchu, Taiwan 300*

Abstract: The flow of two-phase mixtures across sudden expansions and contractions is relevant in many applications such as chemical reactors, power generation units, oil wells and petrochemical plants. As the two-phase mixture flows through the sudden area changes, the flow might form a separation region at the sharp corner and introduce an irreversible pressure loss. This loss occurs in practical pipeline connections and in the heat exchangers. The small and narrow channels are widely adopted in compact heat exchangers. Also, flow in small rectangular channels is an integral part of CPU cold plate using the liquid cooling with or without phase change. Predictions of these pressure drops had been made using correlations developed for the conventional tubes, extrapolations of these correlations to small diameter tube are questionable.

In this study, the authors first give a short overview on the single-phase flow across sudden contraction and expansion, followed by a thorough review of the relevant literature for two-phase flow across sudden contraction and expansion. The applicability of the existing model/correlations for sudden contraction and expansion is then examined with the available data from literature. Comparisons for the pressure change data with the predictions of existing model/correlations indicate that none of them can accurately predict the data.

For the two-phase contraction, it is found that the influence of surface tension and outlet tube size, or equivalently the Bond number plays a major role for the departure of various models/correlations. Among the models/correlations being examined, the homogeneous model shows a little better than the others. Hence by taking into account the influences of gas quality, Bond number, Weber number and area ratio into the homogeneous model, a modified homogeneous correlation is proposed that considerably improves the predictive ability over existing correlations with a mean deviation of 30% to the 503 data.

For the two-phase expansion, most of the correlations highly over predict the data with a mini test section which has a Bond number being less than 0.1 in which the effect of surface tension dominates. Also, some of the correlations significantly under predict the data for very large test sections. Among the models/correlations being examined, the homogeneous model shows a poor predictive ability than the others, but it is handy from the engineering aspect. Hence by taking account the influences of Bond number, Weber number, Froude number, liquid Reynolds number, gas quality and area ratio into the original homogeneous model, a modified homogeneous model is proposed that considerably improves the predictive ability over existing correlations with a mean deviation of 23% and a standard deviation of 29% to the 282 data with wider ranges for application.

Keywords: Two-phase flow, pressure change, sudden contraction and expansion.

INTRODUCTION

The flow of two-phase mixtures across sudden expansions and contractions is commonly seen among

*Address correspondence to Ing Youn Chen:** Mechanical Engineering Department, National Yunlin University of Science and Technology, Yunlin, Taiwan 640; Fax: 886-55312062; E-mail: cheniy@yuntech.edu.tw

Lixin Cheng and Dieter Mewes (Eds)

connection piping as well as relevant to many applications such as chemical reactors, power generation units, oil wells and petrochemical plants. It is well known that the gas-liquid interactions in sudden flow area changes such as pipeline connections and heat exchangers are a complex function of the flow rates of the two phases, their physical properties and pipe geometry. As the two-phase mixture flows through the sudden area changes, the flow might form a separation region at the sharp corner and results in an appreciable pressure loss due to irreversibility. Two-phase flow studies having constant cross-sectional pipes had been widely studied in the literatures, however, frictional performance arisen from singularities such as expansion and contraction are among the least studies of two-phase system [1]. On the other hand, the small and narrow channels are widely adopted in compact heat exchangers; or act as an integral part of CPU cold plate using the liquid cooling with or without phase change. A detailed physical description of two-phase flow across the abrupt flow area change in smaller rectangular channels is still not available.

Though, there are several correlations for the two-phase flow across sudden contractions and expansions available in the literature. Most of the correlations could only predict their own database, and extrapolating their correlations outside their database was uncertain. In this sense, one of the objectives of this study is to examine the applicability of existing correlations subject to sudden contractions and expansions. In the following, a short overview on the single-phase flow across sudden contraction and expansion is firstly given, followed by a thorough review of the relevant literatures for two-phase flow across sudden contractions and expansions. Based on the available data collected from the literature, the applicability for each cited correlation/model is tested. It will be shown later that none of them is able to predict the entire database. Hence, rationally based two-phase pressure change correlations across sudden contractions and expansions incorporating with all the significant parameters are proposed to encapsulate the much larger database for proposing handy and reliable correlations of sudden contraction and expansion for engineering application.

REVIEW OF LITERATURE

Single-phase Pressure Loss Coefficient

For single-phase flow, this pressure loss is mathematically given as a function of the contraction loss coefficient (K_c) or enlargement loss coefficient (K_e) and the kinetic energy of the flow:

$$\Delta P_c = K_c \rho u^2/2 = K_c\, G^2/(2\rho), \Delta P_e = K_e \rho u^2/2 = K_e\, G^2/(2\rho) \tag{1}$$

where σ_A is the passage cross section area ratio, σ_A = (smaller cross section area)/(larger cross section) and $\sigma_A < 1$, mass flux (G) and velocity (u) are calculated based on the smaller cross sectional area of the inlet and outlet tubes. The values of K_c and K_e with a very small cross area ratio ($\sigma_A < 0.05$) are closed to 0.5 and 1.0, respectively [2]. Both values of K_c and K_e, depending on the Reynolds number, decrease as σ_A increases, but the enlargement loss coefficient, K_e, is decreased further more than that of contraction coefficient, K_c. As σ_A approaches 0, K_e becomes 1. The enlargement pressure difference becomes a simple correlation by Delhaye [3] from a simplified momentum balance equation.

$$\Delta P_e = \frac{G^2 \sigma_A (1 - \sigma_A)}{\rho} \tag{2}$$

Fig. **1** shows a typical change of static pressure along the axis for flow across the expansion. Due to the deceleration of the flow in the transitional region, the static pressure initially increases at the expansion area. After the pressure reaches the maximum, the pressure gradient merges with the downstream pressure gradient line. The pressure change at the sudden expansion is defined as the pressure difference for upstream and downstream fully developed pressure gradient lines extended to the expansion position, *i.e.*, ΔP_{EXP}.

For flow through a sudden contraction, the pressure variation along the axis is shown in Fig. **2**. The sudden pressure drop, ΔP_{CON}, at the contraction is defined as the pressure difference for upstream and downstream fully developed pressure gradient lines extended to contraction position. Figs. **1** and **2** were originally given by Schmidt, L. Friedel [4] and [5], respectively.

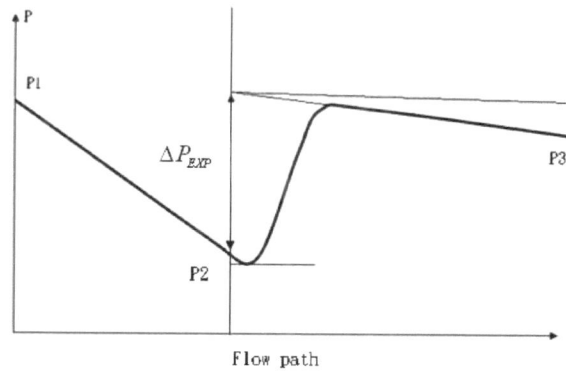

Figure 1: Idealized pressure variations at sudden expansion.

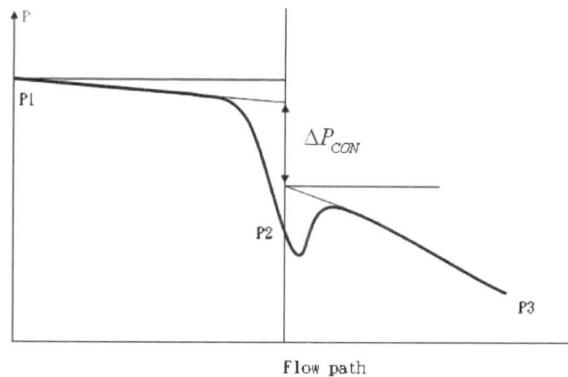

Figure 2: Idealized pressure variations at sudden contraction.

In our previous investigations, a single pressure differential transducer was used to measure directly the difference across the sudden contraction. However, this kind of measurements cannot actually reflect the pressure loss and pressure recovery across the abrupt area change. For an accurate measurement, several pressure transducers are utilized for measuring the local pressures in the upstream and downstream parts of the test sections. The measured axial pressures vs. the pressure tap positions are plotted in a figure, and then the fully developed pressure gradient lines in the upstream and downstream of the test section can be extended to the area change position for obtaining the corresponding pressure change at the sudden expansion.

The enlargement loss coefficient was estimated to become simple correlations by Collier and Thome [6] from a simplified momentum balance equation.

$$K_e = -2\sigma_A (1 - \sigma_A) \tag{3}$$

The negative sign shown in Eq. (3) indicates that there is a pressure gain due to the velocity reduction at sudden enlargement.

For the contraction, as the flow approaches the contraction area, the flow separates from the inner wall, and eddy zones are developed just at the front of transitional cross section. The flow contracts to a small jet for entering the small tube. The narrowest cross section of the jet, the so called vena-contracta, is located immediately after the transition cross section. In the contracted flow region the static pressure decreases rapidly than the fully developed flow at the upstream and downstream. After the ventral-contracta, the pressure gradually increases and reaches its maximum, and then merges into the fully developed pressure gradient line.

For single-phase flow through the contraction, Chisholm [7] combined the static pressure drop to the vena contracta (ΔP_{iv}) and the pressure recovery downstream of the vena contracta (ΔP_{vc}) to give the total static pressure drop at the contraction (ΔP_c):

$$\Delta P_{iv} = (G^2/2\rho_L)[(\sigma_A C_C)^{-2} - 1] \tag{4}$$

$$\Delta P_{vc} = (G/\sigma_A)^2/\rho_L)(C_C^{-1} - 1) \tag{5}$$

$$\Delta P_c = \Delta P_{iv} - \Delta P_{vc} = \Delta P_{cl} = \left(\frac{G^2}{2\rho_L}\right)\left[\frac{1}{(\sigma_A C_C)^2} - 1 - \frac{2(C_C^{-1} - 1)}{\sigma_A^2}\right] \tag{6}$$

Comparing Eqs. (1) with (6), the contraction loss coefficient can be found as:

$$K_c = 1/(\sigma_A C_C)^2 - 1 - 2(C_C^{-1} - 1)/\sigma_A^2 \tag{7}$$

The contraction coefficient C_C was estimated by Chisholm [7] as

$$C_C = \frac{1}{\left[0.639(1-\sigma_A)^{0.5} + 1\right]} \tag{8}$$

Another contraction coefficient C_C given by Geiger [8] is:

$$C_C = 1 - \frac{(1-\sigma_A)}{\left[2.08(1-\sigma_A) + 0.5371\right]} \tag{9}$$

Bejan [9] considered the sharpness of the abrupt change in cross section area may cause the flow separation of the flow upstream and downstream at the step change. The eddy flow formed in the separation region is a clear sign of the irreversibility of the flow. In the absence of irreversibility, the total pressure, $P + \rho u^2/2$, would be conserved from the upstream to downstream across the abrupt flow area change. However, an energy loss would be induced in an irreversible flow. This quantity can be evaluated as:

$$(P_u + \rho u_u^2/2) - (P_d + \rho u_d^2/2) = K\rho u^2/2 \tag{10}$$

where u_u and u_d are the fluid velocities at upstream and downstream, respectively, u is the velocity in the smaller cross sectional area of the inlet and outlet tubes, and K is the loss coefficient. For incompressible flow, the mass conservation requires that $u_u A_u = u_d A_d$, then the pressure change across the sudden expansion can be expressed as:

$$\Delta P_e = P_d - P_u = (1 - \sigma_A^2)\,\rho u_u^2/2 - K_e\rho u_u^2/2 = (1 - \sigma_A^2)\rho u^2/2 - K_e\rho u^2/2 \tag{11}$$

where K_e is the enlargement loss coefficient.

For the sudden contraction, the pressure drop is

$$\Delta P_c = P_u - P_d = (1 - \sigma_A^2)\rho u_d^2/2 + K_c\rho u_d^2/2 = (1 - \sigma_A^2)\rho u^2/2 + K_c\rho u^2/2 \tag{12}$$

where K_c is the contraction loss coefficient. K_c and K_e are varied with the flow and geometric conditions. Examples of the numerical values of K_c and K_e are shown in Figs. **3** and **4** which were originally given by Kay and London [2].

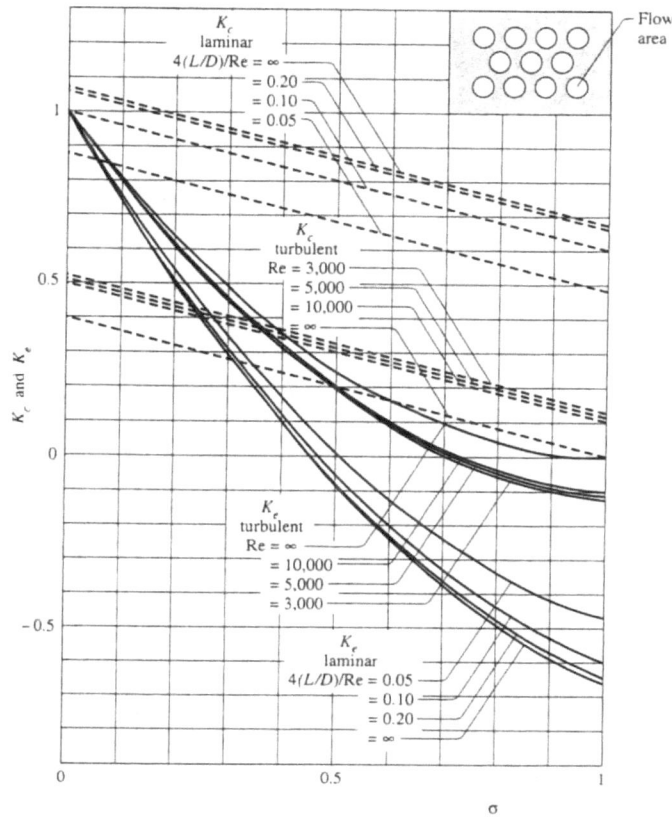

Figure 3: Sudden contraction and expansion loss coefficients for a heat exchanger core with multiple circular-tube passages (Kay and London [2]).

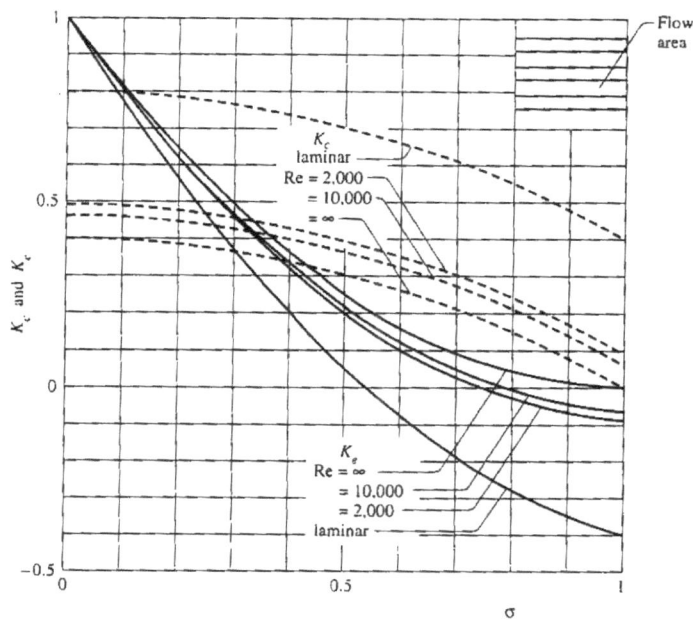

Figure 4: Sudden contraction and expansion loss coefficients for a heat exchanger core with multiple parallel-plate passages (Kay and London [2]).

Two-Phase Pressure Change across Sudden Contraction

Geiger [8] measured pressure drops for steam-water mixtures flowing through sudden contraction with area ratios (σ_A) of 0.398, 0.253, and 0.144. His test conditions are summarized in Table **1**. The data were compared with the homogeneous model, momentum equation and mechanical energy equation across the contractions. The homogeneous model gave the best predictions of the data.

McGee [10] had also measured the steam-water mixtures flowing through sudden contraction using the same test rig as Geiger [8], but with different test sections and conditions ($\sigma_A = 0.608, 0.546$). His test conditions are also listed in Table **1**. The predictions by homogeneous model against the test data were fairly accepted. The predictions by the momentum and mechanical energy equations were much lower than the test data. This deviation is believed due to the assumption of no mechanical energy loss for the acceleration of the fluids at the downstream of the contraction.

For two-phase flow, the change in static pressure at a sudden contraction can be estimated using a homogeneous flow model as recommended by Collier and Thome [6]:

$$\Delta P_c = \left(\frac{G^2}{2\rho_L}\right)\left[\left(C_C^{-1}-1\right)^2+\left(1-\sigma_A^2\right)\right]\left[1+x\left(\frac{\rho_L}{\rho_G}-1\right)\right] \tag{13}$$

where x is the gas quality and ρ_G is the gas density.

Chisholm [7] introduced a constant B coefficient for flow through a discrete interval in evaluating the contraction pressure change:

$$\Delta P_c = \Delta P_{cl}\left[1+\left(\frac{\rho_L}{\rho_G}-1\right)\left(Bx(1-x)+x^2\right)\right] \tag{14}$$

$$B = \frac{\left\{\frac{1}{K_O}\left(\frac{1}{(\sigma_A C_C)^2}-1\right)-\frac{2}{\left(K_O C_C \sigma_A^2\right)}+\frac{2}{\left(\sigma_A^2 K_O^{0.28}\right)}\right\}}{\frac{1}{(\sigma_A C_C)^2}-1-\frac{2}{\left(C_C \sigma_A^2\right)}+\frac{2}{\sigma_A^2}} \tag{15}$$

where ΔP_{cl} is the contraction pressure drop for total flow assumed liquid across the same sudden contraction, and K_O is given as

$$K_O = \begin{cases} \left(1+x(\frac{\rho_L}{\rho_G}-1)\right)^{0.5} & \text{for } X > 1 \\ \\ \left(\frac{\rho_L}{\rho_G}\right)^{0.25} & \text{for } X \leq 1 \end{cases} \tag{16}$$

where X is the Martinelli parameter.

Based on the momentum and mass transfer balance, Schmidt and Friedel [5] developed a new pressure drop model for sudden contraction which incorporates all of the relevant boundary conditions. In this model all the relevant physical parameters were also included in their sudden expansion paper [4]. In addition, the influence of liquid entrainment (α_E) in the gas stream is included along with relevant parameters like area ratio (σ_A), mass flux (G), gas quality (x), mean void fraction (α), surface tension (σ), the viscosity and density of the gas and liquid phases (μ_G, μ_L, ρ_G, ρ_L), and slip ratio (S) to calculate the effective two-phase density (ρ_{eff}). Their test sections incorporated with inlet tube diameters in the range of 44.2~72.2 mm and with outlet tubes in the range of 17.2 ~ 44.2 mm. The model predicts several experimental data sets with different physical properties. The comparison of the model and test results is fair with 80% of the data sets being predicted within ±30%. The test conditions of water-air data for the test section with 72.2 mm inlet diameter and 17.2 mm outlet diameter conducted at 25°C and 5 bar were given by Schmidt and Friedel [5] and listed in Table **1**. This data set is used for the comparison in this study.

Table 1: Available literature data for two-phase flow across sudden contraction

Researchers	Geiger [8] 3 tubes	McGee [10] 2 tubes	Abdelall *et al.* [11] 1 tubes	Chen *et al.* [13] 4 tubes	Schmidt-Friedel [5] 1 tube
Mass flux(kg/m²s)	Max : 6538 Min : 705	Max : 1973 Min : 542	Max : 4679 Min : 2092	Max : 5358 Min : 226	Max : 4000 Min : 500
Quality	Max : 0.265 Min : 0.001	Max : 0.323 Min : 0.004	Max : 0.0168 Min : 0.0019	Max : 0.8027 Min : 0.0008	Max : 0.9 Min : 0.005
Working fluid	Steam-Water at 194-241 °C	Steam-Water at 141-195 °C	Air-Water at 25 °C	Air-Water At 25 °C	Air-Water at 25 °C, 5 bar
Hydraulic Diameter, D_h (mm)	①9.70-25.53 ②12.88-25.53 ③16.10-25.53	①8.64-11.68 ②11.68-14.99	①0.84 -1.6	①2-2.67 ②2-4 ③2-3 ④2-4.8	①17.2-72.2
Area ratio	① 0.398 ② 0.253 ③ 0.144	① 0.608 ②0.546	①0.2756	① 0.3927 ② 0.2618 ③ 0.1963 ④ 0.1309	① 0.0568
Bo	Max : 71.6 Min : 20.4	Max : 30.7 Min : 13.4	0.0953	Max : 0.55 Min : 0.53	39.025
We	Max : 95831 Min : 774	Max : 12016 Min : 332	Max : 1723 Min : 251	Max : 79031 Min : 3.94	Max : 18694 Min : 261
Test points	211	44	26	156	77
Mean Deviation by Homogenous Model	21%	73%	223%	861%	50%

For incompressible and adiabatic flow, the equation for pressure drop across a sudden contraction is given as [5]:

$$\Delta p_c = \frac{G^2[\frac{1}{\rho_{eff}} - \frac{\sigma_A}{\rho_{eff}} + f_{con}\rho_{eff}(\frac{x}{\rho_G \alpha} - \frac{1-x}{\rho_L(1-\alpha)})^2(1-\sigma_A^{1/2})^2]}{1+\Gamma_{con}(\frac{1}{\sigma_A}-1)} \tag{17}$$

$$\text{where } \frac{1}{\rho_{eff}} = \frac{x^2}{\rho_G \alpha} + \frac{(1-x)^2}{\rho_L(1-\alpha)} + \rho_L(1-\alpha)\left(\frac{\alpha_E}{1-\alpha_E}\right)\left[\frac{x}{\rho_G \alpha} - \frac{1-x}{\rho_L(1-\alpha)}\right]^2 \tag{18}$$

$$\alpha = 1 - \frac{2(1-x)^2}{1-2x+\sqrt{1+4x(1-x)\left(\frac{\rho_L}{\rho_G}-1\right)}} \tag{19}$$

$$\alpha_E = \frac{1}{S}\left[1 - \frac{1-x}{1-x\left(1-0.18We^{0.27}\,Re^{0.5}\right)}\right] \tag{20}$$

$$S = \frac{x}{1-x}\frac{(1-\alpha)}{\alpha}\frac{\rho_L}{\rho_G} \tag{21}$$

$$We = G^2 x^2 \frac{d}{\rho_G \sigma}\frac{(\rho_L - \rho_G)}{\rho_G} \tag{22}$$

$$Re = \frac{G(1-x)d}{\mu_L} \tag{23}$$

$$\Gamma_{con} = 0.77\sigma_A(1 - \sigma_A^{0.306}) \tag{24}$$

$$f_{con} = 5.2\times10^{-3}\,x^{0.1}\,(1-x)\left(\sigma_A\frac{\mu_L}{\mu_G}\right)^{0.8} \tag{25}$$

where f_{con} is the total fiction factor for the contraction, α is the void fraction, ρ_{eff} is the effective density, Γ_{con} is the base pressure coefficient for the contraction, Re is the Reynolds number and We is the Weber number (We). In this model all the relevant physical parameters were also included in their sudden expansion paper [4]

Abdelall *et al.* [11] investigated air-water pressure drops caused by abrupt flow area expansion and contraction in a very small test section. The large and small tube diameters were 1.6 and 0.84 mm, respectively. Their test conditions are also listed in Table **1**. Assuming incompressible gas and liquid phases, and assuming x and α remained constant across the sudden contraction, the pressure drop across the sudden contraction was derived as

$$\Delta P_c = G^2\left\{\frac{\rho_h\left(\dfrac{1}{C_c^2} - \sigma_A^2\right)}{2\rho''^2} + \frac{(1-C_c)}{\rho'}\right\} \tag{26}$$

where $\rho_h = \left[\dfrac{x}{\rho_G} + \dfrac{(1-x)}{\rho_L}\right]^{-1}$ $\tag{27}$

$$\rho' = \left[\frac{(1-x^2)}{\{\rho_L(1-\alpha)\}} + \frac{x^2}{(\rho_G\alpha)}\right]^{-1} \tag{28}$$

$$\rho'' = \left[\frac{(1-x)^3}{\{\rho_L^2(1-\alpha)^2\}} + \frac{x^3}{(\rho_G^2\alpha^2)}\right]^{-\frac{1}{2}} \tag{29}$$

The data of two-phase flow pressure change across the sudden contraction were found significantly lower than the predictions of the homogeneous model. It might be attributed to the significant velocity slip at the vicinity of the flow area change. With the assumption of an ideal annular flow regime, the velocity slip ratio was given by Zivi [12] as:

$$S = \frac{u_G}{u_L} = \frac{(1-\alpha)\rho_L x}{\left[(1-x)\rho_G \alpha\right]} = \left(\frac{\rho_L}{\rho_G}\right)^{\frac{1}{3}} \tag{30}$$

where u_G and u_L are the actual gas and liquid velocities of gas and liquid phases, respectively. Since the Abdelall *et al.* correlation [11] requires void fraction during calculation, the above slip ratio equation was thus used to calculate the void fraction. The proposed slip flow model with vena-contracta coefficient (C_C) resulted in a relatively close agreement with a mean deviation of 16.41% between their 26 point data and predictions, but the mean deviation is 32.76% for the all liquid Reynolds number (Re_{LO}) less than 2600. An attempt was also made to correlate the two-phase pressure change data at the contraction point in terms of the Martinelli parameter (X), yielding:

$$X = \left(\frac{u_L}{u_G}\right)^{0.1} \left(\frac{1}{x}-1\right)^{0.9} \left(\frac{\rho_G}{\rho_L}\right)^{0.5} \tag{31}$$

$$\frac{\Phi_{cL}}{X} = 120 \left(X \, Re_L\right)^{-0.7} \tag{32}$$

where $\Phi_{cL} = \Delta P_c / \Delta P_{cL}$ is the two-phase multiplier with all liquid flow through the contraction, and ΔP_{cL} is the pressure drop assuming total flow to be liquid flow. Re_L is the Reynolds number in the smaller tube (regarding total flow to be liquid). Thus, the two-phase pressure change at the sudden contraction can be predicted by

$$\Delta P_c = \Delta P_{cL} X \left[120 \left(X \, Re_L\right)^{-0.7}\right] \tag{33}$$

Based on the same test facility of Abdelall *et al.* [11], Chalfi *et al.* [1] recently reported more data for single-phase and two-phase flow pressure drops caused by flow area expansion and contraction using air and water. Their new data falls within an even lower all liquid Reynolds number ($Re_{LO} < 1020$) than that of Abdelall *et al.*'s data [11] ($1754 < Re_{LO} < 3924$). For flow contraction, the one-dimensional slip flow model along with the slip ratio expression (Eq. 29) agreed with the new data well, provided that no vena-contracta ($C_C = 1$) was considered. The results showed that the homogeneous model over predicted the pressure drop very significantly everywhere, typically by a factor of five. The slip flow model with vena-contracta (C_C) also over predicted the data consistently, typically by a factor of two. However, for the data with higher all liquid Reynolds number, the one-dimensional slip flow model with vena-contracta gave better agreement than that with slip flow model without vena-contracta [11].

From the foregoing review of the two-phase pressure change across the sudden contraction, most of the investigations are related to the abrupt area change for the upstream and downstream having round tube configurations Since the singular configuration variation from rectangular channel to round channel is very common in practice, Chen *et al.* [13] recently had conducted the tests regarding to the influence from rectangular to round singularity. In addition, flow visualization experiment is also carried out to link with certain special pressure drop phenomenon. Their test rig and test sections are given in the following section of experimental setup. Since most of the proposed correlations/model is only applicable to their own database, the effort of their study is to provide a physical modification of the existing model that is capable of handling a much larger database as shown in Table **2**. A modified homogeneous correlation was proposed that considerably improves the predictive ability (with a mean deviation of 30%) over existing correlations with their 156 data and 357 available literature data [13]. The Chen *et al.*'s correlation [13], the related significant parameters and the comparison between predictions and data are given in the following section of results and discussion.

Table 2: Mean deviation for comparison of two-phase contraction data with 5 published correlations

Data \ Methods	Schmidt [5] predictions	Homogenous predictions	Chisholm [7] predictions	Abdelall *et al.* [11] predictions	Chen *et al.* [13] predictions
McGee [10] 44 points	66%	73%	107%	100%	68%
Geiger [8] 211 points	35%	21%	20%	58%	21%
Chen *et al.* [13] 156 points	705%	861%	823%	982%	39%
Abdelall [11] 26 points	58%	223%	223%	15%	31%
Schmidt-Friedel [5] 77 points	21%	50%	41%	24%	18%
Total mean deviation for 514 points	240%	295%	285%	335%	30%

Two-Phase Pressure Change across Sudden Expansion

Romie [14], Richardson [15], Lottes [16], Mendler [17], and McGee [10] were among the first to investigate two-phase flow through sudden expansions. And this topic was continuously investigated by Chisholm and Sutherland [18], Delhaye [3], Wadle [19], Schmidt and Friedel [4], Attou *et al.* [20], Attou and Bolle [21], Abdelall *et al.* [11], and Chalfi *et al.* [1]. In the majority of the previous studies, the void fraction, area ratio and gas quality, as well as the densities of gas and liquid were used to estimate the pressure recovery across the sudden area expansion. The pressure change equations were obtained from the mass and momentum balances without considering the structure of the flow and the frictional effect on the pipe wall. Ahmed [22] had shown that once the flow is fully developed, the flow pattern, void fraction and pressure gradient can be characterized only by the pipe geometry and local flow conditions without memory of its formation. Recently, Ahmed *et al.* [23] considered the influence of wall shear stress in the developing region immediately downstream of the expansion and the wall pressure on the downstream face of the expansion in the flow developing region on the pressure recovery. More recently, Ahmed *et al.* [24] included the length of the developing regime to the area ratio and the upstream liquid Reynolds number in their correlation.

By allowing a relative velocity between the momentum balances among phases, Romie [14] derived an expression for sudden enlargement.

$$\Delta P_e = \frac{G^2 \sigma_A}{\rho_L} \left[(1-x)^2 \left(\frac{1}{(1-\alpha_{in})} - \frac{\sigma_A}{(1-\alpha_{out})} \right) + x^2 \frac{\rho_L}{\rho_G} \left(\frac{1}{\alpha_{in}} - \frac{\sigma_A}{\alpha_{out}} \right) \right] \qquad (34)$$

where the subscript *in* denotes upstream of expansion and *out* represents downstream of expansion.

If the void fraction remains unchanged, Eq. (34) is simplified to [18]:

$$\Delta P_e = \frac{G^2 \sigma_A (1-\sigma_A)}{\rho_L} \left[\frac{(1-x)^2}{(1-\alpha)} + \frac{(\rho_L / \rho_G)x^2}{\alpha} \right] \qquad (35)$$

Table 3: Available literature data for two-phase pressure change across sudden expansions

Researchers	Abdellal *et al.* [11] 1 tube	Chalfi *et al.* [1] 1 tube	Mendler [17] 3 tubes	McGee [10] 3 tubes	Schmidt & Friedel [4] 2 tubes
G (kg/m^{2}s)	3074 - 4212	506 - 664	692 - 5642	544 - 2480	1000 - 3000
x	0.002 - 0.013	0.0285-0.182	0.032 - 0.469	0.001 - 0.304	0.01 - 0.99
Working fluid	Air-Water 25 °C	Air-Water 25 °C	Steam-Water 194 - 252 °C	Steam-Water 141 - 195 °C	Air-Water 25 °C
d_{in} (mm)	0.84	0.84	9.55, 12.9, 17.63	8.64, 11.68	17.2, 19
σ_A	0.276	0.276	0.145,0.264, 0.493	0.332, 0.546, 0.607	0.0568, 0.0937
$Bo = \dfrac{\Delta \rho g d^2}{\sigma}$	0.095	0.095	19,8 - 92.1	13.5-30.7	40 - 48.8
$Fr = \dfrac{G^2}{\rho_m^2 g d}$	9350 -167000	26400-919000	25.7 - 37600	10.3-95700	39.5 - 172000
$We = \dfrac{G^2 d}{\sigma \rho_m}$	312 - 1330	100 - 594	1020 - 45500	185-15500	716 - 8360
$Re_{LO} = \dfrac{G d}{\mu_L}$	2580 - 4530	435 - 571	90200 - 495000	33600-137000	19000 - 63000
ΔP_e (kPa)	0.726 - 4.62	0.801 - 3.71	0.575 - 15.6	0.069 - 14.4	0.2 - 36.0
Data Points	14	24	90	64	90

McGee [10] measured pressure drops for steam–water mixtures flowing through sudden expansions with area ratios (σ_A) of 0.332, 0.546, and 0.607. His test conditions are summarized in Table **3**. The predicted pressure change by Romie's Eq. (34) using the measured upstream and downstream void fractions is in fair agreement with the data. The predictive ability of his correlation against his 64 point data was in the order of ± 40%. A further simplification by neglecting the density change and replacing the void fractions, α_{in} and α_{out} by their average value, the predictive ability of Eq. (35) remains unchanged with a 39% standard deviation as shown in Table **4**.

Table 4: The standard deviations for the predictions of correlations to the data

Data Correlations	Abdellal *et al.* 1 tube [11]	Chalfi *et al.* 1 tube [1]	Mendler 3 tubes [17]	McGee 3 tubes [10]	Schmidt-Friedel 2 tubes [4]	Total 282 Points
Delhaye [3]	139%	28%	36%	39%	59%	54%
Homogenous	468%	207%	65%	138%	31%	143%
Richardson [15]	57%	51%	54%	44%	82%	63%
Chisholm & Sutherland [18]	250%	77%	37%	56%	49%	74%
Wadle [19]	178%	46%	52%	66%	123%	92%
Schmit & Friedel [4]	259%	28%	43%	47%	41%	69%
Attou & Bolle [21]	93%	53%	54%	52%	83%	66%
Abdelall *et al.* [11]	133%	46%	169%	80%	360%	230%

Considering a homogeneous flow from the momentum balance, Eq. (35) is reduced to [3]:

$$\Delta P_e = G^2 \sigma_A (1 - \sigma_A) \left[\frac{(1-x)}{\rho_L} + \frac{x}{\rho_G} \right] \tag{36}$$

However, the homogeneous model gave a very poor predictive ability against McGee's database [10] with a standard deviation exceeding 138% as shown in Table **3**. Attou *et al.* [20] and Abdelall *et al.* [11] had pointed out that the homogeneous model tends to significantly overestimate the experimental results due to the assumption of no slip between the phases.

From the mechanical energy balance for the mixture with neglecting the friction dissipation, Delhaye [3] also derived the following expression:

$$\Delta P_e = \frac{G^2 (1 - \sigma_A^2)}{2} \left[\frac{(1-x)^3}{\rho_L^2 (1-\alpha)^2} + \frac{x^3}{\rho_G^2 \alpha^2} \right] \left[\frac{(1-x)}{\rho_L} + \frac{x}{\rho_G} \right]^{-1} \tag{37}$$

Eq. (37) allows a relative velocity between the phases. For homogeneous flow, the mechanical energy Eq. (37) is simplified to [3]:

$$\Delta P_e = \frac{G^2 (1 - \sigma_A^2)}{2} \left[\frac{(1-x)}{\rho_L} + \frac{x}{\rho_G} \right] \tag{38}$$

Richardson [15] simplified the energy balance model and assumed that the pressure recovery is proportional to the kinetic energies of the phases:

$$\Delta P_e = \frac{G^2 (1 - \sigma_A^2)}{2} \left[\frac{\sigma_A (1-x)^2}{\rho_L (1-\alpha)} \right] \tag{39}$$

Assuming the void fraction remains unchanged, Lottes [16] ignored the gas mass flow rate and assumed that all losses of dynamic pressure head can only take place in the liquid phase, yielding:

$$\Delta P_e = \frac{G^2 \sigma_A (1 - \sigma_A)}{\rho_L (1-\alpha)^2} \tag{40}$$

Mendler [17] also measured the pressure drops for steam–water mixtures flowing through sudden expansions with area ratios (σ_A) of 0.145, 0.264, and 0.693, respectively. His test conditions are listed in Table **3**. Three correlations were proposed by using the least square method to test the database. Since the correlations were empirically obtained from his data, they were only recommended for use at $x < 0.15$, $\sigma_A < 0.5$, and a steam pressure < 41.2 bar. However, the predictions by homogeneous model and the simplified Romie's equation by [3] were fair against Mendler's data with a standard deviation of 65% and 36%, respectively, as shown in Table **4**.

Chisholm and Sutherland [18] also developed a heterogeneous model based on the momentum balance:

$$\Delta P_e = \frac{G^2 \sigma_A (1 - \sigma_A)(1-x)^2}{\rho_L} \left(1 + \frac{C_h}{X} + \frac{1}{X^2} \right) \tag{41}$$

$$\text{where } X = \left(\frac{\rho_G}{\rho_L} \right)^{0.5} \frac{(1-x)}{x} \tag{42}$$

$$\text{and } C_h = \left[1 - 0.5\left(\frac{\rho_L - \rho_G}{\rho_L}\right)^{0.5}\right]\left[\left(\frac{\rho_L}{\rho_G}\right)^{0.5} + \left(\frac{\rho_G}{\rho_L}\right)^{0.5}\right] \tag{43}$$

The model of Chisholm and Sutherland [18] was compared with the air-water bubbly flow data ($\alpha < 0.35$) (Attou et al. [20]). Their predictions remained reasonably good with the test data but with a slightly under estimation. The underestimation of the Chisholm model [18] was also reported by Wadle [19].

Wadle [19] assumed the liquid is decelerated much less than the gas when it passes through the expansion due to its higher inertia. The pressure recovery at the sudden expansion is caused by the bulk deceleration effect and a formula was proposed to describe his pressure recovery data in an abrupt diffuser. The model includes an artificial constant K in connection with different working fluids ($K = 0.667$ for steam–water, $K = 0.83$ for air–water).

$$\Delta P_e = \frac{G^2 K(1-\sigma_A^2)}{2}\left[\frac{(1-x)^2}{\rho_L} + \frac{x^2}{\rho_G}\right] \tag{44}$$

Attou et al. [20] mentioned that Wadle's model significantly overestimated their air-water bubbly flow data ($\alpha < 0.35$). Owen et al. [25] also noticed this phenomenon and claimed that, with $K = 0.22$ (very different from 0.667), Wadle's model agreed quite well with their measurements. The difference of the K values was speculated by the difference of the expansion geometries and flow conditions having a lower mass flow rate. The parameter, K, is an empirical constant that summed up various influencing parameters. For further clarifying the individual effects, an adequate modeling is needed.

Attou and Bolle [21] simplified the jet line emerging from a sudden expansion as a straight line, and then the central flow is confined inside a conical diffuser. By applying the momentum balance within the boundary of the conical jet for an incompressible and adiabatic flow, they obtained a correlation applicable for two-phase flow pressure recovery from a sudden expansion:

$$\Delta P_e = G^2 \sigma_A (1-\sigma_A)[\Phi \theta^r + (1-\theta^r)/\rho_L] \tag{45}$$

where $\Phi = x^2/(\alpha\rho_G) + (1 - x)^2/[(1- \alpha)\rho_L]$, $\theta = 3/[1 + \sigma_A^{0.5} + \sigma_A]$ and r is a correction factor related to the physical properties of the mixture. For a gas quality of $x = 0$, Eq. (45) can be reduced to Eq. (2). The best fitting to the correction factor is $r = 1$ for steam–water mixture and $r = -1.4$ for air–water mixture. The predictions of Eq. (45) had been compared with air-water data and steam-water data. The mean quadratic errors were about 23.4%. The correlation is particularly good for small mass velocities, but it is inapplicable to high gas quality flows ($x > 0.2$).

Based on the momentum and mass transfer balances, Schmidt and Friedel [4] developed a new pressure drop model for sudden expansion which incorporated all of the relevant boundary conditions. Assuming constant properties, the equations for calculating the pressure change across the sudden expansion are simplified as the following.

$$\Delta P_e = \frac{G^2\left[\dfrac{\sigma_A}{\rho_{eff}} - \dfrac{\sigma_A^2}{\rho_{eff}} - f_e \rho_{eff}\left(\dfrac{x}{\rho_G \alpha} - \dfrac{(1-x)}{\rho_L(1-\alpha)}\right)^2 (1-\sigma_A^{0.5})^2\right]}{1-\Gamma_e(1-\sigma_A)} \tag{46}$$

The relevant physical parameters in Eq. (46) were also included in their sudden contraction paper [5] as shown in Eqs. (18) ~ (25).

$$\text{Also } \Gamma_e = 1-\sigma_A^{0.25} \tag{47}$$

$$f_e = 4.9 \times 10^{-3} x^2 \left(1-x\right)^2 \left(\frac{\mu_L}{\mu_G}\right)^{0.7} \tag{48}$$

The expansion model [4] predicted several experimental data sets from 8 test sections with several conventional inlet (17.2~44.2 mm) and outlet (44.2~72.2 mm) diameters. The average of the logarithmic ratios of measured and predicted values was less than -3%, the scatter equaled to 61%. The water–air data for the test sections with area ratio of 0.0568 and 0.0937 conducted at 25°C and 5 bar was shown in their paper [4] which are also listed in Table **3**.

Recently, Abdelall *et al.* [11] investigated air-water pressure drops caused by abrupt flow area expansions in two small tubes. The tube diameters were 1.6 and 0.84 mm with a sudden area change ratio of 0.276. Their measured two-phase pressure difference indicated the occurrence of significant velocity slip. With the assumption of an ideal annular flow regime in accordance with minimum kinetic energy of the flowing mixture, the velocity slip ratio given by Zivi [12] is given in Eq. (30). In practice the slip ratio is actually varying along the flow path. The pressure drop across the sudden expansion (ΔP_e) is the combination of the reversible pressure change (ΔP_{eR}) and the irreversible pressure changes (ΔP_{eI}), i.e.,

$$\Delta P_e = \Delta P_{eR} + \Delta P_{eI}. \tag{49}$$

For an incompressible and adiabatic flow,

$$\Delta P_{e,R} = \frac{G^2}{2}\left(\frac{1}{\rho''^2} - \frac{\sigma_A}{\rho''^2}\right) \tag{50}$$

$$\Delta P_{eI} = 0.5\frac{G^2}{\rho_L}\left[\frac{2\rho_L \sigma_A \left(\sigma_A - 1\right)}{\rho'} - \frac{\rho_h \rho_L \left(\sigma_A - 1\right)}{\rho''^2}\right] \tag{51}$$

The parameters, ρ', ρ'' and ρ_h, had been given in Eqs. (27) ~ (29). For obtaining the two-phase pressure change across sudden expansion by using Eq. (49), an empirical relation between α and x can be solved from the slip ratio for Eq. (30). The predictions of Abdelall *et al.*'s slip flow model are slightly higher than the experimental data, but the simplified homogeneous model, Eq. (5) significantly over predicts the data. The error margins for the predictions of the homogeneous model and Abdelall *et al.*'s slip flow model [11] to their 14 data were not reported in their study.

Based on the same test facility of Abdelall *et al.* [11], Chalfi *et al.* [1] recently reported more data for single-phase and two-phase flow pressure drops caused by expansion and contraction using air and water. Their new data falls within an even lower all liquid Reynolds number ($Re_{LO} < 500$) than that of Abdelall *et al.*'s data [11] ($2578 < Re_{LO} < 3530$). The Zivi slip flow model along with the Armand-type void fraction, $\alpha = 0.5 J_G/(J_L + J_G)$, under predicted their 24 data up to about 30%, where J is the superficial velocity.

Ahmed *et al.* [23] utilized an analytical formulation for the pressure recovery of two-phase flow across a sudden expansion. Their formula covered the pressure change, Eq. (34), for the sudden enlargement from the momentum balance derived by Romie [14] with the change in void fraction across the expansion, and also included the difference between the pressure at the centerline just before the expansion (P_1) and the average pressure of the downstream face at the expansion (P_0), as well as the additional wall shear stress in the developing region downstream of the expansion. Experiments were performed using air–oil two-phase flow to evaluate the relative contribution of the individual terms to the pressure recovery for three area ratios of 0.0625, 0.25 and 0.444. The ranges of the tested mass quality and oil mass flux are $0.0007 - 0.67$ and $20 - 2050$ kg/m²s, respectively. Upstream and downstream flow regimes were identified by a high-speed video camera. The cross-sectional average void fractions at different locations along the test section were obtained using capacitance sensors located along the test section [22]. Four pressure taps were located on the downstream face of the sudden expansion wall to measure the local wall pressures (P_w). The pressure at the centerline of the expansion, P_1, was obtained by extrapolating the fully developed upstream pressure gradient line to the sudden expansion section. The average wall pressure (P_0) was integrated from the measured local wall pressures (P_w).

For the model predictions of Ahmed *et al.* [23], the upstream and downstream flow patterns were identified using the flow pattern map of Taitel and Dukler [26]. The upstream and downstream void fractions were estimated using the appropriate correlations on the local flow pattern. The wall pressure was obtained by extrapolating the measured data to different area ratios and higher mass qualities for all the present air-oil data, while the additional wall shear stress term was estimated using the correlation of Aloui *et al.* [20] for bubbly flow. For annular flow, the wall pressure term rises with the mass quality, and its contribution is more significant than the additional wall shear stress in the developing region. While for the elongated bubbly and intermittent flows, the additional wall shear stress is more significant than the wall pressure. The existing literature data along with their air-oil data (192 points) were compared against the predicted values using their formulation. Most of the data were in a good agreement with the predicted values to within ±35% of the relative standard deviation. However, the data of Schmidt and Friedel [4] at mass qualities higher than 0.5 were highly over predicted with a standard deviation of 50% since the mass quality for most of their data was less than 0.5. The inclusion of the wall pressure term had improved the prediction of the pressure recovery for mass qualities lower than 0.5.

Ahmed *et al.* [23] utilized the pipe geometry and local flow conditions at upstream and downstream to predict the flow patterns, then using the appropriate correlations related to the flow patterns to predict the void fraction. They also proposed a formula to predict the pressure recovery *via* the momentum recovery, the wall pressure at the expansion surface and the additional viscous shear stress in the developing length. The developing length was found to be strongly dependent on the upstream liquid Reynolds number and the sudden expansion area ratio (Ahmed *et al.* [24]) in their correlation. However, for the engineering application, this formula is tedious to use since the flow patterns at the upstream and downstream are required for the prediction of void fractions. Also, the predictions of the pressure change across the sudden expansions entail the empirical wall pressure and the developing length which are not practical for most of the cases. In addition, air and oil are used for the two-phase mixtures in their studies. However, the name of oil and the oil properties were not given in their papers, even not included in Ahmed's Ph.D. Thesis [22]. Therefore, their oil and water data were not included in this study for correlation development.

From the foregoing review of the two-phase pressure change across the sudden expansion, the information of void fraction is required except the homogeneous model, and equations of Wadle [19] and Chisholm and Sutherland [18]. The void fraction may vary in the short length of the sudden expansion due to the flow separation, velocity and geometry changes. Most of the pressure change correlations were based on the inlet conditions to give a constant void fraction. The void fraction can be estimated from measurements, predictions from conventional correlations, or by the individually developed empirical correlations. Some of the investigations included the upstream and downstream void fractions from the measurements (Schmidt and Friedel [4], Abdelall *et al.* [11], Ahmed *et al.* [23]). Ahmed *et al.* [23] also predicted the flow patterns at the upstream and downstream, then using the appropriate correlations to predict the corresponding void fractions.

McGee [10] neglected the density change and replacing the local α by the average value at upstream and downstream, the predictions did not affect the results significantly. Abdelall *et al.* [11] and Chalfi *et al.* [1] utilized Zivi slip flow model for the prediction of the averaged void fraction. There are many void fraction correlations available in the literature. Recently, Dalkilic, *et al.* [27] had surveyed the void fraction correlations and a summary of 35 correlations. The correlations were then compared with data, and it is found that the Thom's void fraction correlation [28] was among the best. Thom obtained an empirical relationship between quality and void fraction by assuming the slip velocity to be dependent on phase viscosities and densities. Thom's correlation is given as

$$\alpha = \frac{\gamma x}{1 + x(\gamma - 1)} \quad \text{where } \gamma = Z^{1.6}, \quad Z = (\frac{\rho_L}{\rho_G})^{0.555} (\frac{\mu_G}{\mu_L})^{0.111} \tag{52}$$

The void fraction α is calculated by the Thom's correlation to predict the pressure change across the sudden expansions in this study. Several void fraction correlations had also been tested, but Thom's correlation gave the overall best results.

In summary of the aforementioned review, most of the correlations highly over predict the data with a mini test section in which the effect of surface tension dominates, and some of the correlations significantly under predict the data for very large test sections. Also, some of the correlations are not handy for the engineering application. Hence by taking into account the influences of Bond number, Weber number, Froude number, liquid Reynolds number, gas quality and area ratio, Chen *et al.* [30, 31] had recently correlated the data listed in Table **3** and proposed a modified homogeneous model that considerably improves the predictive ability over existing correlations with a mean deviation of 23% and a standard deviation of 29% with wider ranges for application. This modified homogeneous correlation [30, 31], the related significant parameters and the comparison between predictions and data are also given in the following section of results and discussion.

EXPERIMENTAL SETUP

The test rig shown in Fig. **3** is designed to conduct tests with air-water mixtures. Air is supplied from an air-compressor and then stored in a compressed-air storage tank. Airflow through a pressure reducer, and depending on the mass flux range, is measured by three Aalborg® mass flow meters for different ranges of flow rates. The water flow loop consists of a variable speed gear pump that delivers water. A mixer provides better uniformity of the air and water mixture. An enlarged view of mixer is shown as the round perspex tube in Fig. **5**. The detail of this mixer had been described by Chen *et al.* [14] in two-phase flow pressure drop tests in small tubes.

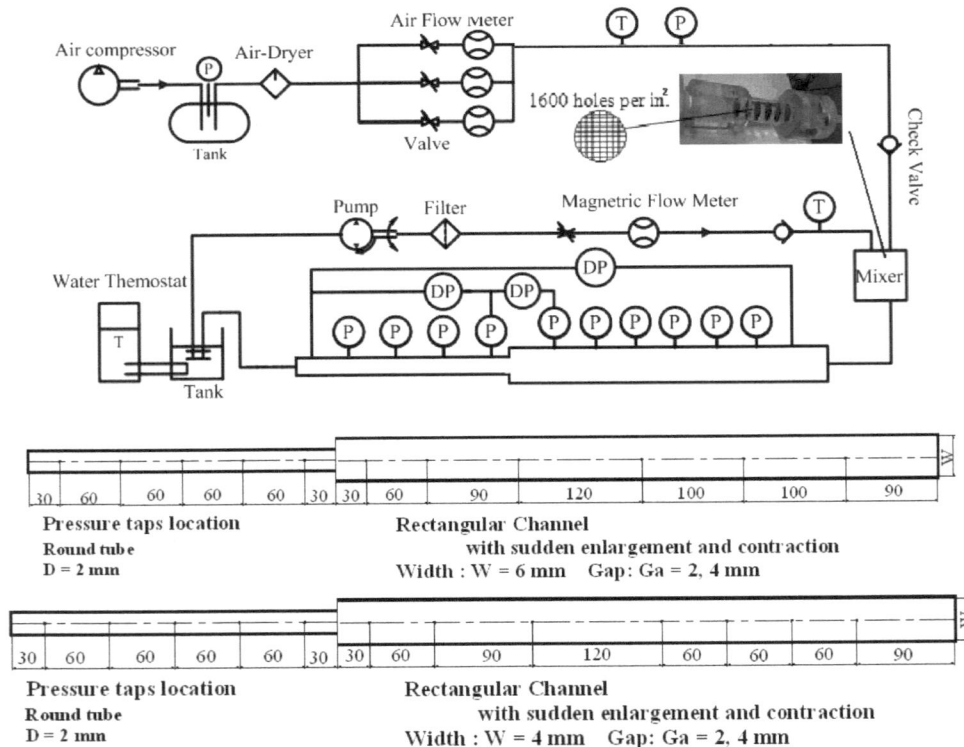

Figure 5: Test rig with schematic diagram of test sections showing pressure taps.

Since the data for two-phase pressure drop in the rectangular channels were taken simultaneously with the sudden contraction tests as shown in Fig. **3**, the mass flux is based on the cross sectional area of rectangular channels. If the test section is reversed, the two-phase flow is firstly through the small round tube and then suddenly entering the rectangular channel for the expansion tests. The total mass flux density (G) of air and water flow rate is ranged from 100 to 700 kg/m^2·s with gas quality (x) being varied from 0.001 to 0.8. However, in order to compare the obtained contraction and expansion data with the available correlations, the mass flux (G) is then changed to the base of the smaller cross sectional area of the round tube [13]. The inlet temperatures of air and water are near 25°C. The pressure measured just after the contraction position is in the range of 101 ~ 350 kPa. The pressure drops of the air-water mixtures are measured by three YOKOGAWA EJ110 differential pressure transducers having an adjustable span of 1300 to 13000 Pa. Resolution of this pressure differential transducer is 0.3% of the measurements. The drilled holes of the pressure taps are perpendicular to the test sections with a diameter of 0.5 mm. Pressure measurements are made in nine locations along the inlet tube and along the rectangular channel of the test sections as shown in Fig. **3**. The resolution of the pressure measurements is 0.1% of the test span. For validation of the present test setup, measurements of the single-phase pressure drops for air and water alone are in terms of friction factors to compare with the friction factor equations for laminar and turbulent flows in rectangular channels. The results are in line with the known correlations having a deviation within ±5%.

Leaving the test section, the air and water mixture flows are separated by an open water tank in which the air is vented and the water is re-circulated. The air and water temperatures are measured by resistance temperature device (Pt100Ω) having a calibrated accuracy of 0.1 K (calibrated by Hewlett-Packard quartz thermometer probe with quartz thermometer, model 18111A and 2804A). Observations of flow patterns are obtained from images produced by a high speed camera of Redlake Motionscope PCI 8000s. The maximum camera shutter speed is 1/8000 second. The high-speed camera can be placed at any position along the rectangular channels or at the side view of the abrupt change of flow area.

The observed two-phase flow patterns across the sudden expansion and contraction in the small rectangular channels were reported by Chen *et al.* [29] and [13], respectively. Fig. **6** shows two figures, having the typical flow pattern across sudden expansion and contraction in small channel, which are depicted from the original photos for semi-annular and plug flow patterns across sudden contraction and expansion, respectively, with quality at 0.01 and mass flux at 300 kg/sm^2. Eddy flow is shown circulated at the up and lower corners of the large intersection for sudden contraction and expansion. However, for the contraction, the liquid like vena-contracta moves toward upstream and even appears in front of the singularity.

(a) Sudden expansion (semi-annular)   **(b) Sudden contraction (plug)**

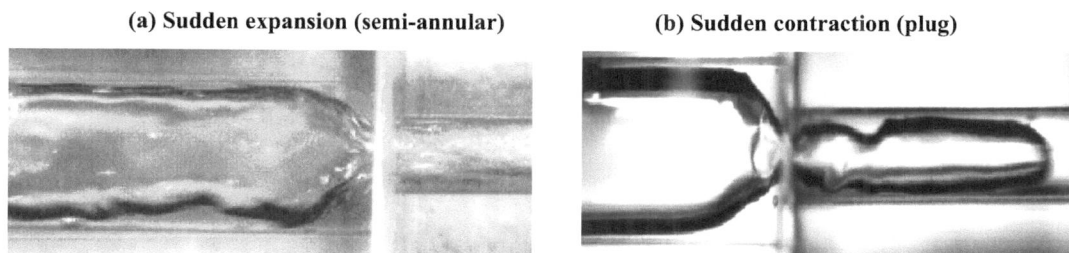

Figure 6: The typical flow patterns across sudden expansion (3 mm inlet tube and a 3 x 6 mm outlet channel) and contraction (2 x 4 mm inlet channel and 2 mm outlet tube).

The dimensions of the test sections are gap (G_a) × width (W) = 2 × 4 mm, 2 × 6 mm, 4 × 4 mm and 4 × 6 mm. All the test sections are connected with a 2 mm diameter glass tube such that the flow will meet the abrupt flow area change at the interconnection between the round tube and the rectangular channel. The area ratio for the abrupt flow area change (σ_A) is ranged from 0.131 to 0.393. The test sections are made of transparent acrylic resin, so that the flow pattern and flow structure at the vicinity of the abrupt cross-sectional area change could be visualized. These test sections are also be arranged in horizontal longitudinal (HL, the wide side is vertical) and the intersection between the rectangular and round tube is well fabricated to avoid any irregularity.

For obtaining the exact pressure change across the sudden contraction or expansion, several pressure transducers are utilized for measuring the local pressures in the upstream and downstream parts of the test sections as shown in Fig. **3**. The measured axial pressures versus the pressure tap positions are plotted in a figure to setup the fully developed pressure gradient lines in the upstream and downstream for further obtaining the corresponding pressure change from the extrapolation of those lines to the point of sudden contraction.

RESULTS AND DISCUSSION

Pressure Change Across Sudden Contraction

To test the validity of the foregoing described models/correlations from the existing literatures, the data (156 points) from Chen *et al.* [13] and the literature data (357 points) listed in Table **1** are compared with the previously described homogeneous model, Eq. (13), and correlations of Chisholm [7], Schmidt and Friedel [5], and Abdelall *et al.* [11], but none of them can accurately predict the entire database. The comparison results for each correlation are summarized in Table **2**. The data sets of Geiger [8] are well predicted by the homogeneous model (mean deviation about 20%) than the other correlations (mean deviation from 58 ~ 447%). The McGee [10] data are fairly predicted by the homogeneous model and the correlation of Schmidt and Friedel [4]. The Schmidt and Friedel data [4] are in good agreement with the Abdelall *et al.* correlation [11] having a mean deviation of 24%, but their correlation gives the best prediction with a mean deviation of 21%. The Abdelall *et al.*'s data [11] are best described by their correlation (15% mean deviation), but are significantly over predicted by the others. The homogeneous model gives the worst predictions to the Abdelall *et al.* data with a mean deviation of 223% which was also reported by themselves. The data of Chen *et al.* [13] along with the micro tube data by Abdelall *et al.* [11] show profound departure to the predictions of the existing correlations. The predictions against present data have mean deviations ranged from 823 ~ 1407%. As a result, a sharp rise of the mean deviation for all the correlations to all the data is encountered. The average mean deviations of the relevant predictions to all the data are 295%, 285%, 737%, 335% by Homogeneous model, Chisholm correlation [7], Schmidt and Friedel correlation [5], and Abdelall *et al.* correlation [11], respectively.

The significant departure of the existing correlations with the present data and Abdelall *et al.* data [11] could be attributed to the very small outlet section, 2 mm and 0.84 mm, respectively. For these mini tubes, surface tension plays essential role. It would be more reasonable to take into account the influence of surface tension force (Tripplet *et al.*, [32]) to have a better predictive ability. The balance of buoyancy force and surface tension can be represented by Bond number (Bo) as:

$$Bo = \frac{\Delta \rho g D^2}{\sigma} \tag{53}$$

where g is the gravity, σ is the surface tension and D is the internal diameter of outlet tube.

When the value of *Bo* is near or less than 1.0, the stratified flow pattern is not able to exist in most of the two-phase flow conditions (Chen *et al.* [33]). Table **1** also tabulated the change of Bond numbers being varied from 0.095 of Abdelall *et al.* [11], 0.53 of present study to 71.6 of Geiger [8] for the related database. Also, considering the effects of total mass flux and gas quality to the surface tension, Schmidt and Friedel [5] had proposed a Weber number (*We*) to correlate the two-phase pressure change across sudden expansions and contractions.

$$We = \frac{G^2 D}{\sigma \rho_m} \tag{54}$$

From the foregoing comparisons, it is found that the homogenous model gives good predictive ability but fails to predict Abdelall *et al.*'s [11] and Chen *et al.*'s data [13]. By examining the tabulated Bond number in Table **1**, it is found that the departure of the predictive ability of homogeneous model is strongly related to the Bond number. The database of Geiger [8] and McGee [10] contains a relatively larger Bond number (> 13), indicating a negligible influence of surface tension, thereby showing a good predictive ability of homogeneous model. The homogeneous model also shows fair predictive ability against Schmidt and Friedel's data [5] with its Bond number around 39.03. Its very small contraction ratio (0.057) may reinforce the influence of contraction (similar to the aforementioned liquid vena contracta). In that regard, an effort is made in this study to include the effects of Bond number, Weber number, gas quality and flow cross-sectional area contraction ratio (σ_A) to extend the applicable range of homogeneous model. Though the Chisholm correlation [7] has a very close average mean deviation to all the data with the homogeneous model, the homogeneous model is much easier to use and to modify. The proposed modification is applicable to the present data (156 points) and those data available from the literatures shown in Table **1** (357 points). By introducing correction factors to the original homogeneous model, Eq. (3), the proposed modification takes the following form:

$$\Delta P_c = \Delta P_{cHom} \times (1+\Omega_1) \times (1+\Omega_2) \times (1+\Omega_3)^{-0.08} \tag{55}$$

$$\text{where } \Omega_1 = -0.99 e^{\frac{-13.1C_1}{C_2}}$$

$$\Omega_2 = -39.4 e^{C_1^{-0.25}} (16.1 C_3^{-1.5} - 13.2 C_3^{-1.8} - 4.2 C_3^{-1})$$

$$\Omega_3 = 0.1 C_3^{-1.3} C_4^{-3} \tag{56}$$

$$C_1 = Bo^{1.1}(1-x)^{0.9}, \ C_2 = 470 e^{-\sigma_A^{-0.2}}$$

$$C_3 = We \times Bo \times \sigma_A \times (1-x)^{-3}, \ C_4 = \sigma_A^{2.5}(1-x)^{-1}$$

The correction factors $\Omega_{i=1,2,3}$ asymptotically approach zero when the Bond number is large enough in which the influence of surface tension is negligible.

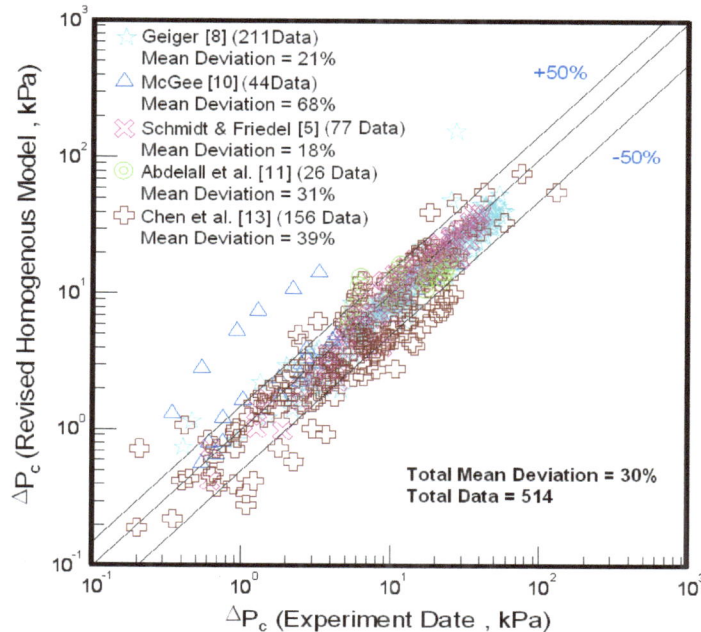

Figure 7: Comparison of the two-phase contraction data with revised homogeneous model, Eq. (55).

For assessment of the proposed correlation, the comparison results with all available data are shown in Fig. 7. The total mean deviation of this modified homogeneous model is 30% which has been greatly improved from 295% of the original homogeneous model. The mean deviation is defined by:

$$\text{Mean deviation} = \frac{\sum_{n=1}^{n} \left| \dfrac{data - prediction}{data} \right|}{n} \tag{57}$$

The mean deviation to the data of Schmidt and Friedel [5] predicted by this proposed correlation is 17%. Though Schmidt and Friedel [11] indicated that 80% of their own data and available literature data were predicted within ±30%, their correlation is rather complicated to use. Furthermore, the mean deviation for the data of Abdelall *et al.* [11] is tremendously reduced from 223% by homogeneous model to 31% by this proposed modification. Although the correlation of Abdelall *et al.* [11] gives a mean deviation of 15 % to their data, their correlation is only valid for their data in a very small test section. In summary, the proposed correlation shows a very good accuracy against the existing data and is capable of handling the effect of surface tension and buoyancy force and is valid for much wider ranges of: $100 < G < 6538$ kg/m^2·s, $0.0008 < x < 0.9$, $0.057 < \sigma_A < 0.608$, $0.84 < D_h < 72$ mm.

Pressure Change across Sudden Expansion

To check the validity of the foregoing described models/correlations from the existing literatures, 90 data from Mendler [17], 64 data from McGee [10], 90 data from Schmidt & Friedel [4], 14 data from Abdellal *et al.* [11] and 24 data from Chalfi *et al.* [1] are collected and their test conditions, as well as the ranges of the momentous parameters are also listed in Table **3**. The data are compared with the previously described correlations/models by [3] for Eq. (4), Homogeneous for Eq. (5), Richardson [15]) for Eq. (8), Chisholm and Sutherland [18] for Eq. (10), Wadle [19] for Eq. (13), Schmidt and Friedel [4] for Eq. (15), Attou and Bolle [21] for Eq. (14) and Abdelall *et al.* [11] for Eq. (25). Table **2** shows the standard deviations for the predictions of correlations with a total of 282 data. The standard deviations of the relevant predictions to the total data are 54%, 143%, 63%, 74%, 92%, 69%, 66%, 230% by correlations of Delhaye [18], Homogeneous, Richardson [15], Chisholm and Sutherland [18], Wadle [19], Schmidt and Friedel [4], Attou and Bolle [21] and Abdelall *et al.* [11], respectively. Also, the mean deviations for the above predictions are obtained as 42%, 83%, 58%, 49%, 65%, 43%, 52% and 146%, respectively. The predictive capability for the models/correlations had being examined; the results are summarized in Table **4**, and None of the existing correlations can accurately predict the entire database. Most of the correlations highly over predict the data with a mini test section which has a Bond number being less than 0.1 in which the effect of surface tension dominates. Also, some of the correlations significantly under predict the Schmidt & Friedel data [4] with very large test sections. The standard deviation is defined by:

$$\text{Standard deviation} = \sqrt{\frac{\sum_{n=1}^{n} \left(\dfrac{data - prediction}{data} \right)^2}{n}} \quad \text{n: number of the data.} \tag{58}$$

The Delhaye [3] correlation seems to have the best overall predictions against the database with a standard deviation of 54%, but it over predicts the Abdelall *et al.* data [11] (139% standard deviation) and under predicts the Schmidt & Friedel data [4] (59% standard deviation), followed by Richardson [15] whose standard deviation is 63%, but it significantly under predicts the Schmidt & Friedel [4] data (82% standard deviation). Attou and Bolle [21] correlation is with a standard deviation of 66%, but it also significantly under predicts the Schmidt & Friedel data [4] (83% standard deviation). The above four correlations require the information of void fraction for the prediction. There are so many correlations available in the literature, it will need further examinations to tell which one shows the best predictive ability. Yet the void fraction is not a constant in the short path of the sudden expansion Therefore, exploitation of the void fraction raises additional uncertainty and inconvenience from the engineer's perspective. In addition, the correlations of Attou and Bolle (1997) and Wadle [19] include an artificial correction factor which is varied for different two-phase flow mixtures.

The highly over predictions by the existing correlations against the Abdelall *et al.*'s data [11] could be attributed to the very small outlet section (d_{in} = 0.84 mm, d_{out} = 1.6 mm). In addition, the significantly under predictions for most of the correlations to the Schmidt & Friedel [4] data could be due to the very large test sections (d_{in} = 17.2 mm, d_{out} = 72.2 mm and d_{in} = 19.0 mm, d_{out} = 56.0 mm). For obtaining a better predictive ability for mini test sections, one should also take into account the influence of surface tension force [32]. The balance of buoyancy force and surface tension force can be represented by Bond number, Bo = $\Delta\rho g d^2/\sigma$. Considering the effects of total mass flux and gas quality to the surface tension, Chen *et al.* [33] had proposed a Weber number, We = $G^2 d/(\rho_h \sigma)$, to correlate the two-phase pressure change across small channels. The Froude number, Fr = $G^2/(\rho_h^2 gd)$, which represents the ratio between the mixture inertia and the buoyancy force was utilized by Friedel [34] for the two-phase frictional pressure drop correlation in conventional straight tubes. In addition, the liquid Reynolds number, Re_{LO} = Gd/μ_L, was used as a significant parameter for correlating the developing length downstream of a sudden expansion by Ahmed *et al.* [24].

Though, homogeneous model shows a poor than the others, but it is handy to use. Hence by taking into account the forgoing parameters of We, Bo, Fr and Re_{LO}, into the original homogeneous model, Eq. (36), the proposed modified Homogeneous model for two-phase flow across sudden expansions takes the form as:

$$\Delta P_{Modify} = \Delta P_{Homogeneous} \times (1 + \Omega_1 - \Omega_2) \times (1 + \Omega_3)$$

$$\Omega_1 = (\frac{WeBo}{Re_{Lo}})^2 \times (\frac{1-x}{x})^{0.3} \times \frac{1}{Fr^{0.8}},$$

$$\Omega_2 = 0.2 \times (\frac{\mu_G}{\mu_L})^{0.4} \tag{59}$$

$$\Omega_3 = 0.4 \times (\frac{x}{1-x})^{0.3} + 0.3 \times e^{1.6/Re_{LO}^{0.1}} - 0.4 \times (\frac{\rho_L}{\rho_G})^{0.2}$$

This proposed modified homogeneous correlation considerably improves the predictive ability over existing correlations with a mean deviation of 23% and a standard deviation of 29% to all the data as shown in Fig. **8**. The standard deviations for the predictions of the Homogeneous model to the data sets of Abdellal *et al.* [11], Chalfi *et al.* [1], Mendler [17], McGee [10] and Schmidt & Friedel [4] are greatly improved by the modified Homogeneous model, from 468%, 207%, 65%, 138% and 31% to 26%, 32%, 31%, 34% and 24%, respectively. In summary, the proposed correlation shows a very good accuracy against the existing data and is capable of handling the effects of gas quality, area ratio, mixture inertia, surface tension and buoyancy force, and is valid for much wider ranges of: 506 < G < 5642 kg/m²·s, 0.002 < x < 0.99, 0.057 < σ_A < 0.607, 0.84 < d_{in} < 19 mm, 0.095 < Bo < 92, 1.03E+1 < Fr < 9.19E+5, 1.0E+2 < We < 8.3E+4, and 4.35E+2 < Re_{LO} < 4.95E+05.

SUMMARY

The flow of two-phase mixtures across sudden expansions and contractions are widely encountered in typical industrial pipe lines and heat exchanging devices. There had been some studies concerning this subject but mostly are applicable for larger channels. Also, most of correlations are only applicable to their data, extrapolating the correlation outside their database is usually not recommended.

In this work, the authors present a thorough review of the relevant literature for two-phase flow across sudden contractions and expansions, and examine the applicability of the existing correlations. Totally, 503 two-phase contraction data and also 282 two-phase expansion data are collected from literature. None of the existing contraction or expansion correlations can accurately predict each entire database. Most of the correlations highly over predict the data for a mini test section with tube diameter around 1 mm. Some of the correlations significantly under predict the data with very large test sections.

For the two-phase flow across sudden contractions, the data of Chen *et al.* [13] along with the micro tube data by Abdelall *et al.* [11] show profound departure to the predictions of the existing correlations. The significant deviation of the existing correlations with the data of Chen *et al.* [13] and the data of Abdelall *et al.* [11] could be attributed to the influence of the surface tension to the very small outlet sections which have bond number less than 1. Among these models/correlations, none of them can accurately predict the entire database. Homogeneous model gives moderate predictive ability than the others, but the mean deviation to the total 503 data is as high as 295%. By introducing the correction factors that includes the influences of gas quality, Bond number, Weber number and area contraction ratio to the original homogeneous model. Through this rational-based modification, the mean deviation of this proposed correlation is dramatically reduced to 30% to the entire database. In summary, the proposed correlation shows a very good accuracy against the existing, and is capable of handling the effects of surface tension and buoyancy force.

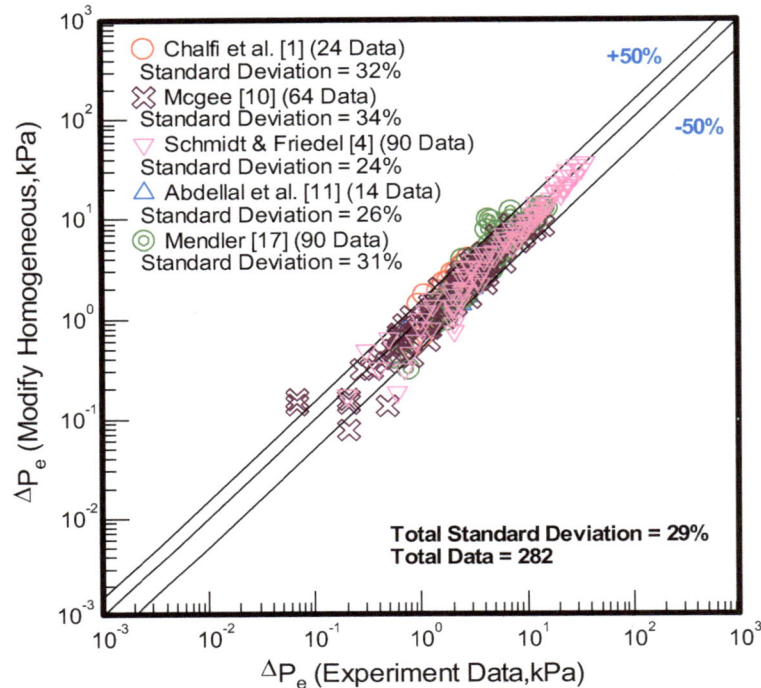

Figure 8: Comparison of the two-phase expansion data with modified Homogeneous model, Eq. (59).

For the two-phase flow across sudden contractions, the homogeneous model though shows a poor predictive ability than the others, but it is handy for the engineering application. Hence by taking account the influences of Bond number, Weber number, Froude number, liquid Reynolds number, gas quality and area ratio into the original homogeneous model for correlating with the data, a modified homogeneous correlation is proposed that considerably improves the predictive ability over existing correlations with a mean deviation of 23% and a standard deviation of 29% to all the data.

The overall objective of study is to have a fundamental understanding of two-phase flow across sudden contractions and expansions in small channels and methods to calculate the corresponding pressure difference across the sudden area changes. A thorough review of the relevant literature for single-phase flow and two-phase flow across sudden contractions and expansions are given. Also, the applicability of for each existing correlation is examined with available data sets, whereas none of them gives satisfactory predictions against the entire database. The homogeneous model is handy from the engineering aspect, hence by taking into account the significant parameters into the original homogeneous models of sudden contraction and expansion, the modified homogeneous correlations for sudden contractions and expansions are individually proposed that considerably improve the predictive ability over existing correlations with a wider range of application.

NOMENCLATURE

A	aspect ratio, gap (G_a)/width (W), $0 < A < 1$
Bo	Bond number
C_C	contraction coefficient
C_h	Chisholm's factor
D, d	internal diameter of circular tube, m
g	gravity, m·s^{-2}
f_e	total friction factor for the expansion given by Schmidt and Friedel [4]
f_{con}	total friction factor for the contraction given by Schmidt and Friedel [5]
Fr	Froude number
G	total mass flux, $\text{kg·m}^{-2}\text{·s}^{-1}$
G_a	channel gap, m
K	loss coefficient
K_O	a factor given by Chisholm [7]
P	pressure, Pa
ΔP_e	pressure change across the sudden expansion, Pa
ΔP_{cL}	pressure drop across the sudden contraction for total flow assumed liquid, Pa
ΔP_{iv}	static pressure drop to the vena contracta, Pa
ΔP_{vc}	pressure recovery downstream of the vena contracta, Pa
ΔP_{eR}	reversible pressure change given by Abdelall *et al.* [11], Pa
ΔP_{eI}	irreversible pressure change given by Abdelall *et al.* [11], Pa
P_o	average wall pressure, Pa
P_w	local wall pressure, Pa
r	correction factor given by Attou and Bolle [21]
Re	Reynolds number
S	slip ratio
u	fluid mean velocity m·s^{-1}
u_G	actual gas velocity, m·s^{-1}
u_L	actual liquid velocity, m·s^{-1}
x	gas quality
X	Martinelli parameter
W	channel width, m
We	Weber number
Z, γ	coefficients given by Tom [28]

Greek Symbols

Ω	correction factor
α	mean void fraction
α_E	mean volumetric entrainment given by Schmidt and Friedel [5]
Φ	characteristic function given by Attou and Bolle [21]
Φ_{cL}	$\Delta P_c/\Delta P_{cL}$, the two-phase multiplier with all liquid flow through the contraction
ρ	density, kg·m^{-3}
ρ_{eff}	mean effect density given by Schmidt and Friedel [5], kg·m^{-3}
ρ'	fictitious mixture density given by Abdelall *et al.* [11], kg·m^{-3}
ρ''	fictitious mixture density given by Abdelall *et al.* [11], kg·m^{-3}
$\Delta\rho$	density difference between liquid and gas, kg·m^{-3}
Γ_{con}	base pressure coefficient for the contraction given by Abdelall *et al.* [11]
Γ_e	base pressure coefficient for the expansion given by Abdelall *et al.* [11]
σ	surface tension, N·m^{-1}
σ_A	flow cross-sectional area expansion ratio, $0 < \sigma_A < 1$
θ	function of the area ratio given by Attou and Bolle [21]
μ	viscosity, N·s·m^{-2}

Subscripts

C	contraction
e	expansion
G	gas phase
h	two-phase homogeneous mixture
in	inlet
L	liquid phase
out	outlet

REFERENCES

[1] T. Y. Chalfi, Y. Toufik, and S. M. Ghiaasiaan. "Pressure drop caused by flow area changes in capillaries under low flow conditions," *Int. J. Multiphase Flow,* vol. 34, pp. 2–12, 2008.

[2] W. M. Kays and A. L. London, *Compact Heat Exchangers*, Third Edition, Mc-Graw-Hill Book Company, New York, Chapter 2, pp. 112, 1984.

[3] J. M. Delhaye, "Singular pressure drops," In: Bergles, A.E. (Ed.), *Two-phase and heat transfer in the power and process industries*, Chapter 3, Hemisphere, Washington, DC, 1981.

[4] J. G. Collier and J. R. Thome, *Convective Boiling and Condensation*, Third Edition, Oxford, New York, Chapter 3, pp. 109, 1994.

[5] D. Chisholm, *Two-phase Flow in Pipelines and Heat Exchangers*, Pitman Press, Bath, England, Chapter 12, pp. 175-192, 1983.

[6] G. E. Geiger, Sudden contraction losses in single and two-phase flow, Ph.D Thesis, University of Pittsburgh, U.S.A. 1964.

[7] A. Bejan and J.S. Jone, *Heat Transfer*, John Wiley & Sons, New York, 1993.

[8] J.W. McGee, Two-phase flow through abrupt expansions and contractions, Ph.D Thesis, University of North Carolina at Raleigh, U.S.A., 1966.

[9] J. Schmidt and L. Friedel, "Two-phase flow pressure drop across sudden contractions in duct areas," *Int. J. Multiphasr Flow*, vol. 23, pp. 283-299, 1997.

[10] J. Schmidt and L. Friedel, "Two-phase flow pressure change across sudden expansions in duct areas," *Chem. Engng. Comm.*, vol. 141-142, pp.175-190, 1996.

[11] F. F. Abdelall, G. Hahm, S. M. Ghiaasiaan, S. I. Abdel-Khalik, S. S. Jeter, M. Yoda, and D. L. Sadowski, "Pressure drop caused by abrupt flow area changes in small channels," *Exp. Thermal & Fluid Science*, vol. 29, pp. 425-434, 2005.

[12] S. M. Zivi, "Estimation of steadt state steam void-fraction by means of principle of minimum entropy production," *ASME Trans. Series. C*, vol. 86, pp. 237-252, 1964.

[13] I. Y. Chen, C. Y. Tseng, Y. T. Lin, and C. C. Wang, "Two-phase flow pressure change subject to sudden contraction in small rectangular channels," *Int. J. Multiphase Flow*, vol. 35, pp. 297-306, 2009.

[14] F. Romie, Private Communication to P. Lottes, American Standard Co., (see Lottes, 1961), 1958.

[15] B. Richardson, "Some problems in horizontal two-phase, two-component flow," Report ANL-5949, 1958.

[16] P. A. Lottes, 1961. "Expansion losses in two-phase flow," *Nucl. Sci. Eng.,* vol. 9, pp. 26–31, 1961.

[17] O. J. Mendler, Sudden expansion losses in single and two-phase flow. Ph.D. thesis, University of Pittsburgh, Pennsylvania, U.S.A, 1963.

[18] D. Chisholm and L.A. Sutherland, "Prediction of pressure gradients in pipeline system during two-phase flow," *Proc. Inst. Mech. Eng.,* vol. 184 (Pt. 3C), pp. 24–32, 1969.

[19] M. Wadle, "A new formula for the pressure recovery in an abrupt diffuser," *Int. J. Multiphase Flow*, vol. 15, pp. 241–256, 1989.

[20] A. Attou, M. Giot, and J. M. Seynhaeve, "Modelling of steady-state two-phase bubbly flow through a sudden enlargement," *Int. J. Heat Mass Transfer*, vol. 40, pp. 3375-3385, 1997.

[21] A. Attou and L. Bolle, "A new correlation for the two-phase pressure recovery downstream from a sudden enlargement," *Chem. Eng. Technol.,* vol. 20, pp. 419–423, 1997.

[22] W. H. Ahmed, Two-phase flow through sudden area expansions, Ph.D. Thesis, McMaster University, Hamilton, Ontarion, Canada, 2005.

[23] W. H. Ahmed, C. Y. Ching, and M. Shoukri, "Pressure recovery of two-phase flow across sudden expansions," *Int. J. Multiphase Flow*, vol. 33, pp. 575–594, 2007.

[24] W. H. Ahmed, C. Y. Ching, and M. Shoukri, "Development of two-phase flow downstream of a horizontal sudden expansion," *Int. J. Heat and Fluid Flow*, vol. 29, pp. 194–206, 2008.

[25] I. Owen, A. Abdou-Ghani, and A.M. Amini, "Diffusing a homogenized two-phase flow," *Int. J. Multiphase Flow*, vol. 18, pp. 531-540, 1992.

[26] Y. Taitel and A. E. Dukler, "A model for predicting flow regime transitions in horizontal and near horizontal gas–liquid flow," *AIChE J.*, vol. 22, pp. 47–55, 1976.

[27] A. S. Dalkilic, S. Laohalertdecha, and S. Wongwises, "Effect of void fraction models on the two-phase friction factor of R134a during condensation in vertical downward flow in a smooth tube," *Int. Comm. Heat and Mass Transfer,* vol. 35, pp. 921–927, 2008.

[28] J. R. S. Thom, "Prediction of pressure drop during forced circulation boiling of water," *Int. J. Heat and Mass Transfer*, vol. 7, 709–724, 1964.

[29] I.Y. Chen, C.C. Liu, K.H. Chien, and C.C. Wang, "Two-phase Flow Characteristics across Sudden Expansion in Small Rectangular Channel," *Exp. Thermal Fluid Science*, vol. 32, pp. 696-706, 2007.

[30] I. Y. Chen, C. Y. Tseng, and C. C. Wang, "Two-phase flow pressure change across sudden expansion in small channels," *7th International Conference on Multiphase Flow*, Tampa, FL USA, May 30-June 4, 2010.

[31] I. Y. Chen, C. Y. Tseng, and C. C. Wang, "A new correlation and the review of two-phase flow pressure change across sudden expansion in small channels," *Int. J. Heat and Mass Transfer* vol. 53, pp. 4287-4295, 2010.

[32] K. A. Tripplet, S. M. Ghiasiaan, S. L. Abdel-Khlik, A. LeMouel, and B. N. McCord, "Gas-liquid two-phase flow in microchannels. Part II: Void fraction and pressure drop," *Int. J. Multiphase Flow*, vol. 25, pp. 395-410, 1999.

[33] I. Y. Chen, K. S. Yang, and C. C. Wang. "An empirical correlation for two-phase frictional performance in small diameter tubes," *Int. J. of Heat and Mass Transfer*, vol. 45, pp. 3667-3671, 2002.

[34] L. Friedel, "Improved friction pressure drop correlations for horizontal and vertical two-phase pipe flow," *European Two-phase Group Meeting*, Ispra, Italy, Paper E2, 1979.

APPENDIX CALCULATION EXAMPLES

TWO-PHASE PRESSURE CHANGE ACROSS SUDDEN CONTRACTION

One set of contraction data was taken from the data sets of Abdelall *et al.* [11]. Air and water were operated at near room temperature and atmospheric pressure. The mass flow rates of air and water are 0.00541 g/s and 1.154 g/s, respectively. The small and large diameters of the test cross section are 0.84 mm and 1.6 mm, respectively, as shown in Fig. **9**.

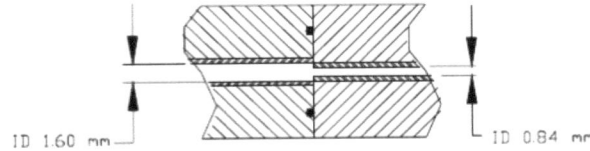

Figure 9: Test section cross-section details of Abdelall *et al.* [11].

Steps of calculation for the pressure change across the sudden contraction:

1. Calculate G, x, σ_A.

$$G = \frac{\dot{m}_L + \dot{m}_G}{A} = \frac{\dfrac{(1.154 + 0.00541)}{1000}}{\dfrac{\pi(0.84/1000)^2}{4}} = 2093(\frac{kg}{m^2 s})$$

$$x = \frac{\dot{m}_G}{\dot{m}_L + \dot{m}_G} = \frac{0.00541}{1.154 + 0.00541} = 0.00467$$

$$\sigma_A = \frac{\dfrac{\pi(0.84/1000)^2}{4}}{\dfrac{\pi(1.6/1000)^2}{4}} = (\frac{0.84}{1.6})^2 = 0.2756$$

2. According to 25°C temperature and atmospheric pressure to find out the properties of air and water, σ, ρ_L, ρ_G, μ_L and μ_G.

$\sigma = 0.07277$ (N/m), $\rho_L = 997$ (kg/m^3), $\rho_G = 1.184$ (kg/m^3)

$\mu_L = 0.001002$ (N·s/m^2) and $\mu_G = 0.00001856$ (N·s/m^2)

3. Calculate the values for the parameters of ρ_m, We, Bo, C_c.

$$\rho_m = \left[\frac{x}{\rho_G} + \frac{1-x}{\rho_L}\right]^{-1} = \left[\frac{0.00467}{1.184} + \frac{1-0.00467}{997}\right]^{-1} = 202.3 \text{ (kg/m}^3)$$

$$We = \frac{G^2 D}{\sigma \rho_m} = \frac{2093^2 \times \dfrac{0.84}{1000}}{0.07277 \times 202.3} = 250$$

$$Bo = \frac{\Delta \rho g D^2}{\sigma} = \frac{(997 - 1.184) \times 9.81 \times (\frac{0.84}{1000})^2}{0.07277} = 0.0947$$

$$C_C = 1 - \frac{1 - \sigma_A}{2.08(1 - \sigma_A) + 0.5371}$$

$$= 1 - \frac{1 - 0.2756}{2.08 \times (1 - 0.2756) + 0.5371}$$

$$= 0.6456$$

4. Calculate the original homogeneous pressure change, $\Delta P_{Homogeneous}$, and the factors, Ω_1, Ω_2 and Ω_3.

$$\Delta P_{Homogeneous} = (\frac{G^2}{2\rho_m}) \left[(\frac{1}{C_C} - 1)^2 + (1 - \sigma_A^2) \right]$$

$$= (\frac{2093^2}{2 \times 202.3}) \times \left[(\frac{1}{0.6456} - 1)^2 + (1 - 0.2756^2) \right] = 13267 (Pa).$$

$$= 13.267 (kPa)$$

$$C_1 = Bo^{1.1}(1-x)^{0.9} = 0.0947^{1.1} \times (1 - 0.00467)^{0.9} = 0.072$$

$$C_2 = 470e^{-\sigma_A^{-0.2}} = 470 \times e^{-0.2756^{-0.2}} = 128.857$$

$$C_3 = We \times Bo \times \sigma_A \times (1-x)^{-3} = 250 \times 0.0947 \times 0.2756 \times (1 - 0.00467)^{-3} = 6.627$$
$$C_4 = \sigma_A^{2.5}(1-x)^{-1} = 0.2756^{2.5} \times (1 - 0.00467)^{-1} = 0.042$$

$$\Omega_1 = -0.99e^{\frac{-13.1C_1}{C_2}} = -0.99 \times e^{\frac{-13.1 \times 0.072}{128.857}} = -0.983$$

$$\Omega_2 = -39.4e^{C_1^{-0.25}}(16.1C_3^{-1.5} - 13.2C_3^{-1.8} - 4.2C_3^{-1})$$

$$= -39.4 \times e^{0.072^{-0.25}} \times (16.1 \times 6.627^{-1.5} - 13.2 \times 6.627^{-1.8} - 4.2 \times 6.627^{-1})$$

$$= 34.972$$

$$\Omega_3 = 0.1C_3^{-1.3}C_4^{-3} = 0.1 \times 6.627^{-1.3} \times 0.042^{-3} = 115.489$$

5. Using the proposed Eq. (55) to calculate the contraction pressure change

$$\Delta P_{Modify} = \Delta P_{Homogeneous} \times (1 + \Omega_1) \times (1 + \Omega_2) \times (1 + \Omega_3)^{-0.08}$$

$$= 13.267 \times (1 - 0.983) \times (1 + 34.972) \times (1 + 115.489)^{-0.08}$$

$$= 5.545 (kPa)$$

6. Comparison

Using the proposed modified homogeneous equation, the predicted ΔP is very closed to the measured data [11] as shown in the following table.

$\Delta P_{Abdelall}$ (kPa)	$\Delta P_{Homogeneous}$ (kPa)	ΔP_{Modify} (kPa)
5.251	13.267	5.545

Based on the measured 26 points of contraction data [11], the mean deviation of the homogeneous predictions is 223%, while the mean deviation of the predictions for the modified homogeneous equation, Eq. (55), is reduced to 31%.

TWO-PHASE PRESSURE CHANGE ACROSS SUDDEN EXPANSION

One set of expansion data was taken from the data sets of Abdelall *et al.* [11]. Air and water were operated at near room temperature and atmospheric pressure. The mass flow rates of air and water are 0.00374 g/s and 1.70 g/s, respectively. The test cross section is shown in Fig. **9**.

Steps of calculation for the pressure change across the sudden expansion:

1. Calculate G, x, σ_A.

$$G = \frac{\dot{m}_L + \dot{m}_G}{A} = \frac{\dfrac{(1.7 + 0.00374)}{1000}}{\dfrac{\pi(0.84/1000)^2}{4}} = 3074(\frac{kg}{m^2 s})$$

$$x = \frac{\dot{m}_G}{\dot{m}_L + \dot{m}_G} = \frac{0.00374}{1.7 + 0.00374} = 0.002195$$

$$\sigma_A = \frac{\dfrac{\pi(0.84/1000)^2}{4}}{\dfrac{\pi(1.6/1000)^2}{4}} = (\frac{0.84}{1.6})^2 = 0.2756$$

2. According to 25°C temperature and atmospheric pressure to find out the properties of air and water, σ, ρ_L, ρ_G, μ_L and μ_G.

$\sigma = 0.07277$ (N/m), $\rho_L = 997$ (kg/m³), $\rho_G = 1.184$ (kg/m³)

$\mu_L = 0.001002$ (N·s/m²) and $\mu_G = 0.00001856$ (N·s/m²)

3. Calculate the values for the parameters of ρ_m, Fr, We, Bo and Re_{Lo}.

$$\rho_m = \left[\frac{x}{\rho_G} + \frac{1-x}{\rho_L} \right]^{-1} = \left[\frac{0.002195}{1.184} + \frac{1-0.002195}{997} \right]^{-1} = 350.3 \text{ (kg/m}^3)$$

$$Fr = \frac{G^2}{\rho_m^2 gD} = \frac{3074^2}{350.3^2 \times 9.81 \times \dfrac{0.84}{1000}} = 9345$$

$$We = \frac{G^2 D}{\sigma \rho_m} = \frac{3074^2 \times \frac{0.84}{1000}}{0.07277 \times 350.3} = 311.4$$

$$Bo = \frac{\Delta \rho g D^2}{\sigma} = \frac{(997 - 1.184) \times 9.81 \times (\frac{0.84}{1000})^2}{0.07277} = 0.0947$$

$$Re_{Lo} = \frac{GD}{\mu_L} = \frac{3074 \times \frac{0.84}{1000}}{0.001002} = 2577$$

4.   Calculate the original homogeneous pressure change, $\Delta P_{Homogeneous}$, and the factors, Ω_1, Ω_2 and Ω_3.

$$\Delta P_{Homogeneous} = \frac{G^2 \sigma_A (1 - \sigma_A)}{\rho_m} = \frac{3072^2 \times 0.2756 \times (1 - 0.2756)}{350.3} = 5027(Pa) = 5.027(kPa)$$

$$\Omega_1 = (\frac{WeBo}{Re_{Lo}})^2 \times (\frac{1-x}{x})^{0.3} \times \frac{1}{Fr^{0.8}}$$

$$= (\frac{311.4 \times 0.0947}{2577})^2 \times (\frac{1 - 0.002159}{0.002159})^{0.3} \times \frac{1}{9345^{0.8}} = 5 \times 10^{-9}$$

$$\Omega_2 = 0.2 \times (\frac{\mu_G}{\mu_L})^{0.4} = 0.2 \times (\frac{0.00001856}{0.001002})^{0.4} = 0.04$$

$$\Omega_3 = 0.4 \times (\frac{x}{1-x})^{0.3} + 0.3 \times e^{\frac{1.6}{Re_{Lo}^{0.1}}} - 0.4 \times (\frac{\rho_L}{\rho_G})^{0.2}$$

$$= 0.4 \times (\frac{0.002159}{1 - 0.002159})^{0.3} + 0.3 \times e^{\frac{1.6}{2577^{0.1}}} - 0.4 \times (\frac{997}{1.184})^{0.2} = -0.853$$

5. Using the proposed Eq. (59) to calculate the expansion pressure change.
$\Delta P_{Modify} = \Delta P_{Homogeneous} \times (1 + \Omega_1 - \Omega_2) \times (1 + \Omega_3)$
$= 5.027 \times (1 - 0.04) \times (1 - 0.853) = 0.709(kPa)$

6. Comparison

Using the proposed modified homogeneous equation, the predicted ΔP is very near the measured data [11] as shown in the following table.

$\Delta P_{Abdelall} (kPa)$	$\Delta P_{Homogeneous} (kPa)$	$\Delta P_{Modify} (kPa)$
0.726	5.027	0.709

Based on the measured 14 points of expansion data [11], the mean deviation of the homogeneous predictions is 468%, while the mean deviation of the predictions for the modified homogeneous equation, Eq. (59) is reduced to 26%.

CHAPTER 3

Coalescence of Drops in Liquid

Pallab Ghosh[*]

Department of Chemical Engineering, Indian Institute of Technology Guwahati, Guwahati–781039, Assam, India

Abstract: Coalescence of drops is an important process for the destabilization of liquid–liquid dispersions. This chapter presents the mechanism of coalescence of drops in dispersion. It also presents the theories of coalescence that are used at present. The film drainage theory and the stochastic theory of coalescence have been discussed explaining their merits and drawbacks. These models are evaluated with numerical examples. The role of van der Waals, electrostatic double layer, steric and hydration forces on the coalescence process has been discussed. The importance of adsorption of surfactant molecules at the liquid–liquid interface on coalescence time has been explained. Possible reasons behind the failure of the film drainage models in predicting the coalescence time in presence of surfactant have been explained. This chapter also presents the calculation of coalescence time in industrial equipment, experimental techniques employed for studying coalescence of a drop at a flat liquid–liquid interface, binary coalescence, and coalescence of two drops in motion. Directions for future research on coalescence have also been presented.

Keywords: Coalescence, disjoining pressure, drop, emulsion, film drainage theory, interfacial force, lubrication flow, surfactant, stochastic theory, thin liquid film.

INTRODUCTION

Coalescence of drops, suspended in liquid, plays a crucial role during preparation and stabilization of emulsions. The rate of coalescence is important in determining the efficiency of emulsifiers, liquid–liquid reactors, phase separators, and polymer blending equipment. Some of the common occurrences of emulsion are in environment and meteorology (*e.g.*, water and sewage treatment emulsions), foods (*e.g.*, milk, butter, cheese and sauces), geology, agriculture and soil science (*e.g.*, sprays), materials science (*e.g.*, polishes, paving asphalt, emulsion and latex paints), biology and medicine (*e.g.*, vitamin and hormone products, blood and cells), petroleum production and mineral processing (*e.g.*, drilling emulsions, reservoir emulsions and transportation emulsions), and home and personal care products (*e.g.*, hair-styling mousse, skin-care creams and lotions). Coalescence is a very important phenomenon in the emulsion-handling equipment used in the petroleum industries [1].

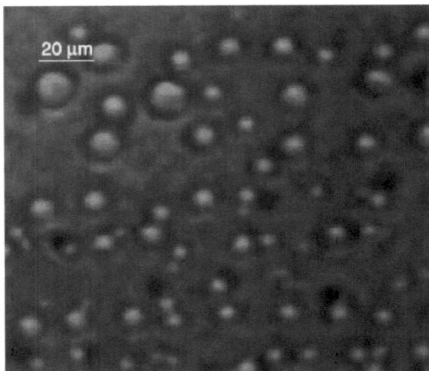

Figure 1: Emulsion drops of paraffin oil in salty water.

*Address correspondence to Pallab Ghosh: Department of Chemical Engineering, Indian Institute of Technology Guwahati, Guwahati–781039, Assam, India; Tel: +91.361.2582253; Fax: +91.361.2690762; E-mail: pallabg@iitg.ernet.in

Lixin Cheng and Dieter Mewes (Eds)

The photograph of an oil-in-water emulsion is shown in Fig. **1**. Emulsions are thermodynamically unstable systems and their free energy of formation $\left(\Delta G_f \right)$ is greater than zero [2]. This instability is a result of the energy associated with the large interfacial area of the drops within the emulsion, given by γA, where A is the total surface area of the drops and γ is the interfacial tension between the aqueous and non-aqueous phases. This energy term outweighs the entropy of formation ΔS_f associated with the formation of the drops from the bulk constituents. The free energy of formation of the emulsion is given by

$$\Delta G_f = \gamma A - T \Delta S_f \qquad\qquad (1)$$

The interfacial tension in emulsions is generally of the order of 1–10 mN/m. The product of interfacial tension and the large interfacial area produces a large positive interfacial energy term. The entropy of formation of emulsions is not very large because the number of drops formed is rather small in entropic terms. These two factors result in $\Delta G_f > 0$, which leads to the thermodynamic instability of the emulsions. Emulsions are, however, kinetically stable due to the presence of an adsorbed layer of surfactant molecules at the oil–water interface. This layer may provide electrostatic (in the case of an ionic surfactant) or steric (for a nonionic surfactant) repulsion when the drops approach each other. These surfactant monolayers therefore act as barriers and prevent the emulsion drops from coming into direct contact. They serve to stabilize the thin film of liquid between two adjacent drops. Coalescence occurs when this thin film becomes unstable and then ruptures. Therefore, the presence of adsorbed surfactants can reduce the likelihood of rupture.

As a result of their thermodynamic instability, emulsions tend to reduce their total free energy through an increase in drop diameter, and hence reduce their total interfacial area. This leads to the degradation of the emulsion. An emulsion may degrade by various mechanisms such as creaming (with or without aggregation), aggregation (with or without creaming), Ostwald ripening and coalescence. These processes are shown schematically in Fig. **2**. Coalescence and Ostwald ripening involve a change in the drop size distribution and can result in complete phase separation into oil and water, as shown in Fig. **2**. Coalescence requires the drops to be in close proximity. Ostwald ripening, on the other hand, does not require the drops to be close, since the process occurs by transport of dissolved matter through the dispersion medium. In the majority of cases, the predominant degradation mechanism is coalescence.

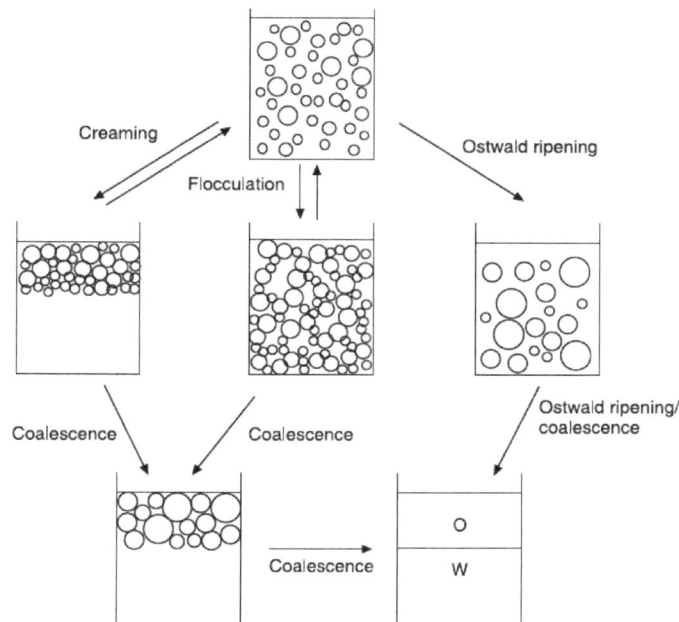

Figure 2: Schematic diagram depicting the role of coalescence in breaking down an emulsion [3] (reproduced by permission from Elsevier Ltd., © 1998).

Coalescence of drops can cause tremendous reduction in the liquid–liquid interfacial area. To illustrate the amount of interfacial area produced by emulsification, let us consider 1000 cm^3 of an oil dispersed in water. The total interfacial area of the drops of 1 μm diameter produced by emulsification can be as large as 6000 m^2. If the drops are reduced to 10 nm diameter, the interfacial area would be 6×10^5 m^2. The work necessary to reduce the size of the drops to the sub-micrometer level is also very high, and work equivalent to several hundreds of joules is necessary to generate drops of size < 10 nm. Since the rate of heat and mass transfer depend upon the interfacial area, coalescence of drops can significantly affect the performance of multiphase contactors. For example, the rate of change of concentration of an organic compound in the aqueous phase can be expressed as [4]

$$\frac{dc}{dt} = k_L \hat{a} \left(c^{\text{sat}} - c \right) \qquad (2)$$

where c is the concentration of the organic compound in the aqueous phase, t is time, c^{sat} is the saturation concentration, k_L is the mass transfer coefficient, and $\hat{a}$ is the interfacial area of the drops of the organic compound per unit volume of the aqueous dispersion.

Scope of the Chapter

Three coupled subprocesses influence the destabilization and phase separation of emulsions. These are flocculation, coalescence, and floc fragmentation. This chapter is focussed on the mechanism of coalescence of drops in liquid–liquid dispersion with emphasis on the film drainage and stochastic theories of coalescence, and the role of intermolecular and surface forces on the coalescence process.

A simplified theory for the coupling of flocculation and coalescence of drops is presented in the section entitled "Flocculation and coalescence of drops". This will provide an overview of the coalescence of drops in emulsions. The theories of van der Waals, electrostatic double layer, hydration, and steric forces are discussed in the section entitled "Role of intermolecular and surface forces on coalescence of drops". Coalescence of two drops occurs when the thin liquid film between them ruptures. The details of thin liquid films is discussed in the section entitled "Thin liquid films in emulsions". The critical film thickness, disjoining pressure model and stability aspects of the thin films are discussed in this section. The lubrication model of film drainage in presence and in absence of surfactants is presented in the section entitled "Film drainage theories of coalescence". The predictions made by these models are evaluated with examples. The stochastic model, its parameters, and their significance are explained in the section entitled "Stochastic theory of coalescence". The procedure of fitting the stochastic model to the experimental coalescence time distributions is illustrated with examples in this section. Some methods used to calculate the droplet growth time in industrial equipment are discussed in the section entitled "Coalescence in industrial equipment". Various experimental techniques have been employed to study coalescence of drops. These methods are discussed in the section entitled "Experimental techniques for studying coalescence of drops".

FLOCCULATION AND COALESCENCE OF DROPS

Two or more drops of the dispersed phase cluster together as aggregates by flocculation. This is a reversible process in most emulsions. Flocculation can cause creaming of emulsion because the aggregates are subject to buoyancy force. Flocculation precedes coalescence. After flocculation occurs, the drops which are in close proximity with each other are separated by a thin film of the continuous phase. Under the various forces acting on the film, the film drains and ultimately ruptures resulting in coalescence. Coalescence is an irreversible process. In 1917, M. V. Smoluchowski [5] proposed the perikinetic theory of flocculation of hydrophobic colloids (described in Appendix 1). This theory has been used for flocculation of emulsion drops as well [6]. The assumptions inherent in the Smoluchowski's theory of Brownian flocculation are also present in these models, e.g., creaming and gravitational flocculation are not accounted for, and hydrodynamic or electromagnetic interactions between the drops are assumed to be absent. The rate constants for Brownian collisions are independent of the size of the drops. Therefore, Smoluchowski's theory is valid for polydispersed aggregates. The scheme of flocculation according to Smoluchowski is depicted in Fig. 3.

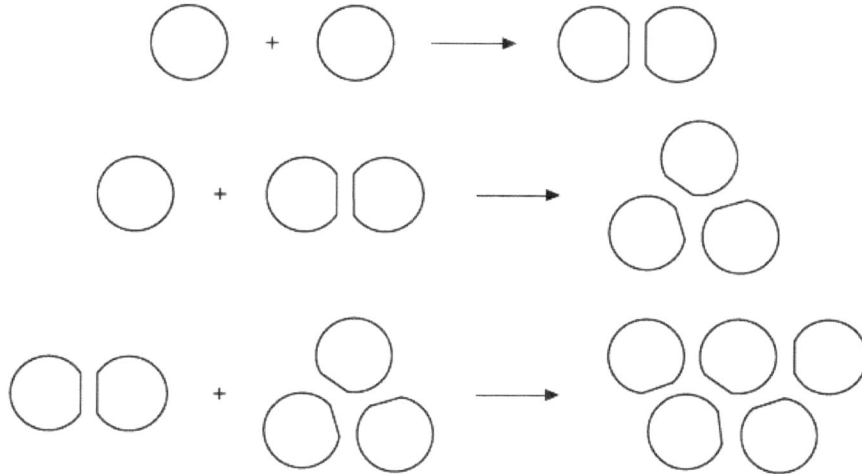

Figure 3: Model of flocculation according to Smoluchowski's scheme.

The rate of formation of aggregates is determined by the diffusion of drops through the continuous phase. If n_0 is the number of primary drops present at time $t = 0$, and n_p is the total number of drops (viz. aggregates and primary drops) at time t, then n_p is related to n_0 by the following equation (see Appendix 1).

$$n_p = \frac{n_0}{1 + \alpha n_0 t} \tag{3}$$

where α is the rate constant for flocculation. The number of primary non-aggregated drops remaining at time t is given by

$$n_t = \frac{n_0}{\left(1 + \alpha n_0 t\right)^2} \tag{4}$$

The number of aggregates n_v at time t is given by

$$n_v = \frac{\alpha n_0^2 t}{\left(1 + \alpha n_0 t\right)^2} \tag{5}$$

The average number of primary drops in an aggregate is given by

$$n_a = \frac{n_0 - n_t}{n_v} = 2 + \alpha n_0 t \tag{6}$$

For perikinetic flocculation, the rate constant α is related to the diffusion coefficient D as

$$\alpha = 8\pi D a \tag{7}$$

where a is the radius of the drop. Since D is inversely related to a, the product Da is independent of the radius of the drop. Therefore, α remains constant even though the size of the aggregates increases. Thus, Smoluchowski's model is independent of the initial size of the drop, and not affected by the polydispersity. Sometimes α is used as a fitted parameter and the model is applicable to the most general case of perikinetic or orthokinetic flocculation, regardless of the nature of interactions between the drops.

In the coupled flocculation and coalescence theory, coalescence proceeds according to the following scheme: the flock composed of i drops can partially coalesce to become an aggregate of j drops $(1 \leq j < i)$ as shown in Fig. **4**.

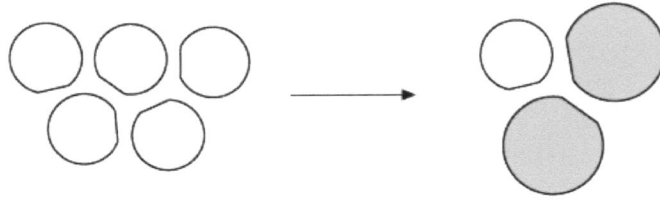

Figure 4: Coalescence in an aggregate of drops. The shaded drops are formed by coalescence.

Suppose that the initial number of drops is n_0, number of primary drops from the initial set remaining at time t is n_t, and number of aggregates (or non-primary drops) at time t is n_v. The average number of drops in an aggregate is $\overline{m}$, which is different from n_a because of coalescence occurring in the aggregates. The number of films in an aggregate is n_f, and the total number of drops in the emulsion after time t is n. Apparently, $n = n_0$ if there is no coalescence.

The aggregate is assumed to be 'linear' so that n_f and $\overline{m}$ are related as

$$n_f = \overline{m} - 1 \tag{8}$$

The lifetimes of the thin liquid films between the drops determine the coalescence rate. Within each aggregate, the rate of increase of the number of drops is equal to αn_0. The decrease in $\overline{m}$ is given by the rate of coalescence, which is assumed to be proportional to the number of films n_f. Thus

$$\frac{\mathrm{d}\overline{m}}{\mathrm{d}t} = \alpha n_0 - K(\overline{m} - 1) \tag{9}$$

where K is the proportionality constant, known as *coalescence rate constant*. The average number of drops in an aggregate at any time t can be calculated by integrating Eq. (9). Using Smoluchowski's theory to calculate the number of aggregates at time t, the total number of drops at time t is given by

$$n = n_t + n_v \overline{m} \tag{10}$$

Substituting n_t from Eq. (4), n_v from Eq. (5), and $\overline{m}$ from Eq. (9) (after integration) in Eq. (10), we get

$$n = \frac{n_0}{(1 + \alpha n_0 t)^2} + \left\{ \frac{\alpha n_0^2 t}{(1 + \alpha n_0 t)^2} \right\} \left[\frac{\alpha n_0}{K} + \left(1 - \frac{\alpha n_0}{K} \right) \exp(-Kt) \right] \tag{11}$$

Equation (11) is known as *van den Tempel* equation. The dimensionless term $\alpha n_0 / K$ determines the relative effects of flocculation and coalescence. If $\alpha n_0 / K \ll 1$, it means that flocculation is much slower than coalescence and therefore, it is the controlling step. On the other hand, if $\alpha n_0 / K \gg 1$, flocculation is much more rapid than coalescence, and thus coalescence is the rate-controlling step.

Several modifications of this coalescence model have been suggested in the literature [6]. For example, the rate of increase of the number of drops in an aggregate was assumed to be proportional to n_0. However, Eq. (6) was obtained from the Smoluchowski's theory, which only accounts for the flocculation of drops (*i.e.*, coalescence does not occur). In an emulsion, the size of an aggregate increases due to the addition of individual primary drops as well as other aggregates, which is accounted for in Eq. (6). However, these incoming aggregates have themselves undergone coalescence, and this equation does not account for the coalescence that has occurred in the incoming aggregates. As a result, the theory of van den Tempel overestimates the rate of increase in the aggregate size. Therefore, Eq. (11) does not represent the correct physical picture of the flocculation–coalescence process, and it cannot predict the change of the rate-controlling step during the lifetime of the emulsions. Borwankar *et al.* [6] adopted a similar approach, but employed a somewhat different formulation. Instead of taking a balance on each aggregate, they took an overall balance on all drops in the emulsion. For linear aggregates, the total number of films in the emulsion is given by

$$n_f n_v = (\bar{m} - 1) n_v \tag{12}$$

The differential equation for n is given by

$$-\frac{dn}{dt} = K'(\bar{m} - 1) n_v \tag{13}$$

where K' is a proportionality constant analogous to K in Eq. (9). The overall balance is expressed by Eq. (10). Note that $\bar{m}$ can be expressed through n using Eq. (10). The variation of n with time is given by the solution of Eq. (13), using Eqs. (4), (5) and (10). This model contains two parameters: the flocculation-rate constant α and the coalescence-rate constant K'. The former depends on the drop interactions as well as on the diffusion coefficient (hence, on the viscosity of the medium), whereas the latter is dependent on the lifetimes of the thin liquid films trapped between the drops. Although the modification of Borwankar *et al.* [6] can be considered as an improvement over the van den Tempel model, there is a common drawback underlying both the theories, which is caused by the use of Smoluchowski's equation for calculating n_t: coalescence changes n_t and correspondingly n_v.

Danov *et al.* [7] have presented a kinetic model in which flocculation and coalescence are interrelated, and take place simultaneously. The flocculation scheme shown in Fig. **3** and the coalescence scheme shown in Fig. **4** are interdependent because the aggregate (after some of the drops have coalesced) is further involved in the flocculation scheme. They have presented a kinetic scheme accounting for the possibilities of flocculation and coalescence in and between the aggregates. They have assigned a rate constant to each elementary act and developed differential equations describing the emulsion system, which are similar to parallel chemical reactions in some respects.

Let us define Smoluchowski time, τ_s, as the time between the drop collisions. Let us denote doublet fragmentation time due to the Brownian movement as τ_d, and coalescence time as τ_c. The quantity τ_c is equal to $1/K'$. If the time between two collisions is shorter than τ_d, a doublet can transform into a triplet before it spontaneously disrupts. In the opposite case (*i.e.*, $\tau_s \gg \tau_d$), the probability for a doublet to transform into a triplet is very low because the disruption of the doublet occurs much earlier than its collision with a singlet. The rate of multiplet formation is very low for $\tau_d / \tau_s \ll 1$. The general rule of physicochemical kinetics states that the slowest step is the rate controlling step. If the flocculation step is rate controlling, flocs composed of three, four or larger number of drops cannot be formed because of rapid coalescence within the floc. Nandi *et al.* [8] observed a reduction in the coalescence rate with increasing shear rate in surfactant-stabilized emulsions sheared in a tangential Couette apparatus. With increase in the shear rate, the time of contact between drops during a collision reduced, and this caused decrease in the rate of film rupture. Hence the coalescence rate decreased with the increase in shear rate.

ROLE OF INTERMOLECULAR AND SURFACE FORCES ON COALESCENCE OF DROPS

The coalescence process can be divided into two segments: (i) hydrodynamic drainage of the liquid film when the drops approach each other, and (ii) rupture of the thin liquid film trapped between the drops. These steps are schematically shown in Fig. **5**.

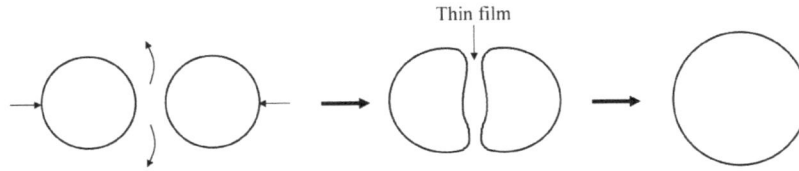

Figure 5: Coalescence of two drops *via* the drainage of the film, and its rupture.

Coalescence can occur only when the film separating the two drops is very small (< 100 nm), and it ruptures so that the liquids contained inside the drops can merge to form a single larger drop. Some of the interfacial forces are active when the separation between two interfaces is less than 100 nm. Therefore, these forces can play a significant part in the stability of the liquid film when the thickness of the film becomes less than 100 nm. The nature and magnitude of these forces depend upon the adsorbed species present at the liquid–liquid interfaces, and the physicochemical properties of the system (*e.g.*, surfactant and salt concentration, pH, viscosity of the liquids, size of the drops, and temperature). The stability of the thin film will be discussed in the next section. In this section, van der Waals, electrostatic double layer, hydration, and steric forces are discussed.

van der Waals Force

In 1873, J. D. van der Waals pointed out that real gases do not obey the ideal gas law. He suggested that two correction terms should be included to improve the accuracy of the ideal gas law. In ideal gas law, intermolecular attraction and the space occupied by the gas molecules were ignored. According to van der Waals, there should be correction terms for pressure (due to the attraction between the molecules, which is valid for polar as well as nonpolar molecules), and volume (due to the finite size of the molecules, which act as hard spheres during collision). Two types of forces were proposed by van der Waals: (i) the short-range repulsive forces which give rise to the excluded volume, and (ii) the long-range attractive forces which lead to the pressure correction. These two correction terms are related to the interaction energy between the gas molecules.

One of the common ways of expressing the interaction energy (ϕ) is the Lennard-Jones model [9], as shown below.

$$\phi = -\frac{\tilde{A}}{s^6} + \frac{\tilde{B}}{s^{12}} \tag{14}$$

where s is the separation between the molecules, and $\tilde{A}$ and $\tilde{B}$ are constants. The interaction energy given by this equation is also known as *L-J potential* or *6-12 potential*. Eq. (14) was developed based on the hypothesis that the pair of molecules is subject to two distinct forces in the limits of large and small separations: an attractive force at the long ranges, and a repulsive force at short ranges. The attraction is due to the dispersion interactions. The short-range repulsion term is due to the overlap of the molecular orbitals (known as *Pauli repulsion* or *Born repulsion*). The force of interaction can be obtained by differentiating ϕ with respect to the distance (*i.e.*, $F = -\mathrm{d}\phi/\mathrm{d}s$).

The total van der Waals interaction between molecules has three components: (i) interaction between two induced dipoles, which is known as *London dispersion force*, (ii) interaction between two permanent dipoles, which is known as *Keesom orientation force*, and (iii) interaction between one permanent dipole and one induced dipole, which is known as *Debye induction force*. Interestingly, each of these contributions to the interaction energy varies with s^{-6}. The dispersion forces were explained by Fritz London in 1930 [10]. These forces exist between all atoms and molecules, even the non-polar molecules. At any given instant, a non-polar molecule will have a dipole moment because of the fluctuations in the distribution of the electrons in the molecule. This dipole creates an electric field that polarizes another molecule located nearby, and an induced dipole results. The interaction between these dipoles leads to the attractive energy.

The time-averaged dipole moment of each molecule is zero but the time-averaged interaction energy is finite due to these temporarily interacting dipoles. The molecules of hydrocarbons and liquefied gases are held together mainly by these forces.

In addition to the London force, additional interactions exist between polar molecules. The Keesom interaction involves interaction between permanent dipoles, and the Debye interaction involves permanent dipole–induced dipole interaction. Since each of these three interactions has energy which varies with the inverse sixth power of the distance, the total van der Waals interaction energy for two dissimilar polar molecules is the sum of these three interaction energies.

The dispersions forces are probably the most important of the three forces which constitute the total van der Waals force. They are always present, but the presence of the other two types depends upon the properties of the molecules. The dispersion forces play very important roles in the stability of thin films. They are often called *long-range forces* because they can operate up to ~10 nanometers. If the distance between two molecules is large, the time taken by the electric field issued from the instantaneously-polarized molecule to reach the second molecule can be longer than the time period of the fluctuating dipole. The oscillating dipole induced by the second molecule re-radiates an electromagnetic field that is propagated back to the first molecule. Therefore, when the latter field reaches the first molecule, it may find that the orientation of the instantaneous dipole of the first molecule has changed from the original, and may be unfavorable for attractive interaction. Therefore, the dispersion energy may decay at a rate that is faster than $1/s^6$. The dispersion force between molecules at large separations is known as *retarded force*, and the effect is known as *retardation effect*. In free space, the retardation effect becomes important when the distance between the molecules is about 5 nm. In media where the speed of light is slower, the retardation effects can occur at smaller separations. Only the dispersion force encounters such retardation, the orientation and induction forces are not affected.

The dispersion interaction between two bodies is influenced by the other bodies present nearby. Two molecules interact directly as well as indirectly through the reflection from other molecules since they are also polarized by the electric field. However, we can calculate an approximate value of the interaction energy by summing up all pair-potentials of a molecule to obtain its net interaction energy with the other molecules. This approach was adopted by H. C. Hamaker [11] to compute the interaction energy between two macroscopic bodies. The interaction energy per unit area between two blocks having planar surface but infinite extension (*i.e.*, plane-parallel half-spaces) is given by

$$\phi = -\frac{A_H}{12\pi h^2} \tag{15}$$

where h is the separation between the two surfaces and A_H is the Hamaker constant (see Appendix 2 for more details of the Hamaker constant). The film between two drops is not flat, as shown in Fig. **5**. A dimple exists at the central part of the film due to the trapped liquid of the continuous phase. If the film is assumed to be flat (which is often done for the sake of simplicity), Eq. (15) can be used to compute the interaction energy. The values of Hamaker constant of some liquids are presented in Table **1**. Note that the Hamaker constant for air is zero.

When two interacting bodies (represented by the superscripts 1 and 2) are separated by a medium (represented by the superscript 3), the Hamaker constant can be calculated by the following equation [11,12] (see Appendix 2 for its derivation):

$$A_H^{1,3,2} = A_H^{1,2} + A_H^{3,3} - A_H^{1,3} - A_H^{3,2} \tag{16}$$

where $A_H^{i,j}$ represents the Hamaker constant for the interacting materials i and j *in vacuo*. The presence of a third medium does not change the distance-dependence of the van der Waals force, but its magnitude is affected by the modified value of the Hamaker constant $\left(A_H^{1,3,2} \right)$. From Eq. (16) is apparent that

$A_H^{1,3,2} < A_H^{1,2}$ if $A_H^{3,3} < A_H^{1,3} + A_H^{3,2}$. This condition holds quite often, which indicates that the Hamaker constant is likely to be reduced in presence of the medium. Several combining relations are available which can be used to calculate the approximate value of the Hamaker constant in terms of the known values. For example

$$A_H^{i,j} = \sqrt{A_H^{i,i} A_H^{j,j}} \ , \ i \neq j \tag{17}$$

From Eqs. (16) and (17), we can derive the following equation for $A_H^{1,3,2}$.

$$A_H^{1,3,2} = \left(\sqrt{A_H^{1,1}} - \sqrt{A_H^{3,3}}\right)\left(\sqrt{A_H^{2,2}} - \sqrt{A_H^{3,3}}\right) \tag{18}$$

These combining relations give good estimates when the dispersion force dominates the interactions. However, they may not give satisfactory results when applied to media of high dielectric constant (*e.g.*, water).

Table 1: Hamaker constants of some liquids interacting *in vacuo* (or air)

Liquid	$A_H \times 10^{20}$ J	Liquid	$A_H \times 10^{20}$ J
Acetone	4.17	$n\text{-}C_{10}H_{22}$	4.82
Benzene	5.00	$n\text{-}C_{11}H_{24}$	4.88
Carbon tetrachloride	5.50	$n\text{-}C_{12}H_{26}$	5.04
Cyclohexane	5.20	$n\text{-}C_{13}H_{28}$	5.05
$n\text{-}C_5H_{12}$	3.75	$n\text{-}C_{14}H_{30}$	5.10
$n\text{-}C_6H_{14}$	4.07	$n\text{-}C_{15}H_{32}$	5.16
$n\text{-}C_7H_{16}$	4.32	$n\text{-}C_{16}H_{34}$	5.23
$n\text{-}C_8H_{18}$	4.50	Toluene	5.40
$n\text{-}C_9H_{20}$	4.66	Water	3.70

The *pairwise-additivity* approach of Hamaker has been quite popular among the scientists because of the simplicity involved in its calculations. However, this approach ignores the influence of the neighbor molecules on the interaction between a pair of molecules. The effective polarizability of a molecule changes when it is surrounded by other molecules. As discussed earlier in this section, a neighbor molecule is also polarized when the electric field of a molecule polarizes a second molecule and the electric field of the first molecule reaches the second molecule directly as well as through the reflection from this neighbor molecule. Because of the contribution from these forces, the pairwise-additivity method does not remain straightforward, and the calculations become very complicated. The pairwise-additivity method is applicable in gases. However, in condensed phases, this method can be inaccurate. The second most important drawback of the Hamaker-approach is the neglect of the retardation effect. The retardation effect becomes important when surfaces interact in a liquid medium.

The equations developed by Hamaker are based on the assumption that the summation over all interacting pairs of molecules may be replaced by an integration procedure. For this assumption to be valid, the separation between the bodies must be large enough so that the interacting materials may be treated as continuous media, and not arrays of discrete molecules. Therefore, these equations do not apply at separations less than a few molecular diameters where the 'graininess of matter' can be an important factor.

The Lifshitz theory of van der Waals interactions was derived entirely from the considerations of the macroscopic properties of the media. Pairwise additivity was completely avoided in this theory. The attraction was assumed to be due to a fluctuating electromagnetic field in the gap which arises due to the spontaneous electric and magnetic polarizations within the media. It was implicit in this theory that the gap must be larger than molecular dimensions. The media were treated as continuous and the force between the macroscopic bodies was derived in terms of the dielectric constants and refractive indices. This theory is particularly suitable when we want to calculate the Hamaker constant for the interaction of the media 1 and 2 across medium 3. As per this theory, the Hamaker constant can be calculated from the following equation [13]:

$$A_H^{1,3,2} = \frac{3kT}{4}\left(\frac{\varepsilon_1 - \varepsilon_3}{\varepsilon_1 + \varepsilon_3}\right)\left(\frac{\varepsilon_2 - \varepsilon_3}{\varepsilon_2 + \varepsilon_3}\right) + \frac{3\tilde{h}\nu_e}{8\sqrt{2}}\left[\frac{\left(\eta_1^2 - \eta_3^2\right)\left(\eta_2^2 - \eta_3^2\right)}{\left\{\left(\eta_1^2 + \eta_3^2\right)\left(\eta_2^2 + \eta_3^2\right)\right\}^{1/2}\left\{\left(\eta_1^2 + \eta_3^2\right)^{1/2} + \left(\eta_2^2 + \eta_3^2\right)^{1/2}\right\}}\right] \quad (19)$$

In Eq. (19), $\varepsilon_i\,(i=1,2,3)$ are the dielectric constants, $\eta_i\,(i=1,2,3)$ are the refractive indices, and ν_e is the main electronic absorption frequency in the ultra-violet region. Its value usually lies between 2×10^{15} s^{-1} and 3×10^{15} s^{-1}. k is Boltzmann's constant ($\equiv 1.381\times10^{-23}$ J/K), T is temperature and $\tilde{h}$ is Planck's constant ($\equiv 6.626\times10^{-34}$ J s). The values of dielectric constant, refractive index, and UV absorption frequency of some liquids are presented in Table **2**.

Table 2: The values of ε, η and ν_e of some liquids at 298 K

Liquid	ε	η	$\nu_e \times 10^{-15}$ s^{-1}
Acetone	20.7	1.36	2.9
Benzene	2.3	1.50	2.1
Carbon tetrachloride	2.2	1.46	2.7
Cyclohexane	2.0	1.43	2.9
Ethanol	24.3	1.36	3.0
n-C$_5$H$_{12}$	1.8	1.35	3.0
n-C$_8$H$_{18}$	2.0	1.39	3.0
n-C$_{12}$H$_{26}$	2.0	1.41	3.0
n-C$_{14}$H$_{30}$	2.0	1.42	2.9
n-C$_{16}$H$_{34}$	2.1	1.42	2.9
Water	78.5	1.33	3.0

The first term in Eq. (19) is known as *zero-frequency contribution* and the second term is known as *non-retarded dispersion energy contribution*. The first term includes contributions from the Debye and Keesom interactions. From Eq. (19) we can observe that for two identical bodies (*i.e.*, $\varepsilon_1 = \varepsilon_2$ and $\eta_1 = \eta_2$), the Hamaker constant is positive and the van der Waals force is always attractive. However, for two different bodies ($\varepsilon_1 \neq \varepsilon_2$ and $\eta_1 \neq \eta_2$), the Hamaker constant can be positive or negative, depending on the medium between them. Therefore, it is apparent that the van der Waals force is not always attractive but it can be repulsive also. The values of dielectric constant and refractive index *in vacuo* or in air are unity, *i.e.*, $\varepsilon_3 = 1$ and $\eta_3 = 1$. Therefore, the van der Waals force between two bodies will always be attractive *in vacuo* or in air. The following example illustrates the calculation of Hamaker constant.

Example 1. Calculate the Hamaker constant for interaction between two cyclohexane drops separated by water at 298 K using Hamaker and Lifshitz theories.

Solution: Let us represent the system as cyclohexane(1) ꞉ water(3) ꞉ cyclohexane(2). From the Hamaker theory we have

$$A_H^{1,3,2} = \left(\sqrt{A_H^{1,1}} - \sqrt{A_H^{3,3}}\right)\left(\sqrt{A_H^{2,2}} - \sqrt{A_H^{3,3}}\right) = \left(\sqrt{A_H^{1,1}} - \sqrt{A_H^{3,3}}\right)^2$$

From Table **1**, $A_H^{1,1} = 5.2\times10^{-20}$ J, and $A_H^{3,3} = 3.7\times10^{-20}$ J. Therefore,

$$A_H^{1,3,2} = \left(\sqrt{A_H^{1,1}} - \sqrt{A_H^{3,3}}\right)^2 = \left(\sqrt{5.2} - \sqrt{3.7}\right)^2 \times 10^{-20} = 1.27 \times 10^{-21} \text{ J}$$

To calculate the Hamaker constant from the Lifshitz theory, the values of ε, η and v_e are collected from Table **2**.

Dielectric constant of cyclohexane $(\varepsilon_1) = 2.0$

Dielectric constant of water $(\varepsilon_3) = 78.5$

Refractive index of cyclohexane $(\eta_1) = 1.43$

Refractive index of water $(\eta_3) = 1.33$

Mean absorption frequency $(v_e) = \dfrac{2.9 + 3.0}{2} \times 10^{15} = 2.95 \times 10^{15} \text{ s}^{-1}$

Boltzmann constant $(k) = 1.381 \times 10^{-23} \text{ J/K}$

Planck's constant $(\tilde{h}) = 6.626 \times 10^{-34} \text{ J s}$

$$A_H^{1,3,2} = \frac{3kT}{4}\left(\frac{\varepsilon_1 - \varepsilon_3}{\varepsilon_1 + \varepsilon_3}\right)^2 + \frac{3hv_e}{16\sqrt{2}}\frac{\left(\eta_1^2 - \eta_3^2\right)^2}{\left(\eta_1^2 + \eta_3^2\right)^{3/2}}$$

Substituting the values in the above equation, we get

$$A_H^{1,3,2} = 5.4 \times 10^{-21} \text{ J}$$

The Hamaker constant is an important parameter for determining the stability of thin liquid films. The intervening medium has a very significant effect on the van der Waals attraction between two bodies. The force of interaction between two bodies is often expressed by "Disjoining pressure". This quantity is often used in film drainage models to balance the pressure in the film. The attractive disjoining pressure due to van der Waals force between two plane parallel half spaces separated by a distance h is given by $\Pi_{vdW} = A_H / 6\pi h^3$.

Electrostatic Double Layer Force

The electrostatic double layer force is one of the major repulsive forces which stabilize the thin liquid films. The surfaces of the drops can be charged by several mechanisms. One of the most common mechanisms of development of charge on an interface is the adsorption of ions (*e.g.*, surfactants) from solution on the initially uncharged interface. The double layer is made of the Stern layer, and the diffuse layer extending into the solution as shown in Fig. **6**.

The Coulomb attraction by the charged surface groups pulls the counterions back towards the interface, but the osmotic pressure forces the counterions away from the interface. This results in a diffuse double layer. The different parts of the electrostatic double layer are shown in Fig. **6**.

The double layer very near to the interface is divided into two parts: the Stern layer, and the Gouy–Chapman diffuse layer. The compact layer of adsorbed ions is known as *Stern layer* in honor of Otto

Stern who proposed the existence of this layer [14]. This layer has a very small thickness (~1 nm). The electrostatic double layer has been discussed at various levels of sophistication [15–19]. Some models divide the Stern layer into two parts: the inner Helmholtz layer, and the outer Helmholtz layer. The counterions specifically adsorb on the interface in the inner part of the Stern layer, which is known as *inner Helmholtz plane* (IHP). The potential drop in this layer is quite sharp, and it depends on the occupancy of the ions. The *outer Helmholtz plane* (OHP) is located on the plane of the centers of the next layer of non-specifically adsorbed ions. These two parts of the Stern layer are named so because the Helmholtz condenser model was used as a first approximation of the double layer very close to the interface. The diffuse layer begins at the OHP. The potential drop in each of the two layers (*i.e.*, from ψ_0 to ψ_1, and from ψ_1 to ψ_d) is assumed to be linear. The dielectric constant of water inside the Stern layer is believed to be much lower (*e.g.*, one-tenth) than its value in the bulk. The value is lowest near the IHP.

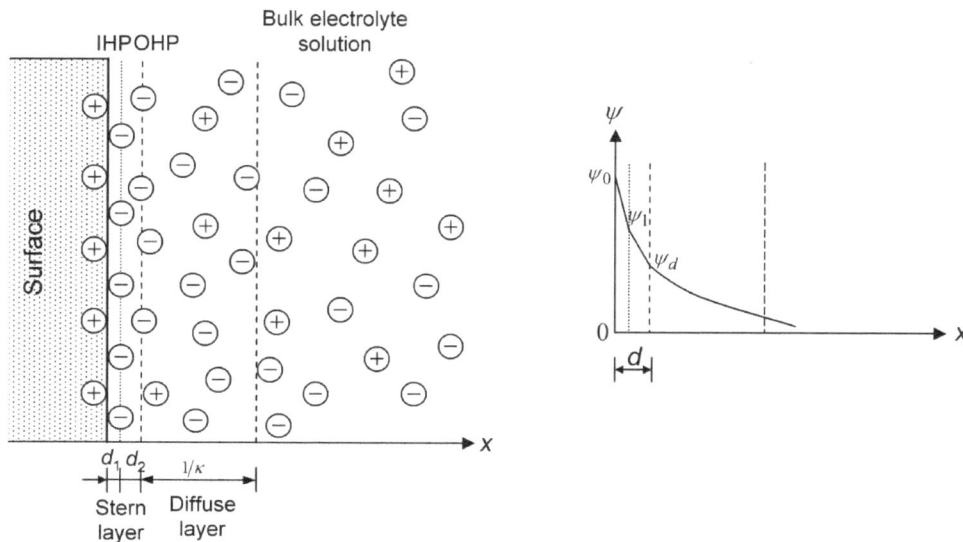

Figure 6: Electrostatic double layer and its parts.

The diffuse part of the electrostatic double layer is known as *Gouy–Chapman layer*. The thickness of the diffuse layer is called *Debye length* (represented by $1/\kappa$). This length indicates the distance from the OHP into the solution up to the point where the effect of the surface is felt by the ions. κ is known as *Debye–Hückel parameter*. The Debye length is significantly influenced by the concentration of electrolyte in the solution. The extent of the double layer decreases with increase in electrolyte concentration due to the shielding of charge at the interface. The multivalent ions are more effective in screening the charge than the monovalent ions.

The diffuse part of the electrostatic double layer begins from the OHP. The thickness of the Stern layer is $d\left(\equiv d_1 + d_2\right)$, as shown in Fig. **6**. Let us express the variation of potential $\left(\psi\right)$ with distance from the OHP (*i.e.*, $x = d$) by the Poisson equation in one dimension

$$\frac{\mathrm{d}^2\psi}{\mathrm{d}x^2} = -\frac{\tilde{\rho}}{\varepsilon\varepsilon_0} \tag{20}$$

where $\tilde{\rho}$ is the charge density (in C/m^3), ε is the dielectric constant of the medium, and ε_0 is the permittivity of the free space. Note that ε is constant in the diffuse layer but it varies significantly with position inside the Stern layer. Outside the OHP, there will be an accumulation of the oppositely charged ions. The work required to bring an ion from infinity to a position where the potential is ψ is $\tilde{z}_i e\psi$ (where, $\tilde{z}_i$ is the valence of the ion i). The ion concentration near the OHP is given by the Boltzmann distribution

$$m_i = m_i^\infty \exp\left(-\frac{\tilde{z}_i e\psi}{kT}\right) \tag{21}$$

where m_i is the number of ions of type i per unit volume near the interface, m_i^∞ is the number of ions of type i per unit volume in the bulk solution, and e is the electronic charge. The charge density $\tilde{\rho}$ can be related to the concentration of ions m_i as,

$$\tilde{\rho} = \sum_i \tilde{z}_i e m_i = \sum_i \tilde{z}_i e m_i^\infty \exp\left(-\frac{\tilde{z}_i e\psi}{kT}\right) \tag{22}$$

Combining Eqs. (20) and (22) we obtain

$$\frac{d^2\psi}{dx^2} = -\frac{e}{\varepsilon\varepsilon_0}\sum_i \tilde{z}_i m_i^\infty \exp\left(-\frac{\tilde{z}_i e\psi}{kT}\right) \tag{23}$$

Equation (23) is known as the *Poisson–Boltzmann equation*. The solution of this equation gives the potential ψ at any distance x.

When the potential is small, the variation of ψ with x is given by [17]

$$\psi = \psi_d \exp\left[-\kappa(x-d)\right] \tag{24}$$

This simplification is known as *Debye–Hückel approximation*. The *Debye length* κ^{-1} can be calculated from the following equation.

$$\kappa^{-1} = \left[\frac{N_A e^2}{\varepsilon\varepsilon_0 kT}\sum_i \tilde{z}_i^2 c_i^\infty\right]^{-1/2} \tag{25}$$

where c_i^∞ is the concentration of ions of type i, expressed in mol/m^3 and N_A is Avogadro's number. In aqueous medium at 298 K, $\dfrac{N_A e^2}{\varepsilon\varepsilon_0 kT} = 5.404 \times 10^{15}$ m/mol. Therefore, Eq. (25) can be written as

$$\kappa^{-1} = \left[5.404 \times 10^{15}\sum_i \tilde{z}_i^2 c_i^\infty\right]^{-1/2} \tag{26}$$

A complete solution of the Poisson–Boltzmann equation provides a better description of the variation of potential with distance. For the systems in which the electrolyte is symmetric (*i.e.*, of $\tilde{z}:\tilde{z}$ type, such as NaCl or MgSO$_4$), the variation of ψ with x is given by [17],

$$\tanh\left(\frac{\tilde{z}e\psi}{4kT}\right) = \tanh\left(\frac{\tilde{z}e\psi_d}{4kT}\right)\exp\left[-\kappa(x-d)\right] \tag{27}$$

This equation is known as the *Gouy–Chapman equation*, which describes the variation of potential in the diffuse part of the electrostatic double layer with distance starting from the Stern layer.

The potential of a charged interface can be reduced by the increase in concentration of the electrolyte in the bulk solution. The valence of the ions is an important factor. The effect of trivalent ions is greater than the divalent ions, which have greater effect than the monovalent ions. The charge density at the interface is reduced very effectively by the binding of counterions on the groups adsorbed at the interface. This reduces the surface potential strongly [20]. The binding of counterions has been found to play an important role in the coalescence of drops in ionic surfactant solutions [21].

The repulsive force between two surfaces begins to develop when they approach each other so closely that the double layers on their surfaces overlap. This repulsion opposes the approach of the surfaces. Let us consider two infinitely large planar charged surfaces separated by a distance h with the solution of electrolyte between them. Let us assume that the Stern layers are absent, and the double layers are solely constituted of the diffuse layers. The repulsive pressure generated by the electrostatic double layers (also known as *positive disjoining pressure due to electrostatic double layer*) on the two flat surfaces is given by

$$\Pi_{EDL} = 64 R T c^{\infty} \tanh^2 \left(\frac{\tilde{z} e \psi_0}{4kT} \right) \exp\left(-\kappa h\right) \tag{28}$$

It is evident from Eq. (28) that the strength of double layer repulsion depends on the electrolyte concentration. If binding of counterions takes place, the surface potential ψ_0 is reduced significantly. This would considerably reduce the repulsion. The following example illustrates how the variation of disjoining pressure with separation between the surfaces can be computed by using Eq. (28).

Example 2. Calculate the variation of Π_{EDL} with h between two planar surfaces in 10 mol/m^3 aqueous NaCl solution at 298 K if the surface potential is 50 mV. Develop the profile between 2 nm and 10 nm separations.

Solution: Putting $R = 8.314$ J mol^{-1} K^{-1}, $T = 298$ K, $c^{\infty} = 10$ mol/m^3, $e = 1.602 \times 10^{-19}$ C, and $k = 1.381 \times 10^{-23}$ J/K in the above equation we get

$$\Pi_{EDL} = 158.565 \times 10^4 \tanh^2 \left(9.732 \psi_0 \right) \exp\left(-\kappa h\right)$$

In this equation, ψ_0 is in V, h is in nm, and Π_{EDL} is in Pa. Let us calculate κ^{-1} from the following equation

$$\kappa^{-1} = \left[5.404 \times 10^{15} \sum_i \tilde{z}_i^2 c_i^{\infty} \right]^{-1/2}$$

Now, $\sum_i \tilde{z}_i^2 c_i^{\infty} = (1)^2 \times 10 + (1)^2 \times 10 = 20$ mol/m^3

$$\therefore \kappa^{-1} = \left[5.404 \times 10^{15} \times 20 \right]^{-1/2} = 3.04 \times 10^{-9} \text{ m} = 3.04 \text{ nm}$$

The variation of Π_{EDL} with h is shown in Fig. 7.

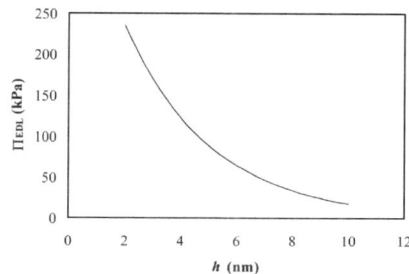

Figure 7: Variation of the disjoining pressure due to electrostatic double layer with separation between two surfaces at T = 298 K, ψ_0 = 50 mV, and c^{∞} = 10 mol/m^3.

Hydration and Steric Forces

The forces due to the van der Waals and electrostatic double layer interactions are the two major forces that play important roles in the stability of thin films. Apart from these forces, hydration and steric forces also play significant roles in coalescence of drops depending on the nature of the adsorbed species present at the interfaces [22,23]. The hydration force has strong repulsive effect below ~5 nm separation. For hydrophilic surfaces, the polar or ionic headgroups are hydrated. A surface which is not inherently hydrophilic can be rendered behaving like a hydrophilic surface by the adsorption of hydrated ions. This is known as *secondary hydration*. The hydration force arises when the adsorbed hydrated ions are prevented from desorbing as two interacting surfaces approach each other [24].

The adsorbed cations retain some of their water of hydration on binding. Dehydration of the adsorbed cations leads to a repulsive hydration force. The strength of the hydration force follows the order: $Mg^{+2} > Ca^{+2} > Li^+ \sim Na^+ > K^+ > Cs^+$. The hydrated radius of the ions and the hydration number (*i.e.*, the number of water molecules bound by these ions) follow similar sequences. The hydration force stabilizes soap films made of cationic surfactant decyltrimethylammonium decyl sulfate at high concentrations of sodium bromide. This stabilization is a result of the interaction between the hydration layers of very small thickness, which have two flanking surfactant monolayers [25].

Synthetic as well as biopolymeric surfactants are frequently used to stabilize emulsions. The adsorbed layers of the polymer on the surface of two drops repel each other. This type of repulsion between two surfaces is known as *steric repulsion*. The repulsion between surfaces covered with polymeric molecules begins when the segments of the polymer begin to overlap. The adsorbed layers encounter reduction in entropy when confined in a very small space as two surfaces approach each other. Since the reduction in entropy is thermodynamically unfavorable, their approach is inhibited. Let us consider the approach of two flat surfaces on which there are adsorbed or grafted chains of polymer as shown in Fig. **8**.

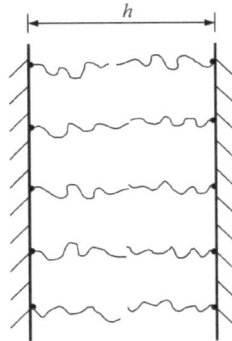

Figure 8: Interaction between polymer layers adsorbed on interfaces leading to steric repulsion.

The polymer is completely soluble in the solvent. The surfaces are well-covered by the polymer and they are brought close to each other without any bridging. The thickness of the polymer brush L can be related to the number of segments of the polymer molecule (n_s) and the length of each segment (l) by the following equation:

$$L = \frac{n_s l^{5/3}}{\bar{s}^{2/3}}, \quad \bar{s} = \left(1/\hat{\Gamma}\right)^{1/2} \tag{29}$$

where $\bar{s}$ is the mean distance between the attachment points and $\hat{\Gamma}$ is the number of adsorbed chains per unit area (assuming that each polymer molecule is grafted at one end to the surface). The polymer brushes come into contact when $h = 2L$. If the separation between the surfaces falls below $2L$, the concentration of polymer inside the brushes increases. This gives two contributions to the force: (i) the osmotic pressure inside each brush increases, and (ii) the elastic restoring force which tends to thin out the brush decreases. The repulsive pressure developed between the surfaces can be calculated from the de Gennes equation [26]:

$$\Pi_s = \frac{kT}{s^3}\left[\left(\frac{2L}{h}\right)^{9/4} - \left(\frac{h}{2L}\right)^{3/4}\right], \quad h < 2L \tag{30}$$

The first term in Eq. (30) arises from the osmotic repulsion between the coils which favors their stretching, and the second term is due to the elastic energy of the chains which opposes stretching.

THIN LIQUID FILMS IN EMULSIONS

Thin liquid films have been a subject of interest since the work of Robert Hooke (*circa* 1672) in which he reported 'holes' in soap films. Later, the works of Isaac Newton and Josiah Willard Gibbs have shown that these so called 'holes' are in fact regions of small thickness in the film where the interference between light reflected from the upper and lower film surfaces leads to nearly-complete extinction of the reflected light. Several aspects of thin liquid films have been studied in the past, such as their geometrical and optical properties, thermodynamics, hydrodynamics, and stability. In this chapter, we will discuss the hydrodynamic and stability aspects of the thin liquid films.

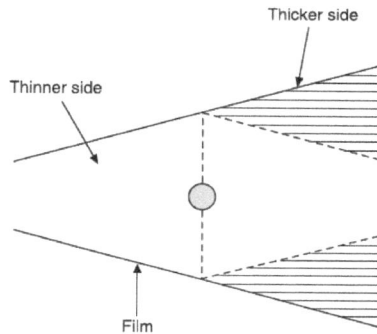

Figure 9: Illustration of how a molecule situated in the thinner part of a liquid film is attracted towards the thicker part of the film by the van der Waals force.

Interfacial tension tends to make the film smooth. However, thermal motions cause a certain amount of roughness. For thin films, the shapes of fluctuations in the two surfaces become correlated because of the molecular interactions in the film. To illustrate, the van der Waals force tends to drive the molecules from the thinner to the thicker parts of the film. Consider the thin film shown in Fig. **9** in which the thickness increases from left to the right.

The molecule located in the middle part of the film is attracted more towards the right by the molecules residing in the hatched portions of the film. Otherwise, the attraction on this molecule from the molecules in the left part of the film is exactly balanced by the molecules in the region enclosed by the dotted lines (*i.e.*, the difference between the entire film on the right side of the molecule and the hatched portion). This, in fact, is the mirror image of the part of the film located on the left side of the molecule. This attraction becomes important when the film is so thin that the range of the attractive force is comparable with the thickness of the film. Therefore, the van der Waals force favors 'corrugation' in the film, which causes disproportionation of a thin liquid film into thinner and thicker parts. Interfacial repulsive forces such as electrostatic double layer, hydration, and steric forces, which are active when surfactant molecules adsorb on the surfaces of the film, favor thicker films. Therefore, they can play an important role in the stability of the film. The stability of thin liquid films is a very important factor in the coalescence of drops [27].

For some deformations, the increase in surface Gibbs energy is more than compensated by the decrease in the van der Waals energy. Such deformations grow spontaneously in amplitude and lead to the rupture of the film. In some cases, rupture does not occur and a stable 'black film' forms if the repulsive force is strong enough. When the film is 'thick' (*e.g.*, 1 μm), its thinning is mainly governed by viscous flow caused by the pressure gradient in the film. When the thickness of the film reduces to such a small value that the rate of growth of fluctuations is faster than the thinning by viscous flow, the film can rupture. This film thickness is known as *critical film thickness* (h_c). Therefore, the total time required for the rupture of the film is comprised of two parts: (i) viscous drainage of the film, and (ii) rapidly-growing fluctuations leading to the rupture. The second part is of much shorter duration than the first part. Vrij and Overbeek [28] have given the following equation for the critical film thickness:

$$h_c = 0.267\left(\frac{a_f A_H^2}{6\pi\gamma\Delta p}\right)^{1/7}$$
(31)

where a_f is the area of the film, A_H is the Hamaker constant, γ is the interfacial tension and Δp is the excess pressure in the film. The value of critical film thickness is less than 50 nm for many emulsion films [29].

To calculate the critical film thickness using Eq. (31), the values of a_f and Δp are required. Let us consider some simple situations which are relevant to the coalescence of drops. Suppose that a drop is resting on a flat undeformable surface placed in a liquid which is immiscible with the liquid constituting the drop Fig. **10a**. In this case, the concerned film lies between the drop and the flat surface. The force acting on it is the gravitational force. The radius of the film is given by [30]:

$$R_f = a^2\left(\frac{2\Delta\rho g}{3\gamma}\right)^{1/2}$$
(32)

where a is the radius of the drop, $\Delta\rho$ is the difference in density between the liquid constituting the drop and the film liquid, and g is the acceleration due to gravity. The area of the film, a_f, is therefore given by

$$a_f = \frac{2\pi a^4 \Delta\rho g}{3\gamma}$$
(33)

If the drop is undeformable but the interface is deformable as shown in Fig. **10b**, the radius of the film is same as that given by Eq. (32). The deformability of the drop can be determined by calculating the Bond number $\left(\equiv \Delta\rho g a^2/\gamma\right)$. If the value of Bond number is much lower than unity, the drop may be considered as undeformable. If both drop and the interface are deformable as shown in Fig. **10c**, the radius of the film is given by

$$R_f = 2a^2\left(\frac{\Delta\rho g}{3\gamma}\right)^{1/2}$$
(34)

However, the area of the film can deviate significantly from the quantity πR_f^2 because of the curvature of the interface. Eqs. (32) and (34) were derived from a balance of gravitational and capillary forces. They give reasonably accurate results for small drops.

The excess pressure in the film, Δp, is given by F_g/a_f, where F_g is the gravitational force given by, $4\pi a^3\Delta\rho g/3$. For a small drop at a deformable interface, the excess pressure is γ/a, in which case the film area is given by $F_g a/\gamma$. Therefore, by calculating a_f and Δp, and with the knowledge of interfacial tension and Hamaker constant, we can calculate the critical film thickness (h_c) from Eq. (31). The calculation of the critical film thickness is illustrated in the following example.

Example 3. Calculate the critical thickness of the aqueous film when a 1 mm radius carbon tetrachloride drop rests on a flat water–CCl₄ interface at 298 K. Given: density of carbon tetrachloride = 1600 kg/m³, density of water = 1000 kg/m³, interfacial tension = 45 mN/m, and Hamaker constant = 6.8×10^{-21} J. Comment on the deformation of the drop.

Solution: The system is represented by Fig. **10a**. The gravitational force, F_g, is given by

$$F_g = \frac{4}{3}\pi a^3 \Delta\rho g = \frac{4}{3}\pi \times \left(1\times10^{-3}\right)^3 \times 600 \times 9.8 = 2.46\times10^{-5} \text{ N}$$

The Bond number is given by

$$\frac{\Delta\rho g a^2}{\gamma} = \frac{\left(1600 - 1000\right)\times 9.8 \times \left(1\times10^{-3}\right)^2}{0.045} = 0.13$$

Since the value of Bond number is small, the deformation of the drop would be small.

$$a_f = \frac{2\pi a^4 \Delta\rho g}{3\gamma} = \left(\frac{2\pi}{3}\right)\frac{\left(1\times10^{-3}\right)^4 \times 600 \times 9.8}{45\times10^{-3}} = 2.74\times10^{-7} \text{ m}^2$$

$$\Delta p = \frac{F_g}{a_f} = \frac{2.46\times10^{-5}}{2.74\times10^{-7}} = 89.8 \text{ Pa}$$

From Eq. (31), the critical film thickness is given by,

$$h_c = 0.267\left(\frac{a_f A_H^2}{6\pi\gamma\Delta p}\right)^{1/7} = 0.267\left[\frac{2.74\times10^{-7} \times \left(6.8\times10^{-21}\right)^2}{6\pi \times 45\times10^{-3} \times 89.8}\right]^{1/7} = 28.7\times10^{-9} \text{ m}$$

Therefore, the critical film thickness is 28.7 nm.

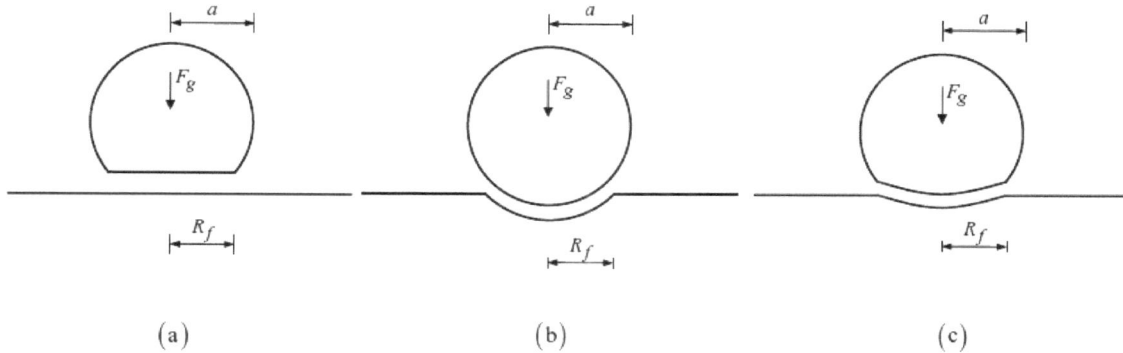

(a)   (b)   (c)

Figure 10: Models for film radius: (a) drop is deformable but the interface is undeformable, (b) drop is undeformable but the interface is deformable, and (c) drop and interface both are deformable.

Experimental studies have confirmed good accuracy of Eq. (31) in predicting the critical film thickness in emulsions. Experiments on the rupture of thin liquid films have shown that a distribution of the values of h_c is always observed, as depicted in Fig. **11** for chlorobenzene films of different radii [31]. In this figure, N is the total number of ruptured films and ΔN is the number of ruptures at thicknesses between $\left(h - \Delta h/2\right)$ and $\left(h + \Delta h/2\right)$.

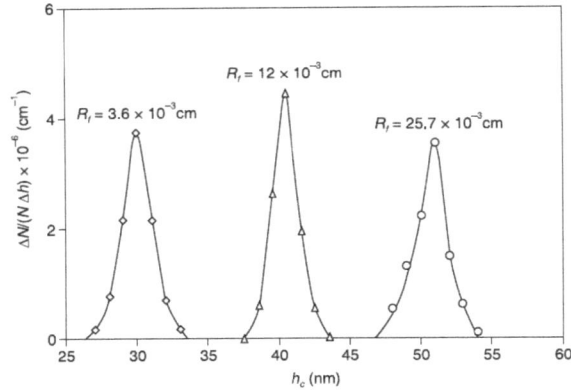

Figure 11: Distribution curves of critical film thickness for chlorobenzene films [31] (adapted by permission from The Royal Society of Chemistry, © 1968).

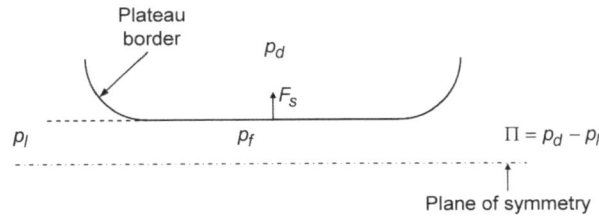

Figure 12: The disjoining pressure model of a thin liquid film.

Disjoining Pressure Model of Thin Liquid Films

Let us consider a thin liquid film formed between two emulsion drops as shown in Fig. **12**. Half of the film is shown in the figure. The film is assumed to be plane-parallel. The liquid in the film exists in contact with the bulk liquid in the Plateau border. The film-liquid would exist in hydrostatic equilibrium with the surrounding liquid if an external force directed perpendicular to the surfaces balances the internal force [32,33]. The internal force per unit area is the disjoining pressure (Π). The total disjoining pressure is the sum of disjoining pressures due to the van der Waals, electrostatic double layer, and short-range forces acting between the surfaces of the thin film.

Suppose that the drop-phase pressure is p_d, bulk liquid-phase pressure is p_l, interfacial tension is γ, and the radius of the drop is a. In the meniscus region outside the film, we have

$$p_d - p_l = \frac{2\gamma}{a} \tag{35}$$

Since p_d is larger than p_l, the equilibrium at the film surface is ensured by the action of the disjoining pressure as

$$\Pi = p_d - p_l \tag{36}$$

The following relationship also holds at equilibrium

$$p_d - p_f = \Pi \tag{37}$$

The disjoining pressure is a function of the film thickness. The disjoining pressure manifests as a normal surface-excess force F_s as shown in Fig. **12**. It is taken to be positive when it acts to disjoin (*i.e.*, separate) the film surfaces.

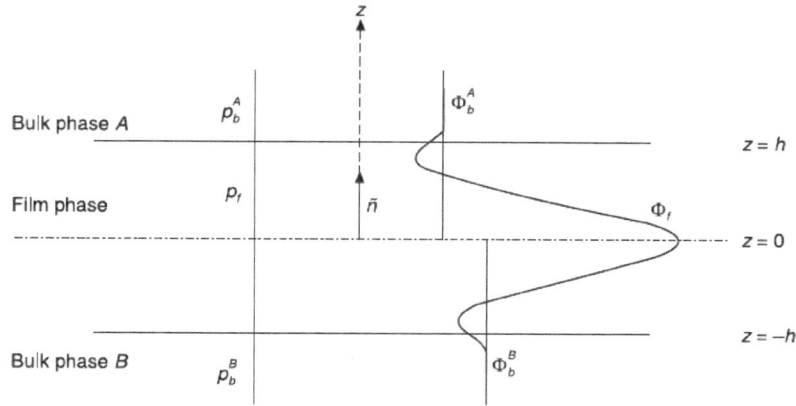

Figure 13: Schematic diagram of pressure and body-force interaction potential distributions in a planar film.

Stability of Thin Liquid Films

The stability analysis of thin liquid films involves the solution of equation of continuity, Maxwell and Navier–Stokes equations. The intermolecular forces in the film are assigned to a body-force acting within the film and incorporated in the Navier–Stokes equation of motion. This approach may be viewed as an alternative to the disjoining pressure model described before. The body-force model can be applied easily to nonsymmetric and nonplanar films [34]. The force is derived from a potential (Φ) which describes the energy of interaction of molecules due to van der Waals force in an infinitesimal volume with respect to the entire ensemble of molecules in the system. This potential can be calculated by a microscopic Hamaker method integrating the intermolecular potential over the system volume. The body-forces are absent within the adjacent bulk phases A and B shown in Fig. **13**. The tangential components of the body-force are significant near the Plateau borders. In equilibrium, the body forces are counterbalanced by pressure gradients and the total equilibrium potentials $(p+\Phi)$ in the film and bulk phases are constant. Therefore, the disjoining pressure can be expressed as

$$\text{at } z = h: \left(p_f + \Phi_f\right) - \left(p_b^A + \Phi_b^A\right) = \Pi \tag{38}$$

and

$$\text{at } z = -h: \left(p_f + \Phi_f\right) - \left(p_b^B + \Phi_b^B\right) = \Pi \tag{39}$$

From the condition of continuity of pressure across the planar surface in equilibrium we have,

$$\text{at } z = h: p_f = p_b^A \tag{40}$$

and

$$\text{at } z = -h: p_f = p_b^B \tag{41}$$

Therefore, we have,

$$\text{at } z = h: \Phi_f - \Phi_b^A = \Pi \tag{42}$$

and

$$\text{at } z = -h: \Phi_f - \Phi_b^B = \Pi \tag{43}$$

The body-force is defined as, $\tilde{F} = -\tilde{n} \left(d\Phi/dz \right)$, where $\tilde{n}$ is a unit vector tangent to the coordinate z and normal to the surfaces of the film (see Fig. **13**). Therefore, the magnitude of the surface contribution of the body-force acting at the film surfaces is $d\Pi/dh$. The disjoining pressure Π can be expressed as the sum of disjoining pressures due to the van der Waals, electrostatic double layer and short-range repulsive forces.

The two modes in which instability occurs in thin liquid films formed between two liquid phases are 'squeezing' and 'stretching' modes as shown in Fig. **14** (when the film is supported on a solid surface, only one surface is corrugated).

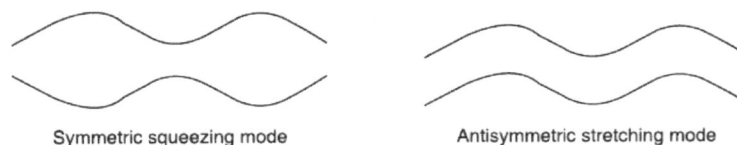

| Symmetric squeezing mode | Antisymmetric stretching mode |

Figure 14: Modes of instability of a thin liquid film: squeezing mode and stretching mode.

In the squeezing mode, the corrugations at one surface of the film are completely out of phase with the corrugations at the other surface. In the stretching mode, this phase difference is zero. The symmetric mode of vibration leads to the thinning of the film causing it to rupture. The anti-symmetric mode of instability usually occurs for the low-tension film surfaces.

The small dynamic fluctuations in the equilibrium thin film engender velocity and pressure fluctuations in the film. The *growth coefficient* of these fluctuations determines the stability. If the growth coefficient is positive, the disturbances grow and the system becomes unstable. If the growth coefficient is zero, the stability is 'marginal'. The disturbances are damped if the growth coefficient is negative. The stability of a thin film depends upon the wavelength of the disturbance. Marginal stability occurs at a 'critical' wavelength. The time required for a small disturbance to destabilize the thin film is of the order of the inverse of the largest growth coefficient.

FILM DRAINAGE THEORIES OF COALESCENCE

The film drainage theories of coalescence deal with the thinning of the film when two drops approach each other, or a drop approaches a flat liquid–liquid interface due to the action of some external force. Models have been developed under various assumptions. A comprehensive review of the film drainage models can be found in the works of Chesters [35], Slattery [36], Edwards *et al.* [37], Rommel *et al.* [38], and Birdi [39].

Models Without Consideration of Surfactant at the Interfaces

Several film drainage models have been developed in the past five decades which have assumed 'clean' interfaces, *i.e.*, the presence of surfactants was ignored in these models [40–58]. Nonetheless, many of these models are frequently used to analyse experimental data on coalescence systems in which surfactants are present. The thinning of a uniform liquid film of thickness h and viscosity μ with time t under the action of a constant force F is given by the Reynolds equation:

$$t = \left(\frac{3\mu\beta^2 a_f^2}{16\pi F} \right) \left(\frac{1}{h^2} - \frac{1}{h_i^2} \right) \tag{44}$$

where a_f is the area of the film of initial thickness h_i and β is the number of immobile surfaces. When $\beta = 2$, the velocity at both the surfaces is zero. On the other hand, when $\beta = 1$, the velocity at one of the surfaces is zero and the velocity gradient at the other surface is zero. Under these conditions, the drainage is very slow. Other values of β are also possible which correspond to different surface velocities and velocity gradients. These may take any value leading to very rapid film thinning or even to film thickening. Values of β less than unity correspond to more mobile interfaces, and values greater than 2 correspond to negative surface velocities.

The time required to reach the critical film thickness can be calculated from Eq. (44) by putting $h = h_c$. According to the film drainage theory, this is the time required for coalescence, or *coalescence time*, represented by $t = t_c$. Charles and Mason [40] developed two models for the calculation of coalescence time. The *parallel-disc model* gives the following equation for coalescence time:

$$t_c = \frac{\mu}{4}\left(\frac{\Delta\rho g a^5}{\gamma^2}\right)\left[\frac{1}{h_c^2} - \frac{1}{h_i^2}\right] \tag{45}$$

where μ is the viscosity of the liquid constituting the film, g is acceleration due to gravity, $\Delta\rho$ is the difference in density between the two liquids, γ is interfacial tension, h_i is the initial film thickness, and h_c is the critical film thickness. Eq. (45) was derived based on the assumption that a drop of radius a approaches a flat undeformable interface by gravity, and deforms by its own weight. The other model, called *spherical–planar model*, considers the drop as a sphere approaching an unbounded plane. According to this model, the coalescence time is given by

$$t_c = \frac{9\mu}{2a\Delta\rho g}\ln\left(\frac{h_i}{h_c}\right) \tag{46}$$

In deriving Eqs. (45) and (46), the electrostatic double layer repulsion and the van der Waals force of attraction were ignored. Furthermore, the theory assumes that the interfaces are rigid. The values of coalescence time predicted by Eqs. (45) and (46) differ considerably, as explained in the following example.

Example 4. Calculate the coalescence time of a 2 mm diameter water drop at a flat toluene–water interface using the parallel-disc and spherical–planar models. Given: density of toluene = 870 kg/m^3, viscosity of toluene = 0.6 mPa s, density of water = 1000 kg/m^3, interfacial tension = 36 mN/m, initial thickness of the film = 1 μm, and the critical thickness of the film = 10 nm.

Solution: Putting the given data, and $g = 9.8$ m/s^2 in Eq. (45), we get,

$$t_c = \frac{\mu}{4}\left(\frac{\Delta\rho g a^5}{\gamma^2}\right)\left[\frac{1}{h_c^2} - \frac{1}{h_i^2}\right]$$

$$= \left(\frac{0.6\times10^{-3}}{4}\right)\left[\frac{(1000-870)\times9.8\times\left(1\times10^{-3}\right)^5}{\left(36\times10^{-3}\right)^2}\right]\left[\frac{1}{\left(10\times10^{-9}\right)^2} - \frac{1}{\left(1\times10^{-6}\right)^2}\right] = 1475 \text{ s}$$

From Eq. (46) we get

$$t_c = \frac{9\mu}{2a\Delta\rho g}\ln\left(\frac{h_i}{h_c}\right) = \frac{9\times0.6\times10^{-3}}{2\times1\times10^{-3}\times(1000-870)\times9.8}\ln\left(\frac{1\times10^{-6}}{10\times10^{-9}}\right) = 9.8\times10^{-3} \text{ s}$$

Therefore, it can be observed that the values of coalescence time obtained from these two models are considerably different.

The model for coalescence time proposed by Jeelani and Hartland [29] incorporates the effects of circulation within the adjacent phases of the film. The coalescence time can be calculated from the following equation assuming $h_c \ll h_i$, and that the circulation lengths in the two liquid phases are nearly equal and similar in magnitude to the radius of the drop.

$$t_c = \left(\frac{9\mu R_f^4}{16a^3 \Delta\rho g h_c^2} \right) \left(\frac{1}{1 + \dfrac{3\mu a}{\mu_d h_i}} \right) \tag{47}$$

where R_f is the radius of the film, and μ_d is the viscosity of the liquid constituting the drop. The role of van der Waals force in coalescence is implicit in these models through the dependence of the critical film thickness (h_c) on the Hamaker constant. The effect of Π_{vdW} on the rate of film thinning has been incorporated in some of the models [48]. For tangentially immobile plane-parallel interfaces, the Reynolds equation can be written as

$$-\frac{dh}{dt} = \frac{2h^3}{3\mu R_f^2} \Delta p \tag{48}$$

where Δp is the driving force per unit area, given by $(p_c - \Pi_{vdW})$, where p_c is the capillary pressure. The capillary pressure is equal to $2\gamma/R_c$, where γ is the interfacial tension and R_c is the radius of the capillary in which the model emulsion film is formed.

To illustrate the film drainage process and the lubrication flow, let us consider the approach of two drops in a liquid under Stokes-flow conditions as shown in Fig. **15a**.

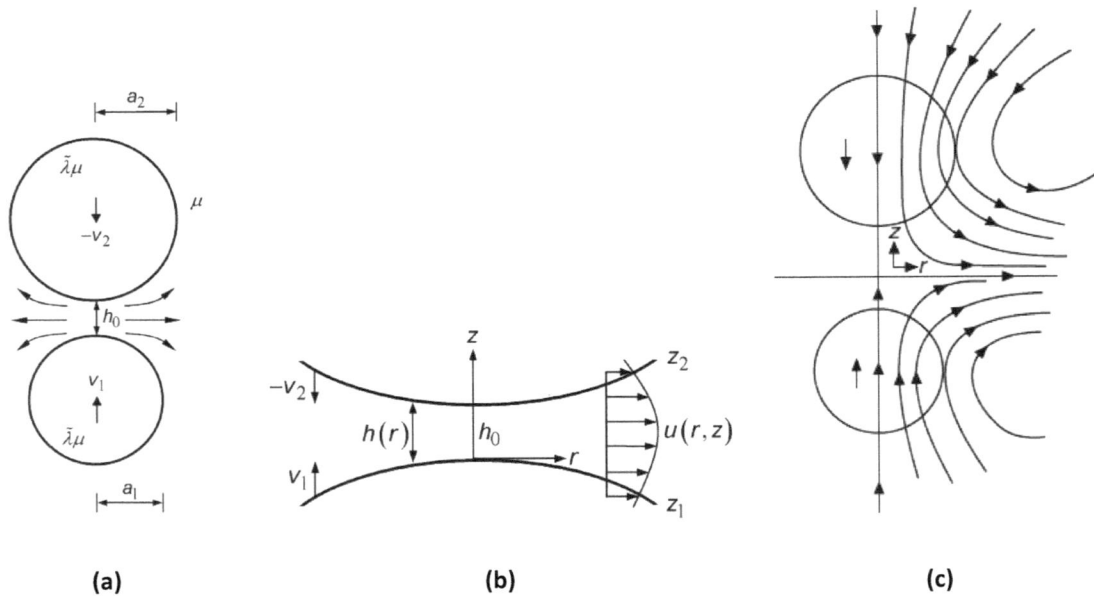

Figure 15: Schematic diagram of the drainage of the liquid film trapped between two drops: (a) drops approach each other, (b) exaggregated view of the near-contact region, and (c) sketches of the streamlines.

As the drops move toward each other along the line of their centers, the hydrodynamic force opposing their motion increases tremendously as the distance between the drops decreases. The exaggerated view of the contact region is shown in Fig. **15b**. A similar phenomenon occurs when a drop moves towards a flat interface. As the two drops approach, the resulting radially-outward flow in the narrow gap separating the drops exerts a tangential shear stress on the drop surfaces. This causes a tangential motion of the drop surfaces and drives a flow inside the drops. A sketch of the streamline-patterns is shown in Fig. **15c**. This flow has a significant effect on the force that resists the approach of the drops. The flow in the narrow gap and the flow within the drops are discussed in this section.

Let the radii of the drops be a_1 and a_2. The viscosity of the continuous phase is μ and that of the drop-liquid is $\tilde{\lambda}\mu$. The relative velocity is $U = v_1 - v_2$. The closest separation between the drop surfaces is h_0. Since we have assumed axisymmetric profile, h_0 is the located at the center of the film. The force that resists the relative motion of the drops and slows down the film-thinning process is dominated by a small region located near the axis of symmetry. This region is known as the *lubrication region*. Our objective is to derive the velocity profile in this region.

The radial velocity profile $u(r, z)$ can be decomposed into two parts as [50]:

$$u(r,z) = u_t(r) + u_p(r,z) \tag{49}$$

where $u_t(r)$ is the tangential velocity, and $u_p(r,z)$ is the velocity for the parabolic portion of the flow driven by the local pressure gradient, which is zero at the drop surfaces (*i.e.*, at z_1 and z_2). The velocity, u_p, is given by

$$u_p = \frac{1}{2\mu}\frac{\partial p}{\partial r}(z - z_1)(z - z_2) \tag{50}$$

The tangential stress exerted by the fluid in the gap on the drop surfaces is

$$f_t = \mu \left.\frac{\partial u}{\partial z}\right|_{z=z_1^+} = -\mu \left.\frac{\partial u}{\partial z}\right|_{z=z_2^-} = -\frac{h}{2}\frac{\partial p}{\partial r} \tag{51}$$

A mass balance on the fluid flowing out of the film gives the following equation:

$$\pi r^2 U = 2\pi r \int_{z_1}^{z_2} u(r,z)\,\mathrm{d}z \tag{52}$$

Substituting $u(r,z)$ from Eq. (49) into Eq. (52), carrying out the integration and substituting $h = z_2 - z_1$, we get

$$\pi r^2 U = 2\pi r \left(hu_t + \frac{h^2 f_t}{6\mu} \right) \tag{53}$$

The magnitude of the hydrodynamic force that resists the relative motion of the drops is given by

$$F_l = 2\pi \int_0^\infty p(r)r\,\mathrm{d}r \tag{54}$$

Since h_0 is small compared to the radius of the drops, $h_0/a \ll 1$, and the pressure is large compared to the z-component of the shear stress on the drop surfaces in the close-contact region. The film thickness h is a function of the radial position r. Therefore, the radial length-scale in the close-contact region is $(\bar{a}h_0)^{1/2}$ where $\bar{a}$ is the reduced radius [$\equiv a_1 a_2/(a_1 + a_2)$], because this is the radial distance for which $h(r)$ changes by $O(h_0)$. The pressure reduces significantly at radial positions far away from the lubrication region, *i.e.*, $p \to 0$ for $r \gg (\bar{a}h_0)^{1/2}$. Therefore, the upper limit of integration in Eq. (54) can be taken as ∞.

The spherical surfaces of the drops may be approximated as paraboloids near the axis of symmetry. Therefore, assuming the drops to be undeformed, the variation of film-thickness h with r can be expressed as

$$h(r) = h_0 + \frac{r^2}{2\overline{a}}$$

(55)

Eq. (55) gives one relationship between the tangential velocity u_t and tangential shear stress f_t. At the drop surfaces, u_t and f_t are significant only within an $O(\overline{a}h_0)^{1/2}$ radius of the axis of symmetry. Since this region is rather small compared with the radii of the drops because $h_0 \ll \overline{a}$, the interfaces may be treated as flat.

Davis *et al.* [50] have used the boundary integral form of the Stokes flow equations to calculate the tangential velocity as

$$u_t(r) = \frac{1}{\tilde{\lambda}\mu} \int_0^\infty \varphi(r',r) f_t(r') dr'$$

(56)

where $\varphi(r',r)$ is an elliptic-type Green's function kernel. It has an integrable (logarithmic) singularity at $r' = r$.

$$\varphi(r',r) = \frac{r'}{2\pi(r^2 + r'^2)^{1/2}} \int_0^\pi \frac{\cos\theta d\theta}{\left[1 - \frac{2rr'\cos\theta}{r^2 + r'^2}\right]^{1/2}}$$

(57)

Eq. (56) gives the second relationship between u_t and f_t. Eq. (53) and (56) are solved simultaneously to obtain the tangential velocity distribution $u_t(r)$ and tangential stress distribution $f_t(r)$. Then Eqs. (51) and (54) are used to find the dynamic pressure distribution $p(r)$ and the hydrodynamic lubrication force F_L. When the viscosity of the drop-liquid is much higher than the viscosity of the continuous phase $\left(\tilde{\lambda} \gg \sqrt{a/h_0}\right)$, the drops behave like rigid spheres. If the viscosity of the drop-liquid is comparable to the viscosity of the continuous phase or smaller $\left(\tilde{\lambda} \ll \sqrt{a/h_0}\right)$ then the drops offer little resistance to the radial flow in the gap. When $\tilde{\lambda}$ is of the order of $\sqrt{a/h_0}$, the drops offer significant resistance to the radial flow in the gap, but does not exhibit the rigid-sphere behavior. In this case, the velocity scales for both the uniform and parabolic portion of the radial flow in the gap are $O\left(U\sqrt{a/h_0}\right)$. The drop flow and gap flow are fully coupled. Davis *et al.* [50] have presented the solutions in these cases. The mobility considered here is purely a hydrodynamic effect involving the viscosities of the liquids constituting the drop and the dispersion medium. The effects of surfactants are not considered in this model.

Adsorption of Surfactants at Liquid–Liquid Interfaces

Surfactants are chemical compounds having amphipathic structure. A surfactant molecule has two parts: one part is soluble in the water, and the other part is insoluble in water. The part of a surfactant molecule that has unfavorable interaction with water is known as the *hydrophobic* part. On the other hand, the part which has favorable interaction with water is called the *hydrophilic* part. For example, the hydrophilic part of the cationic surfactant cetyltrimethylammonium bromide is the headgroup, $-\text{N}(\text{CH}_3)_3$, and the hydrophobic part is the hydrocarbon chain, $-\text{C}_{16}\text{H}_{33}$. When the surfactant molecules come in contact with water, they disrupt the hydrogen bonds between the water molecules. This increases the free energy of the system. Since this is thermodynamically unfavorable, the surfactant molecules are sent away towards the water–hydrocarbon interface. This encourages the surfactant molecules to adsorb at the water–hydrocarbon interface putting their hydrophobic tails in the hydrocarbon phase, and hydrophilic headgroups in water. If a sufficient number of surfactant molecules is present in the medium, the interface becomes covered with a closely-packed monolayer.

When the interface is completely occupied by the surfactant molecules, they undergo self-assembly to form clusters in which the hydrophobic part is directed inside the cluster, and the hydrophilic part is directed towards water. These clusters are known as *micelles*. The self-assembly process is known as *micellization*, and the concentration of surfactant in the aqueous solution at which this phenomenon takes place is known as *Critical Micelle Concentration* (CMC). Therefore, micellization occurs when the surface is saturated and the surfactant molecules have a sufficiently long hydrophobic part that disrupts the structure of water. For ionic surfactants, the hydrophilic headgroups of the surfactant molecules repel each other electrically whereas the hydrophobic groups attract each other by the hydrophobic attraction. Therefore, two opposing forces act in the interfacial region: one tends to increase and the other tends to decrease the headgroup area. The optimal area depends upon these interactions. Depending upon the type of surfactant, and the solution properties, the micelles can have various shapes [59].

Surfactants adsorbed on the film surfaces stabilize the film by two mechanisms: they slow down the hydrodynamic drainage process, and also exert repulsive interfacial forces. Therefore, adsorption of surfactants at liquid–liquid interfaces plays an important role in coalescence. A lot of work has been reported in the literature on surfactant adsorption under equilibrium and dynamic conditions. Most of these works are on air–water interfaces. Prosser and Franses [60], and Chang and Franses [61] have reviewed these works. The adsorption of surfactants at water–hydrocarbon interfaces follows a similar pattern [21,23,62]. Very low interfacial tension (< 1 mN/m) at the water–hydrocarbon interface can be produced by the addition of inorganic salts in ionic surfactant systems, which facilitates the formation of emulsions with fine drops. Many industrial formulations contain surfactants in large concentrations (much above their CMCs) in order to stabilize the drops and prevent them from coalescence. Sometimes a mixture of surfactants is used to enhance the stability of the emulsions. Additives such as salts and alcohols are frequently added to reduce the surfactant requirement and prevent the coalescence of drops.

For a symmetric univalent surfactant (*e.g.*, sodium dodecyl sulfate and cetyltrimethylammonium bromide), which is fully dissociated in the aqueous solution, the Gibbs adsorption equation gives the surface excess concentration $\left(\Gamma\right)$ as

$$\Gamma = -\frac{1}{2RT}\frac{d\gamma}{d\ln c} \tag{58}$$

where γ is the interfacial tension and c is the concentration of surfactant in the bulk. According to the Langmuir isotherm, the surface excess concentration is related to the concentration of surfactant in the bulk solution as

$$\Gamma = \frac{\Gamma_{\infty}K_{L}c}{1+K_{L}c} \tag{59}$$

where Γ_{∞} is the adsorption capacity. Its value depends upon the minimum surface area per adsorbed molecule. The equilibrium constant, K_{L}, is the ratio of the rate constants for adsorption and desorption. From Eqs. (58) and (59), we obtain

$$d\gamma = -2RT\Gamma_{\infty}\left(\frac{K_{L}c}{1+K_{L}c}\right)d\ln c \tag{60}$$

If the interfacial tension for the pure (*i.e.*, surfactant-free) liquids is represented by γ_{0}, integration of Eq. (60) gives

$$\gamma = \gamma_{0} - 2RT\Gamma_{\infty}\ln\left(1+K_{L}c\right) \tag{61}$$

Eq. (61) is known as Szyskowski equation. The difference in interfacial tension, $\gamma_{0}-\gamma$, is known as *surface pressure*. Eq. (61) gives a simple surface EOS which can be used to describe the variation of interfacial tension with the concentration of surfactant in the solution. For a nonionic surfactant, Gibbs adsorption equation defines the surface excess concentration as

$$\Gamma = -\frac{1}{RT}\frac{d\gamma}{d\ln c} \tag{62}$$

The Szyskowski equation in this case is given by [63]

$$\gamma = \gamma_0 - RT\Gamma_\infty \ln(1 + K_L c) \tag{63}$$

The adsorbed monolayer at the water–hydrocarbon interface can be viewed as a two-dimensional lattice in which the total number of sites represents the maximum number of surfactant molecules that can fit on the interface as per the geometry. All sites are supposed to have equal area. In practice, however, all of these sites are unlikely to be occupied by the surfactant molecules, which will be filled up by the molecules of the pure fluids. From the value of Γ_∞, minimum surface area per adsorbed molecule (A_{min}) can be obtained as

$$A_{min} = \frac{1}{\Gamma_\infty N_A} \tag{64}$$

where N_A is Avogadro's number. The value of A_{min} depends upon the properties of the surfactant and the interface. The fit of the surface EOS to the interfacial tension data are illustrated in Fig. **16**.

Interfacial rheological properties can be calculated from the surface excess concentration. For example, the interfacial Gibbs elasticity is given by

$$E_G = -\frac{d\gamma}{d\ln\Gamma} \tag{65}$$

The Gibbs elasticity has been used as an important parameter in some film drainage models of coalescence [37]. The quantity Γ is a very important parameter in the stochastic theory of coalescence, which is discussed in a later section in this chapter.

The surface EOS for ionic surfactants is usually much more complicated than that given by Eq. (61). This is mainly due to the electrostatic interactions. Several EOSs have been developed for the ionic surfactants [64,65] which have incorporated the electrostatic effects.

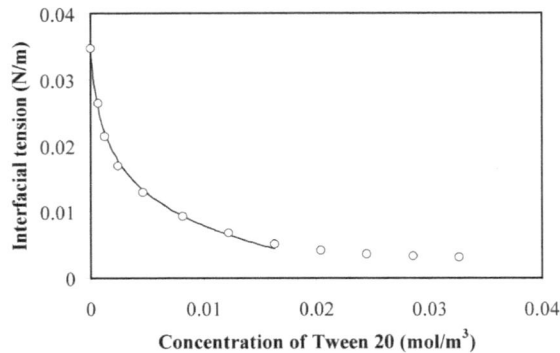

Figure 16: Variation of interfacial tension (γ) with the concentration of the nonionic surfactant Tween 20 in the water–toluene system. The line indicates the fit of Eq. (63). $\Gamma_\infty = 2.96 \times 10^{-6}$ mol/m^2, $K_L = 3800$ m^3/mol and $A_{min} = 56 \times 10^{-20}$ m^2.

Film Drainage Models Considering the Presence of Surfactant at the Interfaces

The film drainage process becomes more complicated in the presence of surfactants because of the effects associated with electrostatic double layer, solvation and steric forces. The surfactant monolayer at the interface significantly alters the rheological properties of the interface as well. The mobility of the interfaces is therefore altered. The interfacial tension gradient can also have a significant effect on film drainage. Several film drainage models have been developed incorporating the effects of the surfactant molecules on the hydrodynamic drainage [29,36,37,66–77].

The film drainage models presented by Slattery [36] are based on several assumptions. The two interfaces bounding the draining liquid film were assumed to be axisymmetric. The deformation of the flat interface was assumed to be small. The drop was assumed to be small such that the Bond number $\left(\equiv \Delta\rho g a^2 / \gamma\right)$ was much smaller than unity. Thus, the deformation of the drop was assumed to be negligible. The Reynolds lubrication approximation was applied. It was assumed that the surfactant molecules adsorbed at the water–hydrocarbon interfaces in a manner such that the resulting interfacial tension gradients were sufficiently large, and the tangential components of velocity were zero. The effects of mass transfer were neglected. The film liquid was assumed to be Newtonian. All inertial effects were neglected. The effects of electrostatic double layer and other repulsive forces in the thin film (such as the hydration or polymeric steric forces) were neglected. However, the van der Waals attraction between the fluid surfaces was taken into account. Slattery [36] has given seven film drainage models which are presented in Table **3**.

Table 3: Film drainage models for calculating coalescence time of drops [36]

Equation for coalescence time	Based on the model of
$t_{c,1} = 1.07 \dfrac{\mu a^{3.4} \left(\Delta\rho g\right)^{0.6}}{\gamma^{1.2} B^{0.4}}$	Chen *et al.* [68]
$t_{c,2} = 0.705 \dfrac{\mu a^{3.4} \left(\Delta\rho g\right)^{0.6}}{\gamma^{1.2} B^{0.4}}$	
$t_{c,3} = 1.046 \dfrac{\mu a^{4.5} \Delta\rho g}{\gamma^{1.5} B^{0.5}}$	Mackay and Mason [78]
$t_{c,4} = 0.37 \dfrac{\mu a^{4.5} \Delta\rho g}{\gamma^{1.5} B^{0.5}}$	
$t_{c,5} = 5.202 \dfrac{\mu a^{1.75}}{\gamma^{0.75} B^{0.25}}$	Hodgson and Woods [79]
$t_{c,6} = 0.79 \dfrac{\mu a^{4.06} \left(\Delta\rho g\right)^{0.84}}{\gamma^{1.38} B^{0.46}}$	Slattery [36]
$t_{c,7} = 0.44 \dfrac{\mu a^{4.06} \left(\Delta\rho g\right)^{0.84}}{\gamma^{1.38} B^{0.46}}$	

The value of the modified Hamaker constant, B, is 1×10^{-28} J m [36]. The equations given in Table **3** were obtained by a stability analysis. The references of the basic models on which these stability analyses were performed are presented in the second column of Table **3**. An interesting aspect of these equations is that they predict widely different values of coalescence time even for the same system. This is illustrated in the following example.

Example 5. The coalescence time of a 2.1 mm diameter cyclohexane drop at flat water–cyclohexane interface in presence of 0.02 mol/m³ CTAB and 0.706 mol/m³ SDS is 30.3 s. The interfacial tension is 15 mN/m, density of cyclohexane is 779 kg/m^3, density of water is 1000 kg/m^3, and the viscosity of water is 1×10^{-3} Pa s. Calculate the coalescence times predicted by the film drainage models given in Table **3**.

Solution: The density difference between the two liquids is, $\Delta\rho = 221$ kg/m^3, $a = 1.05\times10^{-3}$ m, $g = 9.8$ m/s^2, $\gamma = 0.015$ N/m, $\mu = 1\times10^{-3}$ Pa s, and $B = 1\times10^{-28}$ J m. Substituting these values in the equations given in Table **3**, the following results are obtained.

$$t_{c,1} = 1.07\frac{\mu a^{3.4}(\Delta\rho g)^{0.6}}{\gamma^{1.2}B^{0.4}} = 1.07\times\frac{(1\times10^{-3})(1.05\times10^{-3})^{3.4}(221\times9.8)^{0.6}}{(0.015)^{1.2}(1\times10^{-28})^{0.4}} = 196.2 \text{ s}$$

$$t_{c,2} = 0.705\frac{\mu a^{3.4}(\Delta\rho g)^{0.6}}{\gamma^{1.2}B^{0.4}} = 0.705\times\frac{(1\times10^{-3})(1.05\times10^{-3})^{3.4}(221\times9.8)^{0.6}}{(0.015)^{1.2}(1\times10^{-28})^{0.4}} = 129.3 \text{ s}$$

$$t_{c,3} = 1.046\frac{\mu a^{4.5}\Delta\rho g}{\gamma^{1.5}B^{0.5}} = 1.046\times\frac{(1\times10^{-3})(1.05\times10^{-3})^{4.5}(221)(9.8)}{(0.015)^{1.5}(1\times10^{-28})^{0.5}} = 4856.9 \text{ s}$$

$$t_{c,4} = 0.37\frac{\mu a^{4.5}\Delta\rho g}{\gamma^{1.5}B^{0.5}} = 0.37\times\frac{(1\times10^{-3})(1.05\times10^{-3})^{4.5}(221)(9.8)}{(0.015)^{1.5}(1\times10^{-28})^{0.5}} = 1718 \text{ s}$$

$$t_{c,5} = 5.202\frac{\mu a^{1.75}}{\gamma^{0.75}B^{0.25}} = 5.202\times\frac{(1\times10^{-3})(1.05\times10^{-3})^{1.75}}{(0.015)^{0.75}(1\times10^{-28})^{0.25}} = 7.4 \text{ s}$$

$$t_{c,6} = 0.79\frac{\mu a^{4.06}(\Delta\rho g)^{0.84}}{\gamma^{1.38}B^{0.46}} = 0.79\times\frac{(1\times10^{-3})(1.05\times10^{-3})^{4.06}(221\times9.8)^{0.84}}{(0.015)^{1.38}(1\times10^{-28})^{0.46}} = 1005.9 \text{ s}$$

$$t_{c,7} = 0.44\frac{\mu a^{4.06}(\Delta\rho g)^{0.84}}{\gamma^{1.38}B^{0.46}} = 0.44\times\frac{(1\times10^{-3})(1.05\times10^{-3})^{4.06}(221\times9.8)^{0.84}}{(0.015)^{1.38}(1\times10^{-28})^{0.46}} = 560.3 \text{ s}$$

From these results, it can be easily observed that these film drainage models predict widely different values from one another. Also, they differ significantly from the experimental value of 30.3 s.

The model for coalescence time proposed by Jeelani and Hartland [29] incorporates the effect of circulation within the adjacent phases of the film. It also considers the interfacial tension gradient created by the nonuniform distribution of the surfactant molecules at the interfaces. According to their model, the coalescence time can be calculated from the following equation:

$$t_c = \left(\frac{9\mu R_f^4}{16a^3\Delta\rho g h_c^2}\right)\left\{1+\left(\frac{3\mu a}{\mu_d h_i}\right)\left[1+\left(\frac{3R_f^3}{8a^3\Delta\rho g h_i}\right)\left(\frac{\partial\gamma}{\partial r}\right)_{fi}\right]\right\}^{-1} \qquad (66)$$

where R_f is the radius of the film, μ_d is the viscosity of the liquid constituting the drop, and $(\partial\gamma/\partial r)_{fi}$ is the initial value of the interfacial tension gradient at the periphery of the film. It is difficult to predict the value of this parameter. Apparently, this term is zero in absence of surfactant. Jeelani and Hartland [29] calculated the values of $(\partial\gamma/\partial r)_{\text{fi}}$ from Eq. (66) using the experimental values of coalescence time. However, these values differed significantly from one system to another.

As we have noted before, the surfactants alter the rheological properties of the water–hydrocarbon interfaces. Some works on film drainage theory have suggested that the surface shear and dilatational viscosities play an important role in the coalescence of drops [37,71,80–82]. Aderangi and Wasan [81] have observed that coalescence time of drops correlated more favorably with the interfacial shear viscosity than interfacial tension. They defined a dimensionless interfacial viscosity number, $\mu_s/(\mu R_f)$, where μ_s is the interfacial shear viscosity. They observed that the coalescence time increased linearly with the interfacial viscosity number. Similar observations have been reported by Wasan *et al.* [82].

Curvature of the interface plays an important role in the film drainage [83]. The film formed between two drops in an emulsion is actually not flat. The thickness of the film and curvature of the interfaces change with time. When two drops approach each other, a dimple forms as shown in Fig. **5**. Due to the trapped liquid in the central region of the film, the film is thicker in the central part as compared to the peripheral region. The interfaces deform during the drainage of the film and a narrow region is developed near the rim of the film, which is known as the *barrier ring* (because it causes constriction to the outward flow of the film liquid, which slows down the drainage process). The shape of the film changes with time as the film thins. The presence of surfactants is believed to influence the film drainage process. The shape of the film has been studied by interferometry and video microscopy [79,84]. The various stages of thinning of a film formed between a drop and an initially-flat water–organic interface are shown in Fig. **17**.

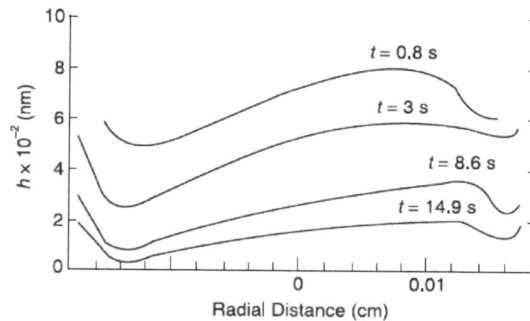

Figure 17: Asymmetric drainage of a film for 3 mm diameter anisole drop at water–anisole interface in presence of 2×10^{-5} kg/m^3 sodium dodecyl sulfate and 10 mol/m^3 KCl [79] (adapted by permission from Elsevier Ltd., © 1969).

The drainage can be symmetric as well as asymmetric. The dimple sometimes flattens with time, and finally a near-flat thin film can be obtained. The theoretical analyses of the thinning process often assume axisymmetric interfaces and flat film. Asymmetric drainage in foam films has been analysed by Joye *et al.* [74].

The interfacial forces play an important role in coalescence in the presence of surfactants [85]. There are some reports in the literature which have suggested that the oil–water interface becomes charged depending upon the pH of the aqueous phase even in the absence of any surfactant [86]. The disjoining pressures due to van der Waals, electrostatic double layer and short-range forces have been included in the film drainage theory developed by Joye *et al.* [73].

STOCHASTIC THEORY OF COALESCENCE

In an experiment of the measurement of coalescence time of drops, a wide distribution of coalescence time is always observed. The stochasticity in the coalescence process is well known for a long time [40,87–91]. A similar phenomenon is observed in the coalescence of bubbles as well [92]. Gillespie and Rideal [88] attempted to explain the stochasticity by employing the theory of surface waves induced by temperature fluctuations [93], and other forms of shocks and vibrations. Lang and Wilke [90] suggested that various disturbances initiate Taylor instabilities, which cause rupture of the film leading to coalescence. They also suggested that the disturbances are randomly distributed in time, frequency and intensity. These factors lead to the distribution of the coalescence time.

The coalescence time distributions are usually very broad. In many systems, some drops have extremely short coalescence times (< 0.1 s) whereas some drops have long coalescence times (> 100 s). This observation indicates that the time required for the film thinning process could be very short in these systems. Since the approach of the film drainage model is deterministic, it cannot account for the stochastic nature of the drop coalescence time. Most workers, however, have ignored this distribution and, instead, used the mean value of the coalescence time distribution for comparison with the model predictions [29].

It was shown by Ghosh and Juvekar [22] that when a drop strikes at a flat liquid–liquid interface, both the drop and the interface undergo an oscillatory (up-and-down) motion before attaining the state of rest. Some of drops coalesce during this period, especially if little or no surfactant is present in the system to stabilize the film. This oscillatory behavior can be explained only if we assume that a repulsive force arises between the drop and the flat interface when the drop is in the close proximity of the interface. This repulsive force can arise only at the rim of the film (*i.e.*, the 'barrier ring') where the distance between the drop and the interface is shortest. This force must be capable of supporting the weight of the drop, and thereby instantaneously arrest the relative motion between the drop and the interface. It was proposed by Ghosh and Juvekar [22] that this repulsive interfacial force, which arrests the relative motion between the drop and the interface be generated by the surfactants adsorbed at the drop surface and at the flat interface.

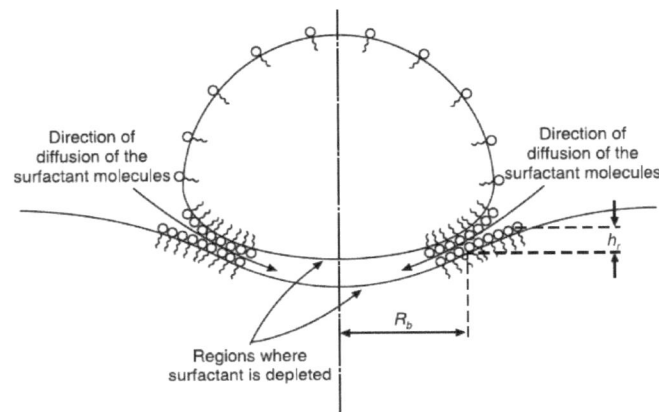

Figure 18: Illustration of the interfacial diffusion of surfactant molecules when a drop rests at a liquid–liquid interface.

During the hydrodynamic drainage of the film, the drag force generated beneath the drop displaces the surfactant molecules in the outward direction. This causes a high concentration of these molecules to build up at the barrier ring, as shown in Fig. **18**. A fraction of the total surfactant molecules present initially in the film may desorb into the bulk phase(s) during the descent of the drop. However, when the drop strikes the interface, the remaining molecules cannot escape, and they are concentrated at the barrier ring, as shown in Fig. **18**. If the repulsive force between the molecules adsorbed on the upper and the lower surfaces of the thin film at the barrier ring is strong enough to balance the weight of drop, the drop stays at the interface, otherwise the drop coalesces at the impact. The repulsive force can be either electrostatic double layer force, hydration force, steric force, or a combination of these, depending on the type of the surfactant present at the barrier ring. Depending upon the pH and electrolyte concentration in the solution, the repulsive force becomes significant at a separation of 100 nm or less. This repulsive force varies inversely with the distance between the drop and the interface. This force, therefore, also determines the separation, h_r, between the drop and the interface at the barrier ring. It allows the drop to actually rest at the interface. The interfacial repulsive force must decay with time, otherwise the drop will never coalesce. After a certain interval of time, it becomes too weak to support the weight of the drop due to this decay, and the drop coalesces. This period of rest is counted as the *coalescence time*. The motion of the drop is rapid before it strikes the flat interface, and therefore this period does not constitute the coalescence time.

The decay of the repulsive force is attributed to the diffusion of the surfactant molecules away from the barrier ring, towards the center of the film. Although the repulsive force is contributed by all the molecules in the two surfaces of the film, only the contribution from the molecules at the barrier ring is significant. This is because the film thickness increases from the rim of the film to its center, and the repulsive force decreases rapidly with increase in the film thickness. Thus the surfactant molecules that diffuse away from the barrier ring would not be able to generate the same repulsive force they would have generated had they been at the barrier ring.

One of the important parameters that determine the coalescence time, is the Gibbs surface excess concentration of the surfactant. This excess amount is displaced towards the barrier ring by the descending drop. If the magnitude of surface excess concentration is large, the concentration of surfactant at the barrier ring will also be large, and this will cause a strong repulsive force. Consequently, the coalescence time will be long. The fluctuation in the coalescence time is caused by the variation in the surface excess concentration at the location where the drop strikes the interface. Ghosh and Juvekar [22] have provided several experimental evidences to support the hypothesis that the fluctuation in the coalescence time is caused by a random variation in Γ_i.

When the drop approaches the flat liquid–liquid interface, the surfactant molecules on the drop surface and those on the flat interface are displaced towards the barrier ring. These molecules concentrate at the barrier ring, and form two concentrated ring sources, one on each face of the barrier ring. Mutual repulsion between these sources is strong enough to bear the weight of the drop. If R_b is the radius of the barrier ring, and if Γ_i is the surface excess concentration of the surfactant before displacement, then the amount of surfactant displaced towards the barrier ring is $\pi R_b^2 \Gamma_i$. A part of the displaced surfactant may escape to the surrounding liquid in which it is soluble. Let ξ be the fraction that remains at the barrier ring. Then the quantity of the surfactant retained on the barrier ring at the beginning of the coalescence process will be, $\pi R_b^2 \Gamma_i \xi$. The effect of non-uniformity in the distribution of surfactant is incorporated in the distribution of Γ_i. It is assumed that Γ_i is distributed in a Gaussian manner, so that its probability density is described by the following equation:

$$f(\Gamma_i) = \frac{1}{\sigma_\Gamma \sqrt{2\pi}} \exp\left[-\frac{1}{2}\left(\frac{\Gamma_i - \bar{\Gamma}_i}{\sigma_\Gamma}\right)^2 \right] \qquad (67)$$

where $\bar{\Gamma}_i$ is the mean value of Γ_i and σ_Γ is the standard deviation.

During the period the drop rests on the interface, the surfactant molecules from the barrier ring diffuse to the regions of low concentration, *i.e.*, towards the center of the film, as illustrated in Fig. **18**. The mode of diffusion is assumed to be mainly the surface diffusion, due to the fact that the repulsion between the opposite faces of the barrier ring does not allow the surfactant molecules to enter the bulk of the film. For simplicity, we neglect the curvature of the surface of the film, and view it as a plane circular disc of radius R_b. The diffusion of the surfactant molecules will occur from the rim of the disc to the center. We describe this diffusion by the Fick's law:

$$\frac{\partial \Gamma}{\partial t} = \frac{D_\Gamma}{r}\frac{\partial}{\partial r}\left(r\frac{\partial \Gamma}{\partial r}\right) \qquad (68)$$

where D_Γ is the surface diffusion coefficient of the surfactant. Eq. (68) is valid for both the drop and the flat interface. Therefore, we can use a single equation to describe the diffusion process except for the fact that the probability desnity $f(\Gamma_i)$ includes the variations of Γ_i at both the interfaces. The combined distribution is a Gaussian distribution because the individual distributions are Gaussian.

It should be noted here that the diffusion of the surfactant molecules either in the drop phase or in the mother phase was neglected in this model. This is true only if the surfactant is insoluble in these phases.

Another important assumption is that the Marangoni flow contribution arising out of the interfacial tension gradient is negligible. This is for two reasons. Firstly, the shear stress that compresses the monolayer during the approach of the drop is so large that the compressed monolayer is in the liquid condensed state. Very slow evaporation of the condensed layer with time will produce a small interfacial tension gradient. Secondly, the nature of the convective flow arising out of the Marangoni stress depends upon the geometry of the region under consideration. In the present case, the region is a thin film which is almost closed at the barrier ring. Further, the interfacial tension gradient is inwardly directed. This is a dead-end system and no net convective flow can occur in this system. The Marangoni stress will therefore generate a local circulating flow, similar to the roll cells described by Sternling and Scriven [94], with dimensions equal to half the thickness of the film. This circulating flow will enhance desorption of the surfactant, and this will increase the bulk diffusion. However, since the scale of the circulation is very small as compared to the radius of the barrier ring, no convective transport of the adsorbate will occur in the r-direction, and hence the form of the diffusion equation is not altered by the Marangoni stress. However, it can alter the effective surface diffusion coefficient. The diffusion coefficient can be a function of both time and location. From this viewpoint, D_Γ may be considered as the space and time-averaged effective diffusion coefficient.

The following intial condition was used for Eq. (68) :

$$\Gamma = 0 \quad \text{at} \quad t = 0 \tag{69}$$

This condition implies that the entire surface excess is depleted from the film during the pre-rest period. This is a reasonable assumption in view of the high intensity of the interfacial shear stress experienced by the film during the approach of the drop towards the interface.

The radial symmetry yields the following boundary condition at the center of the disc.

$$\frac{\partial \Gamma}{\partial r} = 0 \quad \text{at} \quad r = 0 \tag{70}$$

The balance of the surfactant molecules over the disc yields

$$\int_0^{R_b} \Gamma 2\pi r \mathrm{d}r = \pi R_b^2 \Gamma_i \xi \tag{71}$$

Eq. (68) can be solved with the help of the auxiliary conditions given by Eqs. (69)–(71) to yield the following solution:

$$\Gamma(r) = \Gamma_i \xi \left[1 + \sum_{i=1}^{\infty} \exp\left(-\frac{\lambda_i^2 D_\Gamma t}{r^2} \right) \right] \tag{72}$$

where λ_i are the zeros of J_1, the Bessel function of the first kind and of order one.

The concentration of the surfactant at the barrier ring can be obtained by substituting $r = R_b$, into Eq. (72). Thus,

$$\Gamma(R_b) = \Gamma_i \xi \left[1 + \sum_{i=1}^{\infty} \exp\left(-\frac{\lambda_i^2 D_\Gamma t}{R_b^2} \right) \right] \tag{73}$$

If Γ_m represents the minimum value of the adsorbate concentration at the barrier ring required to support the weight of the drop, then we can write

At $t = t_c$, $\Gamma(R_b) = \Gamma_m$ (74)

where t_c is the coalescence time of the drop. Applying the condition given by Eq. (74) in Eq. (73), we obtain,

$$\Gamma_m = \Gamma_i \xi \left[1 + \sum_{i=1}^{\infty} \exp\left(-\lambda_i^2 \tau\right) \right]$$ (75)

where τ is the dimensionless coalescence time which is defined as, $\tau = t_c / \bar{t}$, where $\bar{t}$ is the characteristic diffusion time. It is defined as, $\bar{t} = R_b^2 / D_\Gamma$. It is approximately equal to the time required by the surfactant molecules to diffuse from the barrier ring to the center of the film. There is hardly any experimental data reported in the literature on the diffusion coefficient, D_Γ, at the liquid–liquid interface. Agrawal and Neuman [95] have presented the values of surface diffusion coefficient at the air–water interface at different states (viz. gaseous, liquid-expanded, and liquid-condensed) of the monolayer. A typical value of D_Γ is 1×10^{-10} m^2/s. If the radius of the barrier ring is 1 mm, the value of $\bar{t}$ would be 10^4 s.

We can deduce from Eq. (75) that τ increases with a decrease in Γ_m. This is understandable, since a longer time is then required for the concentration of the surfactant at the barrier ring to decay to a lower value of Γ_m. To achieve a finite coalescence time, $\Gamma_i \xi$ should be less than Γ_m. As $\Gamma_i \xi \to \Gamma_m$, $\tau \to \infty$. This means that the drops for which $\Gamma_i \xi \geq \Gamma_m$, will never coalesce. The reason is, if the drop rests at the interface for infinite time, then the quantity of the surfactant, which was initially concentrated at the barrier ring, will spread uniformly over the entire film, and the concentration of the surfactant will be $\Gamma_i \xi$ everywhere, including the barrier ring. Therefore, for all finite times, the concentration at the barrier ring $\Gamma(R_b)$ will be greater than $\Gamma_i \xi$. Thus, $\Gamma_i \xi \geq \Gamma_m$ implies that $\Gamma(R_b)$ will be greater than Γ_m at all times. Hence, those drops for which this condition is satisfied will never coalesce. This situation can arise in the case of very small drops and also when a water-soluble polymeric surfactant is present. For finite coalescence time, the value of P_Γ must be greater than unity.

The magnitude of Γ_m can be estimated as follows. If w_b is the width of the barrier ring, and if f_r is the repulsive force generated by one mole of the surfactant at the barrier ring, the total repulsive force generated at the barrier ring at the point of coalescence is

$$F_r = 2\pi R_b w_b \Gamma_m f_r$$ (76)

The radius of the barrier ring, R_b, is related to the radius of the drop, a, by Eq. (34), which is applicable when both the drop and the interface are deformable. At the threshold of coalescence, the repulsive force must equal the weight of the drop, viz. $4\pi a^3 \Delta \rho g / 3$. Therefore, Γ_m is given by

$$\Gamma_m = \frac{a}{w_b f_r} \sqrt{\frac{\Delta \rho g \gamma}{3}}$$ (77)

The width of the barrier ring extends over that region for which the contribution to the total repulsive force is significant. Since the repulsive force is very sensitive to film thickness and the film thickness diverges from the rim to the center of the film (see Fig. **5**), the barrier ring spans a very small peripheral part of the film. The divergence of film thickness, and hence the width of the barrier ring is decided by the amount of liquid trapped in the film. The width of the barrier ring would be small if a large amount of liquid is trapped in the film. Drops, which deform as they approach the interface, will trap more liquid and therefore result in a smaller width. Further, as the drop size increases, more liquid will be trapped, and the width of the barrier ring will decrease. A decrease in interfacial tension increases the deformation, and hence should cause a reduction in the width of the barrier ring.

Since Γ_i varies randomly from drop to drop, τ also varies randomly. The probability density of τ can be related to that of Γ_i, using the following equation:

$$f(\tau) = f(\Gamma_i)\frac{d\Gamma_i}{d\tau}$$

(78)

The quantity $d\Gamma_i/d\tau$ can be obtained by differentiation from Eq. (75). Using $f(\Gamma_i)$ from Eq. (67), the following equation for $f(\tau)$ is obtained:

$$f(\tau) = \frac{P_\Gamma}{S_\Gamma \sqrt{2\pi}} \frac{\sum_{i=1}^{\infty} \lambda_i^2 \exp(-\lambda_i^2 \tau)}{\left[1 + \sum_{i=1}^{\infty} \exp(-\lambda_i^2 \tau)\right]^2} \exp\left[-\frac{1}{2S_\Gamma^2} \left\{ \frac{P_\Gamma}{\left[1 + \sum_{i=1}^{\infty} \exp(-\lambda_i^2 \tau)\right]} - 1 \right\}^2 \right]$$

(79)

where

$$P_\Gamma = \frac{\Gamma_m}{\overline{\Gamma}_i \xi} = \frac{a}{w_b f_r \overline{\Gamma}_i \xi}\left(\frac{\Delta\rho g\gamma}{3}\right)^{1/2}$$

(80)

and

$$S_\Gamma = \frac{\sigma_\Gamma}{\overline{\Gamma}_i}$$

(81)

The parameter P_Γ represents the *dimensionless coalescence threshold*. For finite coalescence time, P_Γ should be greater than unity. If $P_\Gamma \leq 1$, at least some drops (for which $\overline{\Gamma}_i \xi \geq \Gamma_m$) will never coalesce. The parameter S_Γ represents the normalized standard deviation of the surface excess concentration, Γ_i.

For comparison with the experimental data on coalescence time, it is desirable to obtain the cumulative probability distribution, $F(\tau)$, of the coalescence time, which is defined as

$$F(\tau) = \int_0^\tau f(t)\,dt$$

(82)

Using Eq. (79), we obtain $F(\tau)$ as

$$F(\tau) = \frac{1}{2}\left[\text{erf}\left\{ \frac{1}{S_\Gamma \sqrt{2}}\left(\frac{P_\Gamma}{1 + \sum_{i=1}^{\infty} \exp(-\lambda_i^2 \tau)} - 1 \right)\right\} + \text{erf}\left(\frac{1}{S_\Gamma \sqrt{2}} \right)\right]$$

(83)

where 'erf' represents the error function.

The stochastic theory of coalescence described in this section was developed for the coalescence of a drop at a flat liquid–liquid interface. However, it can be applicable for coalescence between two drops as well. A few works have reported the use of this model with minor modifications in binary coalescence of drops [23,62,96]. The procedure of fitting Eq. (83) to the experimental cumulative distribution of coalescence time is explained in Example 6.

Example 6. Coalescence of 2 mm diameter water drops was studied at a flat oil–water interface. The coalescence times for 50 observations are as follows (in seconds): 3.7, 6.0, 4.4, 16.9, 11.0, 10.5, 12.5, 9.8, 8.5, 13.8, 8.0, 12.8, 16.9, 3.6, 10.4, 11.2, 11.6, 3.7, 13.8, 14.6, 15.0, 14.4, 8.0, 13.0, 18.1, 18.1, 18.1, 23.2, 6.8, 16.6, 9.0, 5.6, 11.4, 2.5, 8.5, 10.4, 18.7, 12.8, 19.7, 12.4, 2.5, 20.1, 10.2, 11.0, 7.2, 8.0, 17.0, 3.9, 20.1 and 26.4. The density of the oil is 800 kg/m^3 and the interfacial tension is 25 mN/m. Obtain the parameters of the stochastic model using these data. Show your results graphically.

Solution: The mean and standard deviation of the coalescence time distribution are 11.85 s and 5.61 s, respectively. The radius of the barrier ring, R_b, is given by

$$R_b = 2a^2 \left(\frac{\Delta \rho g}{3\gamma} \right)^{1/2} = 2 \times \left(1 \times 10^{-3}\right)^2 \times \left[\frac{(1000-800) \times 9.8}{3 \times 0.025} \right]^{1/2} = 3.235 \times 10^{-4} \text{ m}$$

The characteristic diffusion time, $\overline{t}$, is given by,

$$\overline{t} = \frac{R_b^2}{D_\Gamma}$$

Using $D_\Gamma = 1 \times 10^{-10} \text{ m}^2/\text{s}$, we get

$$\overline{t} = \frac{R_b^2}{D_\Gamma} = \frac{\left(3.235 \times 10^{-4}\right)^2}{1 \times 10^{-10}} = 1046.5 \text{ s}$$

The cumulative distribution of coalescence time is presented in Fig. **19**.

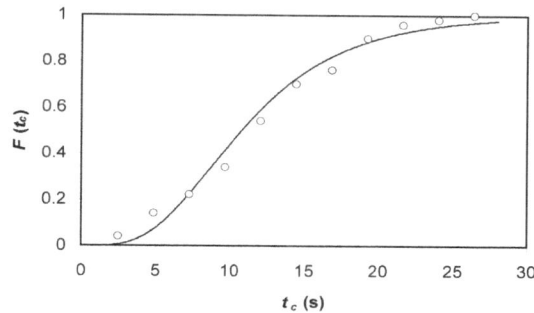

Figure 19: Cumulative distribution of coalescence time for the data given in Example 6 (represented by circles), and the fit of the stochastic model.

The fit of the stochastic model to the experimental data is shown by the line. The optimum values of the two parameters of the model are

$$P_\Gamma = 4.05 \text{ and } S_\Gamma = 0.17.$$

The stochastic model has been used to explain the experimental data on coalescence time in a wide range of surfactant systems [21–23,62,96]. The variation of the model parameter P_Γ with the properties of the systems can be explained from Eq. [80]. Some of the results are presented in this section.

Increase in the surfactant concentration almost always increases the coalescence time (see Fig. **20**). In some systems, drops coalesce rapidly in absence of surfactant, or when the surfactant concentration is low. As the concentration of surfactant is increased, the adsorbed amount of the surfactant at the liquid–liquid interface increases. This leads to an increase in the value of $\overline{\Gamma}_i$. However, the interfacial tension, γ, decreases with the increase in surfactant concentration. From Eq. [80], we observe that $P_\Gamma \propto \sqrt{\gamma}/\overline{\Gamma}_i$. Therefore, P_Γ decreases with increase in surfactant concentration.

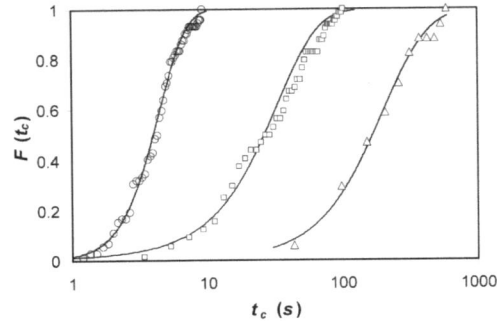

Figure 20: Effect of concentration of sodium dodecyl sulfate (SDS) on the coalescence time of CCl_4 drops at water–CCl_4 interface [22]. The circles represent the coalescence time distribution for 0.18 mol/m^3 SDS concentration, and the line represents the fit of the stochastic model ($a = 1.3$ mm, $\overline{t} = 6889$ s, $P_\Gamma = 12$ and $S_\Gamma = 0.23$). The squares represent the coalescence time distribution for 1 mol/m^3 SDS concentration, and the line represents the fit of the stochastic model ($a = 1.3$ mm, $\overline{t} = 7056$ s, $P_\Gamma = 4.9$ and $S_\Gamma = 0.37$). The triangles represent the coalescence time distribution for 3.5 mol/m^3 SDS concentration, and the line represents the fit of the stochastic model ($a = 1.1$ mm, $\overline{t} = 7225$ s, $P_\Gamma = 2.2$ and $S_\Gamma = 0.33$) (adapted by permission from Elsevier Ltd., © 2002).

The effects of the interfacial forces on coalescence are manifested when a salt is added to the ionic surfactant systems. When the salt is added, two opposing effects come into play. The electrostatic repulsion between the charged headgroups of the surfactant molecules is reduced in the presence of salt. This favors adsorption of surfactant molecules at the interface, and an increase in $\overline{\Gamma}_i$ results. On the other hand, the interfacial tension γ reduces with the increase in salt concentration. The electrostatic double layer repulsion between the drop and the interface (or between two drops) is reduced in presence of salt. This reduces the value of f_r. The relative magnitudes of these effects determine whether coalescence time should increase or decrease with the addition of salt. The variation of P_Γ depends on the variation in γ, $\overline{\Gamma}_i$ and f_r with the addition of salt. An example of the effect of a 1:1 salt on coalescence time in presence of an ionic surfactant is shown in Fig. **21**.

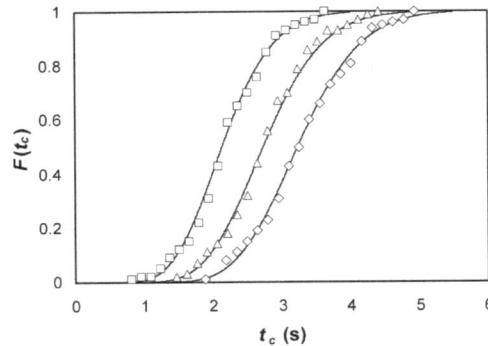

Figure 21: Coalescence time distributions of toluene drops at water–toluene interface at different concentrations of NaBr for the cationic surfactant cetyltrimethylammonium bromide (concentration = 0.003 mol/m^3) [21]. Drop radius = 1.5 mm. The squares represent the coalescence time distribution in absence of NaBr, and the line represents the fit of the stochastic model ($\overline{t} = 2701.5$ s, $P_\Gamma = 11.3$ and $S_\Gamma = 0.12$). The triangles represent the coalescence time distribution in 1 mol/m^3 NaBr, and the line represents the fit of the stochastic model ($\overline{t} = 3323.4$ s, $P_\Gamma = 11.1$ and $S_\Gamma = 0.11$). The rhombi represent the coalescence time distribution in 10 mol/m^3 NaBr, and the line represents the fit of the stochastic model ($\overline{t} = 4865.7$ s, $P_\Gamma = 12.2$ and $S_\Gamma = 0.1$) (adapted by permission from Elsevier Ltd., © 2009).

The salts containing divalent ions are more effective in reducing the electrostatic double layer repulsion than the salts containing monovalent ions. The ionic strength of the solution, $\sum \tilde{z}_i^2 c_i^\infty / 2$, is larger in presence of a salt having higher valence. Sometimes, the binding of counterions reduces the electrostatic repulsion significantly. In the system shown in Fig. **21**, the coalescence time increased with the increase in concentration of the salt. However, the coalescence time did not reduce at the higher salt concentrations. However, in the presence of K_2SO_4, coalescence time increased at 1 mol/m^3 concentration, and then decreased at 10 mol/m^3 concentration (see Fig. **22**), which clearly demonstrates the interplay of the two effects discussed before.

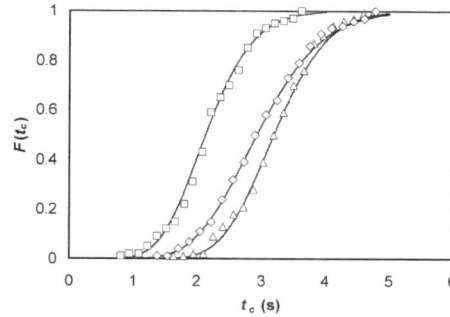

Figure 22: Coalescence time distributions of toluene drops at water–toluene interface at different concentrations of K_2SO_4 for the cationic surfactant cetyltrimethylammonium bromide (concentration = 0.003 mol/m^3) [21]. Drop radius = 1.5 mm. The squares represent the coalescence time distribution in absence of K_2SO_4, and the line represents the fit of the stochastic model ($\bar{t} = 2701.5$ s, $P_\Gamma = 11.3$ and $S_\Gamma = 0.12$). The triangles represent the coalescence time distribution in 1 mol/m^3 K_2SO_4, and the line represents the fit of the stochastic model ($\bar{t} = 4134.7$ s, $P_\Gamma = 11.4$ and $S_\Gamma = 0.09$). The rhombi represent the coalescence time distribution in 10 mol/m^3 NaBr, and the line represents the fit of the stochastic model ($\bar{t} = 4659.8$ s, $P_\Gamma = 12.5$ and $S_\Gamma = 0.12$) (adapted by permission from Elsevier Ltd., © 2009).

COALESCENCE IN INDUSTRIAL EQUIPMENT

In industrial equipment, coalescence is time dependent. In the dispersions of two immiscible liquids, immediate coalescence seldom occurs when two drops collide. If the drop pair is exposed to turbulent pressure fluctuations, and the kinetic energy of the oscillations induced in the coalescing drop pair is larger than the energy of adhesion between them, the drops come out of contact before coalescence can take place.

Experiments with gravity settlers indicate that the time for a drop to grow due to coalescence can be estimated by the following equation [1]:

$$t = \frac{\pi}{6}\left[\frac{a^i - (a_o)^i}{\psi_s K_s}\right] \qquad (84)$$

where a_0 is the initial drop radius, a is the final drop radius, ψ_s is the volume fraction of the dispersed phase, K_s is an empirical parameter, and t is the time required to grow a droplet of radius a. The value of the index i is greater than 3. It depends on the probability that the drops will bounce apart before coalescence occurs. When the energy of oscillations is very low so that the drops hardly bounce, the value of i approaches 3.

The time required for growing drops of size much larger than the initial drops (*i.e.*, $a \gg a_0$) is given by

$$t \approx \frac{a^i}{2\psi_s K_s} \qquad (85)$$

Therefore, Eq. (85) predicts that a doubling of residence time increases the radius of drop grown in a gravity settler by ~19% (for $i = 4$). If $i > 4$, the rate of growth in drop size will be slower. For this reason, after an initial short coalescence period, enforcing additional retention time is not very effective. Coalescence occurs more rapidly in concentrated dispersions. The requirement of residence time increases with decrease in concentration of the dispersed phase. In the gunbarrels and treaters in petroleum refineries, oil is washed with water by entering the treating vessel below the oil–water interface. Flocculation and coalescence occurs most effectively at the interface zone between oil and water.

EXPERIMENTAL TECHNIQUES FOR STUDYING COALESCENCE OF DROPS

The experimental techniques for studying coalescence can be divided into three categories: coalescence of a drop at a flat liquid–liquid interface (*i.e.*, *drop–interface coalescence*), coalescence between two drops (*i.e.*, *drop–drop coalescence*), and coalescence of two drops in motion. Most of the experimental studies have used coalescence cells of various designs for studying the drop–interface coalescence [21,22,40,81,87–91,97,98]. In these cells, precise arrangements are made for forming the flat liquid–liquid interface, and releasing the drop such that it rests on the interface until coalescence. Small quantities of surface-active materials can affect the coalescence time significantly. Such materials are often present even in high-purity liquids. They are difficult to remove. Various techniques were employed in the above-mentioned works for cleaning the interface, controlling the temperature, and generating and releasing the drop with precise control of size. Hodgson and Lee [97] introduced a novel method for cleaning the interface. Using a probe wetted partially by oil and partially by the aqueous phase, the oil–water mixture at the interface was intermittently withdrawn by applying suction to the probe.

Chatterjee *et al.* [99] developed a technique for detecting drop–interface coalescence by using a piezotransducer. When coalescence occurs, the observed piezotransducer signal has two distinct parts. The first part is due to the interfacial instability, while the second part of the signal is due to a transient jet which is ejected into the homophase during coalescence.

A few works have reported experimental techniques for studying binary coalescence. Chen and Pu [100] have used an optical bench with a micro-video camera. Two burettes were mounted on the optical bench to measure the coalescence time of two drops having size in the millimeter range. A 100 microliter syringe was used for forming very small drops. The drops were contacted side-on. The experimental arrangement used by Chen and Pu [100] is shown in Fig. **23a**.

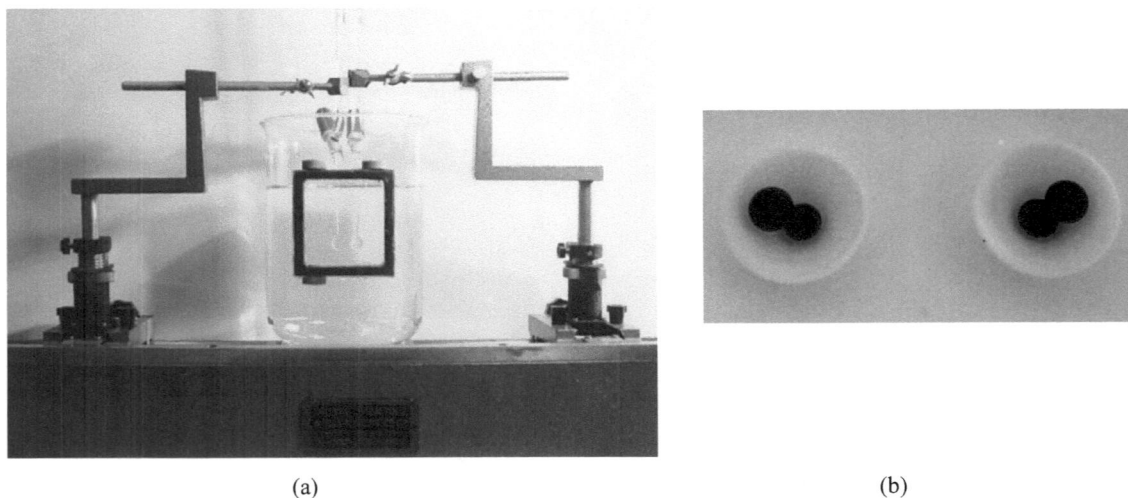

       (a)   (b)

Figure 23: Experimental arrangements for studying binary coalescence of drops: (a) side-on contact of drops [100] (reproduced by permission from Elsevier Ltd., © 2001), and (b) one water drop resting on another water drop inside a Teflon conical cell immersed in an organic liquid [96]. The drops were colorized for illustration (adapted by permission from Taylor and Francis, © 2007).

Mitra *et al.* [96] have developed a setup for studying binary coalescence of water drops in organic liquids (see Fig. **23b**). In this setup, the drops are contacted by gravity inside a small conical cell made of Teflon, which is immersed inside an immiscible organic liquid. The coalescence time is measured by video photography. The main advantage of this method is that the drops always remain in contact with each other until they coalesce, and long coalescence times can be measured in this setup.

Some works have studied coalescence of drops when they are in motion [101,102]. Aul and Olbricht [101] studied the coalescence of two freely-suspended, neutrally-buoyant drops of unequal size in low-Reynolds-number flow through a capillary. They observed that the interaction between a pair of drops resulted in coalescence only when the upstream drop was sufficiently large. They observed large variations in the coalescence times among runs with similar experimental conditions. Zhang *et al.* [102] studied hydrodynamic interactions between two drops during their gravity-induced relative motion. They measured the relative trajectories of two drops, their relative velocities, and the time taken by them to flow around each other. They observed that the hydrodynamic interactions significantly reduced the relative velocity of two nearby drops, and caused them to flow around each other with curved trajectories.

SUMMARY

This chapter presents the mechanism of coalescence of drops in a dispersion. Coalescence is the main process by which a liquid–liquid dispersion becomes unstable and the two liquids separate into distinct phases. The drops grow bigger through binary coalescence. When a flat interface is formed between the two liquids, the bigger drops coalesce at the flat interface. The theories of flocculation and coalescence are discussed in the beginning of this chapter.

The coalescence process can be divided into two steps. In the first step, the film between two drops (or between the drop and the flat liquid–liquid interface) thins by hydrodyamic drainage. When the film becomes thinner than ~100 nm, interfacial forces influence the subsequent drainage of the film. In the final stage of coalescence, the film becomes very thin and the van der Waals forces cause rupture of the film. The liquids in the drops merge to form a single drop. The major forces in the thin films, viz. the van der Waals forces, electrostatic double layer force, hydration, and steric forces are discussed and explained with examples. The critical thickness of rupture of a thin liquid film, its stability, and the models for calculating the radius of the film are discussed. The disjoining pressure model and the body-force model of the thin liquid film are presented.

Two theories have been developed for calculating the coalescence time of drops: the film drainage theory, and the stochastic theory. The film drainage theories are based on various assumptions. These assumptions are discussed, and the models are elucidated with numerical examples. The mechanism of lubrication flow is explained. The adsorption of surfactants at the liquid–liquid interfaces plays an important role in the stability of the drops in an emulsion. The surface equations of state for adsorption are presented. The stochastic theory of coalescence is presented and the parameters of the stochastic model are explained. The procedure to fit this model to experimental data on coalescence of drops is explained with example. Next, methods used to compute the time required for growth of drops in industrial gravity settler is discussed. The last topic discussed in this chapter is the experimental techniques adopted by various researchers for studying coalescence of drops.

From the discussion of the theories and experimental results presented in this chapter, it is evident that coalescence is a stochastic process. The film drainage models attempt to predict the mean values of the coalescence time distributions. However, the agreement between these models and the experimental data is not good. Sometimes the difference is by several orders of magnitude. The disagreement becomes more pronounced in systems in which surfactants are present. The stability of thin liquid films is likely to be the most important factor in the coalescence process. Although the stochastic model fits the experimental coalescence time distributions well, its parameters, viz. the dimensionless coalescence threhsold, and normalized standard deviation cannot be predicted with the present state of knowledge. Two important parameters of this model are surface diffusivity and interfacial force. However, precise values of these quantities at the liquid–liquid interfaces are unknown. Therefore, although the stochastic theory of coalescence can semi-quantitatively explain the mechanism of coalescence of drops, further development is necessary for the prediction of the coalescence time distributions.

NOMENCLATURE

a	drop radius, m
a_f	area of the film, m^2
$\hat{a}$	interfacial area per unit volume of dispersion, m^{-1}
$\bar{a}$	reduced radius, m
A	total surface area of the drops, m^2
A_H	Hamaker constant, J
A_L	London dispersion force constant, $J\,m^6$
A_{min}	minimum surface area occupied by a surfactant molecule, m^2
$\tilde{A}$	constant in Lennard-Jones model, $J\,m^6$
B	modified Hamaker constant, J m
$\tilde{B}$	constant in Lennard-Jones model, $J\,m^{12}$
c	concentration, mol/m^3
c^{∞}	concentration of molecules (or ions) in the bulk solution, mol/m^3
c^{sat}	saturation concentration, mol/m^3
d	total thickness of the Stern layer, m
d_1	thickness of the inner Helmholtz layer, m
d_2	thickness of the outer Helmholtz layer, m
D	diffusion coefficient, m^2/s
D_{Γ}	surface diffusion coefficient, m^2/s
e	electronic charge, C
E_G	Gibbs elasticity, N/m
$f(\Gamma_i)$	probability density function of Γ_i
f_r	repulsive force generated by one mole of surfactant at the barrier ring, N/mol
f_t	tangential stress, Pa
F	force, N
F_g	gravitational force, N
F_l	lubrication force, N
F_r	total repulsive force at the barrier ring, N
F_s	normal surface-excess force, N
$F(\tau)$	cumulative probability distribution of coalescence time
$\tilde{F}$	body force density vector
g	acceleration due to gravity, m/s^2
h	film thickness, m
h_0	closest separation between the drop surfaces, m
h_c	critical film thickness, m
h_i	initial film thickness, m
h_r	separation between the interfaces at the barrier ring, m
$\tilde{h}$	Planck's constant, J s
J_{ij}	total number of collisions occurring between particles of types i and j in unit volume per unit time
k	Boltzmann's constant, J/K
k_L	mass transfer coefficient, m/s
K	coalescence rate constant in van den Tempel's model, s^{-1}
K'	coalescence rate constant in the model of Borwankar *et al.* [6], s^{-1}
K_L	equilibrium constant for the Langmuir model, m^3/mol
K_s	empirical parameter in Eq. (84), $m^i\,s^{-1}$
l	length of a polymer segment, m
L	thickness of the polymer brush layer, m
m	number of molecules (or ions) per unit volume of solution near the interface, m^{-3}

m^∞	number of molecules (or ions) per unit volume of bulk solution, m^{-3}
$\bar{m}$	average number of drops in an aggregate in presence of coalescence
$\hat{m}$	number of molecules per unit volume, m^{-3}
n	total number of drops in the emulsion at time t per unit volume, m^{-3}
n_0	number of primary drops present at $t = 0$ per unit volume, m^{-3}
n_2	number of doublets at time t per unit volume, m^{-3}
n_a	average number of primary drops in an aggregate
n_f	number of films in an aggregate
n_k	number of k-fold aggregates at time t per unit volume, m^{-3}
n_p	total number of drops at time t per unit volume, m^{-3}
n_s	number of segments of the polymer molecule
n_t	number of primary non-aggregated drops remaining at time t per unit volume, m^{-3}
n_v	number of aggregates at time t per unit volume, m^{-3}
$\tilde{n}$	unit vector normal to the surface of the film
N	total number of ruptured films
N_A	Avogadro's number, mol^{-1}
p	pressure, Pa
p_c	capillary pressure, Pa
P_Γ	dimensionless coalescence threshold
r	radial position, m
R	gas constant, $J\ mol^{-1}\ K^{-1}$
$\hat{R}_{ij}$	collsion radius for a pair of particles, m
R_b	radius of barrier ring, m
R_c	radius of capillary, m
R_f	radius of film, m
s	separation between the molecules, m
$\bar{s}$	mean distance between the attachment points, m
S_Γ	normalized standard deviation
t	time, s
t_c	coalescence time, s
t_f	characteristic flocculation time, s
$\bar{t}$	characteristic diffusion time, s
T	temperature, K
u	radial velocity, m/s
u_p	velocity for the parabolic portion of the flow, m/s
u_t	tangential velocity, m/s
U	relative velocity of drops, m/s
v	approach velocity of drop, m/s
w_b	width of the barrier ring, m
x	distance along the x-axis, m
z	distance along the z-axis, m
$\tilde{z}$	valence of ion

Greek Symbols

α	flocculation rate constant, m^3/s
β	number of immobile surfaces
γ	interfacial tension, N/m
γ_0	interfacial tension for pure liquids, N/m
Γ	surface excess concentration of the surfactant, mol/m^2
$\hat{\Gamma}$	number of adsorbed chains per unit area, m^{-2}
Γ_∞	maximum value of Γ, mol/m^2
Γ_i	initial value of Gibbs surface excess of the adsorbate when the drop strikes at the interface, mol/m^2

$\bar{\Gamma}$	mean value of the distribution of surface excess concentration, mol/m^2
$\bar{\Gamma}_i$	mean value of Γ_i, mol/m^2
Γ_m	minimum value of the surfactant concentration at the barrier ring required to prevent the drop from coalescence, mol/m^2
δ_i	average number of i-mers per unit volume, m^{-3}
ΔG_f	free energy of formation, J
ΔN	number of film-ruptures at thicknesses between $\left(h-\Delta h/2\right)$ and $\left(h+\Delta h/2\right)$
Δp	pressure difference, Pa
ΔS_f	entropy of formation, J/K
$\Delta \rho$	difference in density between two liquids, kg/m^3
ε	dielectric constant of the medium
ε_0	permittivity of free space, C^2 J^{-1} m^{-1}
η	refractive index of the medium
θ	angle, rad
κ	Debye–Hückel parameter, m^{-1}
$\tilde{\lambda}$	scaling parameter for viscosity
λ_i	zeros of the Bessel function of first kind and order one
μ	viscosity of the liquid constituting the continuous phase, Pa s
μ_d	viscosity of the dispersed phase, Pa s
μ_s	interfacial shear viscosity, kg/s
v_e	main electronic absorption frequency in the UV region, s^{-1}
ξ	fraction of the total amount of surfactant at the interface which remains at the barrier ring after the displacement of surfactant molecules to the barrier ring
Π_{EDL}	disjoining pressure due to electrostatic double layer interaction, Pa
Π_s	disjoining pressure due to steric interaction, Pa
Π_{vdW}	disjoining pressure due to van der Waals interaction, Pa
ρ	density, kg/m^3
$\tilde{\rho}$	charge density, C/m^3
σ_Γ	standard deviation in the distribution of Γ_i, mol/m^2
τ	dimensionless coalescence time
τ_c	coalescence time in dispersion, s
τ_d	doublet fragmentation time due to Brownian movement, s
τ_s	Smoluchowski collision time, s
ϕ	interaction energy, J
φ	elliptic-type Green's function kernel
Φ	interaction potential, J/m^3
ψ	potential, V
ψ_0	surface potential, V
ψ_1	potential at the inner Helmholtz plane, V
ψ_d	potential at the outer Helmholtz plane, V
ψ_s	volume fraction of the dispersed phase

ACKNOWLEDGEMENT

This work was supported by a research grant from Department of Science and Technology (DST), Government of India.

REFERENCES

[1]   M. Stewart and K. Arnold, *Emulsions and oil treating equipment*. Oxford: Elsevier, pp. 1–75, 2008.

[2]   P. Ghosh, *Colloid and interface science*. New Delhi: PHI Learning, pp. 319–320, 2009.

[3]   P. Taylor, "Ostwald ripening in emulsions," *Advs. Colloid Interface Sci.*, vol. 75, pp. 107–163, 1998.

[4]   E. L. Cussler, *Diffusion mass transfer in fluid systems*. Cambridge: Cambridge University Press, pp. 215, 1997.

[5]   M. V. Smoluchowski, "Versuch einer mathematischen theorie der koagulationskinetik kolloider losungen," *Zeitschrift für Physikalische Chemie*, vol. 92, pp. 129–136, 1917.

[6]   R. P. Borwankar, L. A. Lobo, and D. T. Wasan, "Emulsion stability–kinetics of flocculation and coalescence," *Colloids & Surfaces*, vol. 69, pp. 135–146, 1992.

[7]   K. D. Danov, I. B. Ivanov, Th. D. Gurkov, and R. P. Borwankar, "Kinetic model for the simultaneous processes of flocculation and coalescence in emulsion systems," *J. Colloid Interface Sci.*, vol. 167, pp. 8–17, 1994.

[8]   A. Nandi, A. Mehra, and D. V. Khakhar, "Suppression of coalescence in surfactant stabilized emulsions by shear flow," *Phys. Rev. Lett.*, vol. 83, pp. 2461–2464, 1999.

[9]   J. E. Jones and A. E. Ingham, "The calculation of certain crystal potential constants, and the cubic crystal of least potential energy," *Proc. the Royal Soc.*, vol. 107A, pp. 636–653, 1925.

[10] F. London, "Theory and systematics of molecular forces," *Zeitschrift für Physik*, vol. 63, pp. 245–279, 1930.

[11] H. C. Hamaker, "The London–van der Waals attraction between spherical particles," *Physica*, vol. 4, 1058–1072, 1937.

[12] J. Gregory, "The calculation of Hamaker constants," *Advs. Colloid Interface Sci.*, vol. 2, pp. 396–417, 1969.

[13] J. N. Israelachvili, *Intermolecular and surface forces*. London: Academic Press, pp. 179–192, 1997.

[14] O. Stern, "The theory of the electrolytic double-layer," *Zeitschrift für Elektrochemie und Angewandte Physikalische Chemie*, vol. 30, pp. 508–516, 1924.

[15] A. W. Adamson and A. P. Gast, *Physical chemistry of surfaces*. New York: John Wiley, pp. 169–175, 1997.

[16] D. C. Grahame, "The electrical double layer and the theory of electrocapillarity," *Chem. Rev.*, vol. 41, pp. 441–501, 1947.

[17] P. C. Hiemenz and R. Rajagopalan, *Principles of colloid and surface chemistry*. New York: Marcel Dekker, pp. 499–530, 1997.

[18] R. J. Hunter, *Foundations of colloid science*. New York: Oxford University Press, pp. 317–327, 2005.

[19] J. Lyklema, *Fundamentals of interface and colloid science (vol. 2)*. London: Academic Press, pp. 3.1–3.225, 1995.

[20] V. V. Kalinin and C. J. Radke, "An ion-binding model for ionic surfactant adsorption at aqueous–fluid interfaces," *Colloids & Surfaces (A)*, vol. 114, pp. 337–350, 1996.

[21] P. K. Bommaganti, M. Vijay Kumar, and P. Ghosh, "Effects of binding of counterions on adsorption and coalescence," *Chem. Eng. Res. Des.* vol. 87, pp. 728–738, 2009.

[22] P. Ghosh and V. A. Juvekar, "Analysis of the drop rest phenomenon," *Chem. Eng. Res. Des.*, vol. 80, pp. 715–728, 2002.

[23] K. Giribabu and P. Ghosh, "Adsorption of nonionic surfactants at fluid–fluid interfaces: importance in the coalescence of bubbles and drops," *Chem. Eng. Sci.*, vol. 62, pp. 3057–3067, 2007.

[24] R. M. Pashley, "DLVO and hydration forces between mica surfaces in Li^+, Na^+, K^+, and Cs^+ electrolyte solutions: a correlation of double-layer and hydration forces with surface cation exchange properties," *J. Colloid Interface Sci.*, vol. 83, pp. 531–546, 1981.

[25] J. S. Clunie, J. F. Goodman, and P. C. Symons, "Solvation forces in soap films," *Nature*, vol. 216, pp. 1203–1204, 1967.

[26] P. G. de Gennes, "Polymers at an interface; a simplified view," *Advs. Colloid Interface Sci.*, vol. 27, pp. 189–209, 1987.

[27] A. D. Nikolov and D. T. Wasan, "Effects of surfactant on multiple stepwise coalescence of single drops at liquid–liquid interfaces," *Ind. Eng. Chem. Res.*, vol. 34, pp. 3653–3661, 1995.

[28] A. Vrij and J. Th. G. Overbeek, "Rupture of thin liquid films due to spontaneous fluctuations in thickness," *J. Am. Chem. Soc.*, vol. 90, pp. 3074–3078, 1968.

[29] S. A. K. Jeelani and S. Hartland, "Effect of interfacial mobility on thin film drainage," *J. Colloid Interface Sci.*, vol. 164, pp. 296–308, 1994.

[30] H. M. Princen, "Shape of a fluid drop at a liquid–liquid interface," *J. Colloid Interface Sci.*, vol. 18, pp. 178–195, 1963.

[31] A. Scheludko and E. Manev, "Critical thickness of rupture of chlorobenzene and aniline films," *Trans. Faraday Soc.*, vol. 64, pp. 1123–1134, 1968.

[32] B. V. Derjaguin, "Definition of the concept of and the magnitude of the disjoining pressure and its role in the statics and kinetics of thin layers of liquids," *Colloid J. (Russia)*, vol. 17, pp. 191–200, 1955.

[33] A. Sheludko, "Thin liquid films," *Advs. Colloid Interface Sci.*, vol. 1, pp. 391–464, 1967.

[34] C. Maldarelli, R. K. Jain, I. B. Ivanov, and E. Ruckenstein, "Stability of symmetric and unsymmetric thin liquid films to short and long wavelength perturbations," *J. Colloid Interface Sci.*, vol. 78, pp. 118–143, 1980.

[35] A. K. Chesters, "The modelling of coalescence process: a review of the current understanding," *Trans. IChemE (Part A)*, vol. 69, pp. 259–270, 1991.

[36] J. C. Slattery, *Interfacial transport phenomena*. New York: Springer-Verlag, pp. 385–421, 1990.

[37] D. A. Edwards, H. Brenner, and D. T. Wasan, *Interfacial transport processes and Rheology*. Boston: Butterworth-Heinemann, pp. 281–302, 1991.

[38] W. Rommel, W. Meon, and E. Blass, "Hydrodynamic modeling of drop coalescence at liquid–liquid interfaces," *Separation Sci. Tech.*, vol. 27, pp. 129–159, 1992.

[39] K. S. Birdi, Ed., *Handbook of surface and colloid chemistry*. Boca Raton: CRC Press, pp. 333–494, 1997.

[40] G. E. Charles and S. G. Mason, "The coalescence of liquid drops with flat liquid–liquid interfaces," *J. Colloid Sci.*, vol. 15, pp. 236–267, 1960.

[41] S. P. Frankel and K. J. Mysels, "On the 'dimpling' during the approach of two interfaces," *J. Phys. Chem.*, vol. 66, pp. 190–191, 1962.

[42] S. Hartland, "The approach of a liquid drop to a flat plate," *Chem. Eng. Sci.*, vol. 22, pp. 1675–1687, 1967.

[43] S. Hartland, "The profile of the draining film beneath a liquid drop approaching a plane interface," *Chem. Eng. Prog. Sym. Ser. No.* 91, vol. 65, pp. 82–89, 1969.

[44] E. Matijević, Ed., *Surface & colloid science (vol. 3)*. pp. 167–239, New York: Wiley, 1971.

[45] S. Hartland and J. D. Robinson, "A model for an axisymmetric dimpled draining film," *J. Colloid Interface Sci.*, vol. 60, pp. 72–81, 1977.

[46] D. S. Dimitrov and I. B. Ivanov, "Hydrodynamics of thin liquid films. On the rate of thinning of microscopic films with deformable interfaces," *J. Colloid Interface Sci.*, vol. 64, pp. 97–106, 1978.

[47] A. F. Jones and S. D. R. Wilson, "The film drainage problem in drop coalescence," *J. Fluid Mech.*, vol. 87, pp. 263–288, 1978.

[48] E. D. Manev, S. V. Sazdanova, and D. T. Wasan, "Emulsion and foam stability–the effect of film size on film drainage," *J. Colloid Interface Sci.*, vol. 97, pp. 591–594, 1984.

[49] B. K. Chi and L. G. Leal, "A theoretical study of the motion of a viscous drop toward a fluid interface at low Reynolds number," *J. Fluid Mech.*, vol. 201, pp. 123–146, 1989.

[50] R. H. Davis, J. A. Schonberg, and J. M. Rallison, "The lubrication force between two viscous drops," *Phys. Fluids A*, vol. 1, pp. 77–81, 1989.

[51] E. P. Ascoli, D. S. Dandy, and L. G. Leal, "Buoyancy-driven motion of a deformable drop toward a planar wall at low Reynolds number," *J. Fluid Mech.*, vol. 213, pp. 287–311, 1990.

[52] C. Pozrikidis, "The deformation of a liquid drop moving normal to a plane wall," *J. Fluid Mech.*, vol. 215, pp. 331–363, 1990.

[53] S. G. Yiantsios and R. H. Davis, "On the buoyancy driven motion of a drop towards a rigid surface or a deformable interface," *J. Fluid Mech.*, vol. 217, pp. 547–573, 1990.

[54] X. Zhang and R. H. Davis, "The rate of collisions due to Brownian or gravitational motion of small drops," *J. Fluid Mech.*, vol. 230, pp. 479–504, 1990.

[55] R. J. Hunter, *Foundations of Colloid Science (vol. II)*. Oxford: Clarendon Press, pp. 874–907, 1992.

[56] S. A. K. Jeelani and S. Hartland, "Effect of velocity fields on binary and interfacial coalescence," *J. Colloid Interface Sci.*, vol. 156, pp. 467–477, 1993.

[57] A. Sabotini, C. Gourdon, and A. K. Chesters, "Drainage and rupture of partially mobile films during coalescence in liquid–liquid systems under a constant interaction force," *J. Colloid Interface Sci.*, vol. 175, pp. 27–35, 1995.

[58] M. A. Rother, A. Z. Zinchenko, and R. H. Davis, "Buoyancy-driven coalescence of slightly deformable drops," *J. Fluid Mech.*, vol. 346, pp. 117–148, 1997.

[59] D. F. Evans and H. Wennerström, *The colloidal domain: where physics, chemistry, biology, and technology meet*. New York: Wiley-VCH, pp. 5–17, 1999.

[60] A. J. Prosser and E. I. Franses, "Adsorption and surface tension of ionic surfactants at the air–water interface: review and evaluation of equilibrium models," *Colloids & Surfaces (A)*, vol. 178, pp. 1–40, 2001.

[61] C. -H. Chang and E. I. Franses, "Adsorption dynamics of surfactants at the air/water interface: a critical review of the mathematical models, data, and mechanisms," *Colloids & Surfaces (A)*, vol. 100, pp. 1–45, 1995.

[62] M. K. Kumar, T. Mitra, and P. Ghosh, "Adsorption of ionic surfactants at liquid–liquid interfaces in the presence of salt: application in binary coalescence of drops," *Ind. Eng. Chem. Res.*, vol. 45, pp. 7135–7143, 2006.

[63]   K. Giribabu and P. Ghosh, "Binary coalescence of air bubbles in viscous liquids in presence of non-ionic surfactant," *Can. J. Chem. Eng.*, vol. 86, pp. 643–650, 2008.

[64]   R. P. Borwankar and D. T. Wasan, "Equilibrium and dynamics of adsorption of surfactants at fluid–fluid interfaces," *Chem. Eng. Sci.*, vol. 43, pp. 1323–1337, 1988.

[65]   P. A. Kralchevsky, K. D. Danov, G. Broze, and A. Mehreteab, "Thermodynamics of ionic surfactant adsorption with account for the counterion binding: effect of salts of various valency," *Langmuir*, vol. 15, pp. 2351–2365, 1999.

[66]   C. -Y. Lin and J. C. Slattery, "Thinning of a liquid film as a small drop or bubble approaches a solid," *AIChE J.*, vol. 28, pp. 147–156, 1982.

[67]   J. -D. Chen and J. C. Slattery, "Effects of London–van der Waals forces on the thinning of a dimpled liquid film as a small drop or bubble approaches a horizontal solid plane," *AIChE J.*, vol. 28, pp. 955–963, 1982.

[68]   J. -D. Chen, P. S. Hahn, and J. C. Slattery, "Coalescence time for a small drop or bubble at a fluid–fluid interface," *AIChE J.*, vol. 30, pp. 622–630, 1984.

[69]   P. -S. Hahn, J. -D. Chen, and J. C. Slattery, "Effects of London–van der Waals forces on the thinning and rupture of a dimpled liquid film as a small drop or bubble approaches a fluid–fluid interface," *AIChE J.*, vol. 31, pp. 2027–2038, 1985.

[70]   A. K. Malholtra and D. T. Wasan, "Effect of film size on drainage of foam and emulsion films," *AIChE J.*, vol. 33, pp. 1533–1541, 1987.

[71]   A. K. Malholtra and D. T. Wasan, "Effects of surfactant adsorption–desorpion kinetics and interfacial rheological properties on the rate of drainage of foam and emulsion films," *Chem. Eng. Comm.*, vol. 55, pp. 95–128, 1987.

[72]   D. E. Tambe and M. M. Sharma, "Hydrodynamics of thin liquid films bounded by viscoelastic interfaces," *J. Colloid Interface Sci.*, vol. 147, pp. 137–151, 1991.

[73]   J. -L. Joye, C. A. Miller, and G. J. Hirasaki, "Dimple formation and behavior during axisymmetrical foam film drainage," *Langmuir*, vol. 8, pp. 3083–3092, 1992.

[74]   J. -L. Joye, G. J. Hirasaki, and C. A. Miller, "Asymmetric drainage in foam films," *Langmuir*, vol. 10, pp. 3174–3179, 1994.

[75]   Li, D., "Coalescence between two small bubbles or drops," *J. Colloid Interface Sci.*, vol. 163, pp. 108–119, 1994.

[76]   K. P. Velikov, O. D. Velev, K. G. Marinova, and G. N. Constantinides, "Effect of the surfactant concentration on the kinetic stability of thin foam and emulsion films," *J. Chem. Soc., Faraday Trans.*, vol. 93, pp. 2069–2075, 1997.

[77]   V. Cristini, J. Blawzdziewicz, and M. Loewenberg, "Near contact motion of surfactant covered spherical drops," *J. Fluid Mech.*, vol. 366, pp. 259–287, 1998.

[78]   G. D. M. Mackay and S. G. Mason, "The gravity approach and coalescence of fluid drops at liquid interfaces," *Can. J. Chem. Eng.*, vol. 41, pp. 203–212, 1963.

[79]   T. D. Hodgson and D. R. Woods, "The effect of surfactants on the coalescence of a drop at an interface. II.," *J. Colloid Interface Sci.*, vol. 30, pp. 429–446, 1969.

[80]   Z. Zapryanov, A. K. Malhotra, N. Aderangi, and D. T. Wasan, "Emulsion stability: an analysis of the effects of bulk and interfacial properties on film mobility and drainage rate," *Int. J. Multiphase Flow*, vol. 9, pp. 105–129, 1983.

[81]   N. Aderangi and D. T. Wasan, "Coalescence of single drops at a liquid–liquid interface in the presence of surfactants/polymers," *Chem. Eng. Comm.*, vol. 132, pp. 207–222, 1995.

[82]   D. T. Wasan, S. M. Shah, N. Aderangi, M. S. Chan, and J. McNamara, "Observations on the coalescence behavior of oil droplets and emulsion stability in enhanced oil recovery," *SPE J.*, vol. 18, pp. 409–417, 1978.

[83]   J. C. Lee and T. D. Hodgson, "Film flow and coalescence–I: basic relations, film shape and criteria for interface mobility," *Chem. Eng. Sci.*, vol. 23, pp. 1375–1397, 1968.

[84]   K. A. Burrill and D. R. Woods, "Film shapes for deformable drops at liquid–liquid interfaces–II: the mechanism of film drainage," *J. Colloid Interface Sci.*, vol. 42, pp. 15–34, 1973.

[85]   O. Mondain-Monval, F. Leal-Calderon, and J. Bibette, "Forces between emulsion droplets: role of surface charges and excess surfactant," *J. Physique II*, vol. 6, pp. 1313–1329, 1996.

[86]   S. R. Deshikan and K. D. Papadopoulos, "London–vdW and EDL effects in the coalescence of oil drops," *J. Colloid Interface Sci.*, vol. 174, pp. 302–312, 1995.

[87] E. G. Cockbain and T. S. McRoberts, "The stability of elementary emulsion drops and emulsions," *J. Colloid Sci.*, vol. 8, pp. 440–451, 1953.

[88] T. Gillespie and E. K. Rideal, "The coalescence of drops at an oil–water interface," *Trans. Faraday Soc.*, vol. 52, pp. 173–183, 1956.

[89] L. E. Nielsen, R. Wall, and G. Adams, "Coalescence of liquid drops at oil–water interfaces," *J. Colloid Sci.*, vol. 13, pp. 441–458, 1958.

[90] S. B. Lang and C. R. Wilke, "A hydrodynamic mechanism for the coalescence of liquid drops. II. Experimental studies," *Ind. Eng. Chem. Fund.*, vol. 10, pp. 341–352, 1971.

[91] T. M. Dreher, J. Glass, A. J. O'Connor, and G. W. Stevens, "Effect of rheology on coalescence rates and emulsion stability," *AIChE J.*, vol. 45, pp. 1182–1190, 1999.

[92] P. Ghosh, "Coalescence of bubbles in liquid," *Bubble Sci. Eng. Tech.*, vol. 1, pp. 75–87, 2009.

[93] A. Scheludko, "Über die zerreißwahrscheinlichkeit von schaumfilmen aus isoamylalkohollösungen," *Zeitschrift für Elektrochemie*, vol. 61, pp. 220–222, 1957.

[94] C. V. Sternling and L. E. Scriven, "Interfacial turbulence: hydrodynamic instability and the Marangoni effect," *AIChE J.*, vol. 5, pp. 514–523, 1959.

[95] M. L. Agrawal and R. D. Neuman, "Surface diffusion in monomolecular films. II. Experiment and theory," *J. Colloid Interface Sci.*, vol. 121, pp. 366–380, 1988.

[96] T. Mitra and P. Ghosh, "Binary coalescence of water drops in organic media in presence of ionic surfactants and salts," *J. Dispersion Sci. Tech.*, vol. 28, pp. 785–792, 2007.

[97] T. D. Hodgson and J. C. Lee, "The effect of surfactants on the coalescence of a drop at an interface. I.," *J. Colloid Interface Sci.*, vol. 30, pp. 94–108, 1969.

[98] X. Chen, S. Mandre, and J. J. Feng, "An experimental study of the coalescence between a drop and an interface in Newtonian and polymeric liquids," *Phys. Fluids*, vol. 18, pp. 092103-1–092103-14, 2006.

[99] J. Chatterjee, A. D. Nikolov, and D. T. Wasan, "Study of drop–interface coalescence using piezoimaging," *Ind. Eng. Chem. Res.*, vol. 35, pp. 2933–2938, 1996.

[100] D. Chen and B. Pu, "Studies on the binary coalescence model. II. Effects of drops size and interfacial tension on binary coalescence time," *J. Colloid Interface Sci.*, vol. 243, pp. 433–443, 2001.

[101] R. W. Aul and W. L. Olbricht, "Coalescence of freely suspended liquid drops in flow through a small pore," *J. Colloid Interface Sci.*, vol. 145, pp. 478–492, 1991.

[102] X. Zhang, R. H. Davis, and M. F. Ruth, "Experimental study of two interacting drops in an immiscible fluid," *J. Fluid Mech.*, vol. 249, pp. 227–239, 1993.

APPENDIX 1: SMOLUCHOWSKI'S THEORY OF KINETICS OF COAGULATION

Smoluchowski's theory of coagulation kinetics is based on the assumptions that the collisions are binary and the fluctuations in density are small. An aggregate formed from i identical particles is called an i-mer. The average number of i-mers per unit volume is δ_i. The coagulation of two clusters of the kind i and j is given by the following relation:

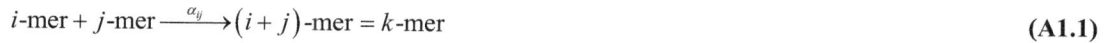

$$i\text{-mer} + j\text{-mer} \xrightarrow{\alpha_{ij}} (i+j)\text{-mer} = k\text{-mer} \tag{A1.1}$$

where α_{ij} is the concentration-independent coagulation constant (or kernel). Physically, it means that the coagulation rate between all kinds of i-mers and j-mers is the same. The equation describing the temporal evolution of the cluster of the kind k is given by

$$\frac{d\delta_k}{dt} = \frac{1}{2} \sum_{i=1, j=k-1} \alpha_{ij} \delta_i \delta_j - \delta_k \sum_i \alpha_{ik} \delta_i \tag{A1.2}$$

where α_{ij} depends on the details of the collision process between i-mers and j-mers. The coagulation constant embodies the dependence on i and j of the meeting of an i-mer and a j-mer, including effects such as the volume dependence of the collision cross-section and the diffusion constant.

The first term in Eq. (A1.2) describes the increase in δ_k owing to coagulation of an i-mer and a j-mer and the second term describes the decrease of δ_k owing to the coagulation of a k-mer with other aggregates. Eq. (A1.2) is valid for irreversible aggregation – no account is taken of the breakup of aggregates, which would require a third term on the right side. It is also assumed that each collision results in the formation of an aggregate. In other words, the collision efficiency and the stability ratio are both unity.

It is possible to use Eq. (A1.2) to derive the concentrations of all aggregate types at any time provided the values of the rate constants are known. The rate constants depend on aggregate size, shape as well as the particle transport mechanism. In certain cases, Eq. (A1.2) is converted into an integral expression and continuous aggregate size distribution is assumed rather than discrete numbers. Such approaches may not be physically realistic but can provide some insight into the way aggregate size distributions evolve during the flocculation process.

Particles in a dispersion undergo Brownian motion. As a result, particles collide. Smoluchowski's approach considers a stationary central particle and calculates the number of particles colliding with it in unit time. Allowance can then be made for the fact that the central particle itself is one of many similar particles undergoing Brownian motion and so the appropriate collision frequency can be derived. The total number of collisions occurring between particles of types i and j in unit volume per unit time, J_{ij}, is given by

$$J_{ij} = 4\pi \hat{R}_{ij} \left(D_i + D_j \right) n_i n_j \tag{A1.3}$$

where D_i and D_j represent the diffusion coefficients, and n_i and n_j represent the number concentrations, respectively. The term $\hat{R}_{ij}$ is the collision radius for the pair of particles. It represents the center-to-center distance at which the particles may be assumed to be in contact. In many cases, the collision radius can be taken simply as the sum of the particle radii. However, if there is long range attraction between the particles, the effective collision radius will be somewhat larger. For spherical particles of radii a_i and a_j

$$\hat{R}_{ij} = a_i + a_j \tag{A1.4}$$

The diffusion coefficients are calculated by the Stokes–Einstein equation:

$$D_i = \frac{kT}{6\pi a_i \mu} \tag{A1.5}$$

where μ is the viscosity of the liquid, k is Boltzmann's constant, and T is temperature. Substituting $\hat{R}_{ij}$ from Eq. (A1.4) and diffusion coefficient from Eq. (A1.5) into Eq. (A1.3), we get

$$J_{ij} = \frac{\left(\frac{2kT}{3\mu}\right) n_i n_j \left(a_i + a_j\right)^2}{a_i a_j} \tag{A1.6}$$

By comparison with Eq. (A1.2), the rate constant α_{ij} can be written as,

$$\alpha_{ij} = \frac{\left(\frac{2kT}{3\mu}\right)\left(a_i + a_j\right)^2}{a_i a_j} \tag{A1.7}$$

For an initially monodisperse dispersion of particles, the initial rate of collision can be calculated easily from Eq. (A1.2) because only one type of collision (*i.e.*, 1–1) is involved. The initial rate of decrease of the total particle concentration, n_p, follows directly from the collision rate because each collision reduces the number of particles by one (*i.e.*, two primary particles lost, and one aggregate gained). Therefore,

$$-\frac{dn_p}{dt} = \left(\frac{4kT}{3\mu}\right) n_p^2 = \alpha n_p^2 \tag{A1.8}$$

where α is the flocculation rate constant. Its value is 6.13×10^{-18} m^3/s for aqueous dispersions at 298 K. Eq. (A1.8) does not involve the particle size. As the particle size increases, the diffusion coefficient decreases, but the collision radius increases. These have opposing effects on the collision rate. For particles of equal size, they balance exactly.

Eq. (A1.8) can be integrated with the initial condition that at $t = 0$, the concentration of the particles is n_0. Upon integration we get

$$n_p = \frac{n_0}{1 + \alpha n_0 t} \tag{A1.9}$$

The *characteristic flocculation time* $\left(t_f\right)$ is defined as the time in which the number of particles is reduced to half of the initial value (*i.e.*, $n_p = n_0/2$). From Eq. (A1.9):

$$t_f = \frac{1}{\alpha n_0} \tag{A1.10}$$

The flocculation time can be viewed as the average time in which a particle experiences one collision. If the initial particle concentration is $n_0 = 1 \times 10^{15}$ m^{-3}, and $\alpha = 6.13 \times 10^{-18}$ m^3/s, Eq. (A1.10) predicts that the flocculation time will be 163 s.

The concentration of the single particles at any time t is given by

$$n_1 = \frac{n_0}{\left(1 + \alpha n_0 t\right)^2} = \frac{n_0}{\left(1 + t/t_f\right)^2} \tag{A1.11}$$

The concentration of doublets is given by

$$n_2 = \frac{\alpha n_0^2 t}{\left(1 + \alpha n_0 t\right)^3} = \frac{n_0 \, t/t_f}{\left(1 + t/t_f\right)^3} \tag{A1.12}$$

The general Smoluchowski expression for k-fold aggregates is

$$n_k = \frac{n_0 \left(t/t_f\right)^{k-1}}{\left(1 + t/t_f\right)^{k+1}} \tag{A1.13}$$

It can be shown that for all aggregates, the concentration rises to a maximum at a characteristic time, and then declines slowly. At all times, the concentration of singlets exceeds that of any other aggregate.

APPENDIX 2: THE HAMAKER CONSTANT

After the appearance of London's treatment of the dispersion forces between atoms, it was realized that the additive nature of these forces would lead to appreciable interaction between bodies, and it would play an important part in the stability of colloidal materials. Among the various approaches adopted by scientists to calculate the interaction between bodies assuming the additlvity of dispersion forces, the equations of Hamaker are the most frequently employed today. For the interaction between two bodies having molecules of type 1 and type 2 *in vacuo*, the Hamaker constant is defined as

$$A_H^{1,2} = \pi^2 \hat{m}_1 \hat{m}_2 A_L^{1,2} \tag{A2.1}$$

where $A_L^{1,2}$ is the London constant for molecules of types 1 and 2, and $\hat{m}_1$ and $\hat{m}_2$ are the number of molecules per unit volumes of the two types of material.

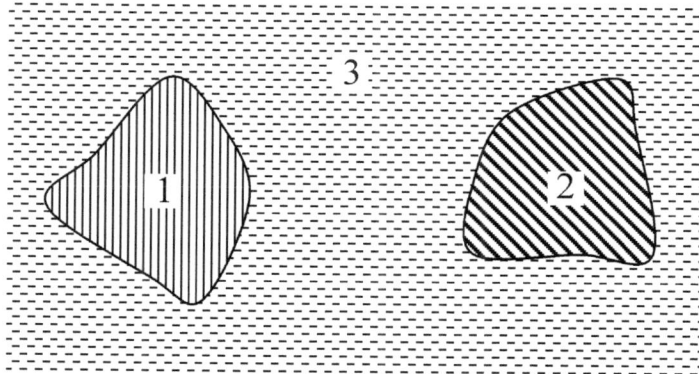

Figure A2.1: Hamaker constant for interaction between two bodies (1 and 2) in a medium (3).

Let us consider two particles constituted of different substances designated by 1 and 2, and embedded in a medium, which is designated as 3 as depicted in Fig. **A2.1**.

In evaluating the energy variations in this system, we have to take into account the two particles 1 and 2, and also the particles of same size constituted of medium 3. Let us denote the interaction energy between the particles 1 and 2 *in vacuo* as ϕ_{12}, the interaction energy between particle 1 and particle 3 *in vacuo* as ϕ_{13}, the interaction energy between particle 2 and particle 3 *in vacuo* as ϕ_{32}, and the interaction energy between the two particles of medium 3 *in vacuo* as ϕ_{33}. These energies will be functions of the distance between the particles. If ϕ_1 represents the energy of particle 1 in the medium at infinity, this particle, when brought in the neighborhood of the particle 2 will possess an energy $\left(\phi_1 + \phi_{12} - \phi_{13}\right)$. While bringing the particle 1 towards the particle 2, we have at the same time to remove a particle of medium 3 towards infinity. This will correspond to a change in energy from $\left(\phi_3 + \phi_{32} - \phi_{33}\right)$ to ϕ_3 when ϕ_3 is the energy of the particle of medium 3 at infinity. Since ϕ_1 and ϕ_3 are constants, the energy changes associated with the variations in the distance between the particles 1 and 2 will be,

$$\phi = \phi_{12} - \phi_{13} - \left(\phi_{32} - \phi_{33}\right) = \phi_{12} + \phi_{33} - \phi_{13} - \phi_{32} \tag{A2.2}$$

This expression is independent of the nature of the forces of interaction. It is, however, inherent in this argument that the energy of interaction of one particle with the medium shall be unaffected by the presence or absence of the other particle. This can be a severe limitation if the interaction between the particles and medium 3 is accompanied by an orientation of the molecules of medium 3. In such cases, we may consider the total energy to be made up of two parts: a part independent of the orientation of the medium molecules, and an additional amount due to this orientation. Whenever the latter part is only a small fraction of the total, it is justified to assume that the conclusions drawn from Eq. (A2.2) will be correct.

Let us represent the interaction energy between two particles of same substance *in vacuo* as,

$$\phi = -A_H \phi_y \left(x\right) \tag{A2.3}$$

where $\phi_y \left(x\right)$ is a function of the geometrical data (*i.e.*, diameter and distance), and A_H is equal to $\pi^2 \hat{m}^2 A_L$. If the two particles are composed of two different substances 1 and 2, the Hamaker constant will be given by Eq. (A2.1). If these two particles are embedded in medium 3, from Eqs. (A2.2) and (A2.3) we get

$$A_H^{1,3,2} = \pi^2 \left[\hat{m}_1 \hat{m}_2 A_L^{1,2} + \hat{m}_3^2 A_L^{3,3} - \hat{m}_1 \hat{m}_3 A_L^{1,3} - \hat{m}_3 \hat{m}_2 A_L^{3,2} \right] = A_H^{1,2} + A_H^{3,3} - A_H^{1,3} - A_H^{3,2} \tag{A2.4}$$

Advances in Multiphase Flow and Heat Transfer, Vol. 3, 2012, 135-150

Carbon Capture and Storage with a Focus on Capillary Trapping as a Mechanism to Store Carbon Dioxide in Geological Porous Media

Stefan Iglauer*

Curtin University, Department of Petroleum Engineering, 26 Dick Perry Avenue, Kensington, Perth, Australia

Abstract: Carbon Capture and Storage (CCS) is a feasible short-to-medium term method to dispose carbon dioxide (CO_2) which would otherwise be emitted into the atmosphere and cause potentially massively damaging climate change. In CCS, CO_2 is captured, compressed and injected deep underground into geological formations. There are four main CO_2 trapping mechanisms, namely stratigraphic or structural trapping, dissolution trapping, capillary trapping anwd mineral trapping. In this text we discuss all these trapping mechanisms with focus on capillary trapping, which has recently been identified as a rapid and reliable CO_2 storage method.

Keywords: Carbon capture and storage (CCS), capillary trapping, multi-phase flow through porous media, carbon dioxide, climate change.

INTRODUCTION

Climate change caused by anthropogenic greenhouse gas (GHG) emissions is one of the primary problems of the 21st century [1, 2].

As summarized by the IPCC [1], observed climate change effects include:

- Substantial increase in radiative forcing (RF) by 1.6 W/m^2 due to anthropogenic influences. RF is a measure of the influence a parameter has on changing the balance of in- and outgoing energy in the Earth-atmosphere system.

- Eleven of the last twelve years considered (1995-2006) rank among the 12 warmest since instrumental temperature records began (*i.e.* 1850).

- Global average sea level rise by 75.6 mm since 1961 (until 2003), this rate is increasing.

- Global average temperature rise by 0.74° C (from 1906 to 2005).

- Permafrost areas defrosting (frozen areas decreased by 7% in the Northern Hemisphere, 15% in spring time).

- Precipitation changes: significantly increased precipitation and flood risk in eastern parts of North and South America, northern Europe, northern and central Asia; drying and longer and more intense droughts have been observed in the Sahel, the Mediterrean, southern Africa, parts of southern Asia – this is however highly variable spatially and temporarily.

- More frequent heat waves.

These effects are very likely (> 90% probability) due to anthropogenic GHG.

*Address correspondence to Stefan Iglauer: Curtin University, Department of Petroleum Engineering, 26 Dick Perry Avenue, Kensington, Perth, Australia; E-mail: stefan.iglauer@curtin.edu.au

Lixin Cheng and Dieter Mewes (Eds)

Predicted future climate changes as summarized by the IPCC [1] include:

- the global average temperature will increase by 1.1-$6.4°$ C until the years 2090-2099.

- increased acidification of the ocean, a drop in pH value of 0.14-0.35 is predicted (in addition to the present decrease of 0.1 since pre-industrial times). Such a pH value drop could have a serious impact on sea ecology with unknown consequences.

- more frequent hot extremes including heat waves and heavy precipitation.

- more frequent and more intense tropical cyclone (typhoons and hurricanes) acitivity.

- changes in wind and temperature patterns.

- further rise in sea level.

- further permafrost thawing.

- anthropogenic warming continues for centuries even if the carbon dioxide (CO_2) concentration would be stabilized.

The largest fraction of anthropogenic CO_2 has a life-cycle of several centuries, but 20-60% of the CO_2 remains airborne for >1000 years. Complete depletion of the anthropogenic CO_2 will take hundreds of thousands of years [3].

Paleoclimatic studies using ice core data [1] indicate that the last half century has been the warmest for at least 1,300 years while around 120,000 years ago the polar regions were 3-$5°$ C warmer than present, which led to 4-6 m sea level rise due to melted polar ice alone. It is also interesting to note that 40 Ma ago in the Eocene during the optimum warm climate the sea level was 70 m higher than today [3,4].

If these climate changes occur, then they will most likely have a dramatic socio-economic impact, including changes in disease vectors, land masses flooded by the sea, mass migration, and food supply shortages.

The most important anthropogenic GHG are CO_2, methane (CH_4) and dinitrogen oxide (N_2O). Their concentrations all massively increased in the last 100 years (cp. Fig. **1**), significantly increasing RF.

Further GHG include halogenated hydrocarbons, sulphur hexafluoride, nitrogen trifluoride and fluorinated ethers. Though the radiative efficieny (= RF per ppb GHG) of these compounds is far higher than that of CO_2 (*e.g.* compare 0.32 $Wm^{-2}ppb^{-1}$ for Halon-1301 ($CBrF_3$) with 1.4 x 10^{-5} $Wm^{-2}ppb^{-1}$ for CO_2), these substances have less of an impact because of much lower emitted quantities. Indeed CO_2 is clearly by far the most important GHG, and it accounts for 1.66 Wm^{-2} in RF, while CH_4 contributes 0.48 Wm^{-2}, N_2O 0.16 Wm^{-2} and the halocarbons 0.34 Wm^{-2} [1]. 26.4 Gt of CO_2 per year were emitted on average between 2000-2005 [1], and this gigantic number presents an equally gigantic engineering challenge. Most of the CO_2 emissions stem from fossil fuel consumption [1,5], which again is closely related to global economic growth, and a feasible way of CO_2 disposal has to be found to keep the planet's ecology intact while guaranteeing economic growth, especially in undeveloped and take-off countries such as India, China, Brazil – these countries are likely to use coal as a major energy source, and burning coal releases substantial amounts of CO_2 (among other pollutants).

Several solutions to mitigate climate change and reduce these anthropogenic CO_2 emissions have been suggested; such processes are called CO_2 sequestration or Carbon Capture and Storage (CCS), and storage options include injection and storage into geological formations, biological, mineral or deep ocean disposal [6]. In this text we will only discuss storage in geological reservoirs (*i.e.* deep saline aquifers, depleted oil or gas reservoirs or unmineable coal seams (cp. Fig. **2**)), which is believed to be feasible [7,8]; I will refer to this method in the following as CCS.

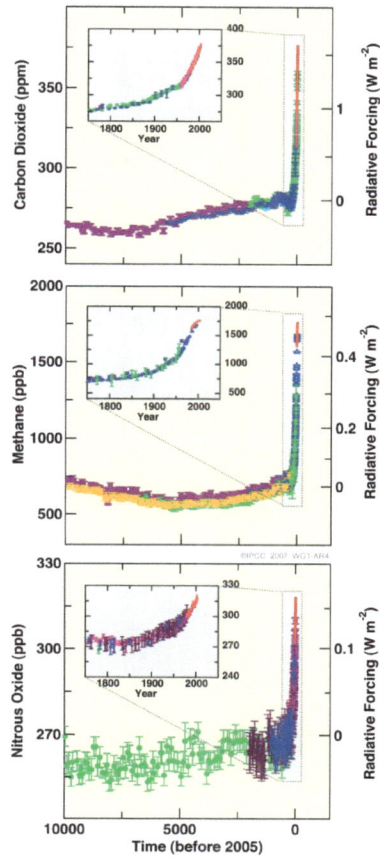

Figure 1: Concentration changes of atmospheric CO_2, CH_4, N_2O in the last 10000 years and their radiative forcings. Ice core measurements and atmospheric samples (red lines) are displayed. The inner panels show the changes since 1750 in more detail (from [1] with permission from the IPCC).

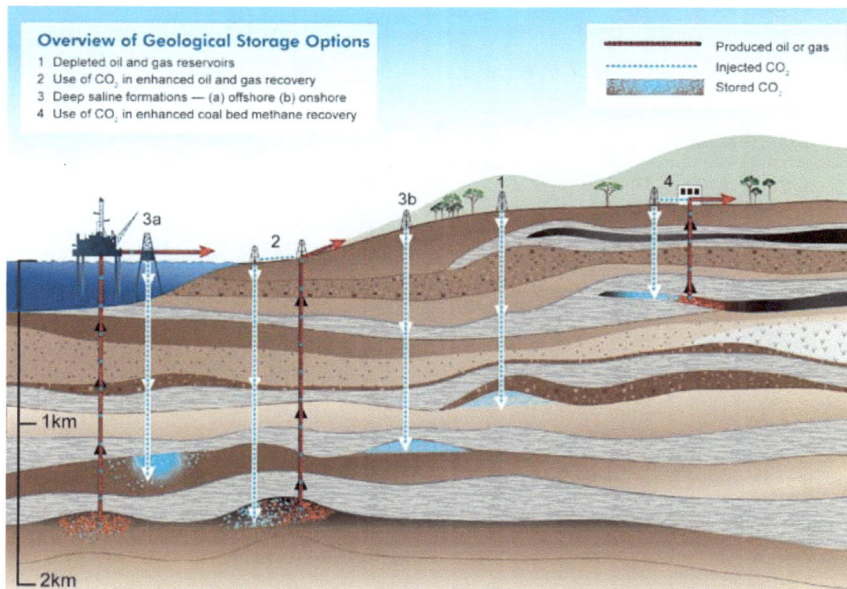

Figure 2: Schematic showing CCS options. These include storage in deep saline aquifers, depleted oil reservoirs or unmineable coal seams (from [9] with permission from the IPCC).

Depleted oil and gas reservoirs have an impermeable caprock which prevents the CO_2 (or originally the oil and/or gas) from flowing upwards, this is the main storage mechanism proposed and we discuss it in more detail in section 1 (Stratigraphic /structural trapping). Such a caprock is not necessarily available in all deep saline aquifers, however storage capacity and geographical spread of aquifers is much higher than for depleted oil and gas reservoirs [9].

The storage mechanism in coal seams is different, here the CO_2 is chemically adsorbed onto the coal surface. We will not discuss this storage method in this text.

Carbon capture means collection and purification of CO_2. Several types of CO_2 capture have been proposed, including post-combustion, pre-combustion and oxyfuel combustion [9]. For example in post-combustion, CO_2 is separated from flue gas, while in oxyfuel combustion, oxygen is separated from air, and the combustion is driven with pure oxygen. The purified CO_2 is then compressed and injected deep into the subsurface. CCS is possible [10,11] and good storage sites will contain the CO_2 for tens of thousands of years [8]. In CCS, CO_2 purification is the expensive step, while injection and storage are relatively cheap. Cost reduction for all these processes is topic of current research.

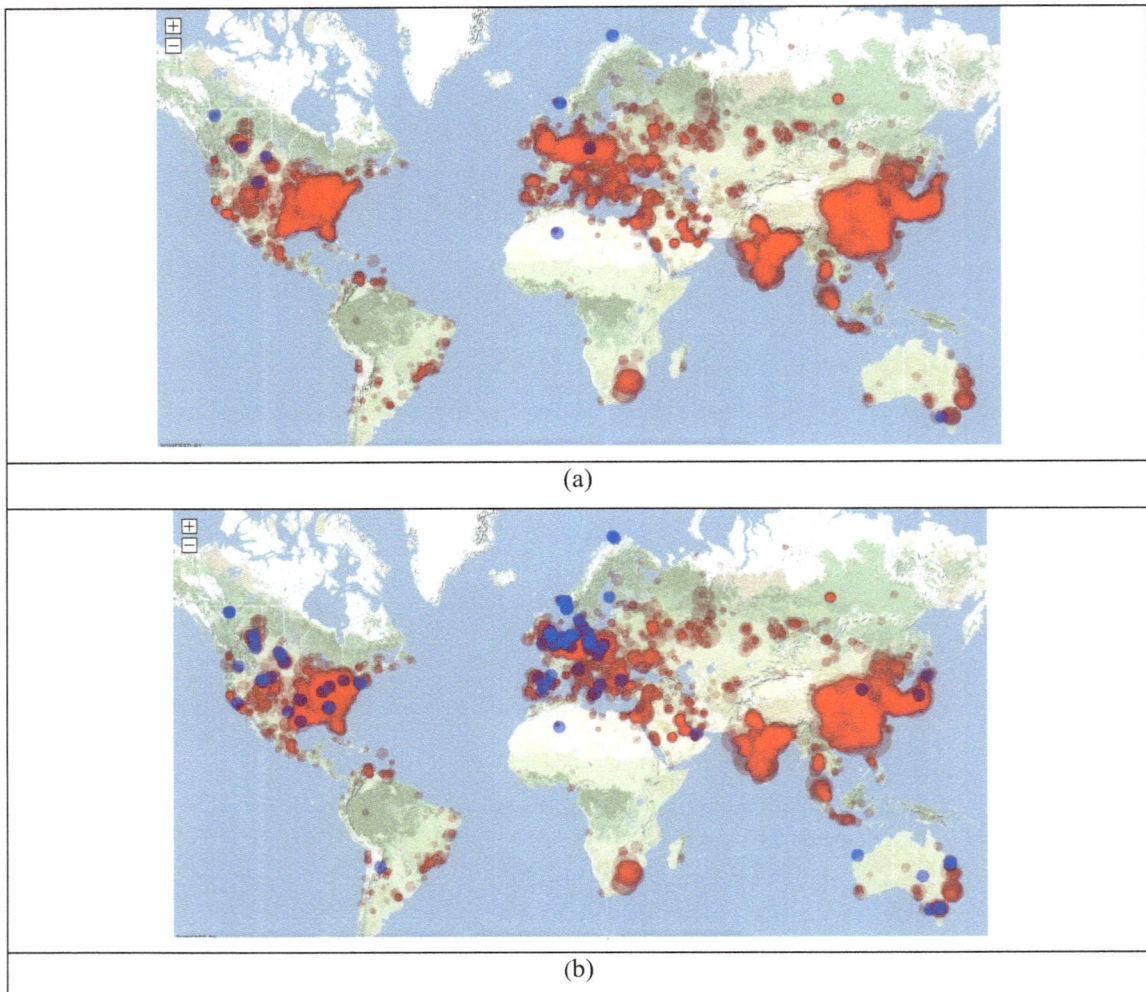

(a)

(b)

Figure 3: Red areas represent magnitude of CO_2 emissions; blue points are CO_2 injection projects; a) life CCS projects, b) planned CCS projects (from [14] with permission from the Bellona Foundation).

Now the likely scenario is that compressed CO_2 is not injected into extreme depths, because this would be too expensive. This means that the CO_2 will be more buoyant than brine [12], even though the CO_2 is in a supercritical (sc) state (critical temperature of CO_2 = 31.3° C and critical pressure of CO_2 = 7.39 MPa) and therefore tends to migrate upwards. This is indeed the main risk of CCS – leakage of CO_2 to the surface [13]. This risk leads to uncertainty and public disapproval – indeed opposition has stopped several possible storage sites in Europe [8].

In order to evaluate the leakage potential and to reduce uncertainty, research and industrial projects are currently ongoing, including theoretical, laboratory and pilot-scale level programs. In this case pilot-scale means injecting 10^4-10^6 t CO_2/a. Several of such pilot projects have been in operation for some time now (Fig. **3a** and Table **1**), while many more have been scheduled (Fig. **3b**). However, because of public concern, many projects are delayed.

Table 1: Current CCS projects marked blue in Fig. **3a**.

Project	Name	Country	Start Date	Injection Rate
1	Sleipner	Norway	1996	1.01 Mt/y
2	Fenn Big Valley	Canada	1998	17.32 Kt/y
3	Weyburn	Canada	2000	1.80 Mt/y
4	Spectra	Canada	2003	190.00 Kt/y
5	In Salah	Algeria	2004	1.21 Mt/y
6	Salt Creek	USA	2006	2.09 Mt/y
7	Schwarze Pumpe	Germany	2008	100.00 Kt/y
8	Otway	Australia	2008	104.72 Kt/y
9	Snøhvit	Norway	2008	665.00 Kt/y

In the following we will discuss only CO_2 storage, and outline the four main trapping mechanisms for CO_2. These include 1) stratigraphic/structural trapping, 2) dissolution trapping, 3) capillary trapping and 4) mineral trapping.

STRATIGRAPHIC/STRUCTURAL TRAPPING

This is the traditional mechanism proposed by geologists and engineers [15]. It stems from well-established petroleum engineering and hydrology science and it relies on impermeable geological seals, which disallow the CO_2 from flowing upwards. Such barriers are a pre-condition of geologic oil or gas reservoirs and are therefore fairly well understood. To illustrate this mechanism, we will discuss the Norwegian CCS project Sleipner (Fig. **4**). Here CO_2 has been injected deep underground since 1996, and injection is ongoing.

Fluid movement is observed by seismic measurements (Fig. **4**); CO_2 is shown in a more intense darker tone (from yellow to blue), while rock and brine are visualized in light green and blue. CO_2 was continuously injected deep into the reservoir at point IP; then over the years it rose upwards due to buoyancy forces and accumulated beneath the geological seal ("top of Utsira sand"). It appears that the caprock stops CO_2 upwards flow, and based on the seismic images and reservoir models the CO_2 then flows horizontally (Fig. **4**, [16]).

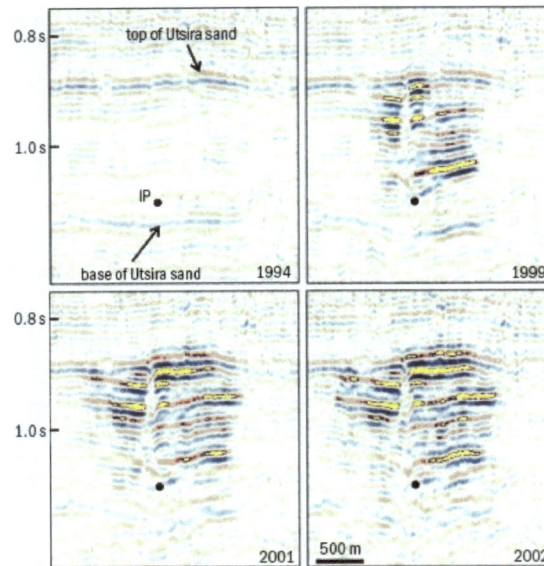

Figure 4: Sleipner project in Norway: CO_2 was injected into the Utsira formation since 1996. The seismic images show CO_2 accumulation beneath the caprock in more intense yellow and blue tones (courtesy Dr. Andy Chadwick, BGS).

Overall, this trapping mechanism is believed to be highly reliable, but there are a few unanswered questions, including a) is the diffusion of CO_2 through the caprock significant? b) is it possible that the CO_2 phase reaches a pressure which exceeds the capillary entry pressure of the caprock (which would mean that CO_2 flows through the caprock)? The capillary entry pressure is reduced for scCO_2 as compared to oil because the interfacial tension scCO_2-brine is lower than that for oil-brine [17,18] and the contact angle scCO_2 on quartz in brine is reduced with increasing pressure [19], c) is the dissolution of the caprock by the acidic CO_2-enriched brine significant? It is also not necessarily guaranteed that every site has a continuous caprock to prevent upwards migration towards the surface.

Overall, these physical/physico-chemical failure mechanisms seem unlikely, but of course they need to be tested thoroughly.

Apart from this, there is one major problem associated with this trapping mechanism, namley undocumented wellbores [1,9]. This is particularly problematic in North America, where there are plenty of wellbores without documentation. This of course leads to great uncertainty – and indeed leakage risk – as the CO_2 could simply escape via this route. A major and thorough survey could help here, but the question whether these wellbores can be sealed over long time periods remains (does the sealing cement hold?). And, most importantly, as mentioned above, it is often not clear in many sites that there is indeed a good caprock present.

DISSOLUTION TRAPPING

CO_2 dissolves in brine as it migrates upwards [20,21]. This CO_2-enriched brine has a slightly higher density than the original brine, so that the CO_2-rich brine sinks in the reservoir over hundreds to millions of years [22,23,12] (cp. Figs. **5** and **6**). This is a very safe mechanism, but also a very slow process, which means that it is risky in the short term (= initial several hundreds of years), since the CO_2 may escape before it can dissolve.

Thermodynamically it has been shown that 0.9-3.6 mol% of CO_2 can be dissolved in brine, depending on pressure, temperature and brine composition [24-27]; less CO_2 can be dissolved as the brine salinity increases. Because large brine volumes are available, all anthropogenic CO_2 could theoretically be disposed this way.

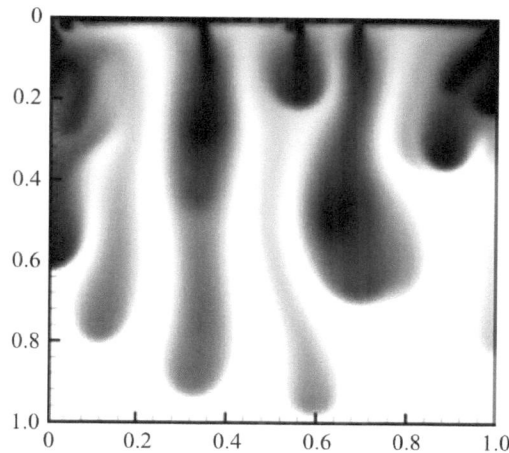

Figure 5: CO_2-enriched brine sinks in a normalized simulated reservoir over hundreds to thousands of years (from [12] with permission from Cambridge University Press). The concentration contours are shown in grayscales. The x- and y-axis are normalized lengths, the corresponding absolute values are in the kilometer range.

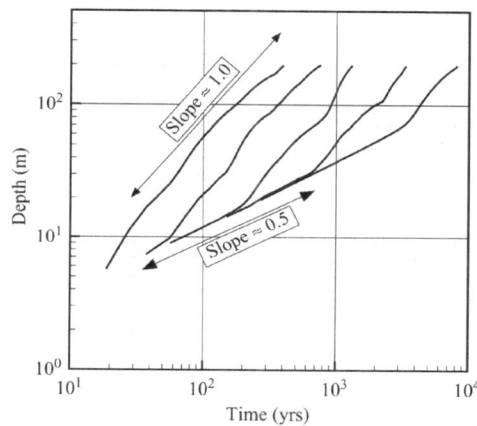

Figure 6: Time required until the CO_2-enriched brine front reaches the bottom of the reservoir (from [12] with permission from Cambridge University Press).

There is a however a detrimental effect in terms of CO_2 dissolving in H_2O in the context of CCS, namely CO_2 reacts with H_2O (scheme 1) which lowers the pH value.

$$CO_2 + H_2O \leftrightarrow H_2CO_3 \leftrightarrow H^+ + HCO_3^- \leftrightarrow 2H^+ + CO_3^{2-}$$

Scheme 1: CO_2 reacting with H_2O.

The reduction in pH value is significant under CCS conditions; pH values between 2.8- 3.2 have been measured in the laboratory [28,29]. These acidic conditions lead to geochemical effects, specifically dissolution of minerals. This has been described for geological diagenetic processes [30] (cp. schemes 2-5), but data for CCS conditions is extremely scarce.

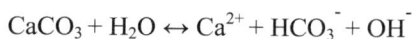

$$CaCO_3 + H_2O \leftrightarrow Ca^{2+} + HCO_3^- + OH^-$$

Scheme 2: Calcium carbonate dissolution [30]

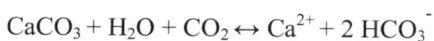

$$CaCO_3 + H_2O + CO_2 \leftrightarrow Ca^{2+} + 2\,HCO_3^-$$

Scheme 3: Accelerated dissolution of calcium carbonate by addition of CO_2 [30,31].

$$Mg_2SiO_4 + 4\ H_2O \leftrightarrow 2\ Mg^{2+} + 4\ OH^- + Si(OH)_4^0$$

Scheme 4: Magnesium silicate dissolution [30].

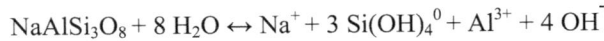

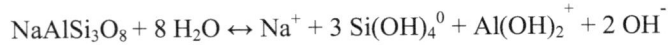

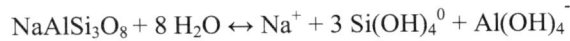

$$NaAlSi_3O_8 + 8\ H_2O \leftrightarrow Na^+ + 3\ Si(OH)_4^0 + Al^{3+} + 4\ OH^-$$

$$NaAlSi_3O_8 + 8\ H_2O \leftrightarrow Na^+ + 3\ Si(OH)_4^0 + Al(OH)_2^+ + 2\ OH^-$$

$$NaAlSi_3O_8 + 8\ H_2O \leftrightarrow Na^+ + 3\ Si(OH)_4^0 + Al(OH)_4^-$$

Schemes 5a-c: Sodium-aluminium silicate dissolution [30].

More recently, an accelerated hydrolysis of a plagiophase [$(Na,Ca)(Al,Si)_4O_8$] under CCS conditions was observed, this was apparently due to the CO_2 [32].

There are several problems associated with such mineral dissolution: a) with the porosity increase, which will be most likely highest in the high permeability areas, the permeability for CO_2 may significantly increase, which increases probability of leakage and complicates predictions *via* reservoir simulations, b) if porosity is increased too much, then structural reservoir integrity may be lost, which could result in land subsidence (as seen in former coal mines) or even earthquakes.

MINERAL TRAPPING

After initial dissolution of host rock the brine and/or the host rock react with the CO_2 (which is then present in the form of aqueous HCO_3^- or CO_3^{2-} anions) and precipitates as a solid, *e.g.* $CaCO_3$, $MgCO_3$, $MgCa(CO_3)_2$ or $CaSO_4$ [28,33,34].

A few additional mechanisms have been summarized by the IPCC [9], however their reaction velocities are probably slow and most importantly the quantities of reactants required are not available (schemes 6-8).

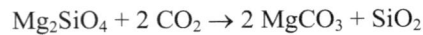

$$Mg_2SiO_4 + 2\ CO_2 \rightarrow 2\ MgCO_3 + SiO_2$$

Scheme 6: Olivine conversion.

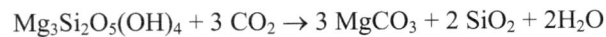

$$Mg_3Si_2O_5(OH)_4 + 3\ CO_2 \rightarrow 3\ MgCO_3 + 2\ SiO_2 + 2H_2O$$

Scheme 7: Serpentine conversion.

$$CaSiO_3 + CO_2 \rightarrow CaCO_3 + SiO_2$$

Scheme 8: Wollastonite conversion.

In addition, alumino-silicate minerals that can act as cation donors are able to trap CO_2 in solid form, possible example reaction mechanisms include [35].

a) feldspar (albite) alteration

$$NaAlSi_3O_8 + CO_2 + H_2O \rightarrow NaAlCO_3(OH)_2\ (dawsonite) + 3\ SiO_2$$

b) alteration of clay minerals, *e.g.* chlorite trapping

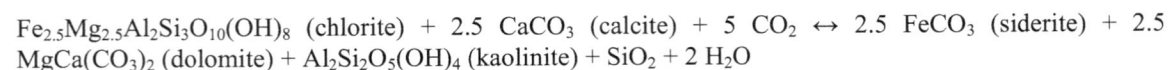

$$Fe_{2.5}Mg_{2.5}Al_2Si_3O_{10}(OH)_8\ (chlorite) + 2.5\ CaCO_3\ (calcite) + 5\ CO_2 \leftrightarrow 2.5\ FeCO_3\ (siderite) + 2.5\ MgCa(CO_3)_2\ (dolomite) + Al_2Si_2O_5(OH)_4\ (kaolinite) + SiO_2 + 2\ H_2O$$

c) alteration of cement; this is a possible process, however its effectiveness is debated as degradation occurs under CO_2 storage conditions.

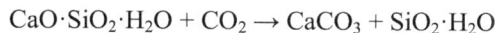

$$CaO \cdot SiO_2 \cdot H_2O + CO_2 \rightarrow CaCO_3 + SiO_2 \cdot H_2O$$

$$Ca(OH)_2 + CO_2 \rightarrow CaCO_3 + H_2O$$

Mineral trapping is the safest mechanism, since the carbon is bound in a solid; however, it is also the slowest one - storage times are estimated to be in the range of thousands to billions of years [9,34,36,37]. These simulations and theoretical conclusions are based on data measured at ambient conditions which was then extrapolated to CCS conditions. It is possible that the chemistry under these conditions (higher pressure, elevated temperature and much lower pH value) is considerably different. This needs clarification and further research.

Apart from this, there is another problem associated with solid precipitation, namely reduced injectivity. On reservoir scale, CO_2 can only be injected until a certain pressure, the fracture pressure, is reached. If more CO_2 is injected above this threshold pressure, the reservoir rock fractures and CO_2 will escape through the formed fractures. The injectivity is however closely related to the permeability of the rock by Darcy's law. If solid mineral is precipitated, then the permeability and with that the injectivity is reduced. This again results in lower possible injection pressures and lower injection flow rates, which will increase the project time and may turn the project into an unfeasible one. This is especially the case when solid is precipitated close to the injection wellbore. This scenario is not only possible by carbonate precipitation, but also by salt precipitation which is caused by drying the near-wellbore area with the dry CO_2 injected [38].

CAPILLARY TRAPPING

CO_2 can be trapped as pore-scale bubbles in the porous matrix due to capillary forces [39]. This is analogous to residual hydrocarbon trapping in waterflooded oil or gas reservoirs, and it is well established that the residual oil cannot be produced without special (tertiary) methods [40]. This mechanism has been identified to be highly reliable and rapid, *i.e.* it can trap the CO_2 in a matter of decades if the injection process is designed appropriately [9,41-45].

The physical process of capillary CO_2 trapping can be divided into three steps:

1. $scCO_2$ is injected into the fully brine saturated porous medium (the reservoir), and brine is displaced because of viscous forces. This is a two-phase flow process (in case of an aquifer; it may be a three-phase flow process in an oil reservoir), and depending on exposure time and location in the reservoir, CO_2 dissolves in and equilibrates with the brine and subsequently reacts with the water to some extend (cp. Section 2 dissolution trapping and scheme 1).

2. The $scCO_2$ migrates upwards due to buoyancy forces.

3. Natural water flow will follow the upward moving CO_2 plume and trap a fraction of the non-wetting phase (CO_2) in snap-off processes. This process can be optimized by chase water injection or water-alternating-gas (WAG) designs so that maximum quantities of CO_2 can be trapped [45].

In Fig. **7** the capillary trapping phenomenon at the pore-level (micrometer-level) is visualized via micro-computed tomography (μ-CT) experiments. These images were recorded at ambient conditions with brine and oil (as the non-wetting phase to represent $scCO_2$). The flow conditions which established this residual condition were representative of flow in an aquifer or oil reservoir. Here the wetting phase is brine and the non-wetting phase is n-octane. The porous medium was a sandstone. In Fig. **7a**, a grayscale image is displayed, where oil is dark, sandstone light gray and brine dark grey; the oil clearly accumulates as pore-scale bubbles in the largest pores. The oil is located in the largest pores because during the drainage phase (oil or CO_2 injection), the largest pores are filled first with the non-wetting fluid because of capillary forces. During the imbibition step (waterflooding), the nonwetting phase is then trapped in such isolated bubbles and hold in place by strong capillary forces – they are capillary trapped. Fig. **7b** shows the same sample in 3D, where the phases were segmented in blue (brine) and red (oil) – rock was removed for better visualization. Fig. **7c** illustrates the distribution of the residual non-wetting phase clusters in a thinner subsection. The nature of these clusters is topic of current research [46-49]. Acquisition of similar images with scCO_2 at reservoir conditions is part of our ongoing research.

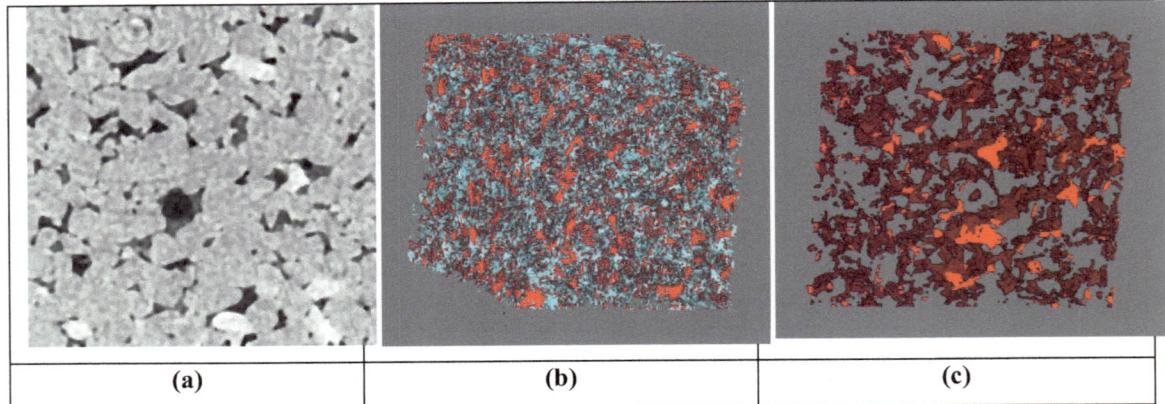

| (a) | (b) | (c) |

Figure 7: (a) 2D grayscale image of a sandstone filled with brine and residual oil drops. The area displayed consists of 300 x 300 pixels = 7.29 mm². An oil drop (black round area) is clearly visible in the middle of a pore, surrounded by brine (dark gray). The sandstone has a light gray color. (b) 3D fluid distribution. Oil is red, brine blue; the volume displayed consists of 300 x 300 x 300 voxels = 19.683 mm³. (c) Residual oil cluster distribution, the displayed volume is 300 x 300 x 30 voxels = 1.9683 mm³.

Such experiments are important in order to verify the spreading and contact angles of scCO_2 on sandstone (or carbonates) in brine, because these wettability factors significantly effect amounts of residual saturation [50] and flow dynamics [51]. In fact there is evidence that CO_2 spreads more on quartz at higher pressure [19]. One result is that the capillary pressure in secondary imbibition can be negative as observed for a brine/scCO_2/quartz sand system [52]. It is therefore possible that less scCO_2 is trapped under CCS conditions; this needs more study.

Capillary trapping has been measured extensively in the context of oil recovery, and a comprehensive summary has been compiled by Pentland *et al.* [53]. One important parameter influencing capillary trapping is the porosity ϕ of the porous medium. An increase in ϕ strongly decreases the amount of residual non-wetting (trapped) phase, $S_{nw,r}$, [54,55] as displayed in Fig. **8**. The scatter in this graph arises from the different fluid systems investigated, different rock samples with different pore morphologies and wettability states and the fact that some authors used different displacement methods which may not be representative of reservoir flow.

Based on this relation the concept of capillary trapping capacity (C_{trap}) can be developed, which states how much CO_2 can be stored per unit rock volume solely by capillary forces [54]. C_{trap} is the product of the porosity ϕ and the trapped non-wetting phase saturation $S_{nw,r}$.

$$C_{trap} = \phi S_{nw,r} \qquad\qquad (1)$$

C_{trap} is the most important parameter in terms of determining CO_2 storage capacities if relying only on capillary trapping. If C_{trap} is plotted against porosity ϕ, then a non-monotonic curve is obtained which has a maximum around $\phi = 20\%$ (Fig. **9**). This indicates that there is an optimal porosity for CCS. The influence of the exact pore morphology on $S_{nw,r}$ is subject of our ongoing research.

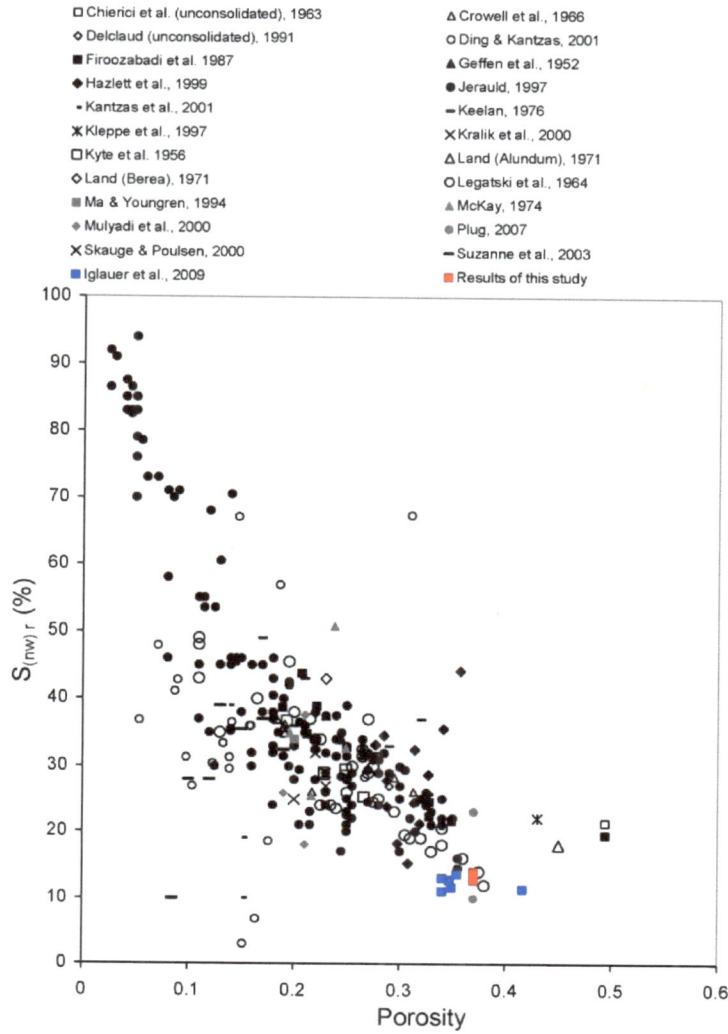

Figure 8: Non-wetting phase fraction trapped by capillary forces plotted versus the porosity of the material; summary of literature data (from [56] with permission from Elsevier).

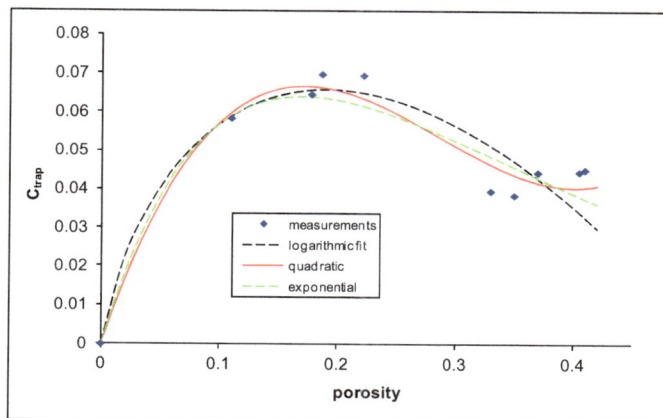

Figure 9: Capillary trapping capacity as a function of porosity (from [54], copyright SPE, with permission from SPE).

C_{trap} is also a function of the initial non-wetting phase saturation $S_{nw,i}$; basically this means that if there is a lot of scCO$_2$ initially (after the drainage step, *i.e.* CO$_2$ injection), then more of it can be trapped. A summary of this concept is shown in Fig. **10**, where C_{trap} is plotted against $S_{nw,i}$ for a range of porous media. The points shown in color were measured by us, we studied an unconsolidated sandpack system ($\phi = 37\%$) and used brine and air (blue) or brine and oil (red) (as an analogue to $scCO_2$) [53,56]. The black points are data measured on consolidated samples; as can be seen, C_{trap} depends on $S_{nw,i}$ and is higher for consolidated media.

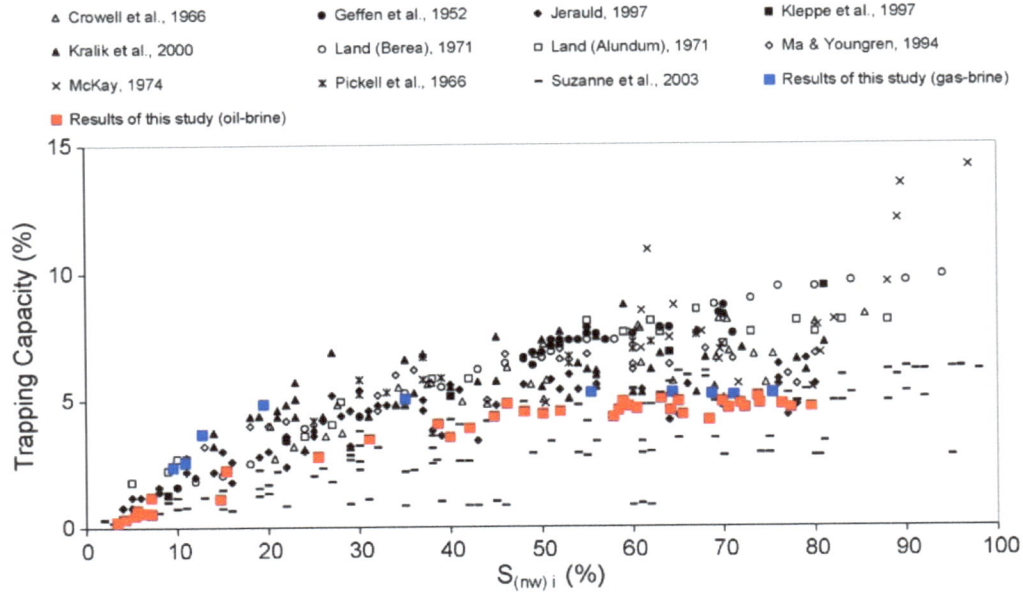

Figure 10: C_{trap} as a function of initial non-wetting phase saturation $S_{nw,i}$. The colored points represent an unconsolidated sandpack ($\phi = 37\% = const.$) with brine/air (blue) or brine/oil (red) fluid systems. The black points represent consolidated media (from [56] with permission from Elsevier).

The influence of wettability, contact angles, high pressure and rock chemistry are topic of our current research – we expect that these parameters can also significantly render C_{trap} and thereby CO_2 storage capacity and reliability. For example, in a potentially CO_2-wet reservoir, there would be less capillary trapping of CO_2; the CO_2 would then be able to migrate even after waterflooding.

All the pore-scale and core-scale phenomena described above are highly relevant at the reservoir (km) scale. For example, Qi *et al.* [45] showed in a reservoir simulation, that 95% of the CO_2 injected can be trapped by capillary forces in a matter of decades if the process is optimized by chase brine injection (Fig. **11**). Fig. **11a** shows the trapped $scCO_2$ saturation, and Fig. **11b** the $scCO_2$ which is still mobile after the chase brine injection. As displayed, most of the CO_2 has been immobilized.

An important engineering parameter to consider is the mobility ratio (M) $scCO_2$ to brine. M is defined as the ratio of CO_2 and brine mobilities λ_{CO2} and λ_{brine} [40].

$$M = \lambda_{CO2}/\lambda_{brine}$$

(2)

with

$$\lambda_i = k_i/\mu_i$$

(3)

where k_i is the permeability of fluid i and μ_i its viscosity. k_i is a function of the fraction of fluid i in the pore space.

Considering possible brine and CO_2 viscosities under CCS conditions [1], M may be in the range 10-25, and certainly above 1. However, if M > 1, then unstable flow occurs, causing viscous fingering which leads to the quick migration of CO_2 upwards through high permeability channels [57], and low sweep efficiency, *i.e.* the average initial $scCO_2$ saturation over the whole reservoir volume is comparatively low. In fact all reservoirs have some form of heterogeneity incorporated (porosity and permeability distribution in the reservoir; the permeability distribution can span several magnitudes caused for example by fractures, this is especially the case in carbonate reservoirs). A low $S_{nw,i}$ results in low average $S_{scCO2,r}$ as discussed above. The solution to this problem suggested by Qi *et al.* [45] is to use WAG injections (alternating CO_2-brine injections), which render the effective mobility ratio between the injected mixture of CO_2 and brine and brine alone favorable (*i.e.* M < 1). Another theoretical solution to this problem would be to increase the viscosity of the CO_2 by additives, such as surfactant to produce a foam [58].

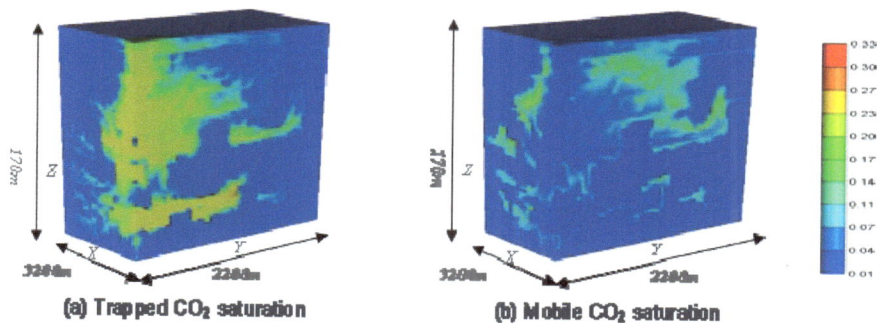

(a) Trapped CO₂ saturation   **(b) Mobile CO₂ saturation**

Figure 11: (a) CO_2 trapped in a model reservoir after 20 years of CO_2 injection followed by 2 years of chase brine injection, (b) only a small amount of CO_2 is still mobile (from [45] with permission from Elsevier).

In Otway, Australia, a pilot field project (CO2CRC) has started where the feasability of this capillary trapping idea is tested. Capillary trapping capacities will be measured and CO_2 migration monitored in test wells. These data will be tremendously valuable in order to improve risk assessment associated with CCS projects.

A similar test project is planned in Northern California (Solano County), where 6000 t of CO_2 will be injected into a saline formation which is structurally a syncline in the subsurface. There is no structural trap for the CO_2, so here CO_2 storage soley relies on capillary trapping, dissolution trapping and mineral trapping [59].

CONCLUSIONS

Overall it can be said that CCS is a feasible short-to-medium term solution in terms of reducing CO_2 emissions and thereby mitigating climate change. In the medium-to-long term, energy sources should be switched to renewable which do not emit GHG, and improvements in energy efficiency will also help significantly.

As $scCO_2$ has been used for enhanced oil recovery for more than 50 years, there is sufficient understanding of the engineering and scientific problems associated with CCS. Of course, as outlined in this text, several open questions remain and further research should be undertaken to address them, including theoretical, laboratory and most importantly additional field-scale pilot projects, mainly to satisfy public concerns. The best thing about CCS is that we can implement it now, all the technology is there, while waiting for potentially massively damaging climate change sounds pretty risky.

There is one last question: Who is going to pay for all this?

We should keep in mind that climate change including floods, droughts, hurricanes, land loss, *etc.* could be far more expensive than implementing CCS.

NOMENCLATURE

CCS	carbon capture and storage (of carbon dioxide)
CO_2	carbon dioxide
GHG	greenhouse gas
RF	radiative forcing [W/m^2]
ppb	parts per billion
a	year
Gt	Gigatons = 10^9 tons
$S_{CO2,i}$	initial CO_2 saturation
$S_{CO2,r}$	residual CO_2 saturation
$S_{nw,i}$	initial non-wetting phase saturation
$S_{nw,r}$	residual non-wetting phase saturation
ϕ	porosity [-]
k	permeability [m^2]
M	mobility ratio [-]
μ	viscosity [Pa.s]
C_{trap}	capillary trapping capacity [-]

REFERENCES

[1]   Intergovernmental Panel on Climate Change (IPCC), "Climate Change 2007: The Physical Science Basis. Working Group I Contribution to the Fourth Assessment Report of the Intergovernmental Panel on Climate Change," Cambridge University Press, 2007.

[2]   M. Sengul, A.E. Pillay, C.G. Francis, and M. Elkadi, "Climate change and carbon dioxide (CO_2) sequestration: an African perspective," *Trade & Ind. Monitor*, vol. 69, pp. 69-78, 2008.

[3]   D. Archer and V. Brovkin, "The millennial atmospheric lifetime of anthropogenic CO_2", *Climatic Change*, vol. 90, pp. 283-297, 2008.

[4]   R.B. Alley, P.U. Clark, P. Huybrechts, and I. Joughin, "Ice-sheet and sea-level changes," *Science*, vol. 310, pp. 456-471, 2005.

[5]   M.I. Hoffert, K. Caldeira, A.K. Jain, E.F. Haites, L.D. Harvey, S.D. Potter, M.E. Schlesinger, S.H. Schneider, R.G. Watts, T.M. Wigley, and D.L. Wuebbles, "Energy implications of future stabilization of atmospheric CO_2 content," *Nature*, vol. 395, pp. 881-884, 1998.

[6]   K.S. Lackner, "A guide to CO_2 sequestration," *Science,* vol. 300, 5626, pp. 1677-1678, 2003.

[7]   F.M. Orr, "Storage of carbon dioxide in geologic formations", *J. Petroleum Tech.*, vol. 56, 9, pp. 90-97, 2004.

[8]   R.S. Haszeldine, "Carbon capture and storage: how green can black be?," *Science*, vol. 325, pp. 1647-1652, 2009.

[9]   Intergovernmental Panel on Climate Change (IPCC), "IPCC Special Report on Carbon Dioxide Capture and Storage. Prepared by Working Group III of the Intergovernmental Panel on Climate Change", Cambridge University Press, 2005.

[10]   S. Holloway, "Storage of fossil fuel-derived carbon dioxide beneath the surface of the earth," *Ann. Revision Energy Environment*, vol. 26, pp. 145-166, 2001.

[11]   S. Bachu, "Sequestration of CO_2 in geological media: criteria and approach for site selection in response to climate change," *Energy Conversion & Management*, vol. 41, pp. 953-970, 2000.

[12]   A. Riaz, M. Hesse, H.A. Tchelepi, and F.M. Orr, "Onset of convection in a gravitationally unstable diffusive boundary layer in porous media," *J. Fluid Mech.*, vol. 548, pp. 87-111, 2006.

[13]   D. Silin, T.W. Patzek, and S.M. Benson, "A one-dimensional model of vertical gas plume migration through a heterogeneous porous medium," *Int. J. Greenhouse Gas Control*, vol. 3, pp. 300-310, 2009.

[14]   The Bellona Foundation, Website: "http://www.bellona.org/ccs," Retrieved April 10, 2010.

[15]   S. Bachu, W.D. Gunter, and E.H. Perkins, "Aquifer disposal of CO_2 – hydrodynamic and mineral trapping," *Energy Conversion & Management*, vol. 35, 4, pp. 269-279, 1994.

[16]   M.A. Hesse, F.M. Orr, and H.A. Tchelepi, "Gravity currents with residual trapping," *J. Fluid Dynamics*, vol. 611, pp. 35-60, 2008.

[17] T. Suekane, T. Nobuso, S. Hirai, and M. Kiyota, "Geological storage of carbon dioxide by residual gas and solubility trapping," *Int. J. Greenhouse Gas Control*, vol. 2, pp. 58-64, 2008.

[18] C. Chalbaud, M. Robin, J.-M. Lombard, F. Martin, P. Egermann, and H. Bertin, "Interfacial tension measurements and wettability evaluation for geological CO_2 storage," *Advs. Water Resources*, vol. 32, pp. 98-109, 2009.

[19] P. Chiquet, D. Broseta, and S. Thibeau, "Wettability alteration of caprock minerals by carbon dioxide," *Geofluids*, vol. 7, pp. 112–122, 2007.

[20] K. Pruess and J.Garcia, "Multiphase flow dynamics during CO_2 disposal into saline aquifers," *Environmental Geology*, vol. 42, pp. 282-295, 2002.

[21] S. Bachu and J.J. Adams, "Sequestration of CO_2 in geological media in response to climate change: capacity of deep saline aquifers to sequester CO_2 in solution", *Energy Conversion & Management*, vol. 44, pp. 3151-3175, 2003.

[22] J. Ennis-King and L. Paterson, "Role of convective mixing in the long-term storage of carbon dioxide in deep saline formations," *SPE Journal*, vol. 10, 3, pp. 349-356, 2005.

[23] E. Lindeberg and D. Wessel-Berg, "Vertical convection in an aquifer column under a gas cap of CO_2." *Energy Conversion & Management*, vol. 38, 1, pp. 229-234, 1997.

[24] B. Rumpf, H. Nicolaisen, C. Öcal, and G. Maurer, "Solubility of carbon dioxide in aqueous solutions and sodium chloride: experimental results and correlation," *J. Solution Chem.*, vol. 23, 3, pp. 431-448, 1994.

[25] D. Koschel, J.-Y. Coxam, L. Rodier, and V. Majer, "Enthalpy and solubility data of CO2 in water and $NaCl_{(aq)}$ at conditions of interest for geological sequestration," *Fluid Phase Equilibria*, vol. 247, pp. 107-120, 2006.

[26] S. Bando, F. Takemura, M. Nishio, E. Hihara, and M. Akai, "Solubility of CO2 in aqueous solutions of NaCl at (30 to 60) °C and (10 to 20) MPa," *J. Chem. Eng. Data*, vol. 48, pp. 576-579, 2003.

[27] J. Kiepe, S. Horstmann, K. Fischer, and J. Gmehling, "Experimental determination and prediction of gas solubility data for $CO_2 + H_2O$ mixtures containing NaCl or KCl at temperatures between 313 and 393 K and pressures up to 10 MPa," *Ind. Eng. Chem. Res.*, vol. 41, pp. 4393-4398, 2002.

[28] P. Egermann, B. Bazin, and O. Vizika, "An experimental investigation of reaction-transport phenomena during CO_2 injection," in *Proc.14th Middle East Oil & Gas Show*, Bahrain, 2005, SPE 93674.

[29] H.T. Schaeff and B.P. McGrail, "Direct measurements of pH in H_2O-CO_2 brine mixtures to supercritical conditions," *7th Int. Conf. Greenhouse Gas Control Technologies*, Vancouver, Canada, 2004.

[30] C.D. Curtis, "Geochemistry of porosity enhancement in clastic sediments," in *Petroleum Geochemistry and Exploration of Europe*, J. Brooks Ed., Oxford: Blackwell Scientific Publications, 1983, pp. 113-125.

[31] A.F. Holleman, N. Wiberg, "*Lehrbuch der anorganischen Chemie*", Berlin: De Gruyter, 1985, p. 196.

[32] H. Lin, T. Fujii, R. Takisawa, T. Takahashi, and T. Hashida, "Experimental evaluation of interactions in supercritical CO_2/water/rock minerals system under geologic CO_2 sequestration conditions," *J. Material Sci.*, vol. 43, 7, pp. 2307-2315, 2008.

[33] T. Xu, J.A. Apps, K. Pruess, and H. Yamamoto, "Numerical modelling of injection and mineral trapping of CO_2 with H_2S and SO_2 in a sandstone formation," *Chem. Geology*, vol. 242, pp. 319-346, 2007.

[34] T.F. Xu, J.A. Apps, and K. Pruess, "Reactive geochemical transport simulation to study mineral trapping for CO_2 disposal in deep arenaceous formations," *J. Geophysical Res.*, vol. 108, B2, 2071, 2003.

[35] I. Gaus, "Role and impact of CO_2-rock interactions during CO_2 storage in sedimentary rocks," *Int. J. Greenhouse Gas Control*, vol. 4, pp. 73-89, 2010.

[36] W.D. Gunter, B. Wiwchar, and E.H. Perkins, "Aquifer disposal of CO_2-rich greenhouse gases: extension of the time scale of experiment for CO_2-sequestring reactions by geochemical modeling," *Mineralogy & Petrology*, vol. 59, pp. 121-140, 1997.

[37] W.D. Gunter and E.H. Perkins, "Aquifer disposal of CO_2 rich gases: reaction design for added capacity," *Energy Conversion & Management*, vol. 34, 9-11, pp. 941-948. 1993.

[38] N. Muller, R. Qi, E. Mackie, K. Pruess, and M.J. Blunt, "CO_2 injection impairment due to halite precipitation," *Energy Procedia*, pp. 3507-3514, 2009.

[39] C.H. Pentland, R. El-Maghraby, S. Iglauer, and M.J. Blunt, "Measurements of the capillary trapping of supercritical carbon dioxide in Berea sandstone," *Geophysical Res. Lett.*, vol. 38, L06401, 2011.

[40] D.W. Green and G.P. Willhite, "*Enhanced oil recovery*", Richardson: SPE Publications, 1998.

[41] R. Juanes, E.J. Spiteri, F.M. Orr, and M.J. Blunt, "Impact of relative permeability hysteresis on geological CO_2 storage," *Water Resources Res.*, vol. 42, W12418, 2006.

[42]   A. Kumar, R. Ozah, M. Noh, G.A. Pope, S. Bryant, K, Sepehrnoori, and L.W. Lake, "Reservoir simulation of CO$_2$ storage in deep saline aquifers," *SPE Journal*, vol. 10, 3, pp. 336-348, 2005.

[43]   M. Flett, R. Gurton, and I. Taggart, "The function of gas-water relative permeability hysteresis in the sequestration of carbon dioxide in saline formations," in *Proc. SPE Asia Pacific Oil and Gas Conf. Exh.*, Perth, Australia, 2004, SPE 88485.

[44]   S.T. Ide, K. Jessen, and F.M. Orr, "Storage of CO$_2$ in saline aquifers: effects of gravity, viscous, and capillary forces on amount and timing of trapping," *Int. J. Greenhouse Gas Control*, vol. 1, pp. 481-491, 2007.

[45]   R. Qi, T.C. LaForce, and M.J. Blunt, "Design of carbon dioxide storage in aquifers," *Int. J. Greenhouse Gas Control*, vol. 3, pp. 195-205. 2009.

[46]   M. Kumar, T.J. Senden, A.P. Sheppard, J.P. Middleton, and M.A. Knackstedt, "Visualizing and quantifying the residual phase distribution in core material," in *Proc. Int. Sym. Society of Core Analysts*, Noordwiijk, Netherlands, 2009, SCA2009-16.

[47]   M. Kumar, J.P. Middleton, A.P. Sheppard, T.J. Senden, and M.A. Knackstedt, "Quantifying trapped residual oil in reservoir core material at the pore scale: exploring the role of displacement rate, saturation history and wettability," in *Proc. Int. Petroleum Tech. Conf.*, Doha, Qatar, 2009, IPRC 14001.

[48]   M. Prodanovic, W.B. Lindquist, and R.S. Seright, "3D image-based characterization of fluid displacement in a Berea core," *Adv. Water Resources*, vol. 30, pp. 214-226, 2007.

[49]   S. Iglauer, S. Favretto, G. Spinelli, G. Schena, and M.J. Blunt, "X-ray tomography measurements of power-law cluster size distributions for the nonwetting phase in sandstones," *Phys. Rev. E,* vol. 82, 056315, 2010.

[50]   E.J. Spiteri, R. Juanes, M.J. Blunt, and F.M. Orr, "A new model of trapping and relative permeability hysteresis for all wettability characteristics," *SPE Journal*, vol. 13, 3, pp. 277-288, 2008.

[51]   J. Bear, *"Dynamics of Fluids in Porous Media"*, New York: Dover Publications, 1988.

[52]   W.-J. Plug and J. Bruining, "Capillary pressure for the sand-CO$_2$-water system under various pressure condition. Application to CO$_2$ sequestration," *Adv. Water Resources*, vol. 30, pp. 2339-2353, 2007.

[53]   C.H. Pentland, E. Itsekiri, S.K. Al Mansoori, S. Iglauer, B. Bijelic, and M.J. Blunt, "Measurement of nonwetting-phase trapping in sandpacks," *SPEJ*, vol. 15, pp 270-277, 2010.

[54]   S. Iglauer, W. Wülling, C.H. Pentland, S.K. Al Mansoori, and M.J. Blunt, "Capillary trapping capacity of rocks and sandpacks," in *Proceedings of the SPE EUROPEC/EAGE Ann. Conf. Exh.*, Amsterdam, The Netherlands, SPE 120960, 2009.

[55]   G.R. Jerauld, "Prudhoe Bay gas/oil relative permeability," *SPE Reserv. Eng.*, vol. 12, pp. 66-73, 1997.

[56]   S. Al-Mansoori, E. Itsekiri, S. Iglauer, C.H. Pentland, B. Bijeljic, and M.J. Blunt, "Measurements of non-wetting phase trapping applied to carbon dioxide storage," *Int. J. Greenhouse Gas Control*, vol. 4, pp. 283-288, 2010.

[57]   F.M. Orr, "CO$_2$ capture and storage: are we ready?," *Energy & Environmental Sci.*, 2, pp. 449-458, 2009.

[58]   G.C. Wang, "A laboratory study of CO$_2$ foam properties and displacement mechanism," in *Proceedings of the SPE Enhanced Oil Recovery Symposium*, Tulsa, Oklahoma, USA, 1984, SPE 12645.

[59]   L. Myer, *private comm.*, 2010.

Advances in Multiphase Flow and Heat Transfer, Vol. 3, 2012, 151-187

CHAPTER 5

Discrete Particle Model for Dense Gas-Solid Flows

C. L. Wu[1*] and A. S. Berrouk[2]

[1]*Engineering College, Guangdong Ocean University, Zhanjiang 524088, China and* [2]*Department of Chemical Engineering, The Petroleum Institute, P.O. Box 2533, Abu Dhabi, United Arab Emirates*

Abstract: The discrete particle model (DPM) is a mesoscale method used to study the hydrodynamics of dense dispersed flows. In this approach, the particle motion is described in a Lagrangian framework by directly solving the Newtonian kinetic equations of each individual particle while the gas flow is studied in an Eulerian framework. The constitutive relations for the dispersed phase are not required because the particle-particle interactions are modeled through a two-variant collision-handling algorithm. In this chapter a full understanding of the DPM technique is presented through a detailed description of the numerical model and the results of its applications to gas-solid fluidization systems. We also detail a multi-component numerical strategy developed by the authors to enhance the DPM efficiency.

Keywords: Discrete particle model, gas-solid flow, fluidization system, numerical efficiency.

INTRODUCTION

Gas-solid flows are encountered in the chemical and petrochemical industries for panoply of processes including catalytic cracking, drying, coating, plastic gasification and combustion. In such a two-phase system, small or fine solid particles are dispersed in the continuous gas phase. A dispersed flow can be classified as being either dilute or dense according to the dominant mechanism controlling the motion of the dispersed phase. In a dilute flow regime, the particle motion is controlled by surface and body forces on the particle (such as drag, lift, gravity and/or magnetic forces). In a dense regime, on the other hand, the motion of the dispersed phase is significantly affected by particle-particle collisions or interactions. Two time scales are often used to distinguish the flow regime [1]. One is the particle response time which is defined by

$$t_p = \frac{\rho_p d_p^2}{18\mu}$$

the other is the average collision time

$$t_c = \frac{1}{n\pi d_p^2 v_r}$$

where n is the particle number density, and v_r is the relative particle velocity. d_p and ρ_p are the particle diameter and density respectively. The flow can be considered as dense if $t_c < t_p$. In such a case, the particles will not have sufficient time to respond to the local gas dynamic forces before the next collision. Otherwise the flow can be considered as dilute.

The hydrodynamics of the dense gas-solid flow has dominant effects on the performance of the chemical reactors in terms of, for instance, reactants' mixing rate, contacting period and separation efficiency. The amount of experimental and numerical research conducted to investigate such important effects has risen

*Address correspondence to C. L. Wu: Engineering College, Guangdong Ocean University, Zhanjiang 524088, China; E-mail: chunliangwu@gmail.com

dramatically in the past 20 years which reflects the need for better knowledge of the determinant role played by the flow dynamics in improving industrial chemical reactors' hardware and performance. Failing to understand the mechanisms through which dense gas-solid flows relate to reactors' hardware and performance makes of the design and scale-up of these industrially important gas-solid reactors, a difficult task. In most cases, the design and scale-up of fluidized bed reactors is a fully empirical process based on some preliminary tests on pilot-scale model reactors, which are time consuming and expensive. Computer simulation has become a powerful tool to aid this design and scale-up process. As a consequence, many physical models have been developed and numerical simulations based on these models are used to gain more insight into the complex hydrodynamics and predict the flow behaviors of the gas-particle systems in chemical reactors.

Due to the multi-scale character of the dense gas-particle systems, several models at different scale levels have been formulated [2, 3]. At the macroscopic level, mixture or two-fluid models (TFM) need closure laws to describe the two-phase interaction and the particulate matter rheology. Although these macroscopic models are the most tractable to simulate actual fluid-particle systems that could help their optimization, they are known to be silent about microstructure and they have never acknowledged the very important micro-structural properties that are induced by particle-fluid interaction. This has motivated the recourse to finer numerical descriptions at the microscopic scale. Indeed, at that level of description, direct numerical simulation (DNS) has been used to fully resolve the gas flow details around the particles. DNS models based on Lattice Boltzmann method, immersed boundary method and fictitious domain or arbitrary Lagrangian-Eulerian methods do not require closure relations. However, they are computationally intensive and resource demanding when applied to dense gas-solid systems of practical relevance, which are three-dimensional and contain a large number of solid particles. This has made DNS models unwieldy to apply for practical engineering problems. Notwithstanding this limitation, however, the DNS results are of great value to understand the physical mechanisms through which fluid and particle interact and help derive more accurate closure relations [4, 5].

Between these two levels of numerical description, a mesoscopic approach for dispersed gas-solid flows has been developed under the name of Discrete Particle Model (DPM) or Discrete Element Model (DEM). In this approach, the particle motion is described in a Lagrangian framework by directly solving the Newtonian kinetic equations of each individual particle. The solids constitutive relations are not required because the particle-particle interactions are modeled through a two-variant collision-handling algorithm; the hard-sphere variant or the soft-sphere variant. The hard-sphere algorithm was originated from the molecular dynamics (MD) modeling of a set of molecules interacting *via* hard potentials [6]. The programming description of the MD simulation is widely available in the literature [7-9]. In the hard-sphere DPMs, the inter-particle interactions are modeled by binary collision dynamics. Collisions are assumed instantaneous and handled one by one according to the order of the collision event occurrence. This is the so-called event-driven. Since the first successful attempt of combining the hard-sphere model with computational fluid dynamics [10], the hard-sphere model has been widely used to simulate dense dispersed flows such as those taking place in bubbling fluidized bed [11-15], circulating fluidized bed [16-19], pressurized fluidized bed [20] and bubble columns [21]. This model was also used to validate the solid closure laws of the kinetic theory of granular dynamics [13, 22]. The soft-sphere variant of the DPM model [23] solves the differential equations of particle dynamics to obtain both momentum variations and displacements. In these equations, the inter-particle contact forces, namely, the normal, damping and sliding forces are accounted for using equivalent simple mechanical elements, such as springs, dashpots and sliders. The soft-sphere model is typically time-driven and requires a very small time step (often $< 10^{-6}$s) in particular when the particle-particle normal collisions are stiff. Generally, this character makes it less efficient than the hard-sphere model. This promotes especially in simulation of different size particle systems since the spring stiffness used in the model is dependent on the particle size. The soft-sphere discrete particle model has been employed by many researchers in numerical simulations of gas-solid flows. Examples are solid slug flow in pneumatic transport pipe and granular flow in bubbling fluidized bed [24, 25], particles' motion caused by lateral gas blasting into a bed [26], particle mixing behavior in bubbling fluidized bed [27]. Also, the soft-sphere model was combined with computational fluid dynamics to simulate gas-solid flows in fluidized beds where the event-driven scheme is applied only to precisely determine the first instant of contact [28].

The DPM has become a versatile tool since the pioneering works that focused on its development and validation [10, 12, 14, 23, 25, 28]. The numerical results obtained have demonstrated the powerful capabilities of the method [10-28]. Although the gas flow details around particles are not fully determined, the particle motion is resolved at such a particle scale that many important flow structures related to the particle motion can be reasonably captured. For instance, several investigation based on DPMs have helped to understand that the formation of heterogeneous structures in fluidized beds is attributed to the combination of energy dissipation due to inter-particle interactions and the strong dependence of the inter-phase drag force on void fraction [10, 12, 16, 19, 20, 29].

Numerically, almost all the methods employed to solve the DPM equations are limited to structured regular grids and usually the SIMPLE algorithm based on staggered variables layout is used to discretize and solve the continuous-phase governing equations. This restricts its application to simulation of gas-solid flow in simple and regular geometry. In addition, the high computational cost linked to DPM has limited its use to applications in a laboratory scale with the number of particles typically at the order of 100,000. This shortcoming can be overcome by designing a computationally-efficient multifaceted numerical strategy [30]. It is deemed that the model accuracy and its implementation efficiency depend strongly but not solely on: (1) void fraction computation, (2) inter-phase coupling, and (3) collision event handling through proper data structures.

The objective of this chapter is to describe the hard-sphere discrete particle model for gas-solid flows in arbitrary domains. A methodology of implementing the model is described along with a detailed illustration of the finite volume method for solving the gas-phase transport equations on generic unstructured grids. Issues related to the numerical implementation and the model efficiency are addressed in detail. Several applications of the model in simulation of gas-solid fluidization systems are demonstrated.

GOVERNING EQUATIONS

The governing equations of the gas phase mainly follow the volume-averaged form of the two-fluid model [31]. The particles' motion obeys Newtonian second law of motion while particle collisions are described by a three-parameter hard-sphere model [10, 14, 22]. Interactions between the two phases are accounted for through the momentum exchange terms.

There are two well-established Eulerian-Eulerian models for gas-soild flow, named Model A and Model B respectively [31]. In Model A, the static pressure is shared by both the gas and solid phases; whilst in Model B, the pressure gradient force is only acted on the gas phase. Thus the force due to the static pressure gradient does not appear in the momentum equations of the dispersed phase in Model B. However, its effect should be considered as an additional part of the drag as it has been discussed in the literature [5, 31-33].

Gas Phase Hydrodynamics

The conservation of the gas mass is balanced by the convective mass fluxes without inter-phase mass transfer:

$$\frac{\partial}{\partial t}\left(\varepsilon\rho_g\right)+\nabla\cdot\left(\varepsilon\rho_g\mathbf{u}_g\right)=0 \tag{1}$$

where ρ_g is the density of the gas phase, ε is the void fraction and $\mathbf{u}_g$ is the gas velocity.

The momentum balance equations are similar to the single-phase momentum equations. For Model A, it reads:

$$\frac{\partial}{\partial t}\left(\varepsilon\rho_g\mathbf{u}_g\right)+\nabla\cdot\left(\varepsilon\rho_g\mathbf{u}_g\mathbf{u}_g\right)=-\varepsilon\nabla p+\nabla\cdot\left(\varepsilon\bar{\bar{\mathbf{\tau}}}_g\right)+\mathbf{S}_g+\varepsilon\rho_g\mathbf{g} \qquad (2)$$

with the stress tensor evaluated as:

$$\bar{\bar{\mathbf{\tau}}}_g=\mu\left(\nabla\mathbf{u}_g+\nabla\mathbf{u}_g^T\right)+\frac{2}{3}\mu\nabla\cdot\mathbf{u}_g\bar{\bar{\mathbf{I}}}$$

While it reads for Model B as follows

$$\frac{\partial}{\partial t}\left(\varepsilon\rho_g\mathbf{u}_g\right)+\nabla\cdot\left(\varepsilon\rho_g\mathbf{u}_g\mathbf{u}_g\right)=-\nabla p+\nabla\cdot\left(\varepsilon\bar{\bar{\mathbf{\tau}}}_g\right)+\mathbf{S}_g+\rho_g\mathbf{g}$$

where p is the static pressure. The source term $\mathbf{S}_g$ represents the forces due to particle-fluid interactions, such as drag, lift and added mass forces. If only particle-fluid drag force is considered, two methods can be used to calculate this term. For the first one [10]:

$$\mathbf{S}_g=-\beta(\mathbf{u}_g-\mathbf{u}_{p,c}) . \qquad (3)$$

The second one is an integral form of Eq. (3), it takes the following form [14, 22]:

$$\mathbf{S}_g=-\frac{1}{V}\int_V\sum_{k=1}^{N_P}\frac{V_{p,k}\beta}{1-\varepsilon}(\mathbf{u}_g-\mathbf{u}_{p,k})\delta(\mathbf{x}-\mathbf{x}_{p,k})dV \qquad (4)$$

where V is the integral volume, V_p is the particle volume, N_p is the number of particles in the integral volume, $\mathbf{u}_p$ is the particle velocity, and $\mathbf{u}_{p,c}$ the cell-averaged particle velocity. The δ function ensures that the reaction force acts as a point force at the position of the particle. β is the momentum exchange coefficient correlated by the void fraction. There are a lot of proposed correlation formulas in the literature [34]. The commonly used formula is the one proposed by Gidaspow [31], which reads as

$$\beta=\begin{cases}150\dfrac{(1-\varepsilon)^2\mu}{\varepsilon d_p^2}+1.75(1-\varepsilon)\dfrac{\rho_g}{d_p}\,|\,\mathbf{u}_g-\mathbf{u}_p\,|,\text{if }\varepsilon\le0.8\\[3mm]\dfrac{3}{4}C_D\dfrac{\varepsilon(1-\varepsilon)}{d_p}\rho_g\,|\,\mathbf{u}_g-\mathbf{u}_p\,|\,\varepsilon^{-2.65},\text{ if }\varepsilon>0.8\end{cases} \qquad (5)$$

This formula implies that in dense regimes, the drag force accounted for the presence of neighboring particles according to the Ergun equation for pressure drop prediction in packed bed [35]; while for dilute regimes, Wen & Yu formula is used [36]. C_D is the drag coefficient for a single particle, and reads as follows:

$$C_D=\begin{cases}\dfrac{24}{Re_p}\left(1+0.15Re_p^{0.687}\right) & Re_p\le1000\\[3mm]0.44 & Re_p>1000\end{cases}, \quad Re_p=\frac{\varepsilon\rho_g d_p\,|\,\mathbf{u}_g-\mathbf{u}_p\,|}{\mu}$$

The formula Eq. (5) for computing the momentum exchange coefficient is applicable for Model A. For Model B, it reads as [31]

$$\beta_{ModelB}=\frac{\beta}{\varepsilon}$$

Dispersed Phase Hydrodynamics

Since the gas inertia is much smaller than that of the solid particles, only the drag and the pressure gradient forces are taken into account in the particle momentum equation. For Model A, it reads as follows

$$m_p \frac{d\mathbf{u}_p}{dt} = -V_p \nabla p + \frac{V_p \beta}{1-\varepsilon}(\mathbf{u}_g - \mathbf{u}_p) + m_p \mathbf{g} \tag{6a}$$

where m_p is the particle mass. For Model B, the gradient force due to the static pressure is absorbed in the drag force and it does not appear in the particle momentum equation:

$$m_p \frac{d\mathbf{u}_p}{dt} = \frac{V_p \beta_{ModelB}}{1-\varepsilon}(\mathbf{u}_g - \mathbf{u}_p) + m_p \mathbf{g} \tag{6b}$$

Models A and B are originated from TFM. They have brought out arguments when they were used in conjunction with Lagrangian simulations [28, 32, 37-40]. One should be prudent to extend the governing equations available in the TFM literature To DPM. For instance, the model presented by Xu & Yu [28] will result in an over-estimation of the pressure drop across the fluidized bed due to the ignorance of the pressure gradient in the particle momentum equation, as noticed by Hoomans *et al.* [37]. It is worthy to mention that Model A is the more commonly used model in the literature due to its clear physical statement on the two-phase interaction. Our discussion is limited to Model A.

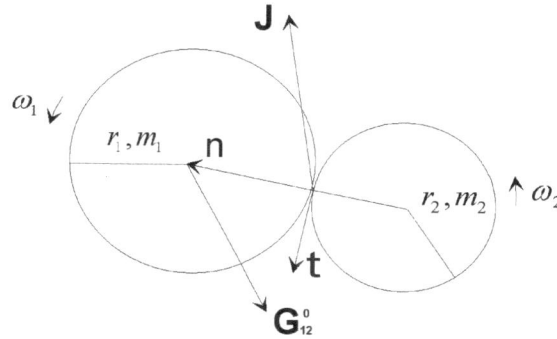

Figure 1: Illustration of collision between two particles with mass m_1 and m_2.

In a hard-sphere approach, collisions among dispersed particles are assumed binary, quasi-instantaneous, and particle contact occurs at one point. During a collision only impulse forces are considered. The relative velocity $\mathbf{v}_{12}$ at the contact point between two particles with velocities $\mathbf{v}_1$ and $\mathbf{v}_2$ is defined as follows:

$$\mathbf{v}_{12} = (\mathbf{v}_1 - \mathbf{v}_2) - (r_1\boldsymbol{\omega}_1 + r_2\boldsymbol{\omega}_2) \times \mathbf{n} \tag{7}$$

where r_1 and r_2 are the radii, and ω_1 and ω_2 are the angular velocities as shown in Fig. **1**.

The normal and tangential unit vectors that define the collision coordinate system are computed as follows:

$$\mathbf{n} = \frac{\mathbf{x}_1 - \mathbf{x}_2}{|\mathbf{x}_1 - \mathbf{x}_2|} \tag{8}$$

$$\mathbf{t} = \frac{\mathbf{v}_{12}^0 - (\mathbf{G}_{12}^0 \cdot \mathbf{n})\mathbf{n}}{|\mathbf{v}_{12}^0 - (\mathbf{G}_{12}^0 \cdot \mathbf{n})\mathbf{n}|} \tag{9}$$

where the superscript 0 denotes pre-collision conditions. The relative velocity of particle centroid is given by:

$$\mathbf{G}_{12} = \mathbf{v}_1 - \mathbf{v}_2 \tag{10}$$

The equations of motions of the two particles are given by the following system:

$$
\begin{aligned}
m_1(\mathbf{v}_1 - \mathbf{v}_1^0) &= \mathbf{J} \\
m_2(\mathbf{v}_2 - \mathbf{v}_2^0) &= -\mathbf{J} \\
\frac{I_1}{r_1}(\boldsymbol{\omega}_1 - \boldsymbol{\omega}_1^0) &= \mathbf{J} \times \mathbf{n} \\
\frac{I_2}{r_2}(\boldsymbol{\omega}_2 - \boldsymbol{\omega}_2^0) &= \mathbf{J} \times \mathbf{n}
\end{aligned}
\tag{11}
$$

where I_1 and I_2 are the moments of inertia and $\mathbf{J}$ is the impulse vector. Using the relation $(\mathbf{n} \times \mathbf{J}) \times \mathbf{n} = \mathbf{J} - \mathbf{n}(\mathbf{J} \cdot \mathbf{n})$, one obtains:

$$\mathbf{v}_{12} - \mathbf{v}_{12}^0 = B_1 \mathbf{J} - (B_1 - B_2)(\mathbf{J} \cdot \mathbf{n})\mathbf{n} \tag{12}$$

The constants B_1 and B_2 for rigid spheres are given by:

$$
\begin{aligned}
B_1 &= \frac{7}{2}\left(\frac{1}{m_1} + \frac{1}{m_2}\right) \\
B_2 &= \frac{1}{m_1} + \frac{1}{m_2}
\end{aligned}
\tag{13}
$$

Some parameters are required to relate the pre-collision and post-collision velocities. The first collision parameter of the model is the coefficient of normal restitution, e_n:

$$\mathbf{v}_{12} \cdot \mathbf{n} = -e_n\left(\mathbf{v}_{12}^0 \cdot \mathbf{n}\right) \tag{14}$$

Combining this equation with Eq. (12) we can get the following expression for the normal component of the impulse vector:

$$J_n = -(1 + e_n)\frac{\mathbf{v}_{12}^0 \cdot \mathbf{n}}{B_2} \tag{15}$$

The second and the third collision parameters are the coefficient of friction μ_f and the coefficient of tangential restitution e_t. These two collision parameters represent two types of collisions in the tangential direction, namely sticking and sliding collisions. If the tangential component of the impact velocity is sufficiently high or the friction coefficient is small by comparison, *i.e.*

$$\mu_f < \frac{(1 + e_t)\mathbf{v}_{12}^0 \cdot \mathbf{t}}{J_n B_1} , \tag{16}$$

gross sliding occurs during the whole duration of the contact. In this case the Coulomb's law applies and the tangential component of impulse is then given by:

$$J_{t,sliding} = -\mu_f J_n \tag{17}$$

For sticking-type of tangential collisions, the relative tangential velocity between the two colliding particles becomes zero. For these conditions, the friction force is sufficiently high:

$$\mu_f \geq \frac{(1+e_t)\mathbf{v}_{12}^0 \cdot \mathbf{t}}{J_n B_1} \qquad (18)$$

In this case, the coefficient of tangential restitution is defined as:

$$\mathbf{v}_{12} \cdot \mathbf{t} = -e_t(\mathbf{v}_{12}^0 \cdot \mathbf{t}) \qquad (19)$$

Thus we get the tangential impulse for sticking collisions:

$$J_{t,sticking} = -(1+e_t)\frac{\mathbf{v}_{12}^0 \cdot \mathbf{t}}{B_1} \qquad (20)$$

With the total impulse vector known the post-collision velocities can now be computed from the system (11).

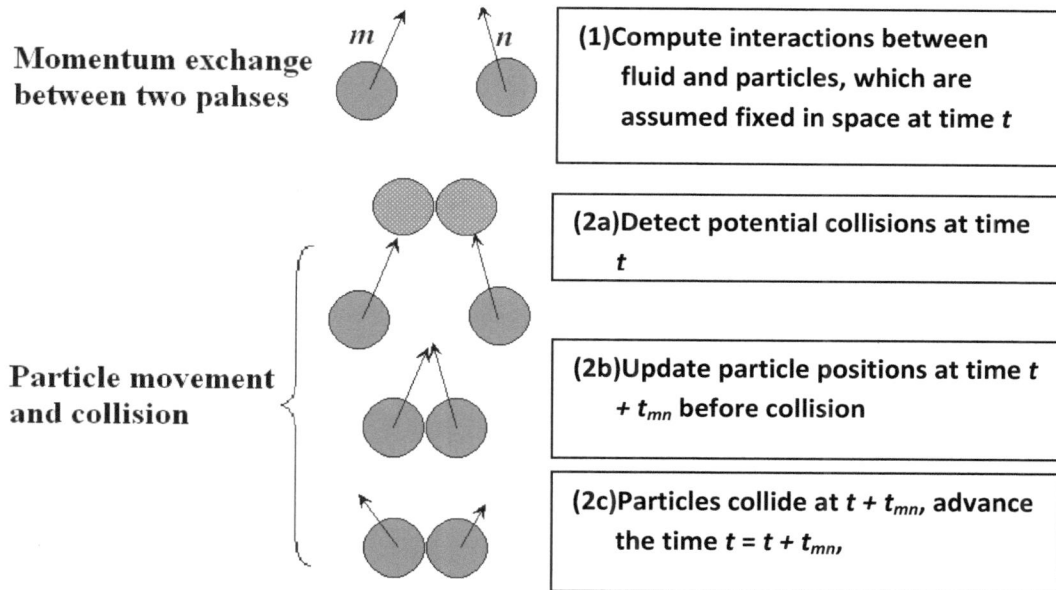

Figure 2: Two steps decoupling the hard-sphere DPM; t_{mn} is the minimum collision time in collision sequence.

NUMERICAL SOLUTION OF DPM

In DPMs, due to the weak gas-phase inertia compared to that of the solid phase, the complex gas-particle coupling system is often de-coupled into two subsystems or processes that are solved separately at each simulation time step, as illustrated in Fig. 2. During the first process or step, the two-phase interactions are taken into account while the particles are assumed fixed in space. Thus, only the momentum exchange between the two phases is accounted for in this step. The gas-phase governing equations are solved using a finite volume/difference method together with the particle momentum equation. The system of two-phase equations is solved using either an explicitly segregated scheme [10, 12] or an implicitly coupled algorithm [14, 30]. The time step Δt for numerical integration should be selected at least one order of magnitude smaller than the particle response time t_p to capture the two-phase interaction. In the second step, all possible collisions between particles are detected and the collision dynamics is computed for each occurring collision. During this step, all collision events are handled one by one in a chronological order and particles are considered in free flight during successive collisions. In this section, we first detail the finite volume method used to solve the gas-phase governing equations, under which the general unstructured mesh is applied. The multi-faceted numerical strategy for an efficient DPM implementation developed by the authors [30, 41, 42] is then discussed. This multi-component strategy includes numerical algorithms designed to: (i) accurately calculate the void fraction, (ii) efficiently locate particles within unstructured grids, (iii) efficiently handle particle collision events, and (iv) ensure a strongly implicit two-phase coupling.

Discretization of Gas Phase Equations

Unstructured mesh is very popular in numerical simulations. Unlike structured grids, unstructured grids have the natural flexibility to mesh complex geometries and they are widely used in numerical simulations of single-phase flow, two-phase flow with free surface, and dispersed gas-solid flows in fluidized beds based on TFM [43-45]. It is almost unavoidable to use unstructured grids to mesh complex solution domains of many real-life chemical reactors where gas-solid flow takes place. These reactors usually contain many complex internals such as immersed tubes, draft tubes or baffles that make the use of the structured grid numerically ineffective. The finite volume method incorporated with the SIMPLE algorithm for unstructured meshes has been commonly applied to pressure-velocity equations of the single-phase flow. In the finite volume method, discretized equations are obtained by integrating the governing transport equations over a finite control volume or cell. In this section, we detail the extension of the finite volume method to solve the gas-phase governing equations based on general unstructured mesh. The difference from its application to single-fluid flow mainly comes from the calculation of the mass flow rate through a cell face.

Let us consider the transport equation of a gas-phase scalar quantity ϕ with diffusivity Γ :

$$\frac{\partial}{\partial x_i}\left(\rho_g \varepsilon u_{gi} \phi\right) = \frac{\partial}{\partial x_i}\left(\Gamma \frac{\partial \phi}{\partial x_i}\right) + S_\phi \tag{21}$$

Here the term representing the local rate of change has been combined with the source term S_ϕ. For convenience, the subscript g will be omitted in this sub-section. The subscript i denotes one of the three orthogonal coordinates x, y, or z. Integration and discretization over the control cell C_0 as shown in Fig. 3 yield:

$$\sum_{f=1}^{NF} \Lambda_f \phi_f = \sum_{f=1}^{NF} D_f + \int_{C0} S_\phi dV \tag{22}$$

where $\Lambda_f = \rho \varepsilon_f \mathbf{A} \cdot \mathbf{u}_f$ is the mass flow rate(defined as positive if flow is leaving C_0), D_f the transport due to diffusion through the face f (denoted by face normal vector $\mathbf{A}$), and NF the number of faces enclosing cell C_0.

For $\phi = 1$ and $S_\phi = -\dfrac{\partial(\rho\varepsilon)}{\partial t}$, Eq. (21) becomes the gas continuity equation and the mass flow rate is discretized as follows:

$$\sum_{f=1}^{NF} \Lambda_f = \rho \Delta V \frac{\varepsilon^n - \varepsilon^{n+1}}{\Delta t} \tag{23}$$

where ΔV is the cell volume and Δt is the time step between two consecutive time levels n and $n+1$.

For $\phi = u_i$, Eq. (21) reduces to the momentum equation: Eq. (2). A second order upwind scheme is used to approximate the transport scalar value in the convection terms at the cell face:

$$\phi_f = \phi_{upwind} + \nabla \phi_{r,upwind} \cdot d\mathbf{r} \tag{24}$$

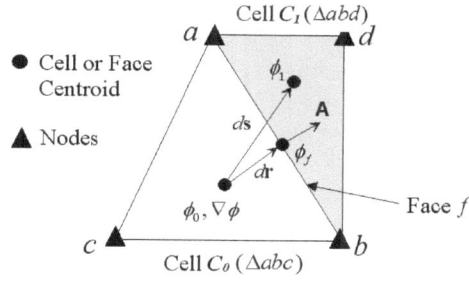

Figure 3: Cell, node, face and scalar discretization over unstructured grid.

where ϕ_{upwind} and $\nabla\phi_{r,upwind}$ are the cell-centered value and its gradient at the upstream cell, respectively. The diffusion D_f through the face f is

$$D_f = \Gamma_f \nabla\phi \cdot \mathbf{A}$$

D_f is further divided into a primary gradient component and a secondary or cross-diffusion component. A second-order discretization scheme that uses only the values of ϕ at the neighboring cells C_0 and C_1 containing face f is often used [46] (see Fig. **3**):

$$D_f = \Gamma_f \frac{(\phi_1 - \phi_0)}{ds} \frac{\mathbf{A} \cdot \mathbf{A}}{\mathbf{A} \cdot \mathbf{e}_s} + \Gamma_f \left(\overline{\nabla}\phi \cdot \mathbf{A} - \overline{\nabla}\phi \cdot \mathbf{e}_s \frac{\mathbf{A} \cdot \mathbf{A}}{\mathbf{A} \cdot \mathbf{e}_s} \right) \tag{25}$$

where $\mathbf{e}_s$ is the unit vector along ds. At face f $\overline{\nabla}\phi$ is taken to be the average of the gradients at the two adjacent cells. The cell gradients are computed by applying the divergence theorem over the control volume, which is second-order accurate:

$$\overline{\nabla}\phi = \frac{\nabla\phi_0 + \nabla\phi_1}{2} \tag{26a}$$

$$\nabla\phi = \frac{1}{\Delta V} \sum_f^{NF} \left(\tilde{\phi}_f \mathbf{A} \right) \tag{26b}$$

$$\tilde{\phi}_f = \frac{\phi_{f,0} + \phi_{f,1}}{2} \tag{26c}$$

$$\phi_{f,c} = \phi_c + \nabla\phi_r \cdot d\mathbf{r} \quad (c = 0, 1) \tag{26d}$$

In addition to the gas-particle interactions, the source term contains the pressure gradient, the gravity, the local momentum change rate and the components of the stress tensor that are not included in the standard diffusion term:

$$\frac{\partial}{\partial x_j} \left(\mu\varepsilon \frac{\partial u_j}{\partial x_i} - \frac{2}{3} \delta_{ij} \mu\varepsilon \frac{\partial u_l}{\partial x_l} \right) - \frac{\partial(\rho\varepsilon u_i)}{\partial t} - \varepsilon \frac{\partial p}{\partial x_i} + \varepsilon\rho g_i$$

By integrating over the control volume, we obtain:

$$\sum_f^{NF} \left(\mu_f \varepsilon_f \frac{\partial u_j}{\partial x_i} - \frac{2}{3} \delta_{ij} \mu_f \varepsilon_f \frac{\partial u_l}{\partial x_l} \right) A_j - \varepsilon \sum_f^{NF} p_f A_i - \frac{\rho\varepsilon^{n+1} u_i^{n+1} - \rho\varepsilon^n u_i^n}{\Delta t} \Delta V + \varepsilon\rho g_i \Delta V \tag{27}$$

Here A_i is the component of vector **A** of the face f and the pressure p_f is calculated by a formula equivalent to Eq. (26c). The time derivatives in the source terms are discretized using a forward-difference scheme. The void fraction is calculated according to the positions of particles located inside the control cell as described as part of the numerical strategy. Its value at the cell face is evaluated using Eq.(26c). Finally the discretized momentum equations are given in the following form:

$$\overline{a}_p u_{i,p} = \sum_{nb} \overline{a}_{nb} u_{i,nb} + \varepsilon \sum_f^{NF} p_f A_i + S_i \tag{28}$$

The first summation in the right side is over all the neighboring cells nb of the cell p. The coefficient $\overline{a}_p$ includes the contribution of the particle-fluid interactions after linearizing the source term for the two-phase momentum exchange. This means that the coupling between phases is accomplished implicitly in the solution of the system of the linear equations. Details about the two-phase momentum coupling will be discussed as part of the numerical strategy.

After solving the gas-phase momentum equations for the velocity vector, the net mass flow rate out of a cell can be obtained from Eq. (23). According to the momentum interpolation [46, 47], the mass flow rate through the face f in Fig. **3** is written as:

$$\Lambda_f = \rho \varepsilon_f \mathbf{A} \cdot \left(\frac{\mathbf{u}_0^* + \mathbf{u}_1^*}{2} \right) - \frac{\rho \varepsilon_f (\Delta V_0 + \Delta V_1)}{(\overline{a}_0 + \overline{a}_1)} \left(\frac{p_1 - p_0}{ds} - \overline{\nabla} p_f \cdot \mathbf{e}_s \right) \frac{\mathbf{A} \cdot \mathbf{A}}{\mathbf{A} \cdot \mathbf{e}_s} \tag{29}$$

where $\mathbf{u}^*$ is the velocity field that satisfies momentum conservation and $\overline{\nabla} p_f$ is the average pressure gradient.

The correction of mass flow rate is defined as:

$$\Lambda_f' = -\frac{\rho \varepsilon_f (\Delta V_0 + \Delta V_1)}{(\overline{a}_0 + \overline{a}_1)} \left(\frac{p_1' - p_0'}{ds} \right) \frac{\mathbf{A} \cdot \mathbf{A}}{\mathbf{A} \cdot \mathbf{e}_s} \tag{30}$$

According to the SIMPLE algorithm, we substitute $\Lambda_f = \Lambda_f^* + \Lambda_f'$ into Eq.(23) yielding the pressure-correction equation:

$$a_p p_p' = \sum_{nb} a_{nb} p_{nb}' + b \tag{31}$$

with the relation $a_p = \sum_{nb} a_{nb}$. The net mass flow rate b is defined as follows:

$$b = \sum_f^{NF} \Lambda_f^* + \rho \Delta V (\varepsilon^{n+1} - \varepsilon^n)/\Delta t \tag{32}$$

The face mass flow rate, cell pressure, and cell velocity are corrected using the same procedure used in the SIMPLE algorithm. The discretized equations for momentum and pressure correction have the same form and they are solved using the Algebraic Multi-Grid (AMG) method. A feature that makes AMG particularly attractive for use on unstructured meshes is that the coarse level equations are generated without the need for re-discretization, in contrast to geometric multi-grid method in which a hierarchy of meshes is required and the discretized equations are evaluated at every level. This means that no fluxes or source terms need to be evaluated at the coarse levels. A detailed description of AMG can be found in the literature [48].

The use of unstructured grids in conjunction with DPM brings about some numerical challenges that should be well addressed in order to do not discount the advantages brought by the use of unstructured grids to mesh fluid-particle systems. These numerical difficulties regard void fraction calculation, particles' tracking, two-phase coupling, and collision events' handling. In the next sub-sections, we detail a multi-component numerical strategy that efficiently handles the above issues for unstructured grids. In addition, this numerical strategy enhances the numerical tractability of DPM for the simulation of real-life dense particulate systems.

Void Fraction Calculation

The void fraction evaluation plays an important role in determining the accuracy of the fluid-particle system simulations based on DPM. Indeed, the momentum exchanges that describe gas-solid interactions are found to be strongly dependent on such an accurate evaluation [34]. Moreover, the solid particles occupy part of the gas phase volume and this is incorporated in the conservation equations of the gas phase through multiplying all the gas properties by the void fraction. Therefore inaccurate computation of the void fraction would adversely affect the general performance of the discrete particle model.

The earlier studies applying DPM were based on the two-dimensional model whilst structured regular grids were often used [10, 12, 13, 16-18]. The voidage ε_{2D} in 2D DPMs is calculated according to the space or area occupied by the particles in the 2D grid cells. This is not consistent with the empirical drag formula in which the correlated porosity ε_{3D} is evaluated in real 3D systems. To correct this inconsistency, two strategies are often used to transform the 2D porosity. The first one is described by the following equation:

$$\varepsilon_{3D} = 1 - \frac{2}{\sqrt{\pi\sqrt{3}}}\left(1 - \varepsilon_{2D}\right)^{3/2} \tag{33}$$

This equation is derived on the basis of a comparison between a 2D hexagonal lattice and a 3D cubic lattice assuming equal inter-particle distances [10, 12].

For the second strategy [28], the void fraction is computed as follows:

$$\varepsilon_{3D} = 1 - \frac{\sum V_i}{\Delta V} \tag{34}$$

where V_i is the volume of particle i, and the summation is taken over all the particles in the cell volume $\Delta V = \Delta x \Delta y d_p$. This means that the 2D domain is regarded as a pseudo 3D one with a thickness of one particle diameter, d_p. The original equation for the first strategy; Eq. (33) can be slightly altered by introducing an empirical parameter containing the maximum experimental solids packing in practice [49]. The numerical results using two-dimensional DPM indicated that the model is sensitive to the porosity estimation strategies [49].

The use of unstructured grids constitutes a challenge when it comes to the estimation of the void fraction for both 2D and 3D DPM, in particular when particles are not fully contained in one cell, which can have any shape (quadrilateral, triangular, wedged, tetrahedral, or hexahedral) and/or can intersect particles through any of its boundaries (node, edge or face). Indeed, when the centre of a particle locates on an edge shared by two 2D cells or on a face shared by two 3D cells, there will be considerable errors in the calculated porosity if its volume is not appropriately shared between these two cells. If the cell volume is 20 times that of the particle, for example, neglecting the share of the particle volume between two cells results in particle volume fractions of 0.05 and 0. This represents a 50% relative error or 2.5% absolute error in the volume fraction. Since the porosity plays a very important role in the local mass and momentum balance of the gas phase, the said errors should be avoided as far as possible.

In case of 2D square cells, the particle area shared by the four neighbouring cells can be calculated approximately as follows (see Fig. **4a**):

$$A_i = \frac{\pi}{4}\left(r_p \pm \delta_1\right)\left(r_p \pm \delta_2\right) \qquad (i = 1,2,3,4) \tag{35}$$

where r_p is the particle radius. We name this approach as the 2D Approximate Method (2DAM).In case of 3D cubic cells, the following formula for the particle volume can be used [50] (see Fig. **4b**):

$$V_i = \zeta_i V_p \tag{36}$$

where V_p is the particle volume. The particle is treated as a cube and ζ_i is the cube volume fraction in the cell under consideration (hereafter referred to as the 3D Approximate Method, 3DAM).

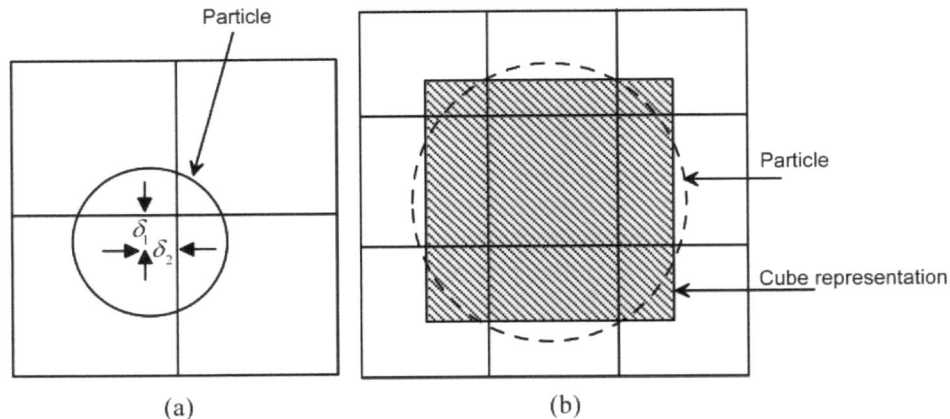

Figure 4: Approximate method to calculate the 2D and 3D porosity.

The above two methods are limited to regular structured grids. They are relatively simple to apply, but not accurate. When these two methods are applied in conjunction with a solution domain discretized by an unstructured grid, the common practice for void fraction calculation is to omit the particle shape and consider the particle as a point. Thus, the split of particles between cells is neglected and particles are considered to belong only to one cell. Hereafter this approach is referred to as the Point Approximate Method, PAM. In this subsection we detail an analytical method to calculate the void fraction for both 2D and 3D general unstructured meshes.

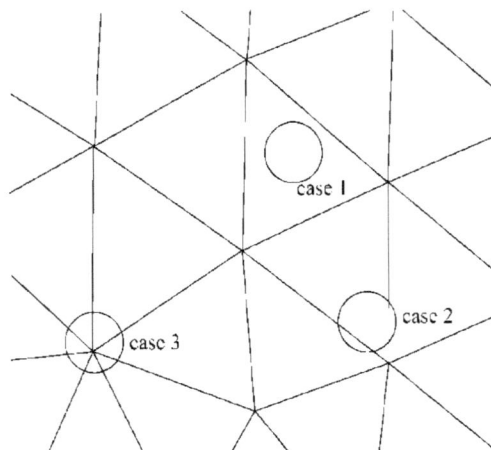

Figure 5: Different particle positions in 2D unstructured mesh.

The 2D unstructured mesh is often composed of triangular, quadrilateral cells or combinations of these two types. All possible particle-cell intersections can be presented by three generic cases as they are shown in

Fig. **5**. For case (1), the covered area is the full particle area while for case (2); the void fraction can be calculated after finding the intersection points between the edges of the cell and the particle circumference. Case (3), however, is more complex. Hereafter we call C_o the cell that hosts the centre (O) of the particle while the other cells that share the particle are denoted by C_n (n =1, 2 ...) as depicted in Fig. **6**. The corresponding areas that the particle contributes to these cells are denoted by A_0 and A_n (n =1, 2 ...), respectively. All geometry information is transformed to the particle local coordinates with the origin placed at the particle centre. The line from the particle centre O to the vertex V shared by the neighbouring cells is defined as the x-axis, and the two points at which the edges of cell C_n intersect the particle boundary are denoted by M and N (see Fig. **6**). The area of the pseudo sector VMN is determined by $A_n = S(l, \theta_2) - S(l, \theta_1)$, where the function S is defined as:

$$S(l, \theta) = \frac{1}{2} r_p^2 \left[\theta - \arcsin\left(\frac{l \sin \theta}{r_p} \right) \right] - \frac{1}{2} l r_p \sin\left[\theta - \arcsin\left(\frac{l \sin \theta}{r_p} \right) \right], \quad -\pi < \theta < \pi \qquad (37)$$

A_n is the difference in area between the pseudo sector VPM or VPN. θ is defined as the angle measured counterclockwise from the local x-axis to the edge intersecting the particle circumference. l is the distance between the particle centre and the common vertex. Finally, A_0 is given by $A_0 = \pi r_p^2 - \sum A_n$.

During the computation, all the particles are involved in a global loop and the case appropriate to each particle is identified. The area (A_p) of any cell C_n covered by one particle is calculated and then the total area $\sum A_p$ of all the particles hosted by this cell C_n is computed. The 2D void fraction of each cell is finally obtained by $\varepsilon_{2D} = 1 - \varepsilon_{p,2D}$, with $\varepsilon_{p,2D} = \sum A_p / A_{cell}$.

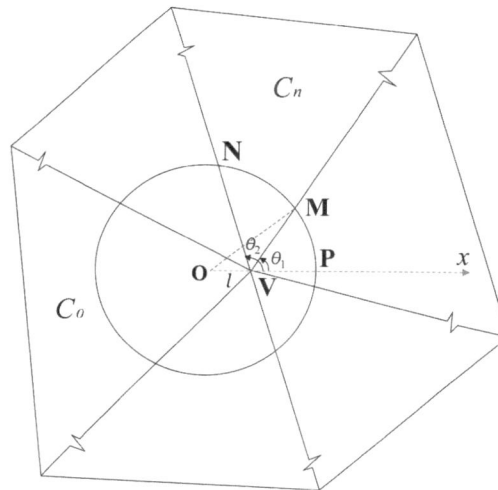

Figure 6: Accurate calculation of 2D void fraction (case 3).

The method presented above for 2D unstructured grids can be extended to calculate the fractional particle volume for the more complicated 3D mesh. Generally a 3D domain is composed of tetrahedral, wedged or hexahedral cells. Four kinds of particle locations are distinguished depending on the distance between the particle and the cell face, the cell edge or the cell node as shown in Fig. **7**. Analogously to the 2D case, the cell in which the centre of the particle is located is denoted by C_0 and the other cells hosting parts of this particle are denoted by C_n. The corresponding fractional volumes are denoted by V_i and V_n, respectively. Obviously if V_n is known, then we get $V_i = 4\pi r_p^2 / 3 - \sum_n V_n$. Four cases are identified for which the volume V_n is accurately computed (see Fig. **7**). These four cases are deemed to cover all possible configurations for the cell-particle intersection.

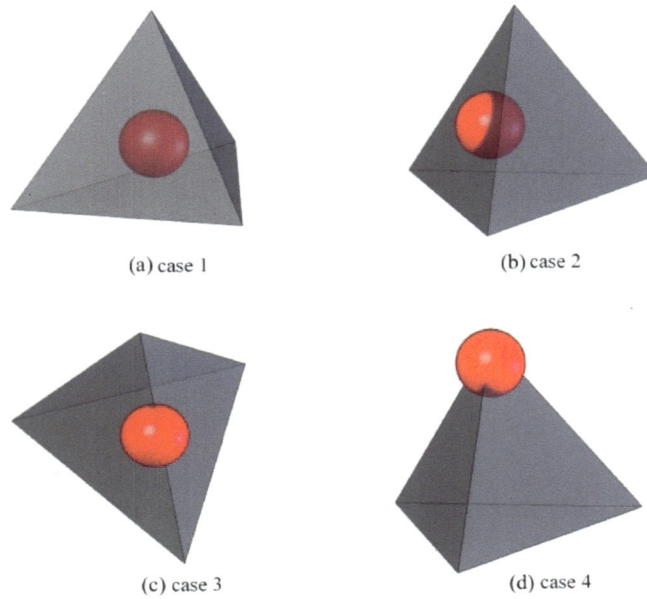

(a) case 1   (b) case 2

(c) case 3   (d) case 4

Figure 7: Different particle positions in an unstructured 3D grid.

Case 1: the sphere is wholly contained in C_0, then $V_n = 0$.

Case 2: a segment of the volume of the sphere is shared by cell C_n totally through one of its faces. The volume of this segment can be calculated by $V_n = \pi r_n^2 \left(r_p - \frac{1}{3} r_n \right)$, where $r_n = r_p - h$ and h is the distance from the particle centre to the face splitting the particle sphere. In this case the particle may be split into several parts by more than one face if the angle between the two adjacent faces of a cell is sharp.

Case 3: one edge (or more than one) of the cell C_i passes through the sphere, so the particle is split into several parts by all the faces sharing this edge. To analyse this case, we specify this edge as VV', the particle centre O as the origin of the local cylindrical coordinate system, the line parallel to VV' as the z-axis, and the line normal to the edge VV' as the x-axis (see Fig. **8a**). The plane at $z = 0$ is shown in Fig. **8b** and the x-z plane in Fig. **8c**. The volume of the sphere cut by the two faces (face 1 and face 2) of the cell C_n can be calculated by:

$$V_n(l, \theta_1, \theta_2) = V(l, \theta_2) - V(l, \theta_1) \tag{38}$$

$$V(l, \theta) = 2\int_0^H S(l, \theta, z)dz \quad (0 \le \theta \le \frac{\pi}{2}, \ 0 < l < r_p)$$

$$V(l, \theta) = \pi s^2 \left(r_p - \frac{s}{3} \right) - 2\int_0^H S(l, \pi - \theta, z)dz \quad (\frac{\pi}{2} < \theta \le \pi, \ 0 < l < r_p)$$

Here the integrand S is defined as:

$$S(l, \theta, z) = \frac{1}{2} r(z)^2 \left[\theta - \arcsin\left(\frac{l \sin \theta}{r(z)} \right) \right] - \frac{1}{2} l \left(\sqrt{r(z)^2 - l^2 \sin^2 \theta} - l \cos \theta \right) \sin \theta$$

with $r(z) = \sqrt{r_p^2 - z^2}$ and $s = r_p - l\sin\theta$. The upper limit of the above integrals is

$$H = \sqrt{r_p^2 - l^2} \ (0 < l < r_p).$$

In the above definition of V_n, l is the distance from the particle centre to VV', θ_1 and θ_2 are the angles from the x-axis to face 1 and face 2, respectively. So $S(l,\theta,z_0)$ represents the area of the pseudo sector VMP at the height $z = z_0$. Integration yields:

$$V(l,\theta) = 2\int_0^H S(l,\theta,z)dz = I_0 + \frac{l\sin\theta}{3}\left(r_p^2 I_1 + I_2\right) \tag{39}$$

$$I_0 = \frac{1}{2}l\sin\theta\int_0^H\left(l\cos\theta - \sqrt{r(z)^2 - l^2\sin^2\theta}\right)dz$$

$$= \frac{1}{2}l^2 H\sin\theta\cos\theta - \frac{1}{2}l\sin\theta\left(r_p^2 - l^2\sin^2\theta\right)\arcsin\left(\sqrt{\frac{r_p^2 - l^2}{r_p^2 - l^2\sin^2\theta}}\right)$$

$$I_1 = \int_l^{r_p}\frac{1}{r^2}\sqrt{\frac{r_p^2 - r^2}{r^2 - l^2\sin^2\theta}}dr^2$$

$$= \frac{2r_p}{l\sin\theta}\arctan\left(\frac{H\tan\theta}{r_p}\right) - 2\arctan\left(\frac{H}{l\cos\theta}\right)$$

$$I_2 = \frac{1}{2}\int_l^{r_p}\sqrt{\frac{r_p^2 - r^2}{r^2 - l^2\sin^2\theta}}dr^2$$

$$= \frac{1}{2}\left(r_p^2 - l^2\sin^2\theta\right)\arctan\left(\frac{H}{l\cos\theta}\right) - \frac{1}{2}Hl\cos\theta$$

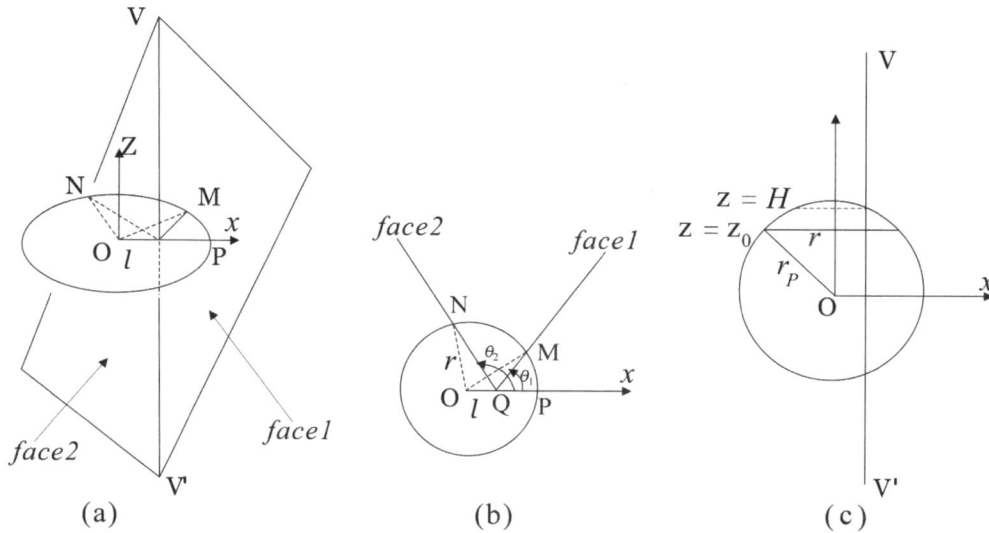

Figure 8: Accurate calculation of 3D void fraction (case 3) (a) perspective view; (b) plane at $z = 0$ and (c) x-z plane.

Once the length l and the angle θ are calculated according to the positions of the particle and the cell faces, the above integrals constitute the sphere volume encompassed by the x-z plane and the face. All the faces sharing the edge VV' are identified first and the angle from local x-axis to each face is calculated in a loop to avoid repetitive computations. Then the volume of a sphere segment cut by any two faces can be calculated using Eqs. (38) and (39). If the two edges of a cell face form a very sharp angle, these two edges may both pass through the particle. For this situation, the procedure for case (3) can be applied twice for both edges to compute the fractional volumes.

Case 4: a node or vertex V that is common to the cell C_0 and its neighbouring cells C_n, is occupied by a particle. The calculation of the volume V_n of the particle shared by C_n is described as follows: firstly the vertex V is identified by checking all the vertices of the cell C_0. The intersection points of each edge (sharing V) with the particle surface are calculated. For cell C_n these intersection points are denoted by P, Q and R, since there are only three edges between the three faces (denoted by f_1, f_2, f_3) for a tetrahedral, wedged or hexahedral cell that are intersected by the particle (see Fig. **9**). Then the particle volume V_k encompassed by the face PQR and f_k ($k=1, 2, 3$) can be calculated by the approach described above for case 3 for the sphere volume surrounded by any two faces. Finally we obtain:

$$V_n = V_t + \left(V_{sg} - \sum_{k=1}^{3} V_k \right)$$
(40)

$$V_t = \frac{1}{6}\left|(PR \times PQ) \times PV\right|$$

V_{sg} is the volume of the sphere segment split by the face PQR.

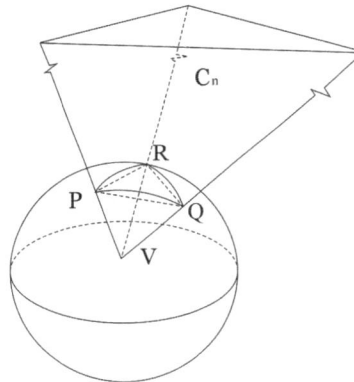

Figure 9: Accurate calculation of 3D void fraction (case 4).

To avoid repetitive computations, all the edges sharing the vertex V are involved in a loop to compute the intersection points with the sphere surface (such as P, Q and R) before calculating V_n. One should be prudent in calculating the angle θ of a given face for case 3 and case 4. The normal vector of a given face plays an important role in deciding which side of the face should be used to calculate the encompassed volume.

The analytical method for the 3D unstructured mesh, though accurate, is relatively time-consuming compared with the point approximate method. This is due to the many arcsine and arctangent expressions that have to be evaluated for case 3 and case 4. To avoid evaluating repeatedly these complex expressions when calculating the void fraction at each time step, we introduced the look-up table. It consists of a pre-computed mapping table to accelerate the computation of the analytical solution [41].

Figs. **10** and **11** show respectively the 2D and 3D volume fraction variations when the relative position between a particle and a structured regular cell changes. In such cases, only two parameters are needed to determine the fractional volume because the two faces of the cell are orthogonal ($\theta_2 = \theta_1 + 90^0$). It can be seen that although both approximate methods (2DAM and 3DAM) give similar trends as the analytical solutions, the errors associated with these two methods are significant over a wide range of particle configurations when the distance from the particle center to the vertex or edge is less than the particle radius. Since the volume fraction is normalized by its maximum value (when the particle is wholly contained in the cell), the accumulated error due to applying the approximate method in a real simulation will be of the same order as shown in Figs. **10** and **11** even if the particle is much smaller than the grid.

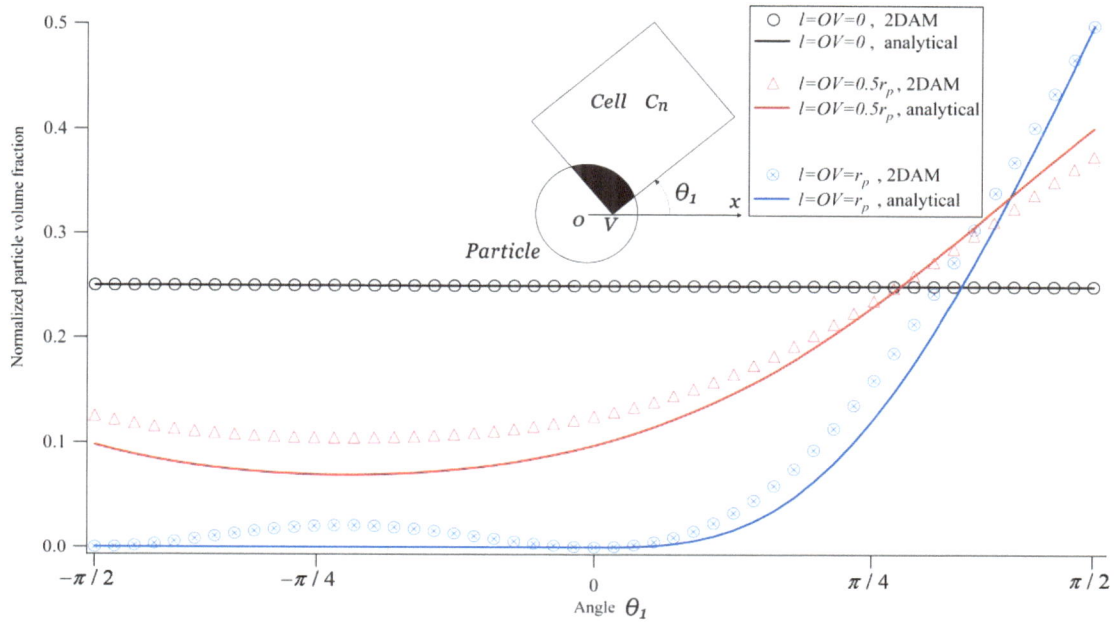

Figure 10: Comparison of analytical and 2DAM computations of void fraction on a 2D unstructured grid (case 3). The void fraction is normalized with the maximum void fraction.

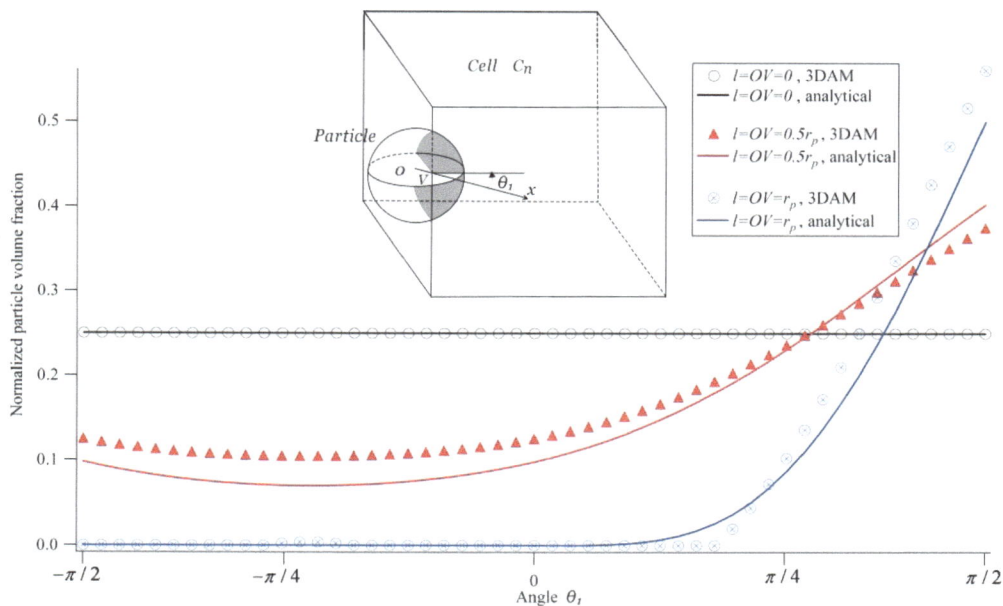

Figure 11: Analytical and 3DAM computations of the void fraction on a 3D structured grid (case 3). The void fraction is normalized with respect to the maximum void fraction.

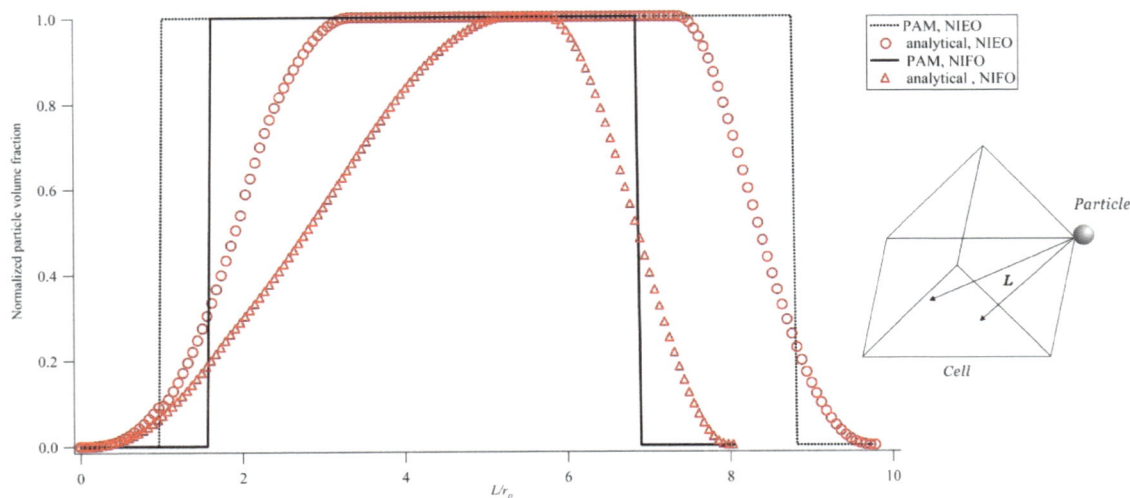

Figure 12: Analytical and PAM computations of the void fraction on a 3D unstructured grid for different particle-cell configurations. The void fraction is normalized with respect to the maximum void fraction. NIEO: node-in-edge-out; NIFO: node-in-face-out.

For unstructured mesh, the point approximate method (PAM) is often used because the approximate methods (2DAM and 3DAM) cannot be applied. In the 3D case, when a particle passes through an irregular grid, applying the PAM will give a step-like particle volume fraction. As shown in Fig. **12**, the error caused by applying the PAM for the case when the particle travels from a node to an edge in a wedged cell is different from that when the particle travels from a node to a face. So, the error incurred in computing the void fraction using the PAM depends on the trajectory of the particle in a cell, and it is usually larger than that incurred in a structured grid, especially when the skewness of the unstructured mesh is very high. The described analytical method gives a very smooth variation of the particle volume fraction when the particle changes its position continuously in space. It is worthy to mention that although the results shown in Fig. **12** are for a wedged cell with the particle entering the cell from a node and leaving it from an edge or a face (NIEO: node-in-edge-out and NIFO: node-in-face-out), they apply also for any cell shape and other particle-cell configurations such as EINO, FINO, EIFO or FIEO.

The application of the described algorithm to simulate bubbling flow in a flat fluidized bed has demonstrated the influence of the void fraction estimation on fluidization. Figs. **13b** and **13c** show the particle distributions as predicted by 3D numerical simulation using the analytical method and PAM for void fraction computation, respectively. The results are recorded 0.2 second after the central gas was injected. The experimental results are shown in Fig. **13a**. The numerical parameters are listed in Table **1**. As expected, the use of the analytical method to compute the void fraction yielded good agreement with the experimental observation: a comparable bed height and similar bubble shape. The use of the PAM has led to an overestimation of the bed height (appreciatively 6%) and a somewhat different shape of the bubble in its upper part. A 2D simulation was also performed for the same. The single bubble simulated by the 2D DPM is shown in Fig. **13d**.

Clearly, the bed expansion and the bubble shape predicted by the 3D DPM are in better agreement with the experiment compared to those predicted by the 2D DPM. In the latter, the void fraction was computed using Eqs.(33&37). The inaccuracy of the 2D discrete particle simulation is partially attributed to the 3D transformation of the voidage achieved by using Eq.(33). This equation holds only for the particle configuration with equal inter-particle distances. Besides the net improvement in accuracy, the 3D analytical method used to evaluate the void fraction is also numerically more efficient. The number of iterations for the solution of the two-phase momentum equations needed to reach convergence was reduced from an average of 35 ± 2 when the PAM was used to 26 ± 2 for the new method (reduction of about 26% in each time step), which saves 15~25% of the CPU time. It is deemed that this enhanced efficiency is brought about by the accurate estimation of the void fraction, which is a very important parameter for the gas mass balance. Thus a smooth and accurate distribution of this parameter ensures numerical stability and enhances convergence. When the lookup-table strategy was used, about 5% of the CPU time was additionally saved.

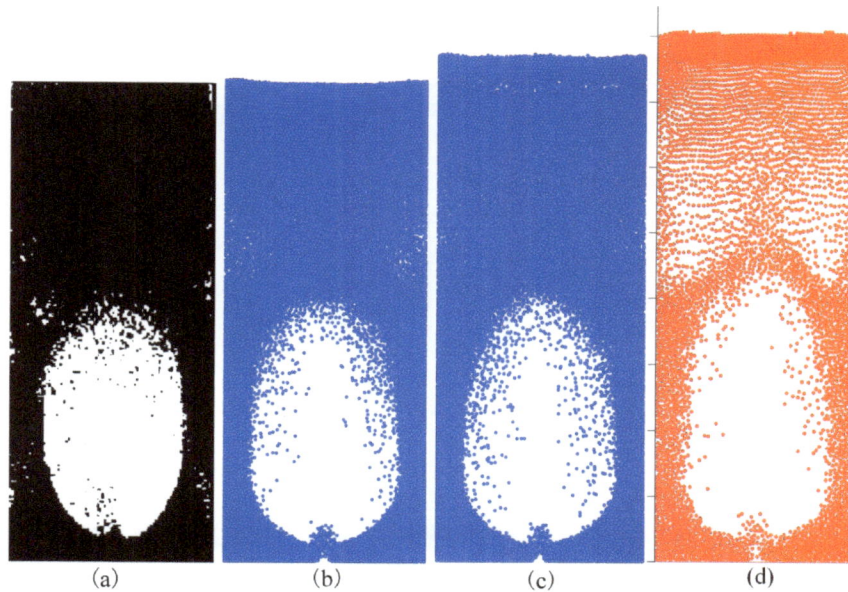

(a)   (b)   (c)   (d)

Figure 13: Single bubble 0.2s after gas injection in a flat fluidized bed: (a) Experiment (from Bokkers *et al.* 2004 [11]); (b) 3D DPM with analytical voidage calculation; (c) 3D DPM using PAM approach; (d) 2D DPM.

Table 1: Parameters for DPM simulation of flat bubbling fluidized bed

Bed dimensions	0.15 m x 0.45 m x 0.015 m
Stagnant bed height	0.2 m
Central jet width	0.01 m
Jet air velocity	20 m/s
Background air velocity	1.25 m/s
Particle diameter	2.5 mm
Particle density	2520 kg/m3
Particle number	31050
Normal restitution coefficient	0.97
Tangential restitution coefficient	0.33
Friction coefficient	0.1
Cell shape	wedge
Cell Number	3780
Volume ratio (V_{cell}/V_p)[a]	32.7
Time step	0.0001s

[a] V_{cell} is the average cell volume and V_p is the particle volume

Particle Locating Algorithm

The cell in which the particle centre is located must be determined for two-phase momentum exchange and void fraction calculation purposes. The development and implementation of a particle locating algorithm is straightforward for regular structured meshes. Many particle locating algorithms for 2D and 3D unstructured grids have been developed in the literature [51-54]. These algorithms are efficient for cases when particles travel long distances and cross several cells as it is the case for particles tracked in dilute homogeneous flows (particles are entrained by fluid eddies and a large time step is used). However, they are not efficient when applied to DPM simulations of dense gas-solid flows. For such cases, particles often alter their positions slightly due to collisions and the small size of the time step usually used in DPM (about 0.0001s). Moreover, many complex vector calculations are involved in these methods, and some algorithm for 3D case is not a straightforward extension of the 2D algorithm.

Here we describe an alternative method to determine the updated cell C_{n+1} to which the particle has moved from the cell C_n in the last time step [55]. As illustrated in Fig. **14**, the particle center is located inside the cell C_{n+1} if and only if:

$$\mathbf{P}_f \cdot \mathbf{A}_f < 0 \tag{41}$$

for each face f of the cell. $\mathbf{P}_f = \mathbf{x}_p^{n+1} - \mathbf{x}_f$ is the vector from the center of face f to the particle. Starting from the first selected cell C_n, the above condition is tested for all the faces of the selected cell. If the above condition is violated for one face of the selected cell, that is, $\mathbf{P}_f \cdot \mathbf{A}_f > 0$, the particle trajectory crosses this face and the neighboring cell sharing this face will be the next-selected cell and the above condition is tested for its other faces. This procedure is repeated until the above condition is met. The last-selected cell will be the one where the particle locates. The method presented here may not give the shortest cell path that the particle travels from cell C_n to C_{n+1}, so it may be less efficient compared with the algorithm by Chordá *et al.* [53] if particle travels a long distance over several cells. But for DPM simulations, it is simpler, more economical and can be applied to both 2D and 3D unstructured grids straightforwardly.

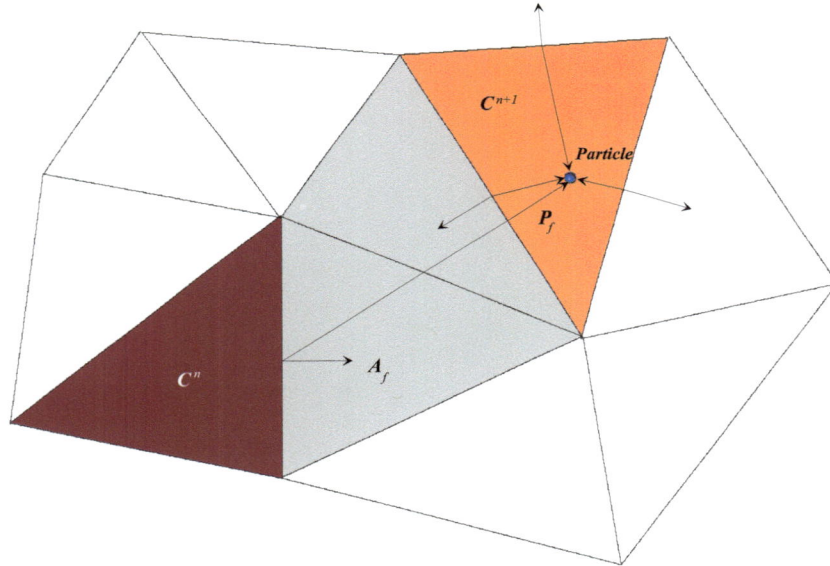

Figure 14: Illustration of particle locating on unstructured mesh for DPM.

Handling Collision Events

The hard-sphere core for particle-particle interactions in DPM is typically event-driven. Detection and handling of potential collision events in hard-sphere DPM often cause a heavy computational burden. Relatively little progress has been achieved in the development of the hard-sphere DPMs regarding the handling of the collision events compared to other aspects such as collision dynamics and fluid-particle interactions. Several optimization strategies for collision event handling in the context of DPM have been discussed in the literature [12]. A reduction in computation time was achieved using these strategies yet the complexity is only of order $O(N_p)$ or $O(N_g)$ because no proper data structure was introduced to handle the event list. Herein, N_p and N_g are the number of particles and cells respectively. Also, the $O(logN_p)$ priority queue algorithms that are widely used in MD simulations [56-58] have not been implemented in the DPM framework to determine the current collision events. Herein, we detail a new efficient algorithm approach to build up a complexity $O(1)$ priority queue suitable for hard-sphere discrete particle modeling [42]. The chained hash-table concept is introduced to speedup both insertion of collision events and location of the current event with the minimum collision time. The crosswise linking data structure for event list is used to determine and delete all invalid events associated with any colliding particle pair.

The basic algorithm for collision event handling is described for a system of N_P particles denoted by $P = \{0,1,...,N_p -1\}$. The system P occupies a computational domain that is discretized into N_g finite volume grids. A collision event involving two particles m and n is denoted by $C_{m,n} = (m, n, t_{m,n})$ where $t_{m,n}$ is the collision time for the particle pair (m, n). The collision time can be calculated according to the positions of the particle pair

$$t_{m,n} = \frac{-\mathbf{x}_{mn} \cdot \mathbf{v}_{mn} - \sqrt{(\mathbf{x}_{mn} \cdot \mathbf{v}_{mn})^2 - \mathbf{v}_{mn}^2 [\mathbf{x}_{mn} \cdot \mathbf{x}_{mn} - (r_m + r_n)^2]}}{\mathbf{v}_{mn}^2} \tag{42}$$

where $\mathbf{x}_{mn} = \mathbf{x}_m - \mathbf{x}_n$ and $\mathbf{v}_{mn} = \mathbf{v}_m - \mathbf{v}_n$ ($m,n \in P$).

A variable t_c is introduced to mark the time point at which the current collision is handled. The algorithm to handle all possible collisions in a simulation time step Δt is summarized as follows:

S1. Scan all particles and build up the collision event list, set $t_c = 0$;

S2. If there is no event in the event list, go to the next simulation time step; else

S3. Determine the current event $C_{m,n}$ having the minimum collision time in the list;

S4. $t_c = t_c + t_{m,n}$, update particle positions: $\mathbf{x}_p = \mathbf{x}_p + \mathbf{v}_p \, t_{m,n}$;

S5. Update velocities of the current pair (m, n) using collision dynamics

S6. Detect future possible events for the current pair (m, n)

S7. Update the event list. Go to S2.

If N_c is the number of collisions handled within one time step, it is evident that S2~S7 will be repeated N_c times in this time step. For the same flow conditions, this frequency should be proportional to the particle number N_p, so at this stage we just assume that N_c is of the same order as N_p. It means that if millions of particles are involved in the computation, steps S2~S7 will be computed millions of times. Thus one should pay close attention to these steps while optimizing the algorithm. Indeed, any loop in these steps must be avoided as far as possible. Though time consuming, all particles must be scanned when initializing the collision list in S1. We introduce the complexity in time to illustrate a rough evaluation of the looping times for a given operation. In the following discussions, the computation complexity of S1 is evaluated for one particle, while those of S2~S7 are evaluated for one collision event or iteration.

By introducing a neighbor list for each particle, only the neighborhood particles are scanned to detect potential collisions. Thus the complexity for S1 and S6 is $O(N_{nb})$. N_{nb} is the average particle number in the neighbor list. A neighbor list is constructed as follows:

$$Partner_m = \{n \mid \forall n \in P, |\mathbf{x}_n - \mathbf{x}_m| < R_s, m \neq n\}$$

It can be compiled for each particle $m \in P$ before initializing the collision list. For a mono-disperse system, R_s can be determined as follows:

$$R_s = d_p + 2|\mathbf{v}|_{max} \Delta t$$

$|\mathbf{v}|_{max}$ is the maximum particle velocity.

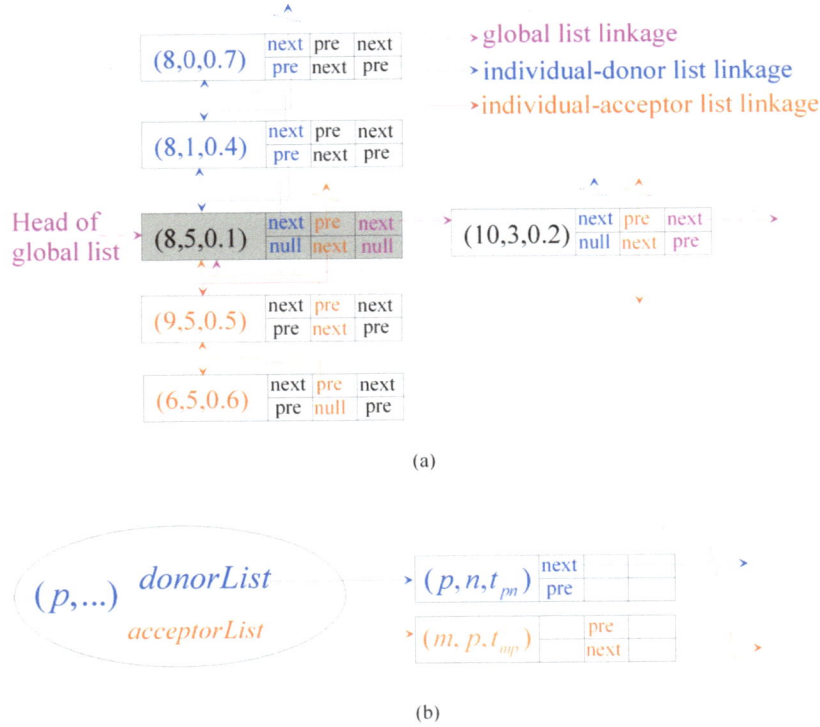

(a)

(b)

Figure 15: Illustration of the crosswise linking of collision events. (a)Three kinds of linkage are maintained for a given collision event (8, 5, 0.1). "Next" and "pre" are pointers to the next and preceding objects, respectively. "null" means a zero pointer without any object pointed to it; (b) two pointers for a given particle p are responsible for maintaining the events in the donor and acceptor lists of particle p.

Since there is no loop involved in S2 and S5, the complexity of these two steps is $O(1)$. By introducing a time stamp for each particle to mark the last updating time point, only the current colliding pair needs updating of their positions. So the complexity of S4 is also $O(1)$. When updating the event lists in S7, all events related to the current pair and their neighboring particles should be updated. If the events in all lists are arranged in an unsorted way, the complexity order of S7 should be $O(N_{nb})$. By recourse to the local minimum algorithm often used in MD studies [56, 57], the individual event list associated with the particle, is constructed as follows:

$$LP_m = \left\{ (m,n,t_{m,n}) \mid \forall n \in Partner_m \right\}$$

A global list having N_p elements is composed of the event with the smallest collision time in LP

$$GLP = \left\{ (p,n,t_{p,n}) \mid \forall p \in P, (p,n,t_{p,n}) \in LP_p \text{ and } \forall (p,m,t_{p,m}) \in LP_p, t_{p,n} < t_{p,m}, m \neq n \right\}$$

If a complete binary tree (CBT) data structure is employed for GLP, the $O(1)$ and $O(logN_p)$ complexity can be achieved for S3 and S7, respectively. Otherwise, only the $O(N_p)$ complexity can be achieved for either S3 or S7, if the events in GLP are stored in a linear linked list.

To avoid additional overhead brought up by the redundant storage of the potential collision events, we first introduce a crosswise linking data structure for events in collision list. Three linkages for a given event $C_{p,q}=(p,q,t_{p,q})$ are built up and maintained when updating relevant collision lists. The C-Language pointer is introduced to access the linked event. The pointer variables for these linkages are all defined in the event data structure, which means that the event object is self-explained and any information related to the collision event can be accessed from the pointer to the event object. This allows the unique storage for each event in the lists said above. As shown in Fig. **15**, the first linkage is the individual-donor list for particle p, LPD_p, which is defined as the set of potential collision events detected when particle p is the host. The second one is the individual-acceptor list for particle q, LPA_q, which is defined as the set of collision events detected when the partner of particle q is the host. The third one is used for the global list. Thus, a given event $C_{p,q}=(p,q,t_{p,q})$ is certainly contained within the lists LPD_p and LPA_q, and may be contained in the global list GLP. The collision event is crosswise linked into these lists through a doubly-linking structure (see Fig. **15a**). Thanks to this type of structure, the previous and the next element can be determined without scanning from the list head during the element deleting and updating operations. Because a given particle p can be a host or an acceptor, it is always associated with both the donor list LPD_p and the acceptor list LPA_p. Two pointer variables associated with the particle p are allocated to store the memory locations of the head elements of these two lists, respectively, as shown in Fig. **15b**. When operating collision events (inserting into or deleting from a list), we always take the first particle in the pair as the host and the second as the acceptor. This makes the updating operation of the donor and acceptor lists straightforward. For instance, when initializing the individual lists in S1, we just need to scan the particle p with its partners whose indexes are smaller than p. This allows us to build the list LPD_p

$$LPD_p = \left\{ (p,n,t_{p,n}) \mid \forall n \in Partner_p, p > n \right\}$$

At this stage, the acceptor lists LPA_n of its partners are built up too. For particle p's partners that their indexes are larger than p, we built the list LPA_p

$$LPA_p = \left\{ (n,p,t_{p,n}) \mid \forall n \in Partner_p, p < n \right\}$$

By taking advantage of the cross linking structure of the collision events, any relevant particle and/or collision event can be easily identified without searching in the corresponding individual lists. Since there is only one entity for any event, there is no additional scanning to update the individual lists for the partners of the colliding pair. Moreover, by sorting the elements in the individual lists, only the first element (local minimum one) in the donor list is added to the global one so that there is no duplicate event in the global list. The additional operation on the global list is avoided too. Thus, in S7, before building up the individual lists for the colliding pair, their corresponding events in the old lists are invalidated and cleaned, resulting in an efficient update of the local event lists and an $O(1)$ complexity of this step.

In computer science, a hash table, or a hash map, is a special data structure that associates keys with records. The primary operation it supports efficiently is a lookup: given a key, find the corresponding record. It works by transforming the key using a hash function into a hash code, a number that is used as an index in an array to spot the desired location where the record should be. Hash tables support the efficient insertion of new entries, in expected complexity of $O(1)$ in time.

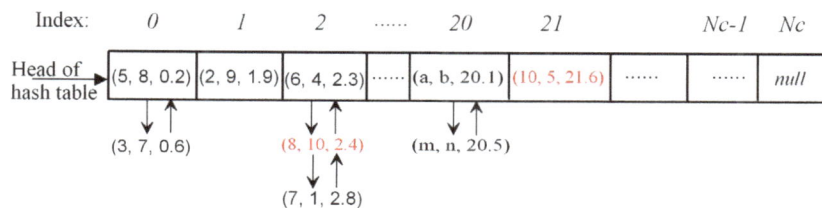

Figure 16: Logical view of event chaining. When two or more events are mapped to the same location in the hash table, they are chained in a doubly-linked list. The collision time is normalized by Δt_c.

Herein a chained hash table is introduced for the storage of the event in the global list *GLP* as shown in Fig. **16**. The one-dimensional array is used to implement the hash table. The key of the hash table is the collision time $t_{m,n}$ and the record is the memory location pointing to the corresponding event in the individual-donor list. We implement the individual-donor list *LPD* as a priority queue so that only its first element is mapped/inserted into the hash table. The individual-acceptor list is implemented as an unsorted linked list since there is no need to pick events from this list (an *O (1)* complexity is achieved for insertion and deletion on this list).

To satisfy the priority queue requirement of having the event with the smallest collision time accessed first in the hash table, the hash function should be well designed to map orderly the collision times to integers or indices *I* and to locate the position where the events are in the table. A simple hash function is defined as:

$$I(t_{m,n}) = INT(t_{m,n} / \Delta t_c) \text{ with } \Delta t_c = \Delta t / N_c , \tag{43}$$

where the function *INT(x)* gives the integral part (lower integer) of *x*. If the particles collide successively in a uniform time interval, the above hash function maps only one element to a table location. When inserting a collision event, its address or index in the global list is determined by the hash function without any loop and the expected computational complexity in time will be reduced to *O(1)*.

However, one cannot determine in a priori manner the number of collisions in a time step. Considering the slow change of a given gas-particle system through the small time steps that are often used in DPM simulations (at least one order less than the particle response time), the number of collision occurrences for successive time steps should change very little. As a direct consequence, the length/capacity of hash table will also vary little. In fact, the number of collisions N_c in Eq.(43) is actually selected as that in the last time step N_c^0. Another fact is that the particles in gas-solid system may never collide in a uniform time interval, which may generate a same index using the hash function Eq.(43). This conflict problem can be solved by chaining [59]. In the chaining methodology, the hash-table looks like a two-dimensional array with a variable length for the second dimension, or like a one-dimensional array of priority queues, as demonstrated in Fig. **16**. Because the length of the hash table is a chosen property and it is a nearly continuous function of the collision number in gas-solid system, chaining events is very rare process (in real simulations, the number of events in the chained lists is often less than 3). Thus the chained hash-table strategy gives exactly an *O(1)* complexity in computational time for event insertion.

As discussed above, there are three types of linkage to be maintained for any given event. The third linkage for the global list *GLP* is used here for chaining in the hash table, as shown in Fig. **16**. The event address/hash code in the hash table is stored with the corresponding event using an integer variable. When the head event $C_{a,b}$ in the donor list LPD_a is inserted to the hash table, its linkages in both the donor list LPD_a and the acceptor list of the particle a are kept. That means for any given event, whenever it is mapped to the hash table, its three linkages are retained. If $C_{a,b}$ becomes the current event, it will be released to the event pool while handling it, together with all other elements in LPD_a. Otherwise if there is an event $C_{b,m}$ occurring before $C_{a,b}$, $C_{a,b}$ will be released together with all other elements in LPA_b because particle b is the acceptor part of the collision when the event $C_{b,m}$ is processed. In the former case, there is no additional operation on the hash table since only the head event $C_{a,b}$ in LPD_a is mapped to the hash table. In the latter case, a null pointer is directly assigned to the location in the hash table indicated by its hash code if any element in LPA_b (such as $C_{a,b}$) has been mapped to the hash table. As shown in Fig. **16**, the events (8, 10, 2.4) and (10, 5, 21.6) are marked invalid since they are associated to the individual lists of the current colliding pair (5, 8). Event (8, 10, 2.4) can be released from the chained hash table by re-linking its previous and next objects maintained in its third linkage pointers. Event (10, 5, 21.6) can be released directly by assigning a null pointer in the location (21) of the hash table. By the crosswise linking data structure, all events in the individual lists can be deleted without any searching. Thus the complexity for event deletion is also *O(1)*. Because there is no duplicate event, the deleting operation is minimized. The simulation is forwarded by scanning the entire hash table from the beginning. This can be done in a simple loop. The loop indicator presents the location of the current event in the hash table. When the last recode of the hash table is null, it means there are no longer events in this time step. Since the capacity of the hash table is chosen close to N_c, the expected complexity to locate the current event is also $O(1)$. In summary, by resorting to the chained hash table methodology, all operations of collision events are performed at the computation complexity of $O(1)$.

Table 2: Diagnostic numerical simulation parameters

Bed diameter	0.2 m
Bed height	0.6 m
Stagnant bed height	0.2 m
Centre jet diameter	0.02 m
Minimum fluidization velocity U_{mf}	0.85 m/s
Background inlet gas velocity	$1.2U_{mf}$
Centre jet gas velocity	$20U_{mf}$
Particle diameter	1.5 mm
Particle density	2400 kg/m^3
Cell number	11,580
Particle number	1,770,000
Restitution coefficients e_n, e_t	0.95, 0.4
Friction coefficient μ_f	0.2
Time step Δt	0.0001s

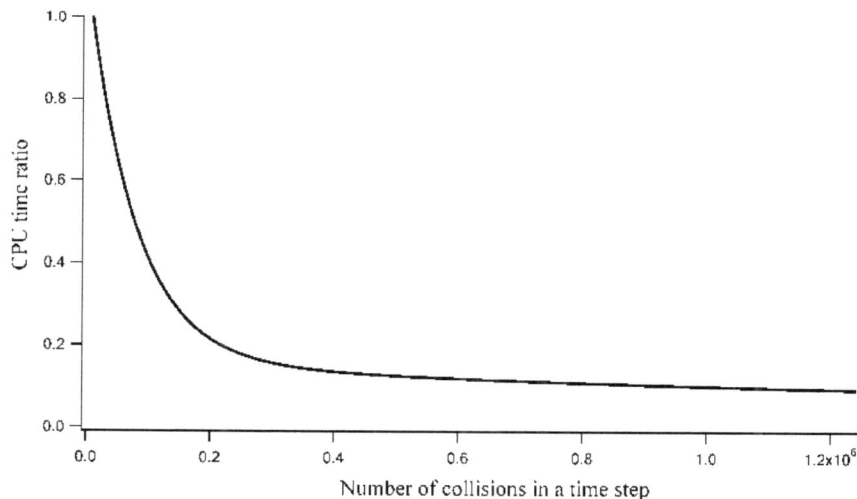

Figure 17: Ratio of the CPU time for computing particle collisions using chained hash table to that using linear linked list (data fitted).

To examine the algorithm efficiency, two diagnostic simulations of a cylindrical bed were conducted. In these numerical experiments, the particle number is large, reaching about 1,800,000. The numerical parameters are listed in Table **2**. The computational cost linked to the particle collision dynamics is shown in Fig. **17**. The result is presented as a ratio of the computational time consumed in a simulation that uses the chained hash table strategy to the one necessary for a simulation without the said strategy. The data of the first 160 simulation time steps is selected for comparison. When a large number of collisions are involved in a time step (especially when it is equal to or larger than the particle number), it can be seen, that about 90% of the CPU time is saved if the hash table is used. In these simulations, the maximum CPU time consumed in a time step is about 256s if the hash table strategy is not applied, but is reduced to less than 20s when it is used. Similar 2D simulations were performed and it was found out that up to 98% of the CPU time is saved if the hash table storage for the collision list is applied. This is because the 2D collision dynamics is simpler than that of the 3D, resulting in a higher computational weight regarding the handling of event lists. Based on these numerical experiments, the time to simulate a 3D dense particulate flow with more than 500,000 particles is not acceptable on a normal desktop computer if the hash table strategy is not used. It should be noted that the collision dynamics implemented herein is a rigorous hard-sphere model without particle overlap, which is different from the one widely cited in the literature [17].

Implicit Two-Phase Coupling

The semi-implicit scheme is used to integrate the particle momentum equation Eq.(6). The momentum exchange coefficient is evaluated explicitly while other terms (such as static pressure gradients, slip velocities) implicitly. This results in a system of linear equations. Integration of Eq.(6) over time reads

$$m_p \left(\mathbf{u}_p^{n+1} - \mathbf{u}_p^n \right) = -V_p \Delta t (\nabla p)_p^{n+1} + \frac{V_p \Delta t \beta_p^n}{1 - \varepsilon_{,p}^{n+1}} (\mathbf{u}_{g,p}^{n+1} - \mathbf{u}_p^{n+1}) + m_p \mathbf{g} \Delta t \tag{44}$$

The Eulerian properties such as volume fraction ε and gas phase velocity $\mathbf{u}_g$ are mapped to the particle's mass centre. Because the collocated finite volume discretization of the gas transport equations is employed, all Eulerian scalar values as well as their gradients are defined at the cell centre. The scalar gradients are computed using the divergence theorem:

$$\nabla \phi_r = \frac{1}{\Delta V} \sum_f^{N_f} \tilde{\phi}_f \mathbf{A}_f \tag{45}$$

The face values $\tilde{\phi}_f$ are the averaged values of ϕ in the two cells adjacent to face f, and $\nabla \phi_r$ are limited so that no new maxima or minima are introduced. Then the scalar value at the particle position is expressed by:

$$\phi_p = \phi_c + \nabla \phi_r \cdot d\mathbf{r} , \tag{46}$$

$$d\mathbf{r} = \mathbf{x}_p - \mathbf{x}_c$$

$\mathbf{x}_c$ and $\mathbf{x}_p$ are the cell centroid and the particle position vectors, respectively. To map the pressure gradient force to the particle position, the gradient for each component of the pressure gradient force is evaluated too. The above mapping method is very simple and more efficient and flexible, when compared to the bilinear or tri-linear interpolations [10, 50]. The latter interpolation methods are only suitable for structured regular grids.

The coupling between the gas and particle phases is accomplished through the inter-phase volume fraction exchange and the inter-phase momentum transfer. The former can be computed precisely since the void fraction is accurately computed according to the particle distribution in the solution domain. For the gas-phase momentum source term due to reactions from dispersed particles, we use the integral form, Eq. (4), and the integration is taken over the grid cell:

$$\mathbf{S}_g = -\frac{1}{\Delta V} \sum_{k=1}^{NPC} \frac{V_{p,k} \beta_k^n}{1 - \varepsilon} (\mathbf{u}_g^{n+1} - \mathbf{u}_{p,k}^{n+1}) \tag{47}$$

NPC is the number of particles overlapping with or inside the cell, and $V_{p,k}$ the corresponding fractional volume shared. Consistent with the interpolation from the Eulerian grid to the particle centre, this formula implies that a particle may be acted on and also react to the fluid in several Eulerian cells. It is worthy to mention that the above source term is evaluated in unit volume.

In 2D cases only overlapped area can be calculated and not overlapped volume. If $V_{p,k}$ is just replaced by $A_{p,k}$ the source term will be over-predicted. But if $V_{p,k}$ is set as the actual volume of a particle; ($\pi d_p^3/6$), the source term will be under-predicted because the volume ΔV is the cell area for 2D mesh. Noting that:

$$1 - \varepsilon = \frac{1}{\Delta V_{3D}} \sum_{k=1}^{NPC} V_{p,k} \tag{48}$$

the source term is computed by the area-weighted summation for 2D case:

$$\mathbf{S}_g = -\frac{1}{\sum_{k=1}^{NPC} A'_{p,k}} \sum_{k=1}^{NPC} A'_{p,k} \beta (\mathbf{u}_g - \mathbf{u}_{p,k}) \tag{49a}$$

$$A'_{p,k} = \frac{A_{p,k}}{\pi r_p^2} \tag{49b}$$

The source term is linearized finally:

$$\mathbf{S}_g = \mathbf{S}_a + B\mathbf{u}_g^{n+1}$$

$$B = -\sum_{k=1}^{NPC} A'_{p,k} \beta_k^n \Big/ \sum_{k=1}^{NPC} A'_{p,k}$$

$$\mathbf{S}_a = \sum_{k=1}^{NPC} A'_{p,k} \beta_k^n \mathbf{u}_{p,k}^{n+1} \Big/ \sum_{k=1}^{NPC} A'_{p,k}$$

From our numerical experiments, the above method gives a correct order for the source term and a correct prediction of pressure drop across a 2D particle bed at the incipient fluidization state. For 3D cases, it is straightforward to calculate the source term, which is linearized as

$$\mathbf{S}_g = \mathbf{S}_a + B\mathbf{u}_g^{n+1} \tag{50a}$$

$$B = -\frac{1}{\Delta V} \sum_{k=1}^{NPC} \frac{V_{p,k} \beta_{p,k}^n}{(1 - \varepsilon^{n+1})} \tag{50b}$$

$$\mathbf{S}_a = \frac{1}{\Delta V} \sum_{k=1}^{NPC} \frac{V_{p,k} \beta_k^n \mathbf{u}_{p,k}^{n+1}}{1 - \varepsilon^{n+1}} \tag{50c}$$

It can be seen from Eqs. (44) and Eq.(47) that all the particle momentum increase gained due to gas drag is feedbacked to the gas for 3D cases. This means a conservation of the momentum exchange due to the two-phase drag interactions. This owes to the combined facts: (i) the nonlinear drag force is linearized and the drag coefficient is evaluated explicitly; (ii) both the particle and gas velocities are calculated in the same time layer. Before iteration in a time step, the drag coefficient for each particle is first calculated according to the two-phase velocities in the last time step along with the gradient of the void fraction. Then, in each iteration on the gas pressure-velocity equations, the gradients of other involved Eulerian variables are calculated according to Eq.(45), and the linearized particle momentum equations are solved. At the same time, the gas source terms due to the inter-phase momentum transfer are evaluated in each iteration step.

The iteration stops only when the solution of both the gas mass and momentum equations converges. The solution procedure is illustrated in Fig. **18a**.

It should be noted that the gas momentum source is calculated implicitly. The linear coefficient including the part due to the two-phase drag interactions, in the source term, is automatically absorbed into the principle diagonal elements of the AMG solver matrix, which physically enhances the two-phase coupling and numerically helps the solution convergence. The particle drag force exerted on the gas is weighted by the particle fractional volume, which has a smooth effect on the source forces and helps numerical stability. The solution of the particle velocity and the gas field is synchronized, resulting in a strongly implicit two-phase coupling. This coupling algorithm differs significantly from the algorithms detailed in the some of the literatures [10, 12] as illustrated in Fig. **18b**. In these algorithms, the two phase momentum equations are solved in a segregated way, which means an explicit two-phase coupling. Moreover, the particle velocities are mapped back to the Eulerian grid and the gas source term is evaluated according to Eulerian properties, which is not quite in line with the discrete feature of the particulate phase.

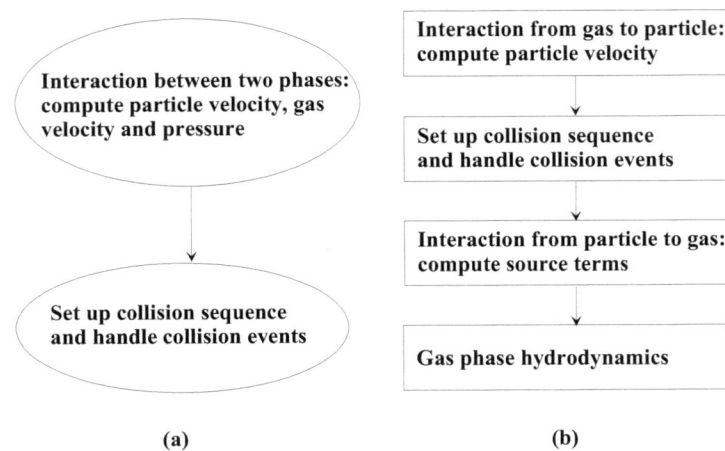

(a)   (b)

Figure 18: The DPM solution procedure in one time step. (a) Implicit two-phase coupling ; (b) explicit two-phase coupling [10, 12].

MODEL'S APPLICATION TO CYLINDRICAL CFB RISER

The gas-solid flow hydrodynamics in CFB risers has attracted great deal of research interest in recent decades because of its wide applications in the chemical and petrochemical industries. The granular flow in a CFB riser is often featured by a high concentration of solids flowing downward near the walls and a core dilute region with upward flowing solids. This is the so-called core-annulus flow structure that has been reported by many experimental studies [60-63]. Although the Eulerian two-fluid model has been the choice to simulate such a flow, the 3D discrete particle model has the potential to clarify many of the flow characteristics in the CFB risers [16-19]. The 2D DPM is known to under-estimates the particle-wall interaction and hence may not be able to simulate accurately the heterogeneous structure in real 3D CFB risers.

A laboratory scale riser with a diameter of 0.1m and a height of 1m is simulated by the developed 3D DPM. Particle properties and numerical parameters are given in Table **3**. It is relatively difficult to analyze the numerical diffusion for DPM than for TFM, since the former is related to the two-phase coupling algorithm. For two-phase flows simulated using TFM or for the general case of single-phase flows simulated by a finite volume method, the numerical diffusion is mainly caused by the mesh quality and the numerical scheme used to calculate the scalar value at the cell faces when integrating the transport equations. For DPM, the source terms in the momentum equations of the continuous phase may cause additional diffusions because the momentum exchange is evaluated not only by the cell scalar values, but also related to the particle velocities in the Lagrangian coordinates. It is well-known that the use of hexahedral cells results in less numerical diffusion than the tetrahedral cells since more faces are involved to evaluate the scalar gradient. Thus, to reduce the numerical diffusion, the unstructured hexahedral grid is adopted to mesh the domain and the second-order upwind scheme is used to calculate the scalar value at the cell face. Gas phase turbulence is not taken into account. At the beginning, the particles are distributed uniformly inside the cylindrical column with an average voidage of 0.982. The particles are introduced to the inlet gradually over several time steps to avoid numerical instabilities after travelling out of the riser outlet. The simulation is run for 12s real time. To reduce the influence of the initial conditions, time-averaged results are obtained for the last 6 seconds of the simulation.

Table 3: CFB simulation parameters

Riser diameter	0.1 m
Riser height	1.0 m
Superficial gas velocity	5.46 m/s
Particle diameter	800μm
Particle density	2520 kg/m^3
Grid number	17,440
Particle number	521,000
Restitution coefficients e_n, e_t	0.97, 0.35
Friction coefficient μ_f	0.1
Time step Δt	0.0001s

Fig. **19** shows the time-averaged axial solids hold-up and the radial profiles of the particle volume fraction at different bed levels. It can be seen that the dense regime exists at the riser bottom below 0.1m with the solids fraction between 0.05 and 0.08; about 2 to 4 times the average solids holdup. Above z=0.2m, the fast fluidization regime can be noticed with an almost constant axial solids holdup. However, the radial particle distribution is not uniform and is characterized by the core-annulus flow structure with dense solids flowing downward near the wall and a more dilute core flowing upward in the riser centre. The particle velocity distribution at two vertical cross sections through the riser center is shown in Fig. **20**. It can be seen that the flow is not axi-symmetry notwithstanding the average of results in 6 seconds and the particle vertical velocity is higher in the core region than in the annulus region. The particle radial profiles at different levels indicate that the higher the cross-sectional average solids holdup, the more non-uniform the radial distribution becomes, resulting in a thicker annular down-flow layer, which agrees well with the experimental observations [60, 61].

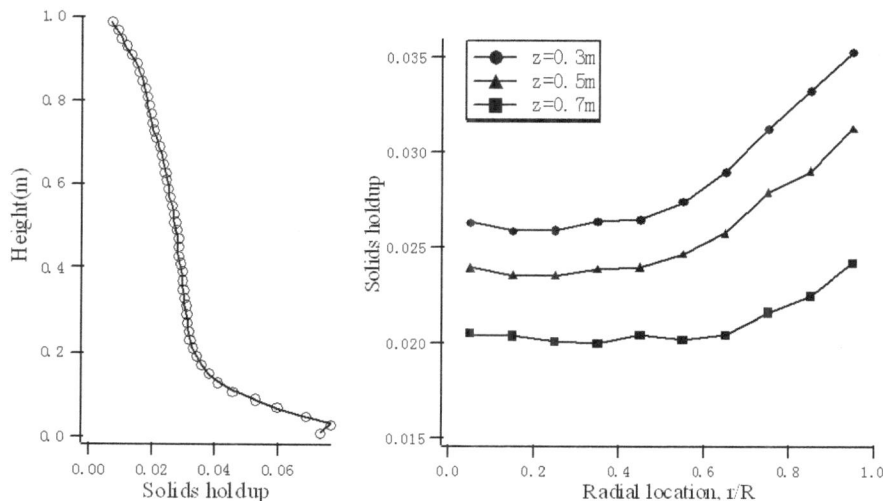

Figure 19: Time-averaged axial distribution and radial profiles of particle holdup.

The numerical model also captures the transient state of the heterogeneous structure such as particle clusters or sheets near the riser wall, as shown in Fig. **21**. The snapshots reveal that the particle clusters or swarms form near the wall at first. Two particle circulations exist below and above the swarm (see Fig. **21b**), which lead to a converged flow regime at the interface. The shear movement of particles in the interface elongates the swarm and forms the particle sheet. Because of the particle-wall interactions the particle sheet twists to form a U-shape. The particle sheet disappears and splits into discrete particles as the particle shear movement continues. The cluster shape and size are comparable to the observations reported in the literature [64]. Since the particle flow behavior can be resolved at the individual particle level transiently, the 3D discrete particle model can be a powerful tool for analyzing such heterogeneous structures taking place in CFB risers.

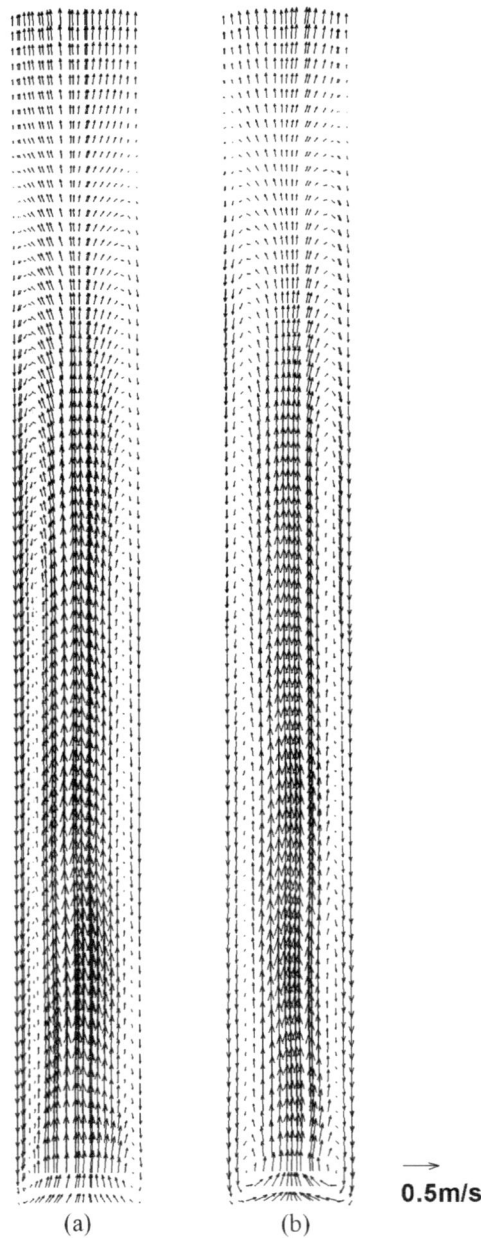

(a)   (b)   0.5m/s

Figure 20: Time-average particle velocity distribution at two cross sections of the cylindrical riser: (a) *x=0* and (b) *y=0*.

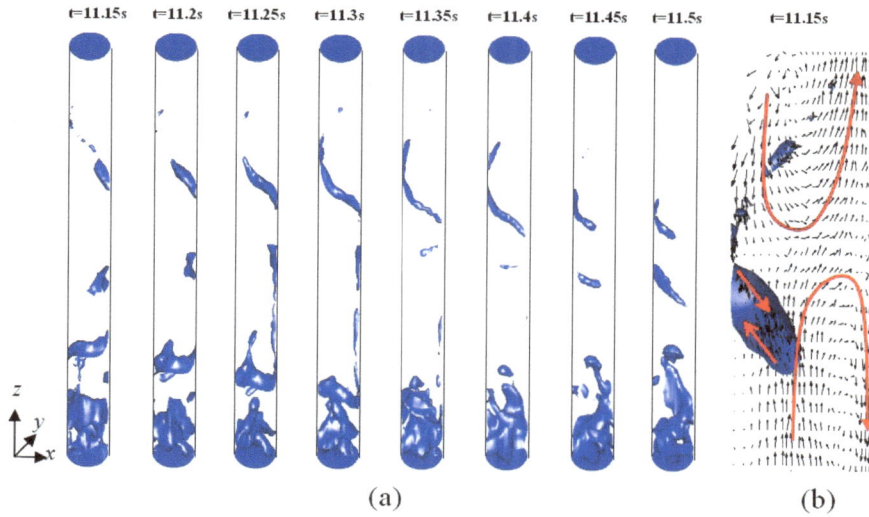

Figure 21: Snapshots of heterogeneous structures in the riser simulated using 3D DPM: (a) particle cluster or swarm elongated to sheet with iso-surface representing dense phase and (b) cell-averaged particle velocity field at $x = 0$ cross section between $z = 0.58$m and $z = 0.88$m.

MODEL'S APPLICATION TO FLAT SPOUTED BED

The gas-solid spouted beds provide good mixing and contact between gas and coarse particles. They are often characterized by a dense region of downward particle flow called the annulus, a dilute core region of upward particle flow called the spout, and the fountain region with center upward and outer downward particle motion. Because the particle regime in the annulus is very dense and the particle-particle interactions are mainly contact-dominated, the discrete or distinct element model, in which the inter-particle interaction is described by the soft-sphere model, is popular in Lagrangian simulations of the spouted bed [65, 66]. In these studies, the structured grid is adopted and the conical part of the bed is meshed in a stair-step way.

Experimental and 2D DEM numerical results on the time-averaged and instantaneous flow behaviors of granular solids in a quasi-2D spouted bed are reported in the literature [65]. We performed a 3D discrete particle simulation of the quasi-2D spouted bed using the proposed DPM model [30]. The bed dimensions are selected the same as those of the experiment in [65] and the collision coefficients are selected as, $\mu_f = 0.15$, $e_n = 0.95$, and $e_t = 0.35$. A hybrid mesh combining hexahedral and wedged cells is used with the total cell number of 6231 and the time step is set to 0.0001s. In the tapered section of the bed, the wedged cells are used to mesh the domain. The bed is composed of glass beads with a diameter of 2.0 mm and a density of 2.38×10^3 kg/m^3. Both physical and numerical experiments were conducted at stagnant bed height of 100 mm, superficial gas velocity of 1.58 m/s (for a minimum spouting velocity $U_{ms} = 0.91$ m/s). Because there are steep gradients of the gas velocity near the interface between the spout and the annulus, the Saffman lift force due to gas shear flow is taken into account as follows:

$$\mathbf{F}_{Saff} = 1.61 d_p^2 (\mu \rho_g)^{1/2} \left| \boldsymbol{\omega}_g \right|^{-1/2} (\mathbf{u}_g - \mathbf{u}_p) \times \boldsymbol{\omega}_g$$

where $\boldsymbol{\omega}_g = \nabla \times \mathbf{u}_g$ is the gas vorticity.

Fig. **22** shows the flow pattern of particles at different points of time. The numerical simulation captures the main features of the spouted bed; the annulus, the spout and the fountain flow regions are distinguished very clearly. The transient granular flow is not steady but appears to be periodic. Also, a denser neck appears near the inlet, flows upward and finally disappears at the end of the spout. That means the particles do not move individually but as a cluster in the spouting region. While the particle cluster in the neck moves upward, it progressively grows bigger and denser with more particles entrained from the annulus. When it reaches the spout end, the spout is almost 'choked'. After that the cluster is injected into the fountain and the particles fall down to the annulus. The above flow behavior agrees well with the experimental observations [65, 67]. The predicted period is about 0.11s, which is slightly less than the experimental one (0.15s).

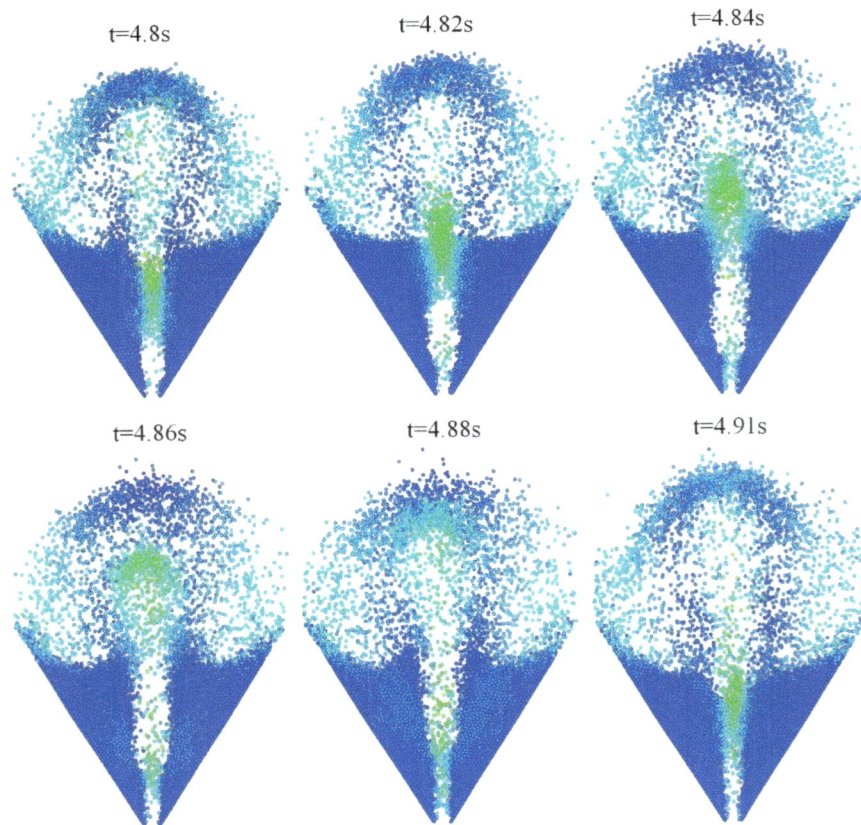

Figure 22: Snapshots of particle distribution in flat spouted bed (color from blue to yellow denotes particle velocity magnitude from 0 to 1.1m/s).

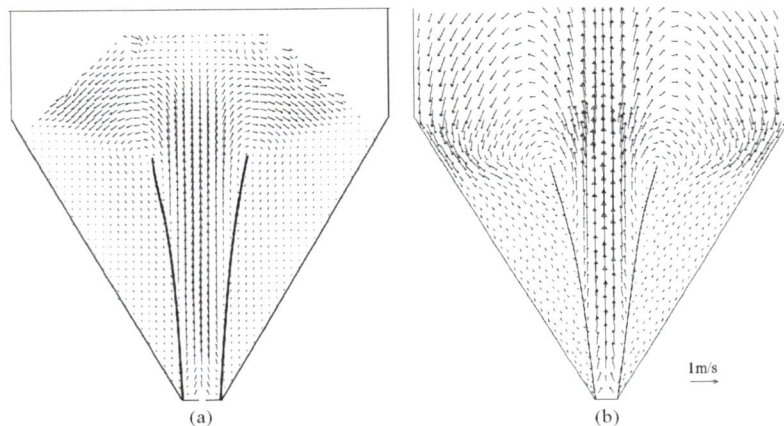

Figure 23: Comparison of time-averaged particle velocity field: (a) experimental results by Liu *et al.* 2008 [67] (the top part of the fountain is not shown because of the small number of particles measured there, see Liu *et al.* 2008 [67]) and (b) 3D DPM results, region between thick lines (zero vertical velocity) is spout.

The time-averaged particle velocity simulated by the present 3D DPM is compared with that of the experiment, as shown in Fig. **23**. Based on the particle velocity field, the interface separating the spout and the annulus can be delineated, where the time-averaged vertical particle velocity is zero. The predicted spout height is about 0.088m, close to the experimental value of 0.08m. The predicted time-averaged flow also agrees well with that of the experiment, but the particle velocity in the upper spout and fountain is over-predicted. This can also be noticed on the axial and lateral profiles of the particle vertical velocity as shown in Figs. **24** and **25**. The particle vertical velocity reaches about 1.1m/s, larger than that of the experiment, at the spout end. The axial profile shows that the particles accelerate more in the lower spout than in the upper spout, which agrees with the experimental observations.

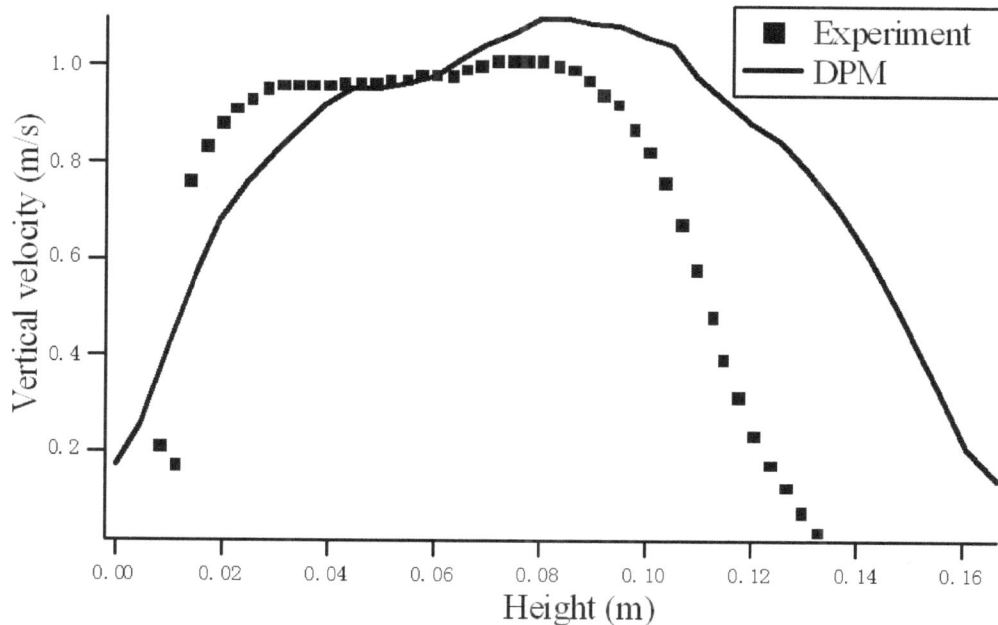

Figure 24: Comparison of vertical particle velocity along the spout axis with that of experiment by Zhao *et al.* (2008) [65].

It can be seen that the particle vertical velocity reaches maximum in the spout center and declines to zero which is defined as the interface separating the spout and the annulus. This is caused by two combined effects: (1) the centre gas jet has a maximum velocity which results in a maximum drag on the particle in the spout centre; 2) the flow is dilute in the spout center and inter-particle frictional interaction (shear force) is less significant than in the interface. The expansion of the radial position of the vertical velocity (see Fig. **25**) reflects the expanding shape of the spout, which is also shown in Fig. **23**. Particles accelerate almost in the whole spout region. This phenomenon is completely different from the traditional conical spouted bed, in which particles only accelerate for a very short period near the inlet orifice and gradually decelerate through the spout [68]. The difference may be caused by the formation of cluster in the spout. In conical spouted bed, no particle cluster or "chocking" was observed and the flow is completely dilute in the spout [68]. The voidage inside the forming cluster is considerably low which results in a higher drag force on the particles (the drag force is correlated to the void fraction), which causes the particles in the cluster to accelerate. The lateral particle velocity profiles predicted by the 3D DPM shows better agreement with those of the experiment than those by the 2D DEM [65]. The latter under-estimated the particle vertical velocity in the spout, which was considered as a result of the simplification of the 2D simulation including erroneous porosity and/or drag force calculations [65]. Nevertheless, the 3D DPM over-estimated the spout height and particle velocity in the fountain, which may be caused by the Gidaspow's drag formula used in the present simulation. This formula over-predicts the two-phase drag force when the heterogeneous flow structure forms in the bed [19].

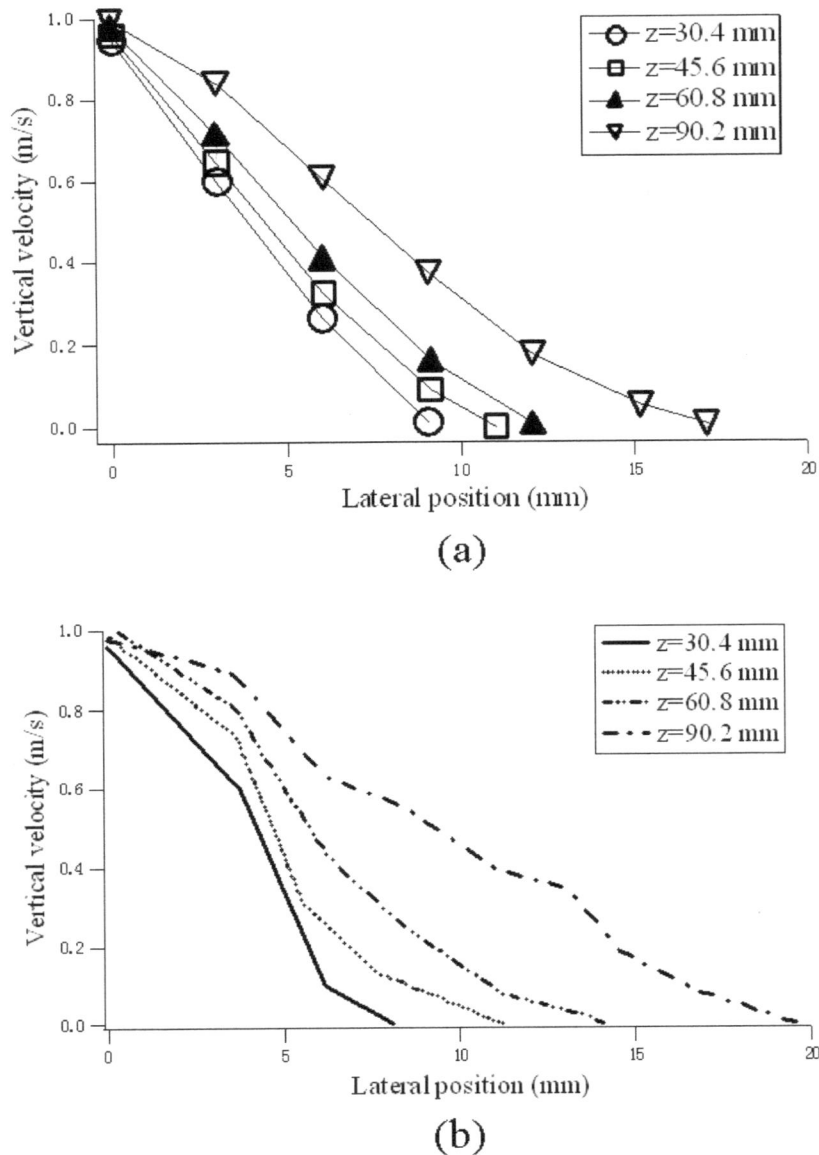

(a)

(b)

Figure 25: Lateral profiles of vertical particle velocity at different levels: (a) Experimental results; (b) 3D DPM results.

REFERENCES

[1]   C. Crowe and M. Sommerfeld, Y. Tsuji, *Multiphase Flows with Droplets and Particles*, CRC Press, Boston, 1998.

[2]   M.A. van der Hoef, M. van Sint Annaland, N. G. Deen, and J.A.M. Kuipers, "Numerical simulation of dense gas-solid fluidized beds: A multiscale modeling strategy," *Ann. Rev. Fluid Mechanics*, vol. 40, pp. 47-70, 2008.

[3]   Y. Tsuji, "Multi-scale modeling of dense phase gas–particle flow," *Chem. Eng. Sci.*, vol. 62, pp. 3410-3418, 2007.

[4] D.L. Koch and R.J. Hill, "Inertial effects in suspension and porous-media flows," *Ann. Rev. Fluid Mechanics*, vol. 33, pp. 619-642, 2001.

[5] M.A. van der Hoef, R. Beetstra, and J.A.M. Kuipers, "Lattice-Boltzmann simulations of low-Reynolds-number flow past mono- and bidisperse arrays of spheres: results for the permeability and drag force," *J. Fluid Mechanics*, vol. 528, pp. 233-254, 2005.

[6] B.J. Alder and T.E. Wainwright, "Phase transition for a hard-sphere system," *J. Chemistry Physics*, vol. 27, pp. 1208-1209, 1957.

[7] M.P. Allen and D.J. Tildesley, *Computer Simulations of Liquids*, Oxford Science Publications, Oxford, UK, 1990

[8] D.C. Rapaport, "The event scheduling problem in molecular dynamic simulation," *J. Comput. Physics*, vol. 34, pp. 184-201, 1980.

[9] M. Marin, D. Risso, and P. Cordero, "Efficient algorithms for many-body hard particle molecular dynamics," *J. Comput. Physics*, vol. 109, pp. 306-317, 1993.

[10] B.P.B. Hoomans, J.A.M. Kuipers, W.J.Briels, and W.P.M. van Swaaij, "Discrete particle simulation of bubble and slug formation in a two-dimensional gas-fluidised bed: a hard-sphere approach," Chem. Eng. Sci., vol. 51, pp. 99–118, 1996.

[11] G.A. Bokkers, M. van Sint Annaland, and J.A.M. Kuipers, "Mixing and segregation in a bidisperse gas-solid fluidized bed: a numerical and experimental study," *Powder Tech.*, vol. 140, pp. 176–186, 2004.

[12] B.P.B. Hoomans, "Granular dynamics of gas-solid two-phase flows," PhD thesis, Twente University, Netherlands, 2000.

[13] H. Lu, S. Wang, Y. Zhao, L. Yang, D. Gidaspow, and J. Ding, "Prediction of particle motion in a two-dimensional bubbling fluidized bed using discrete hard-sphere model," *Chem. Eng. Sci.*, vol. 60, pp. 3217–3231, 2005.

[14] C.L. Wu, J.M. Zhan, Y.S. Li, and K.S. Lam, "Dense particulate flow model on unstructured mesh." *Chem. Eng. Sci.*, vol. 61, pp. 5726-5741, 2006.

[15] C.L. Wu and J.M. Zhan, "Numerical prediction of particle mixing behavior in a bubbling fluidized bed," *J. Hydrodynamics, Ser. B*, vol. 19, pp. 335-341, 2007.

[16] J. Ouyang and J. Li, "Particle-motion-resolved discrete model for simulating gas-solid fluidization," *Chem. Eng. Sci.*, vol. 54, pp. 2077–2083, 1999.

[17] E. Helland, R. Occelli, and L. Tadrist, "Numerical study of cluster formation in gas-particle flow in a circulating fluidized bed," *Powder Tech.*, vol. 110, pp. 210-221, 2000.

[18] H.S. Zhou, G. Flamant, D. Gauthier, and J. Lu, "Lagrangian approach for simulating the gas-particle flow structure in a circulating fluidized bed riser," *Int. J. Multiphase Flow*, vol. 28, pp. 1801-1821, 2002.

[19] E. Helland, H. Bournot, R. Occelli, and L. Tadrist, "Drag reduction and cluster formation in a circulating fluidized bed," *Chem. Eng. Sci.*, vol. 62, pp. 148-158, 2007.

[20] J. Li and J.A.M. Kuipers, "On the origin of heterogeneous structure in dense gas–solid flows," *Chem. Eng. Sci.*, vol. 60, pp. 1251-1265, 2005.

[21] E. Delnoij, F.A. Lammers, J.A.M. Kuipers, and W.P.M. van Swaajj, "Dynamic simulation of dispersed gas-liquid two-phase flow using a discrete bubble model," *Chem. Eng. Sci.*, vol. 52, pp. 1429-1458, 1997.

[22] M.J.V. Goldschmidt, R. Beetstra, and J.A.M. Kuipers, "Hydrodynamic modelling of dense gas-fluidised beds: comparison and validation of 3D discrete particle and continuum models," *Powder Tech.*, vol. 142, pp. 23–47, 2004.

[23] P. A. Cundall and O. D. L. Strack, "A discrete numerical model for granular assemblies," *Geotechnique*, vol. 29, pp. 47-65, 1979.

[24] Y. Tsuji, T. Tanaka, and T. Ishida, "Lagrangian numerical simulation of plug flow of cohesionless particle in a horizontal pipe," *Powder Tech.*, vol. 71, pp. 239-250, 1992.

[25] Y.Tsuji, T. Kawaguchi, and T. Tanaka, "Discrete particle simulation of two-dimensional fluidized bed," *Powder Tech.*, vol. 77, pp. 79-87, 1993.

[26] B.H. Xu, A.B. Yu, S.J. Chew, and P. Zulli, "Numerical simulation of the gas-solid flow in a bed with lateral gas blasting," *Powder Tech.*, vol. 109, pp. 13-26, 2000.

[27] M.J. Rhodes, X.S. Wang, M. Nguyen, P. Stewart, and K. Liffman, "Study of mixing in gas-fluidized beds using a DEM model," *Chem. Eng. Sci.*, vol 56, pp. 2859-2866, 2001.

[28] B.H. Xu and A.B. Yu, "Numerical simulation of the gas-solid flow in a fluidized bed by combining discrete particle method with computational fluid dynamics," *Chem. Eng. Sci.*, vol. 52, pp. 2785-2809, 1997.

[29]   J. Li and J.A.M. Kuipers, "Effect of competition between particle-particle and gas-particle interactions on flow patterns in dense gas-fluidized beds," *Chem. Eng. Sci.*, vol. 62, pp. 3429-3442, 2007.

[30]   C.L. Wu, A.S. Berrouk, and K. Nandakumar, "Three-dimensional discrete particle model for gas-solid fluidized beds on unstructured mesh," *Chem. Eng. J.*, vol. 152, pp. 514-529, 2009.

[31]   D. Gidaspow, *Multiphase Flow and Fluidization–Continuum and Kinetic Theory Descriptions*. Academic Press, San Diego, 1994.

[32]   B.H. Xu and A.B. Yu, "Reply to comments on our paper 'Numerical simulation of the gas–solid flow in a fluidized bed by combining discrete particle method with computational fluid dynamics' by Hoomans *et al.*," *Chem. Eng. Sci.*, vol. 53, pp. 2646–2647, 1998.

[33]   H.P. Zhu, Z.Y. Zhou, R.Y. Yang, and A.B. Yu, "Discrete particle simulation of particulate systems: Theoretical developments," *Chem. Eng. Sci.,* vol. 62, pp. 3378-3396, 2007.

[34]   J. Li and J.A.M. Kuipers, "Gas–particle interactions in dense gas-fluidized beds," *Chem. Eng. Sci.*, vol. 58, pp. 711–718, 2003.

[35]   S. Ergun, "Fluid flow through packed columns," *Chem. Eng. Proc.*, vol. 48, pp. 89–94, 1952.

[36]   C.Y. Wen and Y.H. Yu, "Mechanics of fluidization*," A.I.Ch.E. Series*, vol. 62, pp. 100–111, 1966.

[37]   B.P.B. Hoomans, J.A.M. Kuipers, W.J.Briels, and W.P.M. van Swaaij, "Comments on the paper "Numerical simulation of the gas-solid flow in a fluidized bed by combining discrete particle method with computational fluid dynamics" by B. H. Xu and A. B. Yu," *Chem. Eng. Sci.*, vol. 53, pp. 2646–2647, 1998

[38]   K.D. Kafui, C. Thornton, and M.J. Adams, "Discrete particle-continuum fluid modelling of gas–solid fluidised beds*," Chem. Eng. Sci.*, vol. 57, pp. 2395–2410, 2002.

[39]   Y.Q. Feng and A.B. Yu, "Comments on "Discrete particle-continuum fluid modelling of gas–solid fluidised beds" by Kafui *et al.* [Chem. Eng. Sci., 57 (2002) 2395–2410]," *Chem. Eng. Sci.*, vol. 59, pp. 719–722, 2004.

[40]   K.D. Kafui, C. Thornton, and M.J. Admas, "Reply to comments by Feng and Yu on 'Discrete particle-continuum fluid modeling of gas–solid fluidized beds'," *Chem. Eng. Sci.*, vol. 59, pp. 723–725, 2004.

[41]   C.L. Wu, J.M. Zhan, Y.S. Li, K.S. Lam, and A.S. Berrouk, "Accurate void fraction calculation for three-dimensional discrete particle model on unstructured mesh," *Chem. Eng. Sci.*, vol. 64, pp. 1260-1266, 2009.

[42]   C.L. Wu, A.S. Berrouk, and K. Nandakumar, "An efficient chained-hash-table strategy for collision handling in hard-sphere discrete particle modeling," *Powder Tech.*, vol. 197, pp. 58-67, 2010.

[43]   D. J. Mavriplis, "Unstructured grid techniques," *Ann. Rev. Fluid Mechanics*, vol. 29, pp. 473-514, 1997.

[44]   Y. Zhao, H.H. Tan, and B. Zhang, "A high-resolution characteristics-based implicit dual time-stepping VOF method for free surface flow simulation on unstructured grids," J. Computational Physics, vol. 183, pp. 233-273, 2002.

[45]   M. Jacob, "ProCell technology: modelling and application," Powder Tech., vol. 189, pp. 332-342, 2009.

[46]   S.R. Mathur and J.Y. Murthy, "A pressure-based method for unstructured meshes," *Num. Heat Transfer*, vol. 31, pp. 195–215, 1997.

[47]   C.M. Rhie and W.L. Chow, "A numerical study of the turbulent flow past an isolated airfoil with trailing edge separation," *AIAA J.*, vol. 21, pp. 1525–1552, 1983.

[48]   B.R. Hutchinson and G.D. Raithby, "A multigrid method based on the additive correction strategy," *Num. Heat Transfer*, vol. 9, pp. 511–537, 1986.

[49]   B.G.M. van Wachem, J. van der Schaaf, J.C. Schouten, R. Krishna, and C.M. van den Bleek, "Experimental validation of Lagrangian–Eulerian simulations of fluidized beds," *Powder Tech.*, vol. 116, pp. 155–165, 2001.

[50]   D. Darmana, R.L.B. Henket, N.G. Deen, and J.A.M. Kuipers, "Detailed modelling of hydrodynamics, mass transfer and chemical reactions in a bubble column using a discrete bubble model: Chemisorption of CO2 into NaOH solution, numerical and experimental study," *Chem. Eng. Sci.*, vol. 62, pp. 2556-2575, 2007.

[51]   R. Lohner and J. Ambrosiano, "A vectorized particle tracer for unstructured grids," *J. Computational Physics*, vol. 91, pp. 22–31, 1990.

[52]   R. Lohner and J. Ambrosiano, "Robust, vectorized search algorithms for interpolation on unstructured grids," *J. Comput. Physics*, vol. 118, pp. 380–387, 1995.

[53]   R. Chordá, J.A. Blasco, and N. Fueyo, "An efficient particle-locating algorithm for application in arbitrary 2D and 3D grids," *Int. J. Multiphase Flow*, vol. 28, pp. 1565–1580, 2002

[54]   S.B. Kuang, A.B. Yu, and Z.S. Zou, "A new point-locating algorithm under three-dimensional hybrid meshes," *Int. J. Multiphase Flow*, vol. 34, pp. 1023–1030, 2008.

[55]   G.D. Martin, E. Loth, and D. Lankford, "Particle host cell determination in unstructured grids," *Computers and Fluids*, vol. 38, pp. 101–110, 2009.

[56]   M. Marin, D. Risso, and P. Cordero, "Efficient algorithms for many-body hard particle molecular dynamics," *J. Comput. Physics*, vol. 109, pp. 306-317, 1993.

[57]   M. Marin, and P. Cordero, "An empirical assessment of priority queues in event-driven molecular dynamics simulation," *Computational Physics Comm.*, vol. 92, pp. 214-224, 1995,

[58]   H. Sigurgeirsson, A. Stuart, and W.L. Wan, "Algorithms for particle-field simulations with collisions," *J. Comput. Physics*, vol. 172, pp. 766-807, 2001.

[59]   M.T. Goodrich, R. Tamassia, and D.M. Mount, *Data Structures and Algorithms in C++*, Wiley & Sons Inc., 2004

[60]   D. Bai, E. Shibuya, Y. Masuda, K. Nishio, N. Nakagawa, and K. Kato, "Distinction between upward and downward flows in circulating fluidized beds," *Powder Tech.*, vol. 84, pp. 75-81, 1995.

[61]   H.T. Bi and J.R. Grace, "Flow regime diagrams for gas-solid fluidization and upward transport," *Int. J. Multiphase Flow*, vol. 21, pp. 1229-1236, 1995.

[62]   M.J. Rhodes, M. Sollaart, and X.S. Wang, "Flow structure in a fast fluid bed," *Powder Tech.*, vol. 99, pp. 194-200, 1998.

[63]   S.W. Kim, G. Kirbas, H.T. Bi, C.J. Lim, and J.R. Grace, "Flow structure and thickness of annular downflow layer in a circulating fluidized bed riser," *Powder Tech.*, vol. 142, pp. 48-58, 2004.

[64]   M. Horio and H. Kuroki, "Three-dimensional flow visualization of dilutely dispersed solids in bubbling and circulating fluidized beds," *Chem. Eng. Sci.*, vol. 49, pp. 2413-2421, 1994.

[65]   X.L. Zhao, S.Q. Li, G.Q. Liu, Q. Yao, and J.S. Marshall, "DEM simulation of the particle dynamics in two-dimensional spouted beds," *Powder Tech.*, vol. 184, pp. 205-213, 2008.

[66]   S. Takeuchi, S. Wang, and M. Rhodes, "Discrete element method simulation of three-dimensional conical-base spouted beds," *Powder Tech.*, vol. 184, pp. 141-150, 2008.

[67]   G.Q. Liu, S.Q. Li, X.L. Zhao and Q. Yao, "Experimental studies of particle flow dynamics in a two-dimensional spouted bed." *Chem. Eng. Sci.*, vol. 63, pp. 1131-1141, 2008.

[68]   M.J San José, M. Olazar, S. Alvarez, M.A. Izquierdo, and J. Bilbao, "Local bed voidage in conical spouted beds," *Ind. Eng. Chem. Res.*, vol. 37, pp. 2553-2558, 1998.

CHAPTER 6

Experimental Study of Water Boiling in Microchannel

S.G. Singh[1], R.R. Bhide[2], S.P. Duttagupta[2], Arunkumar Sridharan[2] and Amit Agrawal[2*]

[1]*Indian Institute of Technology, Hyderabad and* [2]*Indian Institute of Technology Bombay, Powai, Mumbai 400076 India*

Abstract: Detailed experimental study involving flow boiling of water in microchannel is discussed in this chapter. The work aims to study the different aspects of the problem such as pressure drop, heat transfer coefficient, pressure instability, and void fraction. Flow visualization has also been performed. Experiments have been conducted *in silico*n microchannels with trapezoidal or rectangular cross-section of hydraulic diameter 45-140 μm, and a microheater fabricated on the reverse side of the silicon wafer to provide well controlled and metered input power. The experimental data is compared with the annular flow model and various empirical correlations, and complemented with clear discussion of the observed phenomenon. In the two-phase regime, the average pressure drop increases with a decrease in the flow rate and reaches a maximum (with a minimum on either side), while in the dryout regime the pressure drop decreases with flow rate. Our results suggest that there are up to four mass flow rate values with same pressure drop penalty and the operating point can be chosen with the maximum heat transfer coefficient. The average pressure drop is found to have a strong dependence on the channel aspect ratio and becomes minimum in rectangular microchannel for width-to-depth ratio of about 1.5. The instability in pressure drop has been quantified and linked to the underlying flow regime. The minimum P_{RMS}/P_{avg} is found to occur when the flow transitions to annular. For the first time, a flow regime map is also obtained for such systems. The flow is found to be predominately annular at high heat flux and flow rate. A breakup of the flow frequency suggests that the flow is bistable in the annular regime. At a fixed location, the flow periodically switches from single phase liquid to annular, and vice-versa. Otherwise, all three regimes: single phase liquid, bubbly and slug are obtained. An image analysis technique has also been developed and utilized to estimate the void fraction as a function of position in the microchannel, heat flux and mass flow rate. The technique has been extended to obtain heat transfer coefficient purely from image analysis. Both void fraction and heat transfer coefficient are found to increase monotonically with position in the microchannel. There are several novel aspects of this study. For example, effect of microchannel aspect ratio on pressure drop is studied for the first time. Some guidelines for choosing the operating point with desired constraints have been proposed. Ways to reduce instability have also been explored. Development of flow regime map and flow visualization technique is not available in the literature currently. Pressure drop data near CHF condition have been presented. The results are interesting from both fundamental and electronic cooling point of view.

Keywords: Electronic cooling, two-phase flow, microchannel, pressure drop, heat transfer coefficient, instability, void fraction, flow regime map, flow visualization, annular flow.

INTRODUCTION

Motivation and Applications

The trend of enhanced IC performance as a combination of higher functional density, larger power density ($> 250 W/cm^2$) and higher operational frequency, has resulted in the possibility of randomly generated very high heat flux (VHHF) transients, amongst other non-ideal effects. For the purpose of illustration, Fig. **1** from Ref. [1] shows a comparison between the heat fluxes experienced by Renewable Launch Vehicles (Space Shuttle) and integrated circuits. Such VHHF transients can severely degrade MOSFET gate dielectric and ohmic contacts of electronics devices including p-n diodes. Accordingly, electronic cooling strategies must take into account current and future device performance trends, and should have in-built

*Address correspondence to Amit Agrawal: Indian Institute of Technology Bombay, Powai, Mumbai 400076 India; E-mail: amit.agrawal@iitb.ac.in

flexibility to be effective under various steady state and transient heat flux scenario. Microchannel heat sink has emerged as a promising on-chip cooling solution due to its large surface area to volume ratio [2]. Tuckerman and Pease first demonstrated heat transfer characteristics in microchannels. Subsequently numerous studies have been performed to understand the cooling mechanism of two phase and single phase fluid flow in microchannels [3-7]. Recently, IBM has developed a prototype device with thousands of 'hair-width' cooling arteries [8]. This cooling technology is being developed in anticipation of 3-D integrated circuits, which will utilize 3-D interconnects to realize even higher IC densities and consequently even higher heat dissipation as compared to the present 2-D ICs [9].

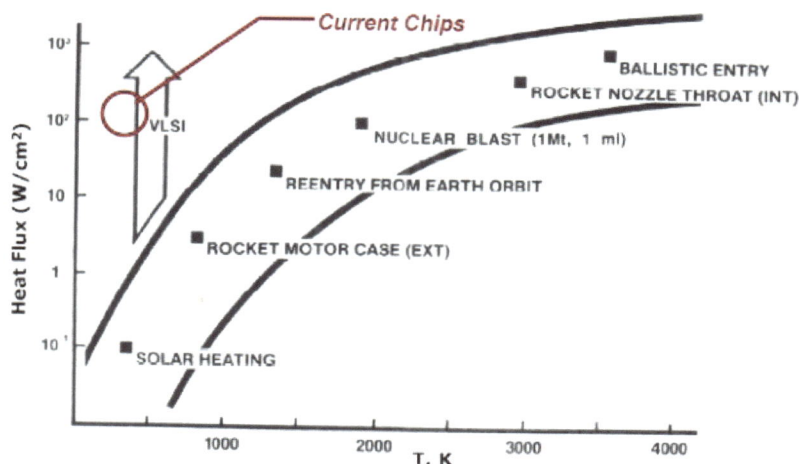

Figure 1: Comparative Heat dissipation from different sources [1].

Microfluidic devices, referred to as micro-thermal-mechanical systems (μ-TMS) are currently being used in radar, aerospace avionics components and in micro-chemical reactors [10], apart from being used as microcooling elements for electronic components. They are also widely used in reactors for modification and separation of biological cells, selective membranes and liquid/gas chromatography [11].

Two phase heat transfer to fluids flowing through microchannels is a promising solution to the problem of cooling integrated circuits with very high power densities. Based on conventional knowledge it is understood that the heat transfer coefficient for boiling is 10-100 times larger than the single-phase heat transfer coefficient. The Nusselt number is constant (~4) for hydro-dynamically and thermally fully developed laminar flow in an enclosed channel, but it varies with the shape of the channel and thermal boundary condition. From $Nu = hd_h/k$, the single-phase heat transfer coefficient for a 100 μm channel with water as the working fluid is 26 kW/m^2-K. Such a high heat transfer coefficients in single phase flow have been experimentally measured, and results because of miniturization of the channel. A further increase in h can be expected with boiling due to additional heat removal during phase change. Another advantage of employing boiling is a smaller axial temperature variation along the length of the microchannel as compared to single phase flow. The penalty, of course, is substantially larger pressure drop and flow instability. However, the possibility of increasing the heat transfer coefficient without a corresponding increase in pressure penalty and with minimal instability needs to be explored.

Note however that the results at the conventional scale need to be applied with care to microchannels suggesting that additional studies at the micro-scale are required. There is clearly a need to generate experimental data for a large range of mass flow rate and heat flux values. Measurement of heat transfer coefficient in microchannel remains a challenge as reliable data is missing. Researchers have reported different trends of heat transfer coefficient in microchannel. The dependence of the two phase heat transfer coefficient in microchannels on the vapor quality, mass flow rate and heat flux is still a matter of debate. Thus there is need to develop alternate techniques to measure this important parameter.

Objectives of the Present Work

The goal of the present study is to design, fabricate, characterize and possibly optimize a microfluidics network for electronic cooling application. Keeping the gaps in literature in mind, the following objectives have been formulated:

- To conduct experiments on single phase and two-phase microchannel heat sinks. This includes flow visualization studies and pressure and temperature measurements.

- To compare the experimental results against prediction from two-phase flow models such as annular flow model and existing empirical correlations.

- To study the effect of microchannel geometry on pressure drop with the aim of arriving at an optimal geometry.

- To characterize the instabilities and explore ways to reduce the same.

LITERATURE SURVEY

Using microscopes and high speed video cameras, attempts have been made to observe boiling in microchannels. Many different flow regimes have been reported by researchers. The bubbly flow regime was observed by Lee and Mudawar [12] in their study of R-134a in parallel microchannels. They also observed elongated slug and annular flow regimes in microchannels of $d_h = 348.95\mu m$ and suggested that transition from slug to annular takes place around quality (x) of 0.55, which is lower than that suggested by Tran *et al.* [13]. Yen *et al.* [4] performed flow visualization studies in square and circular cross section microchannels of diameter 214 μm and 210 μm, respectively. They used HCFC123 as the working fluid and observed bubbly, plug, annular and capillary flow patterns. These flow patterns were periodic in nature. The bubbly and annular flows were predominant in the square channel while the capillary flow was dominant in circular microchannels. They attributed this difference to the evenly distributed liquid film in the case of circular microchannel, which is not the case with square geometry. Revellin and Thome [13] developed a flow pattern map based on the evaporation of refrigerants. They classified the flow into isolated bubble regime, coalescing bubble regime, and annular regime. Qu and Mudawar [14] employed water in parallel microchannels of hydraulic diameter of 348.95 μm and reported that there was an abrupt transition to annular flow regime at the point of zero thermodynamic quality. Lee *et al.* [15] studied the flow patterns in a high aspect ratio, small hydraulic diameter microchannel using de-ionized (DI) water and noticed bubble activity at $q"/q"_{chf} \sim 0.65$. They also observed that the bubble grew rapidly to occupy the entire cross-section. The bubble was stretched in the direction of the fluid flow and departed when the two interfaces merged. Jiang *et al.* [5] performed visualization and measurements of flow boiling *in silico*n microchannels with triangular cross-sections for two different hydraulic diameters of 40 and 80 μm. They did not observe bubbly flow in either instance, unlike in macro-channels. Jiang *et al.* [16] also noted the absence of bubbly regime in their study. Zhang *et al.* [17] reported that mostly annular flow was observed in their experiments with a very thin layer of liquid on the channel walls. However, they did not observe bubbly and slug flows. Kandlikar and Balasubramanian [18] studied the growth of bubbles, formation of slugs, and periodic wetting and re-wetting of the walls [19]. Steinke and Kandlikar [20] reported the occurrence of slug, annular, churn and dryout conditions in their microchannel [21]. Serizawa *et al.* [22] have reported different flow patterns in water-steam flow in very small diameter circular channel. It is to be noted that most of the above studies employ fluid other than water.

Qu and Mudawar [14] presented experimental results for flow boiling in a water-cooled microchannel heat sink. The heat transfer coefficient was found to be in the range of 22-44 kW/m²-K. Contrary to behavior observed in macrochannels, the heat transfer coefficient decreased with increasing vapor quality [23]. Similar dependence of heat transfer coefficient on vapor quality was reported by Hetsroni *et al.* [24] for a heat sink that had 21 parallel triangular microchannels each of hydraulic diameter 129 μm. Vertrel XF (with a much lower saturation temperature of 52 °C) was used as the working fluid. Steinke and Kandlikar

[20] and Yen *et al.* [4] studied flow boiling of water and refrigerants HCFC123 and FC72 in microchannels, respectively. They also observed that the heat transfer coefficient decreases monotonically with an increase in vapor quality, in agreement with the earlier studies. Yun *et al.* [25] studied convective boiling heat transfer to CO_2 in microchannels. They reported that the heat transfer coefficient for CO_2 at a heat flux of 10 kW/m^2 increases with vapor quality, while it remains fairly constant at a heat flux of 15 kW/m^2. Thome [10] referred to data that suggest that microchannel flow boiling heat transfer coefficients are neither a function of vapor quality nor mass velocity but are a function of heat flux and saturation pressure, just like in nucleate pool boiling. Diaz and Schmidt [26] worked with water and ethanol in a single 0.3 mm × 12.7 mm × 200 mm channel made of nickel alloy Inconel 600 and obtained heat transfer coefficient in the range of 10-160 kW/m^2-K, depending on the heat and mass flux applied. Lee and Pan [27] found heat transfer coefficient to vary between 1.8-50 kW/m^2-K for vapor quality between 0.02-0.2, in their 33.7 μm hydraulic diameter channel. With triangular microchannels, Hetsroni *et al.* [28] found the following range for *h* (depending on the boiling number): 9-31 kW/m^2-K (0.00045-0.0009), 12-34 kW/m^2-K (0.0006-0.0010), and 11-28 kW/m^2-K (0.0009-0.0019) for hydraulic diameter of 100, 130, and 200 μm, respectively. Ribatski *et al.* [29] noted that different trends of *h* with respect to quality, mass velocity, and heat flux have been reported in the literature. The most common trend is that *h* decreases with an increase in quality and hydraulic diameter, and increase of *h* with mass velocity for a given quality. They note that nucleate boiling has been incorrectly attributed to be the dominant heat transfer mechanism at the micro-scales. The existing correlations perform poorly in predicting the experimental data; this failure was attributed to large variation in the experimental data itself. Thus, there appears to be lack of consensus over the dependence of heat transfer coefficient on mass flow rate, vapor quality and heat flux. This may partly be because different researchers have used different fluids in their experiments. The heat transfer mechanisms for different fluids may be different and can lead to differences in the results.

Pressure drop is large in microchannel owing to the small flow passage. However, it is widely established that liquid flow in microchannel obeys the macroscale laws. As the flow in microchannel is mostly laminar, we have from [30]:

$$f\,Re = 24(1 - 1.353\beta + 1.9467\beta^2 - 1.7012\beta^3 + 0.9564\beta^4 - 0.2537\beta^5)$$

where β is the aspect ratio of the vapor core. It has been found that when a constant heat flux is applied and the mass flux is decreased, the pressure drop decreases initially. The value at a particular mass flux is less than the pressure drop for no heat flux. This is primarily because the heat flux raises the fluid temperature and lowers the viscosity. However, once the two phase flow starts in microchannel, the pressure drop starts to increase due to increase in vapor quality. Zhang *et al.* [31] reported flow boiling pressure drop in their study. The pressure drop was found between 8-26 k Pa and upto 5K of wall superheat was observed with boiling in their study. The pressure drop was compared with homogeneous and annular flow models; the comparison was found to be particularly good for the homogeneous model. Zhang *et al.* [17] reported pressure drop of 6-25 kPa in 113 μm hydraulic diameter rectangular microchannel for heating rate of 1.8-3.3 W, and 37-42 kPa in a 44 μm microchannel for heating rate of 2.15-2.6 W. Qu and Mudawar [32] and Ribatski *et al.* [29] compared their pressure drop data against correlations in the literature. Wu *et al.* [33] observed pressure drop between 0.27-5.84 bar with water flowing in eight trapezoidal microchannels, each of hydraulic diameter of 72.7 μm and 6 cm length. Lee and Pan [27] found that the frictional pressure drop is the dominant component accounting for about 75% of the overall pressure drop. On the other hand, Bowers and Mudawar [34] report that the acceleration pressure drop is approximately 75% of the total pressure drop in their microchannel (d_h = 510 μm) with R-113 as the working fluid.

It is well known that flow instability can occur with boiling. Owing to low velocities and bubble formation, the fluctuations are likely to be more severe in microchannels as compared to conventional channels [35]. Instability in two phase flow is inherent yet undesirable because it would affect the heat transfer and fluid flow characteristics, possibly resulting in wall temperature and heat flux oscillations or even complete dryout. Hence, flow instability and its mitigation have received a good amount of attention in the literature. The oscillation in the inlet temperature and pressure, outlet temperature, wall temperature, and mass flux through the channel have been systematically measured and documented by Hetsroni *et al.* [35], Muwanga *et al.* [36], Diaz and Schmidt [26], Wu and Cheng [37], among others. Wu and Cheng [37] observed with a successive reduction in mass flow rate and increase in heat flux: liquid/two-phase alternating flow, continuous two-phase flow, and liquid/two-phase/vapor alternating flow. The amplitude of pressure, temperature and mass flux oscillation was reported to be the smallest in the continuous two-phase flow regime. The oscillations of pressure and mass flux are in phase in the continuous two-phase flow regime, and out of phase in the other two regimes. These results have subsequently been corroborated by Wang *et al.* [38]. Hetsroni *et al.* [35] reported an increase in RMS pressure normalized by the average pressure drop with quality. Lee *et al.* [39] reported an increase in the frequency of fluctuation (determined from spectral analysis of pressure-time series) with input power. Qu and Mudawar [14] identified two types of instabilities severe pressure drop oscillations and mild parallel channel instability. Muwanga *et al.* [36] found that the frequency of oscillation reduces with an increase in heat flux and the amplitude increases. The amount of sub-cooling was reported to have considerable influence on both the frequency and amplitude of oscillations. A stability analysis was performed by Chavan *et al.* [40]. They found that the flow can be stable only in a small range of input parameters. Kandlikar *et al.* [41] and Qu and Mudawar [32] have tried to remove flow boiling instability by introducing pressure drop elements upstream of the heated microchannels and addition of nucleation sites on the wall of the microchannels. It was found in [41] that the introduction of nucleation sites in conjunction with a 51% area restriction located upstream of the microchannels could partially reduce the instabilities, but did not completely prevent it. The severe restriction to flow introduced by decreasing the pressure drop elements from 51% area to 4% area could completely eliminate the instabilities, but the pressure drop across the microchannels becomes unacceptably high. Bhide *et al.* [42] has shown that instabilities can be reduced by working with small hydraulic diameter channels and by employing a higher surface roughness.

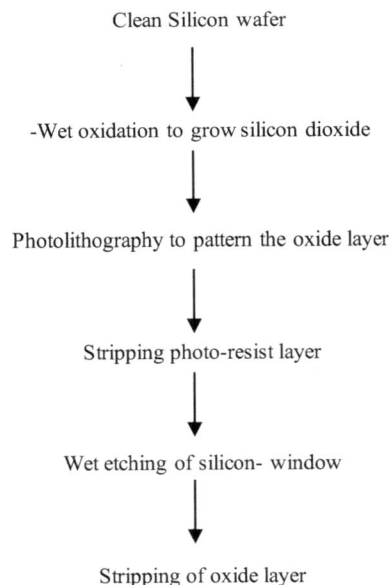

Clean Silicon wafer

↓

-Wet oxidation to grow silicon dioxide

↓

Photolithography to pattern the oxide layer

↓

Stripping photo-resist layer

↓

Wet etching of silicon- window

↓

Stripping of oxide layer

Figure 2: Flow chart for fabrication of trapezoidal <100> and rectangular microchannels.

EXPERIMENTAL SETUP

Fabrication of Microchannel and Microheater

The microchannels are fabricated on a 2 inch, 275±25μm thick, p-type, <100> and <110> double-side-polished silicon wafer. The channel length is 20 mm. Both single and parallel channels have been realized by a sequence of process steps including TMAH (tri-methyl ammonium hydroxide) etching. Fig. **2** shows the fabrication process of microchannels in silicon. Fig. **3** shows the cross-section of a trapezoidal microchannel in <100> silicon where θ is the angle at which etching occur with respect to the wafer surfaces, typically 54.7°. For <110> silicon, $\theta=90°$ and etching results in rectangular cross-section microchannel. At inlet and exit of the channels, reservoirs of size 10 mm × 10 mm × 75 μm were also etched. The complete device produced is shown in Fig. **4**, the etch step and surface roughness are characterized using a profilometer. The surface roughness is less than 0.1 μm for most of the microchannels fabricated. The microchannel is closed with the help of a quartz plate. The bonding of the microchannel with quartz plate is a critical step in the whole process since it is vital that there is no leakage of fluid.

A microheater was fabricated on the other surface of the wafer in order to supply a controlled amount of heat flux. The need for localized heating necessitated the development and fabrication of efficient microheaters. Fig. **5** presents flow chart for fabrication of microheater. A chrome-gold/ titanium-platinum thin film stack is used as the microheater material. The microheater is patterned on the front side of the wafer by using lift-off process. The processed wafer is sputtered with chrome-gold. The thickness of the stack is 50 nm as measured by a profilometer. Fig. **6** shows the step by step processing of the micro-heater and its cross sectional view after each step. The final device is shown in Fig. **7**.

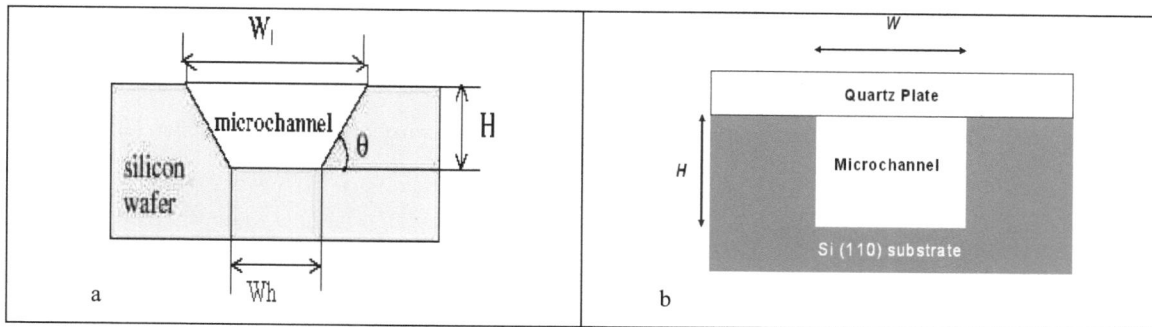

Figure 3: Cross-section of an etched microchannel on a) <100> and b) <110> wafer.

Figure 4: Angular view of test apparatus.

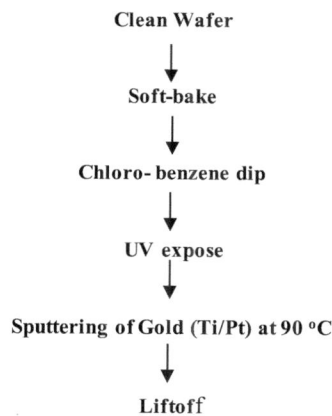

Figure 5: Flow chart of the entire process to fabricate the micro-heater.

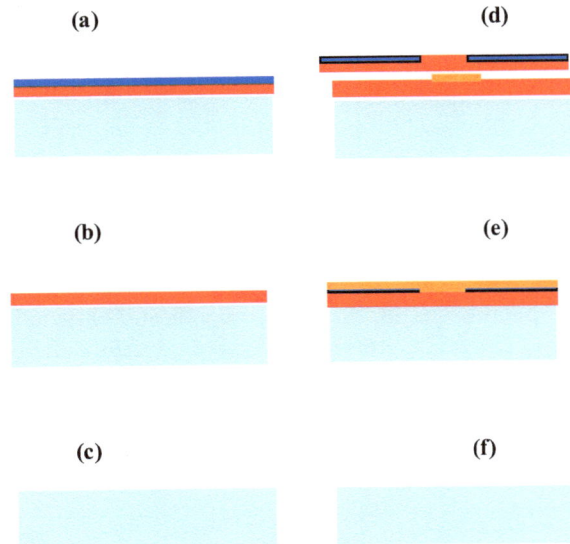

Figure 6: (a) (100) p type Si, **(b)** 1 micron thick oxide over Si, **(c)** PPR over oxide, **(d)** Patterning wafer, **(e)** Sputter the Gold, **(f)** Liftoff the gold.

Figure 7: Top view of micro-heater.

Characterization of Flow in Microchannel

Characterization of microchannel and micro heater is very important before performing any measurements. Fig. **8** presents the experimental set-up employed for pressure drop measurement across the microchannels. The setup consists of DI water reservoir, pump, mass flow rate measurement system, damper (to reduce disturbance from the pump), probe, differential pressure measurement, and data logger. A pre-calibrated peristaltic pump (Master Flex, L/S 2) is used for metering and control of mass flow rate of water, up to a maximum of 1.7 bar (absolute) pressure. A pre-calibrated digital pressure gauge (Keller, Leo 1) with a response time of 1 s and range of -1 bar to 3 bars is used to measure the pressure difference across the channel. Table **1** presents an estimate of the error in quantities measured in the experiments.

The following procedure is adopted for performing the experiments. The water reservoir is filled with de-gassed (boiled for at least 10 min and cooled to 25 ^{0}C) DI water and the micro-pump is set for the desired flow rate. The pressure drop, temperature of inlet and outlet, and flow rate values are measured simultaneously using a data logger. Note that the pressure drop across the entire microchannel is measured, which includes entry and exit loses. The pressure drop due to expansion and contraction (at the entry and exit) is however estimated to be negligibly small (0.3%) as compared to the overall pressure drop.

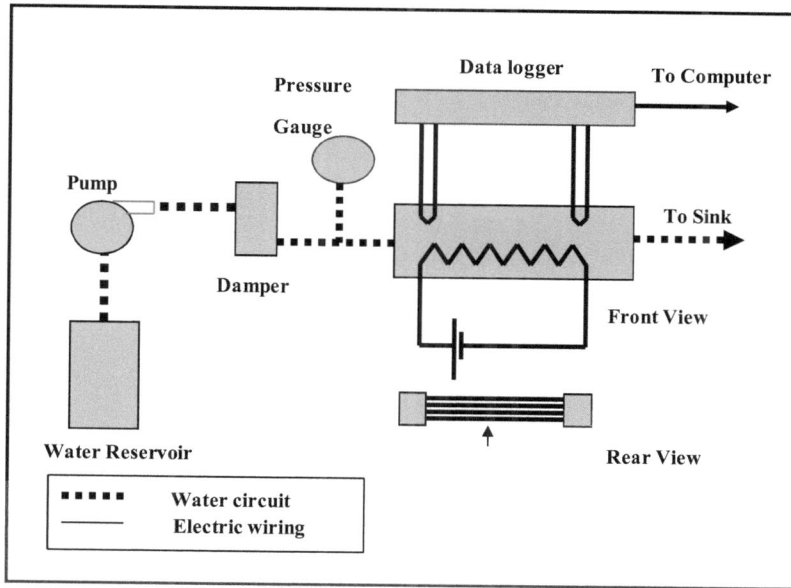

Figure 8: Schematic of the experimental set-up.

The experiments were done to measure the pressure drop across the micro channel with different flow rates at constant temperature. The size of the microchannel used in this section is 115 μm (W_L) ×55 μm (H) ×20 mm, resulting in a hydraulic diameter of 58.8 μm. Fig. **9** shows that as the flow rate increases, the pressure drop across the microchannel increases linearly. Further, as the temperature of the fluid increases the pressure drop decreases. Note that the pressure drop reduces to less than half as the temperature increases from 25°C to 90°C. The reason for this behavior is a decrease in fluid viscosity with an increase in temperature. The significant lowering of pressure drop at higher temperatures has practical implication since pressure drop penalty is regarded as an impediment for application of microchannels for electronics cooling. We now show that on non-dimensionalizing the pressure drop and mass flow rate to friction factor (*f*) and Reynolds number *(Re)*, the conventional behavior is obtained. Fig. **10** show that *f.Re* is independent of both the Reynolds number and the temperature of the fluid. Here *f* and *Re* are calculated as:

$$f = 2\Delta p D_h / (L\rho U^2) \tag{1}$$

$$\mathrm{Re} = \rho U D_h / \mu(T) \tag{2}$$

$$\mu(T) = \mu(T_{ref})(T / T_{ref})^n \exp[B(1/T - 1/T_{ref})] \tag{3}$$

where $n = 8.9$

$B = 4700$ /K,

$\mu(T_{ref}) = 1.005 \times 10^{-3}$ kg/m-s

$T_{ref} = 293$ K

Morini [43] showed that for laminar single phase flow in for trapezoidal microchannels with an aspect ratio of ($W_L/H = 2.18$), *f.Re* is constant at 62.53. The experimentally determined values are within the experimental uncertainty of this theoretical value. This result is important because several contradicting results about the value of friction factor for flow in microchannel have been reported in the literature (Mala et. al. [44], Rahman [45], Wu and Cheng [46], Peng and Peterson [47]). These results help to validate the experimental setup and measurement procedure.

Micro-Heater Characterization

A microheater is used to supply a controlled heat flux to the working fluid. The procedure for characterization of the fabricated microheater is as follows. The resistance of microheater is first measured at room temperature. For this a source meter is connected to the microheater through micromanipulators probes. The resistance is obtained by measuring the current and voltage, and applying the Ohm's law. The procedure is repeated for different heat fluxes (by placing it on a hotplate) while the temperature of the wafer is measured by a thermocouple, with a least count of 0.5°C. The linear relation between current and voltage (Fig. **11a**), suggests that it is indeed obeys the Ohms law and acts as a simple resistor. The resistance of the micro-heater increases from 345.6 Ω (at 30°C) to 382.37 Ω (at 120°C) (Fig. **11b**) – an almost 10% change in resistance over a 90°C temperature range. This suggests that because the resistance of chrome-gold is linear with temperature, it is a good material for microheater. The above test establishes confidence in the microheater fabricated.

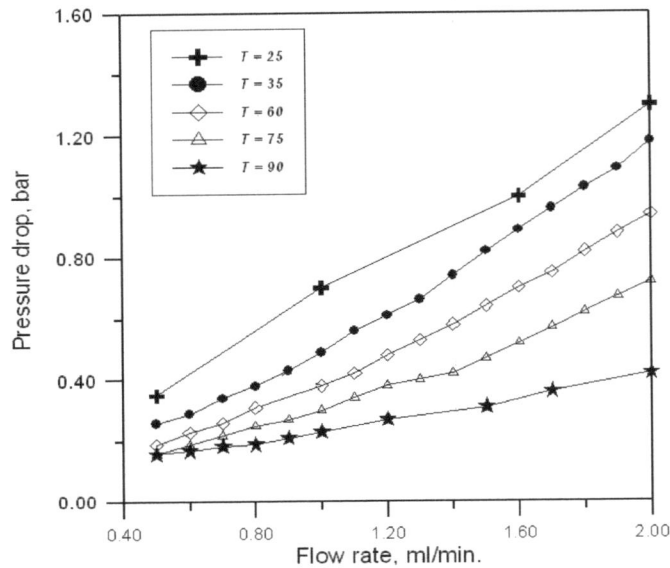

Figure 9: Pressure drop versus mass flow rate, at different temperatures in the single phase regime (four parallel channels of hydraulic diameter 58.8 µm). Note that the data points have been connected by straight line to aid the eye.

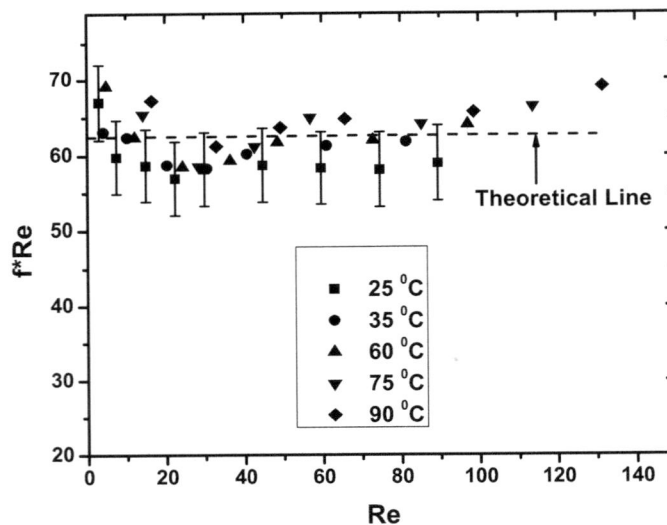

Figure 10: $f \times Re$ versus Reynolds number at different temperatures in single phase regimes (four parallel channels of hydraulic diameter 58.8 µm). The theoretical value as determined by Morini [43] is also plotted.

Table 1: Estimation of error in various parameters measured and derived in the experiments.

Parameter	Maximum error	Parameter	Maximum error
Volume flow	0.01 ml/min	P_{Ch}	1.23%
L	1 µm	A_{Ch}	1.68%
W_l, W_h	1 µm	D_h	3.23%
H	1 µm	M	0.62%
T	0.5 ^{0}C	U	2.31%
ΔP	2 mbar	Re	3.97%
Electrical Power	0.02 W	$f \times Re$	8.40%

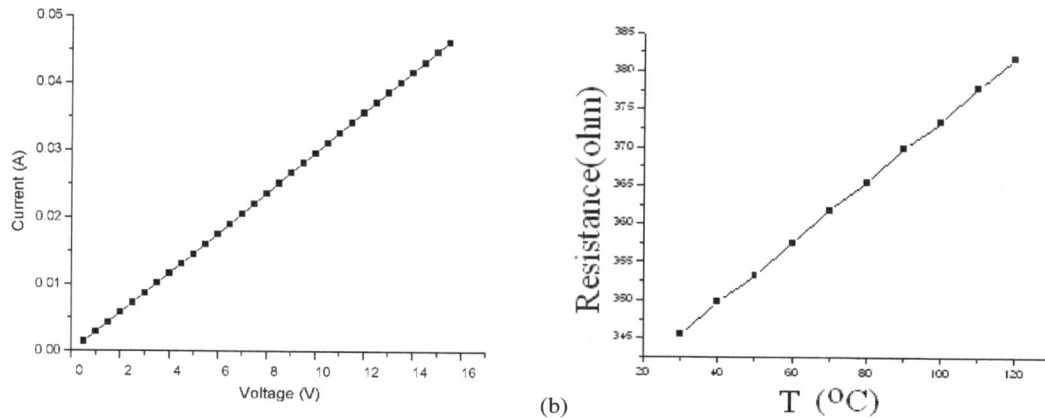

Figure 11: (a) I-V characteristics and (b) variation in electrical resistance with temperature for the microheater fabricated.

Data Reduction

This section deals with the procedure and formula adopted for calculating the heat flux and heat transfer coefficient. The de-ionized water enters the reservoir at room temperature and gets heated before entering the microchannel. The power lost in heating water in the reservoir is

$$P_{reservoir} = \dot{m} C_p (T_{in} - T_{amb}) \tag{4}$$

The power supplied to the microchannel is therefore

$$P_{channel} = P_{supplied}(1 - \lambda) - P_{rersevoir} \tag{5}$$

where the heat loss factor (λ) was carefully determined and found to be about 20 % under the experimental conditions employed in this work. Now the power per unit length and thermodynamic quality can be calculated from the following equations:

$$q' = \frac{P_{channel}}{L} \tag{6}$$

$$x = \frac{q'L - \dot{m} C_p (T_{sat} - T_{in})}{h_{fg} \dot{m}} \tag{7}$$

Note that L is the total length of the channel, while l is the distance from the inlet.

The channel heat flux can be obtained by dividing the power received by the microchannel by the heated surface area. (Note that due to the quartz plate on one side, the microchannel gets heated from the remaining three sides only.)

$$q'' = \frac{P_{channel}}{A_h} \qquad (8)$$

The single phase heat transfer coefficient (required later) can be obtained using the definition of Nusselt number as

$$h_{sp} = \frac{Nu k_f}{d_h} \qquad (9)$$

where the Nusselt number for three side heated rectangular channel is given by [48]

$$Nu = 8.235 \left(\begin{array}{c} 1 - 1.883 / \beta_c + 3.767 / \beta_c^2 \\ -5.814 / \beta_c^3 + 5.361 / \beta_c^4 - 2 / \beta_c^5 \end{array} \right) \qquad (10)$$

where β_c is the microchannel aspect ratio.

Two-Phase Pressure Drop Characteristics in Microchannels

As compared to single-phase flow, a larger amount of heat can be removed by employing two-phase flow in microchannels. Therefore, numerous studies have been carried out to understand the complicated physics of boiling in micron sized channels. But as literature survey confirms, not many studies report the average pressure drop with water boiling in microchannels. The range of heat flux and mass flux covered in the present study is much larger than the earlier works allowing us to see features not reported earlier.

Pressure Drop Characteristics in Trapezoidal Silicon Microchannels

This section presents the operational characteristics of flow in the given microchannel. The experimental data is compared against existing correlations for microchannel and annular flow model. The value of the parameters used in these experiments are wall heat flux in the range of 0-30 W/cm², water flow rate from 0.1 ml/min (44.5 kg/m²-s) to 2.5 ml/min (1114 kg/m-s²), and outlet channel pressure being atmospheric (Table **2**). The inlet Reynolds number covered in the experiments is 3 to 140. It should be noted that the fluid entering the microchannels is subcooled – the amount of subcooling is in the range of 15-46°C in the present experiments, while the exit pressure is atmospheric (1 bar). As apparent from the results presented in the following section, the maximum pressure drop is 105 mbar in two phase, implying that the maximum inlet pressure is about 1.10 bar. The variation in saturation temperature over this pressure range is at most 2.3%.

Fig. **12**(i) shows the pressure drop across the microchannel for various mass flow rates, for six different heat flux values. The pressure drop in the absence of heat flux is also included for comparison. For a given heat flux (Fig. **12**(ii)) with decreasing flow rate, pressure drop is observed to (i) decrease linearly, (ii) reach an onset of boiling point, (iii) pass through a pressure maxima, and finally (iv) increase rapidly at the onset of dryout. For higher heat flux input, the pressure drop maxima observed in (iii) tends to progressively larger values and the maxima point shifts towards right (*i.e.*, corresponding to a higher mass flow rate). The onset of boiling also occurs at a higher mass flow rate, as the heat flux at input is increased. At high mass flow rates the input heat flux is not sufficient to boil the fluid, and the flow is completely in single phase. At (e), vigorous flow boiling and substantial back flow are observed which can cause damage to the microchannel; the experiments are therefore terminated at this point. Qu and Mudawar [49] termed the corresponding heat flux as critical heat flux (CHF) in their study. Thus to the extent possible, the full range for two-phase flow has been covered in these experiments.

Due to the presence of subcooled fluid in the microchannel, single phase flow occurs at the inlet-end of microchannel and switches to two phase flow along the length of the channel. The pressure drop maxima (section b-c-d in Fig. 12(ii)) are due to the following two opposing effects:

1.   There is a decrease in the two-phase length and a corresponding increase in the single-phase length with an increase in the mass flow rate. This effect is expected to reduce the net pressure drop across the channel at larger mass fluxes, since the per unit length two-phase pressure drop is substantially higher than the per unit length single-phase pressure drop.

2.   There is an increased resistance to flow at larger mass flow rates, leading to an increase in pressure drop.

The first effect is dominant at higher flow rates (c-d), while the second effect dominates at lower flow rates (b-c). The decrease in single phase flow length with an increase in heat flux has been visually observed. Additional discussion on existence of a maximum is provided in a later Section through the annular flow model equations.

The data in the two-phase region has been plotted as a function of exit quality in Fig. **13**. It is interesting to note that the maximum exit quality is 0.5 at smaller heat flux input levels and decreases to about 0.1 at higher heat flux levels; the lower exit quality at higher heat flux is owing to an earlier onset of CHF in this case. Our study of pressure drop variation with time confirmed through visual observation that the flow is intermittent in nature at point (e; Fig. **12**(ii)). The back flow observed in the microchannel can be explained as following – at a given heat flux, as the mass flow rate is reduced the amount of vapor generation increases. A part of this vapor flows back into the inlet reservoir leading to an increase in the channel resistance and consequently the pressure drop. Dryout occurs intermittently although the average vapor quality (computed from the overall energy balance) is low. We refer to part d-e of the curve in Fig. **12**(ii) as "high-quality" region – although this term should be taken with a pinch of salt. We note that a systematic study of pressure variation in this regime for microchannel has not been reported earlier. Bergles and Kandlikar [50] mention the difficulty in performing measurements in this regime, highlighting that there is paucity of such data at the micro-scales.

The most significant observation is that for a given channel geometry and the range of heat flux input, there may be up to four flow rates (for example points 1, 2, 3, 4 in Fig. **12**(ii)) with identical pressure drop. The heat transfer coefficient at these points will be different, due to flow conditions being single-phase (a-b), mixed single/two-phase (b-c, c-d), and approaching dryout (d-e) respectively. Since the goal is to identify the most effective operational regime for electronics cooling, we should choose the point where the heat transfer coefficient is the maximum. For this purpose the experiment has been performed to estimate to heat transfer coefficient between points (b-d) in Fig 12(ii) as a function of mass flow rate. Fig. **14** shows that the heat transfer coefficient decreases with an increase in flow rate, suggesting that point 3 in Fig. **12**(ii) is the appropriate operating point. However, additional considerations such as flow stability and safety margin from CHF will be required in a practical situation. Anyhow, we believe that this information will be useful for heat-sink design involving two phase flow in microchannels and has not been reported earlier in literature with respect to microchannel to the best of our knowledge.

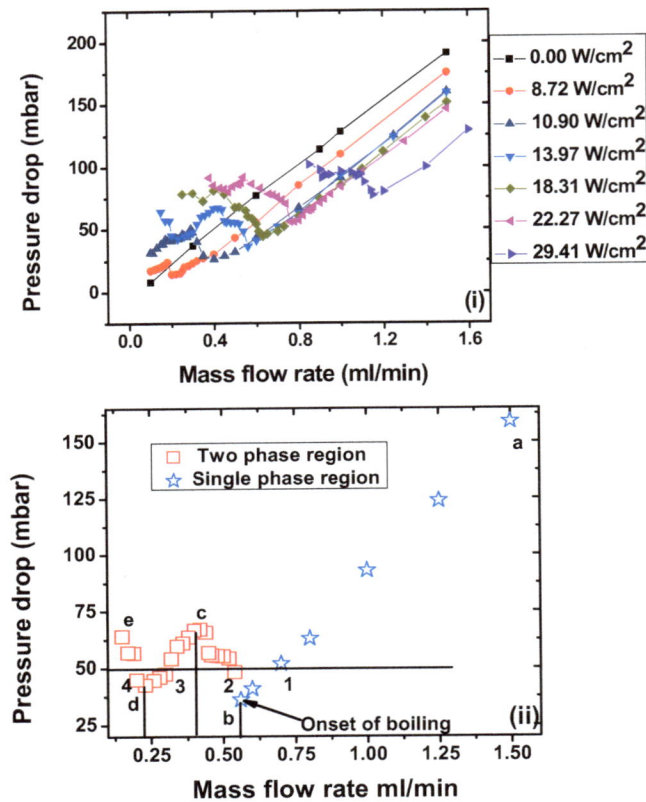

Figure 12: Variation in pressure drop with mass flow rate. A single case (corresponding to heat flux 13.97 W/cm^2) has been shown separately in (ii) for clarity. The data points in (i) have been connected by straight line to aid the eye. (Four parallel channels of hydraulic diameter 108.8 µm. Note that mass flow rate of 0.1 ml/min corresponds to a mass flux of 44.5 kg/m^2-s).

We further note that the present results are for a relatively large range of heat flux (0-30 W/cm^2) and mass flux (44-1114 kg/m^2-s). For example, Qu and Mudawar [32] employed two mass fluxes (135 and 402 kg/m^2-s) and heat flux ranging from 16.25 to 32.25 W/cm^2, while Zhang *et al.* [51] utilized a single mass flux (5.12 kg/m^2-s) and heat flux (2.6 W/cm^2). Hetsroni *et al.* [35] worked with a single mass flux (95 kg/m^2-s) and heat flux ranging from 8-33 W/cm^2. The heat flux and mass flow rate varies over an order of magnitude in the present study.

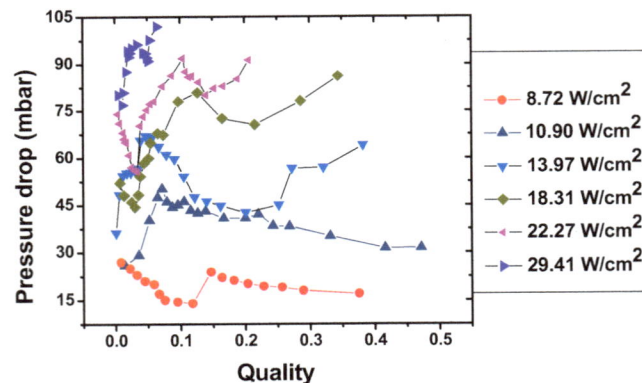

Figure 13: Variation in pressure drop with exit thermodynamic quality for different heat inputs. (Four parallel channels with hydraulic diameter 108.8 µm.) The data points have been connected by straight line to aid the eye. (Single microchannel of hydraulic diameter 140 µm). The data points have been connected by straight line to aid the eye.

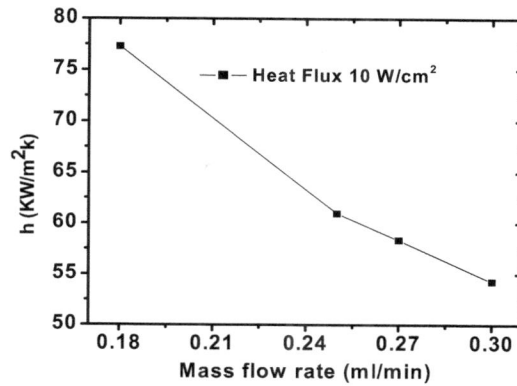

Figure: 14: Variation in heat transfer coefficient versus mass flow rate for fixed heat flux of 10 W/cm^2.

Table 2: Range of heat flux and flow rate over which experiments have been performed

S. No.	Heat flux W/cm^2	Mass flow rate (ml/min)	Mass flux (kg/m^{2}s)	Exit quality
1	8.72	0.10 to 2.5	44.5 to 1114	0 to 0.38
2	10.92	0.10 to 2.5	44.5 to 1114	0 to 0.47
3	13.97	0.15 to 2.5	66.7 to 1114	0 to 0.37
4	18.31	0.22 to 2.5	97.9 to 1114	0 to 0.34
5	22.27	0.38 to 2.5	169 to 1114	0 to 0.20
6	29.41	0.82 to 2.5	364 to 1114	0 to 0.07
7	47.2 to 59.9	0.30	324	0.14-0.182
8	47.8 to 83.7	0.40	433	0.07-0.20

Discussion on Pressure Drop

In this section, we compare the experimental pressure drop in two phase flows with correlations for microchannel and annular flow model. We also attempt to explain the maxima in pressure in the two phase regime using the annular flow model.

Comparison with Correlations

A large number of correlations are available in the literature; however, Qu and Mudawar [32] have already shown that correlations from the macroscale studies are not applicable at the micro-scales. Therefore, the comparison is confined to correlations for microchannel. Table **3** summarizes the correlations that have been considered for comparison in this study. Mishima and Hibiki [52] have considered the combined effect of laminar liquid and vapor flow on pressure drop. The effect of channel size is accounted for by including the hydraulic diameter in the Martinelli-Chisholm constant C. The correlation of Lee and Lee [53] (correlation 2) incorporates the effect of Reynolds number in constant C, which accounts for the channel size and the fluid flow rate. Yu *et al.* [54] (correlation 3) have postulated a fluid flow regime comprising laminar liquid and turbulent vapor for pressure drop calculations. Qu and Mudawar [32] based their correlation on Mishima and Hibiki's correlation with a mass velocity term incorporated in C (correlation 4).

Fig. **15** compares the variation of experimental pressure drop as a function of mass flow rate with the above four correlations, for two heat flux values. The mean and maximum differences in pressure drop between the experiment and correlations are provided in Table **4**. Overall the mean difference is reasonably small (within 36%) for almost all heat fluxes (other than heat flux of 8.72 W/cm^2). Further it is noted that the mean difference is smaller for higher heat flux values. The maximum difference for correlation 3 is about 70% for all heat fluxes used in the experiment, while for other correlations the maximum difference varies substantially with heat flux. Overall, correlation 2 seems the most consistent (*i.e.* the mean and maximum errors are relatively small) but under predicts, while correlations 1 and 4 fit well at higher mass flow rates (Fig. **15**). In summary, the above tested correlations predict the pressure drop reasonably well for higher heat fluxes but show poor comparison at lower heat flux values. This may be because the correlations were developed for different physical conditions.

Table 3: Microchannel two phase pressure drop correlations tested in the present study

Correlation Number	Reference	Pressure drop
1.	Mishima and Hibiki [52]	$\Delta P_{tp,f} = \frac{L_{tp}}{x_{e,out}} \int_0^{x_{e,out}} \frac{2f_f G^2 (1-x_e)^2 v_f}{d_h} \Phi_f^2 dx_e$ $\Phi_f^2 = 1 + \frac{C}{X_{vv}} + \frac{1}{X_{vv}^2}$, $C = 21\left[1 - \exp(-0.319 \times 10^3 d_h)\right]$ $X_{vv} = \left(\frac{\mu_f}{\mu_g}\right)^{0.5} \left(\frac{1-x_e}{x_e}\right)^{0.5} \left(\frac{v_f}{v_g}\right)^{0.5}$ $\Delta P_{tp,a} = G^2 v_f \left[\frac{x_{e,out}^2}{\alpha_{out}}\left(\frac{v_f}{v_g}\right) + \frac{(1-x_{e,out})^2}{1-\alpha_{out}} - 1\right]$ $\alpha_{out} = \cfrac{1}{1 + \left(\frac{1-x_{e,out}}{x_{e,out}}\right)\left(\frac{v_f}{v_g}\right)^{2/3}}$
2	Lee and Lee [53]	$\Delta P_{tp,f} = \frac{L_{tp}}{x_{e,out}} \int_0^{x_{e,out}} \frac{2f_f G^2 (1-x_e)^2 v_f}{d_h} \Phi_f^2 dx_e$ $\Phi_f^2 = 1 + \frac{C}{X_{vt}} + \frac{1}{X_{vt}^2}$, $C = 6.185 \times 10^{-2} \, Re_{f0}^{0.726}$ $X_{vt} = \left(\frac{f_f \, Re_g^{0.25}}{0.079}\right)^{0.5} \left(\frac{1-x_e}{x_e}\right)\left(\frac{v_f}{v_g}\right)^{0.5}$ $\Delta P_{tp,a} = G^2 v_f \left[\frac{x_{e,out}^2}{\alpha_{out}}\left(\frac{v_f}{v_g}\right) + \frac{(1-x_{e,out})^2}{1-\alpha_{out}} - 1\right]$ $\alpha_{out} = \cfrac{1}{1 + \left(\frac{1-x_{e,out}}{x_{e,out}}\right)\left(\frac{v_f}{v_g}\right)^{2/3}}$ $Re_{f0} = \frac{Gd_h}{\mu_f}$, $Re_g = \frac{Gx_e d_h}{\mu_g}$, $Re_f = \frac{G(1-x_e)d_h}{\mu_g}$ $f_f \, Re_f = 24\left(1 - 1.355\beta + 1.947\beta^2 - 1.701\beta^3 + 0.956\beta^4 - 0.256\beta^5\right)$
3	Yu et al [54]	$\Delta P_{tp,f} = \frac{L_{tp}}{x_{e,out}} \int_0^{x_{e,out}} \frac{2f_f G^2 (1-x_e)^2 v_f}{d_h} \Phi_f^2 dx_e$

$$\Phi_f^2 = \frac{1}{X_{vt}^{1.9}}, \quad X_{vt} = \left(\frac{f_f \, \mathrm{Re}_g^{0.25}}{0.046}\right)^{0.5} \left(\frac{1-x_e}{x_e}\right) \left(\frac{v_f}{v_g}\right)^{0.5}$$

$$\Delta P_{tp,a} = G^2 v_f \left[\frac{x_{e,out}^2}{\alpha_{out}}\left(\frac{v_f}{v_g}\right) + \frac{(1-x_{e,out})^2}{1-\alpha_{out}} - 1\right]$$

$$\alpha_{out} = \frac{1}{1+\left(\frac{1-x_{e,out}}{x_{e,out}}\right)\left(\frac{v_f}{v_g}\right)^{2/3}}$$

4   Qu and Mudawar [32]

$$\Delta P_{tp,f} = \frac{L_{tp}}{x_{e,out}} \int_0^{x_{e,out}} \frac{2 f_f G^2 (1-x_e)^2 v_f}{d_h} \Phi_f^2 \, dx_e$$

$$\Phi_f^2 = 1 + \frac{C}{X_{vv}} + \frac{1}{X_{vv}^2}$$

$$C = 21\left[1 - \exp(-0.319 \times 10^3 d_h)\right](0.00418G + 0.0613)$$

$$X_{vv} = \left(\frac{\mu_f}{\mu_g}\right)^{0.5} \left(\frac{1-x_e}{x_e}\right)^{0.5} \left(\frac{v_f}{v_g}\right)^{0.5}$$

$$\Delta P_{tp,a} = G^2 v_f \left[\frac{x_{e,out}^2}{\alpha_{out}}\left(\frac{v_f}{v_g}\right) + \frac{(1-x_{e,out})^2}{1-\alpha_{out}} - 1\right]$$

$$\alpha_{out} = \frac{1}{1+\left(\frac{1-x_{e,out}}{x_{e,out}}\right)\left(\frac{v_f}{v_g}\right)^{2/3}}$$

Table 4: Mean and maximum error (in percent) between the four correlations tested in this study, against present experimental data. (See Table 3 for equations of the various correlations).

Channel heat flux (W/cm²)	8.72		10.90		13.97		18.31		22.27		29.41	
Error in pressure drop (%)	Max	Mean	Max	Mean	Max	Mean	Max	Mean	Max	Mean	Max	Mean
Correlation 1 [52]	272.3	89.2	172.2	38.9	107.4	34.7	58.9	17.4	56.4	24.4	69.1	34.0
Correlation 2 [53]	36.7	16.7	58.6	33.7	55.2	33.3	44.2	23.1	33.1	21.3	75.0	25.2
Correlation 3 [54]	70.2	41.2	77.7	52.7	74.2	50.5	72.2	35.9	65.0	36.9	74.1	32.8
Correlation 4 [32]	233.9	72.6	148.5	31.9	86.1	31.0	44.2	15.4	43.7	20.3	69.1	34.2

Comparison with Annular Flow Model

The annular flow model developed and validated by Patankar [55] has been used for comparison with our experimental data. This model follows Qu and Mudawar [23] and is based on mass and momentum conservation for both liquid and vapor phases (equations 13 and 14 presented later), along with appropriate correlations for closure. The model postulates that vapor phase is present at the core of the microchannel (some liquid may also exist as droplets in the vapor core) while there is uniform liquid flow at the channel walls. Qu and Mudawar [23] showed that their annular flow model is in good agreement with their experimental data.

This flow model was applied on the b-c-d-e part of the curve in Fig. **12**(ii) [55]. Fig. **16** shows comparison between the present experimental results and that predicted by the model, for different heat fluxes. Overall the experimental results match well with the model, difference being less than 10% for almost all the cases. In particular, the peak pressure drop values compare well. The decrease in pressure to the left of the peak pressure point is also captured by the model; however, the local minima point observed in the experiments is not predicted by the model. This is because the back flow of vapor into the inlet reservoir has not been considered in the model.

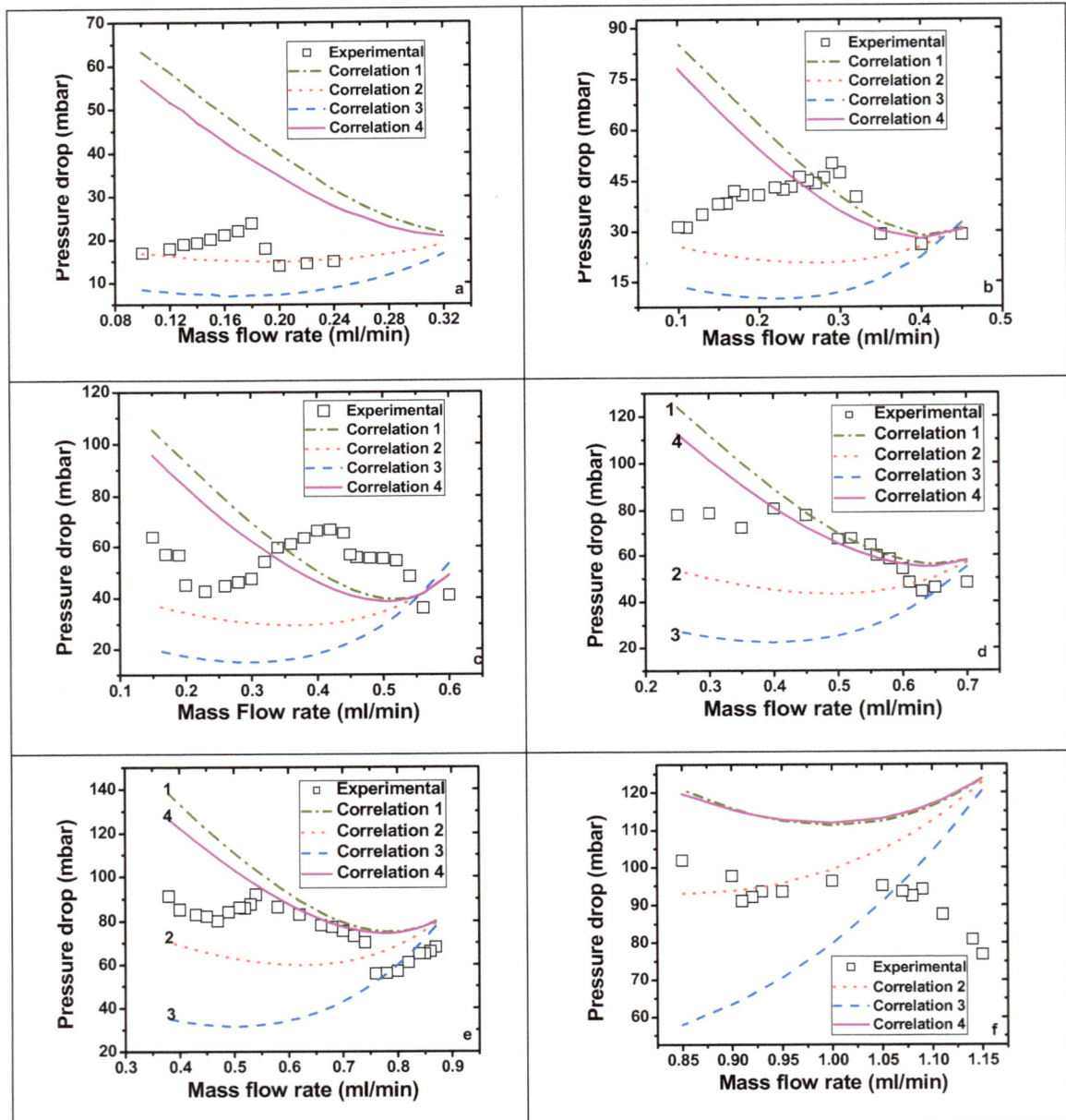

Figure 15: Variation of experimental pressure drop and predicted pressure drop from four different correlations, as a function of mass flow rate. The heat flux in (a-b-c-d-e-f) is 8.72, 10.89, 13.97, 19.31, 22.27 and 29.41 W/cm^2, respectively. See Table **3** for equations of the various correlations. (The mass flow rate of 0.1 ml/min corresponds to a mass flux of 44.5 kg/m^2-s).

Discussion on Pressure Maxima in Two Phase Flow

The reason for pressure maxima in Fig. **12** for pressure drop characteristics with mass flow rate in the two phase regime is explored through the annular flow model. The overall pressure drop across a microchannel is the sum of pressure drop due to contraction and expansion loss at the inlet and outlet reservoirs, the single phase liquid pressure drop, and two phase pressure drop (comprising both acceleration and friction pressure drop components). As already stated, the inlet and outlet pressure drops are negligible for the present case. The pressure drop across the channel due to single phase is

$$\left(-\frac{dP}{dz}\right)_{1\varphi} = \left(\frac{f}{d_h}\right)\left(\frac{\rho U^2}{2}\right).$$

(11)

As per the annular flow model, momentum conservation in the liquid film gives [23]

$$\left(-\frac{dP}{dz}\right)_L = \frac{(\tau - \tau_i) + \dfrac{1}{P_{ch}}(\Gamma_{fg} u_i - \Gamma_d u_c)}{\delta - y}.$$

(12)

Similarly, momentum conservation in the vapor core gives [23]

$$\left(-\frac{dP}{dz}\right)_g = P_c\left\{\tau_i + \frac{1}{A_c}\frac{d}{dz}(\rho_H u_c^2 A_c)\right\} - \frac{1}{A_c}(\Gamma_{fg} u_i - \Gamma_d u_c).$$

(13)

Ignoring the effect of mass evaporation and mass deposition per unit channel length for simplicity, and putting $y = 0$, in the above equation, we obtain the total pressure drop as the sum of eqs. (11), (12) and (13) as

$$\left(-\frac{dP}{dz}\right)_T = \underbrace{\left[\left(\frac{f}{d_h}\right)\left(\frac{\rho U^2}{2}\right) + \frac{1}{A_c}\frac{d}{dz}(\rho_H u_c^2 A_c) - \frac{3\tau_i}{2\delta}\right]}_{A} + \underbrace{\left[\frac{3\mu_f\, m_{Ff}}{P_{ch}\rho_f\delta^3} + \frac{P_c\tau_i}{A_c}\right]}_{B}$$

(14)

where

$$f_i = \frac{\tau_i}{\dfrac{1}{2}\rho_H(u_c - u_i)^2}$$

(15)

$$f_i\,\mathrm{Re}_c = 24(1 - 1.355\beta + 1.947\beta^2 - 1.701\beta^3 + 0.956\beta^4 - 0.256\beta^5)$$

(16)

and β is the aspect ratio of the vapor core.

There are two terms A and B contributing to the overall pressure drop in Eq. 14. As the mass flow rate increases (at constant heat flux), Term A being directly proportional to the square of the mass flux increases; thereby increasing the pressure drop. On the other hand, Term B apparently decreases with an increase in the mass flow rate – note that both numerator and denominator increases. However, the denominator being proportional to the cube of the liquid film thickness (δ^3) increases at a faster rate than the numerator, leading to a decrease in the magnitude of term B, and therefore the overall pressure drop. In summary, A leads to an increase while B leads to a reduction in pressure drop with an increase in mass flow rate. These two opposite effects lead to a maximum in pressure drop with flow rate.

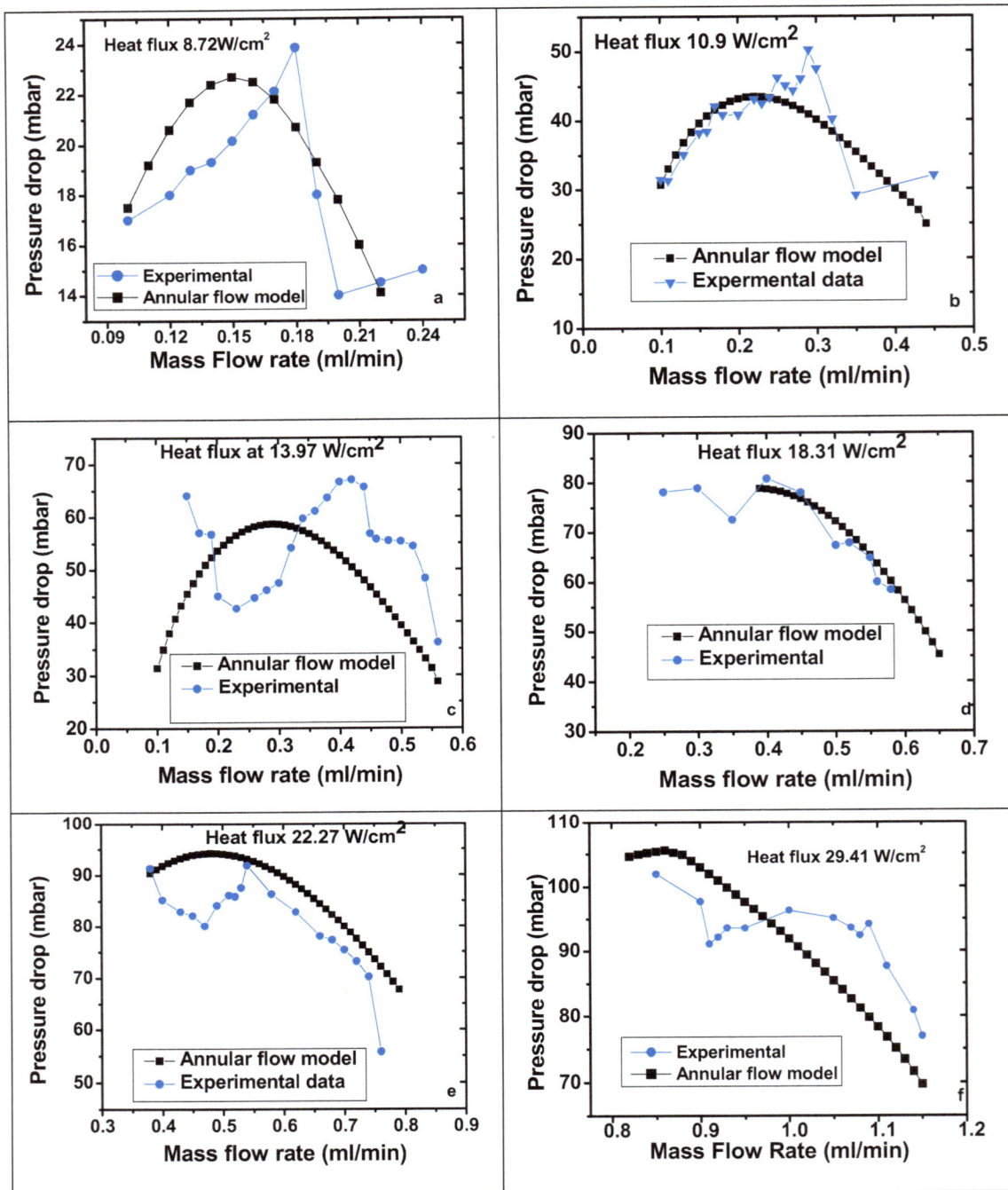

Figure 16: Comparison between experimental data and annular flow model, at heat flux of 8.72, 10.89, 13.97, 19.31, 22.27 and 29.41 W/cm² respectively (a-b-c-d-e-f). Note that mass flow rate of 0.1 ml/min corresponds to a mass flux of 44.5 kg/m²-s.

Effect of Aspect Ratio on Pressure Drop

It is well known that the pressure drop in two-phase flow is dependent on the microchannel geometry. It is therefore apparent that the effect of geometry on pressure drop across the microchannel should be studied. The effect of aspect ratio (width/height) is investigated in this section. Note that in single phase, the aspect ratio has only a marginal role and the pressure drop is primarily determined by the hydraulic diameter. The impact of cross-sectional aspect ratio on single-phase fluid flow and heat transfer characteristics has been reported in Refs. [46,47,56-59]. Specifically, Peng and Peterson [47] and Morini and Spiga [58] performed theoretical and experimental investigations on pressure drop and convective heat transfer for flow in rectangular microchannels, which support the conclusion that the product of friction factor and Reynolds number (*f.Re*) shows a slight, monotonic increase with aspect ratio. To the best of our knowledge, a systematic study of impact of aspect ratio variation on two-phase fluid flow characteristics has not been reported in the literature [60].

The specific aims of the present work are to determine if the pressure drop across microchannels in two phase flows can be minimized by an appropriate choice of aspect ratio (*AR*), and to compare the experimental results with annular flow model predictions. We have studied rectangular channels on (110) silicon wafer. The advantage of choosing rectangular cross-section channels is that comparison studies are easier, as there is only one *AR* (width to depth ratio) value under consideration. (Note that multiple aspect ratio values exist for non-rectangular microchannels.) Also, the calculation of hydraulic diameter for rectangular channels is straight-forward. This is an important consideration, since we have maintained a constant hydraulic diameter value for all the microchannels. Any change observed in fluid flow behavior can therefore be ascribed to aspect ratio. From a practical standpoint, rectangular channels can be fabricated at higher density (lower pitch) levels as compared to trapezoidal channels, which is important in increasing the surface to volume ratio for a given substrate surface area.

For the purpose of these experiments, the hydraulic diameter at 142 µm is maintained constant within ±2 µm, while the aspect ratio is 1.24, 1.43, 1.56, 1.73, 2.56, 3.60 and 3.75 (see Table **5**). The experimental aspect ratio (*AR*) range of 1-4 includes the critical *AR* range of interest and is also compatible with conventional CMOS processing capability.

Table 5: Dimensions of different microchannels fabricated

Channel	Width (µm)	Depth (µm)	Aspect Ratio	Hydraulic diameter (µm)
1	156	127	1.23	140
2	175.5	122	1.44	143.9
3	180	115	1.56	140.4
4	194.5	112	1.73	142.2
5	253	99	2.56	142.3
6	325	90.5	3.60	143
7	337	90	3.75	142

Table 6: Cases investigated in this study (42 in total). The hydraulic diameter for all the microchannels is 142 ± 2 µm

	Mass flow rate (ml/min)	Heat input to Microchannel (W)	Aspect Ratio
Case 1	0.15	0.0	1.23, 1.44, 1.56, 1.73, 2.56, 3.60, 3.75
Case 2	0.15	3.0	1.23, 1.44, 1.56, 1.73, 2.56, 3.60, 3.75
Case 3	0.15	3.5	1.23, 1.44, 1.56, 1.73, 2.56, 3.60, 3.75
Case 4	0.20	0.0	1.23, 1.44, 1.56, 1.73, 2.56, 3.60, 3.75
Case 5	0.20	3.0	1.23, 1.44, 1.56, 1.73, 2.56, 3.60, 3.75
Case 6	0.20	3.5	1.23, 1.44, 1.56, 1.73, 2.56, 3.60, 3.75

Results from Annular Flow Model

An annular flow based model to calculate the pressure drop through microchannels has been developed by Patankar [55] and used in the present study. We have applied the above annular flow model after sufficient validation to test for any departure of pressure drop with respect to aspect ratio of microchannel in both single and two phase flow regimes. The theoretical results predicted a minimum in pressure drop at an aspect ratio range of 1.6 in the two phase flow regime; the location of minima depends weakly upon heat flux and mass flow rate. Note that in above case the hydraulic diameter at 142 μm is maintained constant (Fig. **17**). This suggests that the aspect ratio of microchannels is an important parameter governing two-phase fluid flow and provided the motivation for undertaking the experimental study.

Experimental Results

The experiments are carried out for varying aspect ratios, three different heat flux values and two different mass flow rate values, with deionized water as the working fluid. The values of mass flux and heat flux for a given set of experiment are given in Table **6**. Note that the mass flow rate of water and input power are held constant in these experiments, which implies that the exit quality is the same for all cases. However due to a variation in both cross-sectional area and heating area with aspect ratio, the mass flux and channel heat flux vary with aspect ratio (Table **7**). The implication of these variations is discussed later.

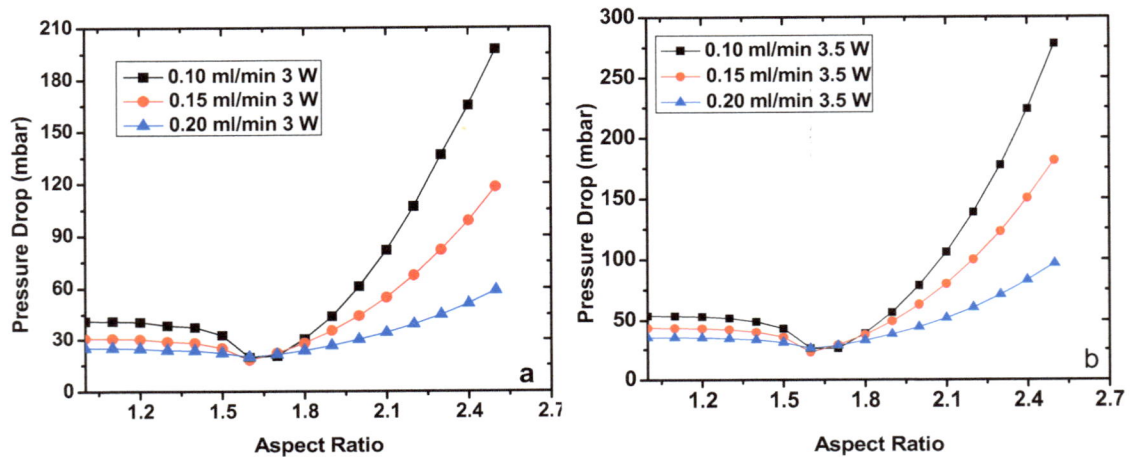

Figure 17: Pressure drop across the microchannel versus aspect ratio at heat input of (a) 3 W and (b) 3.5 W, as obtained from the annular flow model.

Figs. **18** and **19** show the channel pressure drop versus aspect ratio for both single phase as well as two phase flow. Single phase experiments were undertaken to confirm results pertaining to aspect ratio in single phase reported in the literature, and firmly establish the uniformity in hydraulic diameter of the microchannels. The pressure drop remains unchanged for different aspect ratios for a given flow rate and no heat flux. This clearly indicates that the hydraulic diameter of the seven different microchannels is nearly constant for all cases (Fig. **19**). Upon repeating the experiment for a different (higher) mass flow rate, a higher pressure drop is obtained as expected. In both these cases, the pressure drop variation with respect to aspect ratio is within the experimental uncertainty, and correlates very well with the slight variation in hydraulic diameter of the microchannels (Table **5**).

The onset of two-phase flow is for a heating power of 2.5 W for the fluid flow rates under consideration. Therefore, experiments are performed for heating power values of 3 W (Fig. **18**) and 3.5 W (Fig. **19**). The existence of two phase regime was confirmed visually. It can be seen that, for a given mass flow rate and heat flux value, the pressure drop first decreases with an increases in aspect ratio, and then increases with a further increase in aspect ratio. Fig. **19** confirms the trend in Fig. **18** for a different heat flux. Other pertinent observations from Figs. **18** and **19** are:

1. A certain regime is observed where the pressure drop in two phase flow can be comparable or even lower than that of single phase flow. This observation has already been noted earlier over a wide range of heat and mass fluxes for a microchannel of a given hydraulic diameter and aspect ratio.

2. The minimum pressure drop recorded experimentally, corresponding to $AR = 1.56$ is about 1/6 that of pressure drop for $AR = 1.23$ and about 1/4 of pressure drop for $AR = 3.75$. The deep minima at $AR = 1.56$ is therefore noteworthy. The minima in pressure drop at $AR = 1.56$ is beyond the combined uncertainty due to variation in hydraulic diameter and experimental error in the measurements.

3. The minimum pressure drop recorded experimentally is at $AR = 1.56$ for all the four cases (heat flux, mass flow) investigated. However, since experimental AR values are discrete in nature, we stipulate that the pressure-drop minimum is observed over a range rather than any single point.

The above results demonstrate theoretically as well as through experiments, the strong effect of aspect ratio in two phase flow in microchannels. Note that the location of minima as predicted by the annular flow model agrees well with the experiments. Also the value of pressure drop at the minima agrees within 8%. However, the model does not predict the pressure drop well at large aspect ratios, suggesting that the model is good for approximately square cross-sections and poor for large aspect ratio rectangular cross-sections. The result has practical significance in that for a given hydraulic diameter, an aspect ratio range close to 1.56 may be employed to minimize the pressure drop penalty.

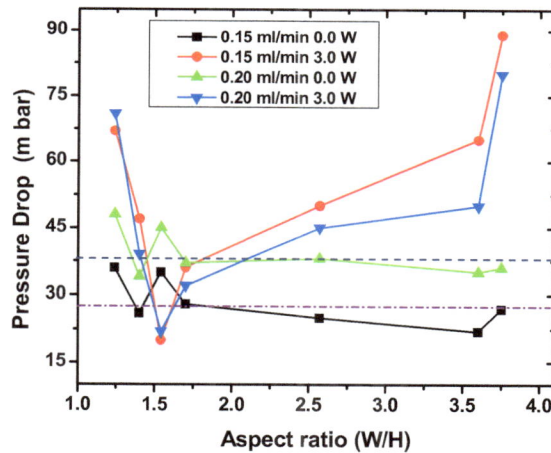

Figure 18: Experimentally determined pressure-drop across the microchannel versus aspect ratio, at a heat input of 3 W. The theoretical single-phase pressure-drop values are shown by dash-dot and dashed lines, for mass flow rates of 0.15 and 0.20 ml/min, respectively.

Discussion on Effect of Aspect Ratio

A discussion on the above interesting behavior is attempted in this section. First, the obvious effect of variation in mass flux and heat flux is considered. As evident from Table **7**, the mass flux reduces monotonically with an increase in AR due to a corresponding increase in the cross-sectional area. Similarly, the channel heat flux reduces with an increase in AR. It is however noted that the quality, at any given point along the length of the channel, is the same for all aspect ratio microchannels. This can be seen from energy balance

$$Q = \dot{m}C_p\Delta T + \dot{m}h_{fg}x_e$$

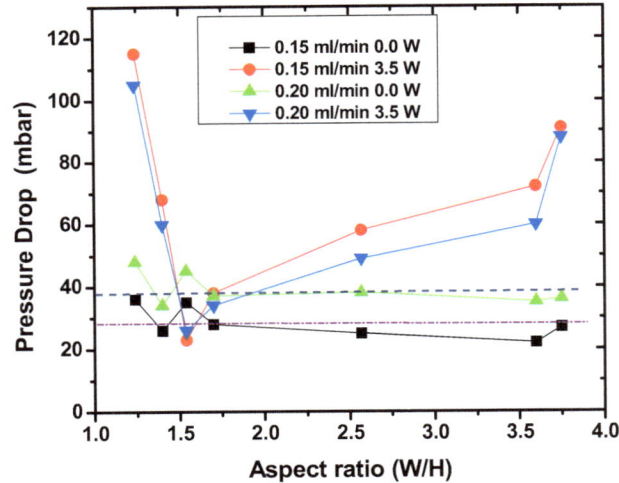

Figure 19: Experimentally determined pressure-drop across the microchannel versus aspect ratio, at a heat input of 3.5 W. The theoretical single-phase pressure-drop values are shown by dash-dot and dashed lines, for mass flow rates of 0.15 and 0.20 ml/min, respectively.

From the above equation, because the mass flow rate, inlet temperature and input power are constant, and the heat loss factor is approximately the same for all *AR* microchannels, the quality at the same length from the inlet will be the same for all *AR* microchannels.

The overall pressure drop across a microchannel is the sum of pressure drop due to contraction and expansion loss at the inlet and outlet reservoirs, single phase pressure drop, acceleration pressure drop, friction pressure drop due to liquid film on the walls of the channel, and frictional pressure drop in vapor. As already stated, the inlet and outlet pressure drop is negligible, while the single phase pressure drop is the same for all *AR* microchannels. The overall pressure drop is therefore governed by the frictional and acceleration pressure drops. The variation of these components with respect to *AR* is the key to the observation of a minimum.

Table 7: Variation of meat flux and mass flux for a given mass flow rate and input power, for different AR microchannels.

AR	Mass flow rate (ml/min)	Input power (W)	Cross-sectional area ($\times 10^{-8}$ m^2)	Heated surface area ($\times 10^{-6}$ m^2)	Mass Flux (kg/m^2-s)	Heat flux (W/cm^2)	Exit quality
1.23	0.15	3.0	2.0	8.2	126.2	36.6	0.29
1.44	0.15	3.0	2.1	8.4	116.8	35.8	0.29
1.56	0.15	3.0	2.1	8.2	120.8	36.6	0.29
1.73	0.15	3.0	2.2	8.3	114.8	35.8	0.29
2.56	0.15	3.0	2.5	9.0	99.8	33.3	0.29
3.60	0.15	3.0	3.0	10.1	85.0	29.6	0.29
3.75	0.15	3.0	3.1	10.3	82.4	29.0	0.29

As per the correlation of Qu and Mudawar [23], which provides the best match without experimental data as noted earlier, the frictional and acceleration pressure drops can be estimated as:

$$\Delta P_{tp,f} = \frac{L_{tp}}{x_{e,out}} \int_0^{x_{e,out}} \frac{2 f_f G^2 (1-x_e)^2 v_f}{d_h} \Phi_f^2 \, dx_e \tag{17}$$

$$\Phi_f^2 = 1 + \frac{C}{X_{vv}} + \frac{1}{X_{vv}^2} \tag{17a}$$

$$C = 21\left[1 - \exp(-0.319 \times 10^3 \, d_h)\right](0.00418G + 0.0613) \tag{17b}$$

$$X_{vv} = \left(\frac{\mu_f}{\mu_g}\right)^{0.5} \left(\frac{1 - x_e}{x_e}\right)^{0.5} \left(\frac{v_f}{v_g}\right)^{0.5} \tag{17c}$$

$$f_f \, \mathrm{Re}_f = 24\left(1 - 1.355\beta + 1.947\beta^2 - 1.701\beta^3 + 0.956\beta^4 - 0.256\beta^5\right) \tag{17d}$$

$$\Delta P_{tp,a} = G^2 v_f \left[\frac{x_{e,out}^2}{\alpha_{out}}\left(\frac{v_f}{v_g}\right) + \frac{(1 - x_{e,out})^2}{1 - \alpha_{out}} - 1\right] \tag{18}$$

$$\alpha_{out} = \frac{1}{1 + \left(\frac{1 - x_{e,out}}{x_{e,out}}\right)\left(\frac{v_f}{v_g}\right)^{2/3}} \tag{18a}$$

From application of Eq. (18), we note that all quantities in Eqs. (19) and (20) are the same for all *AR* microchannels, other than *G* and f_f. Therefore, the acceleration pressure drop is a function of G^2, which reduces with an increase in *AR* (Table 7). Similarly, the frictional pressure drop is directly proportional to *G*, and should again decrease with an increase in aspect ratio. The overall pressure drop should therefore decreases with *AR* as per this correlation. (A similar conclusion is obtained with other correlations.) The failure of these correlations is because they do not explicitly account for the *AR* and therefore cannot explain the minima in our experimental data.

Since the annular flow model also predicts the minima, we resort to this model for a possible explanation. The momentum conservation for the liquid film gives [23]

$$\left(-\tfrac{dP}{dz}\right)_L = \left(\tau - \tau_i + \tfrac{1}{P_{cch}}(\Gamma_{fg} u_i - \Gamma_d u_c)\right) / (\delta - y) \tag{19}$$

while momentum conservation for the vapors core gives [23]

$$\left(-\tfrac{dP}{dz}\right)_v = P_c\left(\tau_i + \tfrac{1}{P_{cch}}\tfrac{(\rho_H u_c^2 A_c)}{A_c}\right) - \tfrac{1}{A_c}(\Gamma_{fg} u_i - \Gamma_d u_c) \tag{20}$$

Ignoring the effect of mass evaporation and deposition mass per unit channel length for simplicity, we obtain the total pressure drop as the sum of eqs. (19) and (20)

$$\left(-\tfrac{dP}{dz}\right)_T = (\tau - \tau_i) / (\delta - y) + P_c(\tau_i) / A_c + \tfrac{1}{A_c P_{cch}}\tfrac{d}{dz}(\rho_H u_c^2 A_c) \tag{21}$$

Putting *y* = 0, in the above equation yields

$$\left(-\tfrac{dP}{dz}\right)_T = (\tau - \tau_i) / (\delta) + P_c(\tau_i) / A_c + \tfrac{1}{A_c P_{cch}}\tfrac{d}{dz}(\rho_H u_c^2 A_c) \tag{22}$$

The first two terms on the right hand side are the frictional pressure drop, while the third term is the acceleration pressure drop. The acceleration pressure drop is proportional to G^2 and this term leads to a decrease in pressure drop as *AR* increases. Apparently, the first term increases with aspect ratio due to a reduction in liquid film thickness (δ). In other words, the frictional pressure drops increases while the acceleration pressure drop reduces, with an increase in *AR*. These two opposite effects lead to a minimum in pressure drop with aspect ratio. Note that an explicit relation of δ with *AR* is required to fully substantiate the above argument, which is not known at present.

Instability in Two Phase Flow

As mentioned above, at high heat flux and low flow rate, the pressure drop across the microchannel is increased sharply because of occasional back flow of water from microchannels to inlet reservoirs and connecting tubes. This suggests that flow instability can occur in two-phase flow regime. The results in this section are reported for both single microchannel ($d_h = 140\mu m$) and multichannel ($d_h = 109$ µm). First we will discuss the single channel, since we observe instabilities to get suppressed in the single channel case under certain condition. Also, the analysis of parallel microchannels is more involved due to the presence of multiple variables.

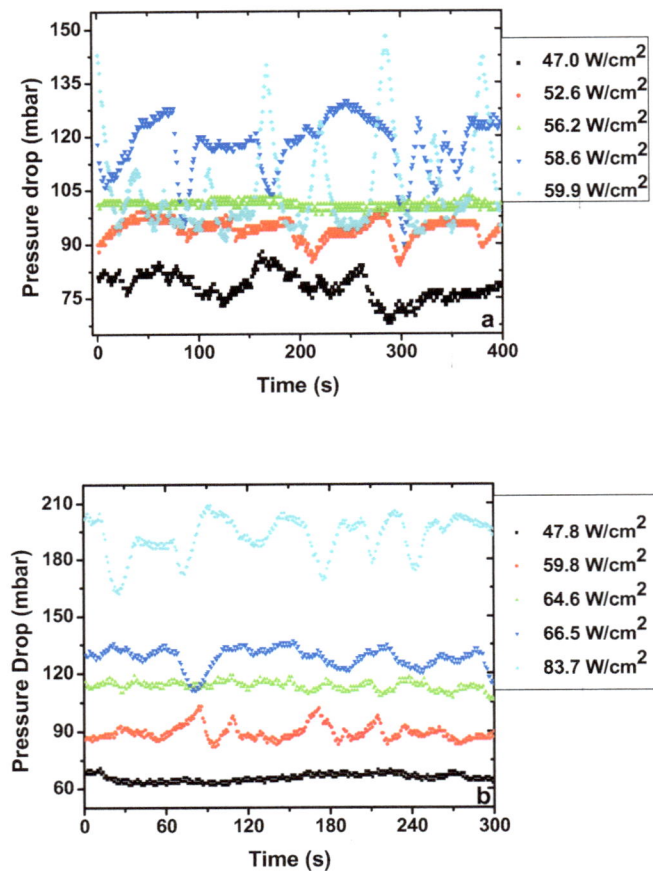

Figure 20: Variation in pressure drop with time, at flow rate of (a) 0.3 ml/min, and (b) 0.4 ml/min for different heat fluxes. (Single microchannel of hydraulic diameter 140 µm).

Fig. **20** shows the pressure drop across the channel as a function of time, for different mass flow rates and heat fluxes. Note that these fluctuations are rather periodic – more so at higher heat flux and mass flow rate. It is worthwhile to note that for constant flow rate, as the heat flux increases the average pressure increases; however the pressure fluctuation first increases, exhibits a minima, and then increases with an increase in the heat flux. For given flow rate the frequency of oscillations appear to increase with an increase in heat flux. Both the amplitude and frequency of oscillation are related to underlying flow pattern discuss later. From the pressure drop versus time characteristics the root mean square pressure deviation (P_{RMS}) is computed as

$$P_{rms} = \frac{\sqrt{\sum_{i=1}^{n}(P_i - P_{avg})^2}}{n}.$$

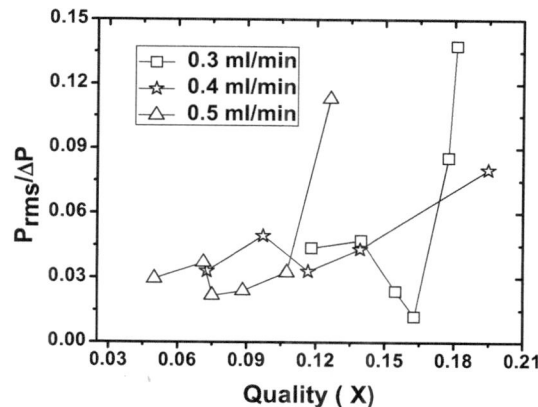

Figure 21: Variations of $P_{rms}/\Delta P$ with exit quality for different mass flow rates.

The fluctuations normalized by the mean value ($P_{RMS}/\Delta P$) lies between 1-14% in our set of measurements (Fig. **21**). The $P_{RMS}/\Delta P$ value shows a slight initial increase, followed by a reduction, and finally it shows a rapid increase, with a successive increase in the exit quality. This minima in pressure fluctuation has not been reported earlier; the earlier researchers [35, 36] observed a monotonic increase in pressure fluctuation with an increase in the vapor quality however Jones *et al.* [61] shows that for 253 μm hydraulic diameter microchannels with 20 μm augmented cavity in base to trigger controlled nucleation activity and help to control large-scale instabilities. The initial increase (as revealed by visual inspection) is due to transition of flow from predominately bubbly to slug regime. The subsequent reduction is due to transition from slug to annular regime. At higher mass flow rates, owing to a direct transition from bubbly to annular, the initial increase in $P_{RMS}/\Delta P$ is smaller.

It is observed that the flow is obstructed when the bubble is nucleating. At the time of departure, the bubble need not always move downstream; we have observed at several instances a slight upstream movement of the bubble. All these events contribute to pressure fluctuations in the bubbly regime. It is clearly observed that although the mean pressure drop is relatively large in annular regime, the fluctuations (at least at the onset of annular flow) are small. With an increase in heat flux, the intermittency between annular flow and single phase increases, leading to an increase in pressure fluctuations in the annular regime. The reason for pressure fluctuations in the dryout regime is as follows – the channel is intermittently blocked due to a large amount of vapor generation in the regime, and the fluid flow in the channel downstream of the blockage reduces. The increase in vapor generation starts to elevate the pressure upstream of the blockage, to a point that it pushes out the blocked vapor completely. This removal of blockage leads to a sudden reduction in pressure drop. The process of blockage formation and its subsequent removal happens rather periodically, as reflected in the pressure time series.

Our results are in qualitative agreement with the literature. However, the behavior of P_{RMS} and its connection to flow visualization as well as observation of a minimum in $P_{RMS}/\Delta P$ has not reported earlier to the best of our knowledge.

Not only instability in single channel case but parallel channel instability was also studied in sufficient detail. Fig. **22** shows the pressure drop across the channel as a function of time, for different mass flow

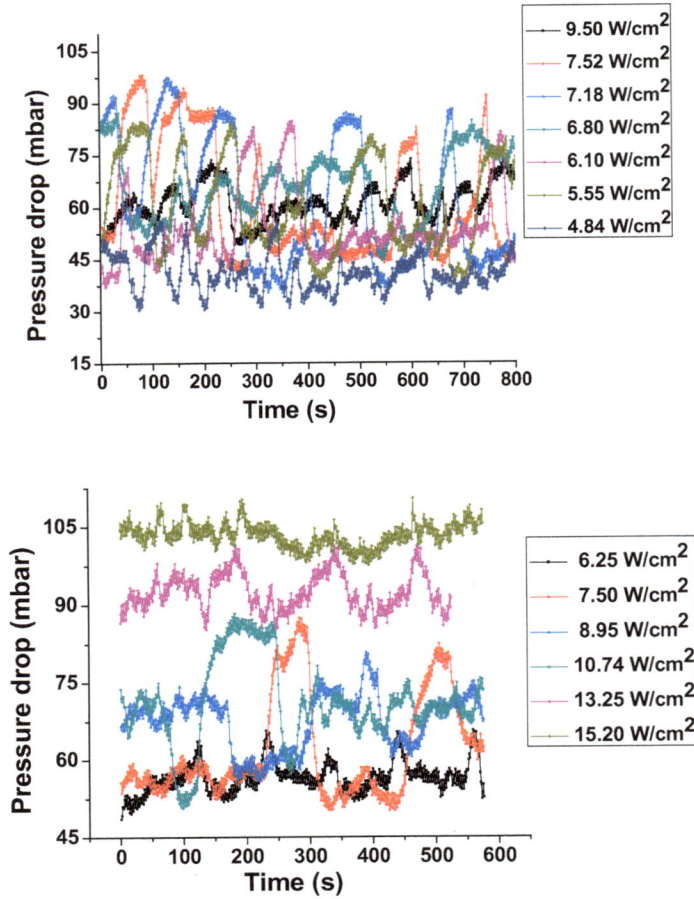

Figure 22: Variation in pressure drop with time, at flow rate of (a) 0.2 ml/min, and (b) 0.3 ml/min for different heat fluxes (Parallel microchannel of hydraulic diameter 108 μm).

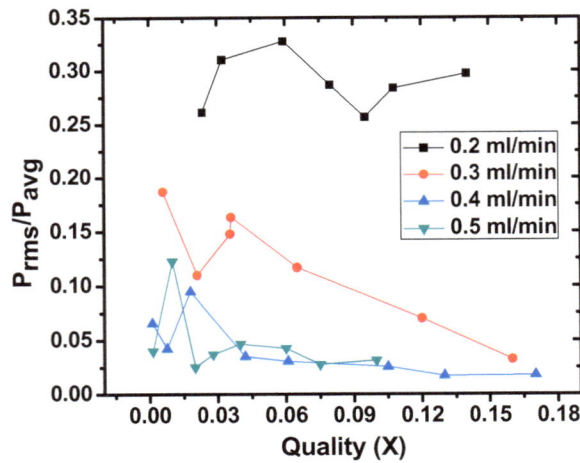

Figure 23: Variations of $P_{rms} / \Delta P$ with exit quality for different mass flow rates (Parallel microchannel of hydraulic diameter 108 μm). The data points have been connected by straight line to aid the eye.

rates and heat fluxes. Note that these fluctuations are not as periodic as noted earlier. There is however an increase in average pressure drop and while the fluctuation passes through a minima as noted earlier. This reinforces the idea of fluctuation being due to formation and departure of bubble leading to periodic variation in single channel however due to alternate passage, the flow is not hampered due growth of bubble in a channel and pressure fluctuation is periodic. Fig. **23** shows the variation $P_{RMS}/\Delta P$ value with quality for different flow rates. For exit quality (0.006 to 0.17) the $P_{RMS}/\Delta P$ is maximum for 0.2 ml/min and keeps decreases with increases with increase flow rate. Note that for constant flow rate, as the boiling number increases, fluctuations exhibit a non-monotonic variation. The reason for initial decrease is a transition in flow pattern as discussed above, while the subsequent monotonic decrease is attributed to fact that although both pressure drop and pressure fluctuation increases, the former increases with faster rate. Further details can be found in Ref. [62].

MEASUREMENT OF VOID FRACTION

Boiling flow patterns are mainly classified as bubbly, slug and annular. Identification and possible control of these flow regimes is critical for practical purposes. To correctly model boiling flow in microchannel, flow regimes in question must be known. Depending upon the application, especially those related to electronics cooling, it is essential to correctly predict the heat transfer coefficient, among other parameters. Numerous researchers [63-65] have utilized flow visualization technique to determine the flow regimes in single phase and two phase flow; however, use of the image processing technique for investigation of boiling water flow in microchannel is rather limited. Revellin and Thome [13] generated a database for two refrigerants (R-134a and R-245fa) and two channel diameters (0.509 and 0.790 mm) using optical techniques. Zhao *et al.* [66] have generated a flow map for the flow of immiscible fluids – dyed de-ionized water and kerosene, in a poly-2-methyl-propenoate microchannel (with 300 μm width and 600 μm depth) by examining the video images.

Although several studies to identify the flow regimes have been undertaken with flow boiling in microchannels, attempts to quantify their occurrence as a function of location, heat flux and mass flow rate has not been undertaken [67]. Further, development of a flow regime map is of practical significance but is not available in the literature for this class of flow. This section attempts to fill some of these voids by employing an image analysis technique. It is possible to study a large amount of video data in a relatively short time through this method. Moreover this method also enables calculations of parameters such as void fraction, flow type percentage, and even the local heat transfer coefficient. Void fraction in combination with the flow regime gives an estimate of the amount of liquid/vapour at a given location, thereby providing important information regarding the nature of the flow. The value of heat transfer coefficient is of practical significance, and all the above can have a direct impact in interpreting the heat transfer characteristics in a microchannel. For example, knowledge of flow regime coupled with void fraction can indicate whether nucleation or evaporation is responsible for the phase change. While nucleation/bubble growth occurs typically in low void fraction (or quality) regime (bubbly or slug), evaporation is the main mode of heat transfer for phase change in annular flow situations. Further, in case critical heat flux situation occurs, the possible mechanism (dryout/departure from nucleate boiling - DNB) can be inferred with reasonable confidence based on knowledge of void fraction and flow pattern. Therefore, measurement of the above mentioned parameters is important for understanding both fundamental and practical aspects of flow boiling in microchannel.

A single microchannel of rectangular cross-section and size 173 μm × 119 μm × 20 mm ($W{\times}H{\times}L$) (hydraulic diameter of 140 μm) has been employed in these measurements. Fig. **24** shows the experimental set-up for recording of images along the microchannel length. The images are recorded using a digital camera (Cannon, Power Shot A560, 7.1 mega pixel) mounted on the eye piece (having a magnification lens of 10x) of the microscope. The images are recorded at 60 Hz, with the image type being 'true color'. The viewing region in the experiments is 2 mm by 2 mm; however, the lateral dimension of the microchannel is smaller than 200 μm and the extra region in the view frame is cropped. The microchannel is resolved by 3200 pixels along the length and 240 pixels along the width, resulting in a spatial resolution of 0.7 μm per pixel.

Image Processing Technique

Identification of Vapor Region

An automated method using an in-house image processing algorithm is employed to calculate the void fraction. For this, the images are first recorded using a video camera. The regions occupied by the vapor are then identified next, and finally, the fraction occupied by the vapor as the ratio of the total image area is computed. Note that the images correspond to the plane normal to the flow direction (and not the cross-sectional view) of the microchannel.

The primary steps in the image processing technique developed are the following [63]. These steps are illustrated through Figs. 25 and 26:

1. Read background image and whiten its boundaries for edge detection. Extract the microchannel from the frame by cropping the borders. The whitening is required to ensure that the bubble (if any) at the boundary of the view frame is fully enclosed by a demarcating edge and its area is well defined.

2. Read the first sample image and crop its borders to extract the microchannel.

3. Remove the background from the sample image in order to reduce the noise. This is accomplished by subtracting the background image from the sample image.

4. Convert the sample image to intensity image (grayscale) from RGB file type. Perform histogram equalization before converting it to a binary image.

5. Remove noise using median filter.

6. Detect edges and fill closed contours (region containing vapors).

7. Small objects (objects with an area less than 40 pixels) are considered noise and are left out while counting the total vapor area.

The vapor film in the case of annular flow is obtained from analysis of the processed image by counting the pixels. Once the algorithm has been developed it can be applied to all video clips with only a slight amount of adjustment for different datasets. Note that the tweaking required is minimal and done to find the best threshold values in steps 5 and 7. However no adjustment was done within a given dataset. Additional details on image processing can be obtained in Jain [63].

Definition of Regimes

From the output of the above program, the area of the vapor can be calculated and this information is used to define the different flow regimes. The void fraction for a given frame is simply the ratio of vapor area in the frame to the total channel area being viewed. Three flow regimes – bubbly, slug and annular are identified through a visual inspection of the frames. The identification of the flow regimes was then automated by comparing the area ratio to a pre-defined threshold, which was carefully chosen and reproduced faithfully in the flow regimes as obtained by visual observations. That is, first a threshold value was arbitrary chosen; the threshold is then adjusted such that the correct flow regime (as determined by visual inspection) is obtained for that threshold value.

The thresholds used in the present work are the following: In a frame, if the area fraction of vapor > 0.7, it is regarded as annular; 0.25 < area fraction < 0.7 is designated as slug; 0.08 < area fraction < 0.25 is designated as bubbly; area fraction < 0.08 is regarded as single phase water. Note that the annular regime primarily comprises a vapor core with liquid water film on the walls of the microchannel. The slug flow comprises an elongated vapor at the center with water on the periphery and part of the same frame filled entirely with water (Fig. 25).

The bubbly flow comprises several vapor bubbles in the view-frame, with the size of each vapor less than the lateral dimension of the microchannel (Fig. **26**). Note that elongated slug regime could also be identified through visual inspection – however, this regime is difficult to distinguish in an automated procedure from both annular and slug; therefore, this regime has not been separately marked.

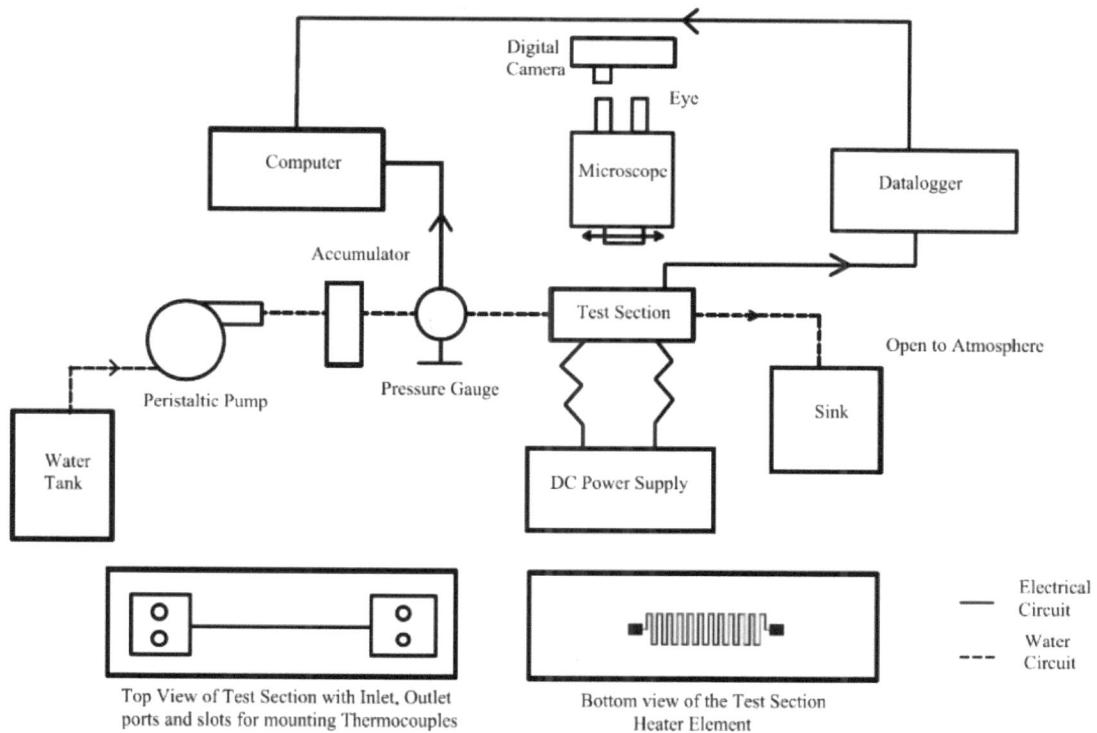

Figure 24: Schematic of the experimental set up for two phase flow visualization in microchannel.

Note that the images are recorded along several axial positions of the microchannel. At each location, three videos are recorded and from each video 1000 frames separated by 0.28 second are extracted. The average value from the entire dataset is being reported here. The percentage of time a particular flow regime occurs is also calculated using the following equations:

$$\text{Annular/Slug/Bubbly \%} = \frac{\text{Number of frames (annular/ slug/ bubbly) flow}}{\text{Total number of frames}} \times 100 \qquad (23)$$

Flow Regime Map

A flow regime map is generated through visual observations by three independent observers, by varying the heat flux and mass flow over a sufficiently large range. A total of 413 points are recorded, and (as noted above) the following regimes are identified: annular, elongated slug/annular, slug, bubbly and single phase water. The flow pattern map has been plotted as heat flux versus quality (Fig. **27**) and volumetric flow rate versus quality (Fig. **28**). Approximate lines to demarcate the different flow regimes have also been drawn. The separation of the flow regimes (with no redefinition at the time of compiling the results) is rather good.

Note that the maximum quality covered in the experiments is 0.3 (see also Fig. **13**) and Bergles and Kandlikar [50], the maximum quality is limited by backflow of vapor into the inlet reservoir, leading to premature CHF (Qu and Mudawar [23]). There is also substantial evidence of subcooled boiling (the minimum quality is negative 0.07). The magnitude of negative quality (or the amount of subcooling) looks rather large and therefore additional experiments to confirm it were undertaken using different amounts of dissolved air in water and even by changing the microchannels (with slightly different surface conditions). A large amount of subcooled boiling has been reported earlier also. For example, Jiang *et al.* [5] found boiling to start at 80 ^{0}C whereas the saturation temperature is 100 ^{0}C, which agrees with the present result. The occurrence of even elongated slug/annular flow in the subcooled boiling regime is however surprising.

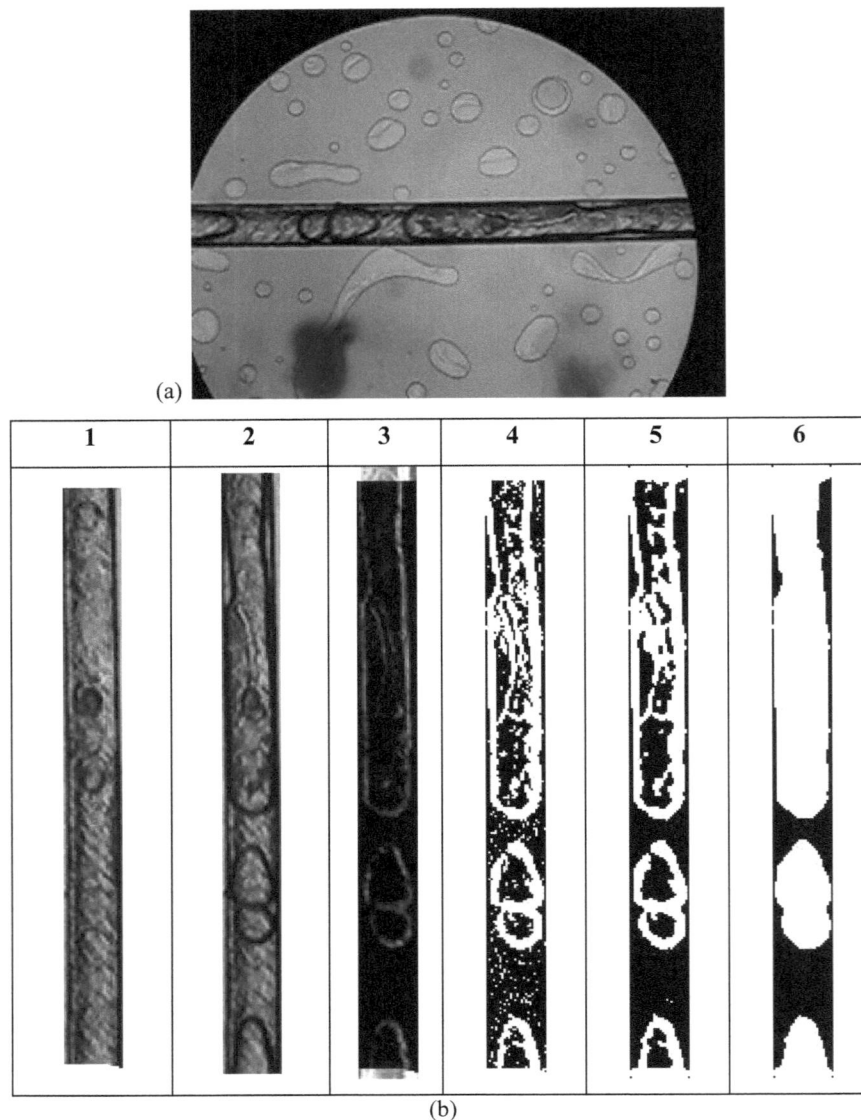

Figure 25: (a) Sample image of slug flow pattern in microchannel, and (b) flow chart used to determine the regions occupied by the vapor. The images in (b) represent: 1. background image and whitening of the boundaries for edge detection; 2. sample image after cropping the sides; 3. sample image with the background subtracted and converted to 'grayscale'; 4. binary version of image obtained after histogram equalization and removal of noise using median filter; 5. image with edges of the vapor regions marked; 6. image with the holes filled. (d_h = 140 μm).

Fig. **28** suggests that at low qualities as the heat flux is increased all the flow regimes are encountered. However, as the quality increases to 0.1 and above, the flow transits predominately from elongated slug to annular flow, with an increase in heat flux. Similarly, at low heat fluxes (less than 15 W/cm^2 in the present experiments), it is difficult to observe the annular flow, whereas above 40 W/cm^2 only elongated slug and annular flow regimes are apparent. Note that this picture applies only on an average – another result in the following section will show that the when the flow is predominately in the annular regime, it comprises both annular and single phase water; else, the flow is a mixture of bubbly, slug and single phase water.

(a)

(b)

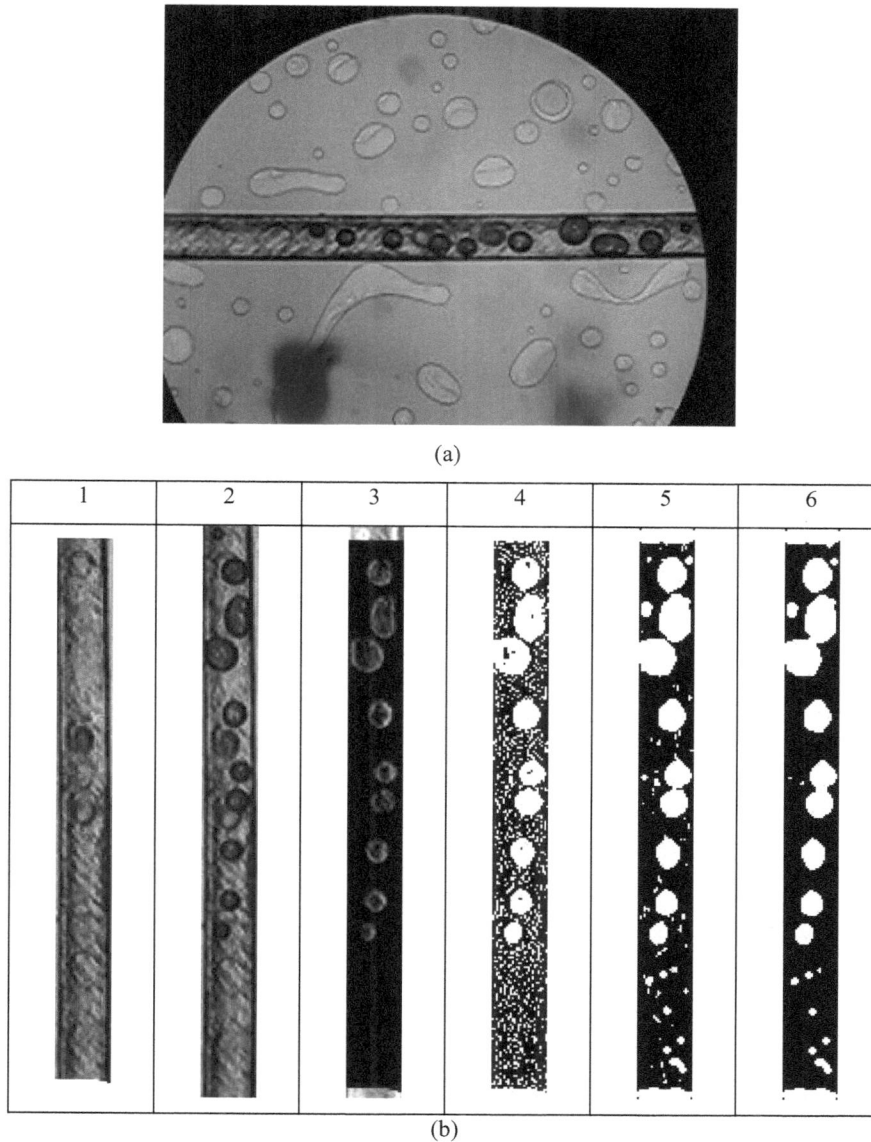

Figure 26: (a) Sample image of bubbly flow pattern in microchannel, and (b) illustrates the sequential steps used to determine the regimes occupied by the vapor. The images in (b) represent: 1. background image and whitening of the boundaries for edge detection; 2. sample image after cropping the sides; 3. sample image with the background subtracted and converted to 'grayscale'; 4. binary version of image obtained after histogram equalization and removal of noise using median filter; 5. image with edges of the vapor regions marked; 6. image with the holes filled. (The hydraulic diameter of the microchannel is 140 μm).

The flow regime map is determined by plotting the flow rate versus quality is qualitatively similar to Fig. **28**, and the observations noted with respect to heat flux versus quality apply. It is worth emphasizing that for volumetric flow rate greater than 0.2 ml/min and quality of 0.1 and above, the flow remains annular.

Note that the above flow map is for water flowing in a microchannel of rectangular cross-section (190 × 110 μm^2) and length of 2 cm, which is heated from three sides (and a roughly insulated top side). To the best of our knowledge, this is the first instance of flow map with water boiling in microchannel. Flow map with refrigerant (R-134a and R-245fa) has been reported by Revellin and Thome [13]. Their flow map results are in qualitative agreement with this earlier flow map. Revellin and Thome [13] however did not report occurrence of subcooled boiling.

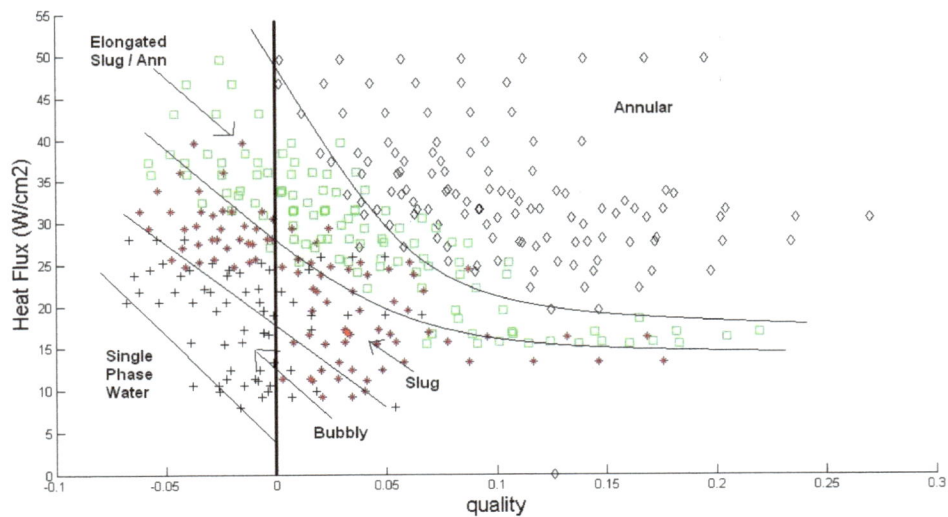

Figure 27: Flow pattern map with experimental transition lines for water flow in microchannel of hydraulic diameter of 140 µm for heat flux versus quality.

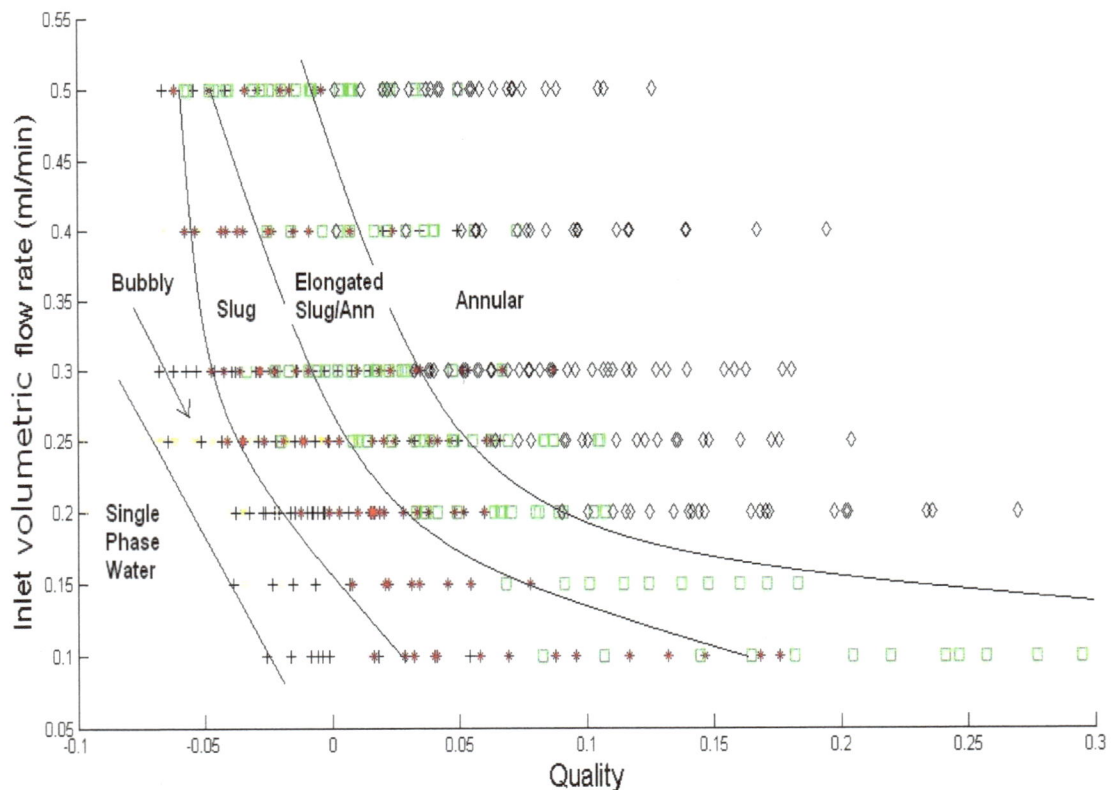

Figure 28: Flow pattern map with water flow in microchannel of hydraulic diameter of 140 µm.

Void Fraction

The void fraction is calculated using the image analysis tool discussed above. For this, a set of images at a maximum of seven different axial location of the microchannel were recorded and analyzed. The adaptors and connecting tubes at the inlet and outlet reservoirs posed a serious challenge in positioning of the microscope, especially towards at the entry and exit of the microchannels making it difficult to extend the measurement to additional locations.

As evident from Fig. **29**, the void fraction increases as we go downstream in the microchannel. The maximum void fractions are 0.4 and 0.15 for the higher and lower heat fluxes, respectively. As we move downstream in the microchannel, the liquid fraction in the microchannel decreases due to evaporation, leading to a monotonic increase in the void fraction. Also, the void fraction increases with a reduction in the volumetric flow rate. This is because, for a given heat flux as the volumetric flow rate is reduced, a larger fraction of liquid is evaporated to vapor, leading to an increase in void fraction. These are perhaps the first measurements of void fraction with boiling water in a microchannel.

Frequency of Occurrence of Flow Regimes

From the database of the images analyzed, the frequency of occurrence of different flow regimes is easily obtained. The results have been presented as a function of position in the microchannel, for two different mass flow rates and heat fluxes.

As it apparent from Figs. **30** and **31** slug/bubbly/single phase liquid occur at lower heat flux (26.6 W/cm^2), for both the mass flow rates and at all locations in the microchannel. At the first station (0.4 cm from inlet), the flow is either bubbly or plain water, with the flow being bubbly for less than 10% of the time (Fig. **30**). The slug regime first appears at 0.6 cm from the inlet. The percentage of both bubbly and slug increases (almost monotonically) along the length of the channel, at the cost of occurrence of single phase liquid. A reduction in the amount and frequency of occurrence of single phase liquid flow with downstream distance is expected. On the other hand, Fig. **31** shows occurrence of annular/single phase liquid at the higher heat flux (33.1 W/cm^2), for both the mass flow rates and at all locations in the microchannel. Very small occurrence of bubbly regime (less than 0.2%) is also found at certain locations. The percentage of single phase liquid is about 72% at the first station (0.8 cm from inlet) for both flow rates. This reduces to about 32% and 28% for the higher and lower mass flow rates, respectively. These results are consistent with the increase in void fraction along the length of the microchannel.

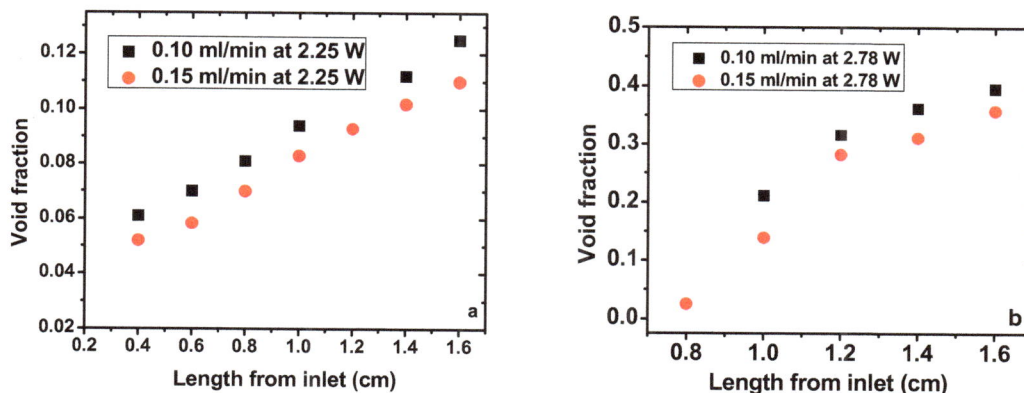

Figure 29: Variation of void fraction with different position of the microchannel of d_h = 140 μm. The heat flux values are (a) 26.6 W/cm^2 (2.25 W) and (b) 33.1 W/cm^2 (2.78 W).

Two typical situations suggested by the results (Figs. **30** and **31**) are: when the flow is predominately in the annular regime, it comprises both annular and single phase liquid; else, the flow is a mixture of bubbly, slug and single phase water. In the latter case, the regime is regarded as either bubbly or slug depending on their respective dominance. Another pertinent observation (which is perhaps difficult to glean from the figures) is that at low heat flux and mass flow rate, all the flow regimes are apparent. However, at high heat flux and mass flow rate, the flow is predominately annular (with absence of bubbly and slug flow regimes). This observation may help to resolve the contradictory observations reported in the literature with respect to occurrence of flow regimes at the micro-scales.

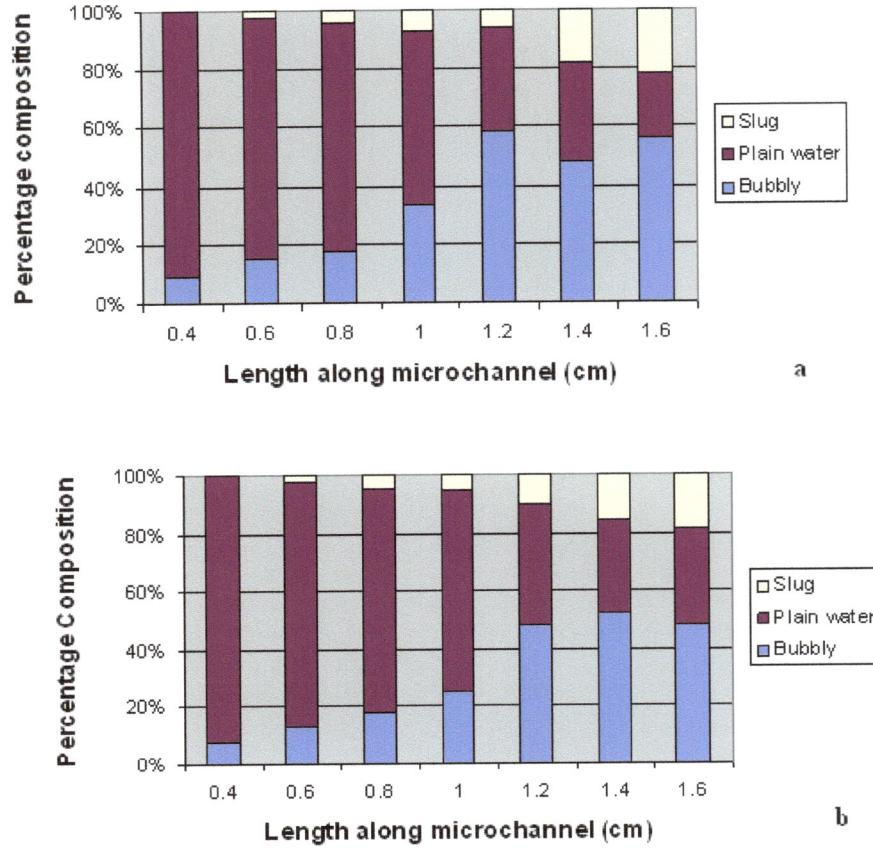

Figure 30: Flow composition versus along the length of microchannel for flow rates of 0.10 ml/min (79 kg/m²s) and 0.15 ml/min (119 kg/m²s) at 2.25 W (26.6 W/cm²) of heating power.

Heat Transfer Coefficient

The two phase heat transfer coefficient at a given position in the microchannel, in the annular flow regime, is obtained from

$$h_{annular}(l) = \frac{k_f}{\delta(l)} \tag{24}$$

where $\delta(l)$ is the corresponding film thickness at axial location. The thickness of liquid film at the wall can be measured using the image analysis technique in the annular flow regime. Therefore, the overall heat transfer coefficient at position l in the microchannel can be calculated from

$$h_{total} = h_{annular}F_{annular} + h_{sp}F_{spw} \tag{25}$$

The uncertainty in the overall heat transfer is estimated as 15%. The above procedure has been employed to obtain the heat transfer coefficient in the annular flow regime, but not the other regimes.

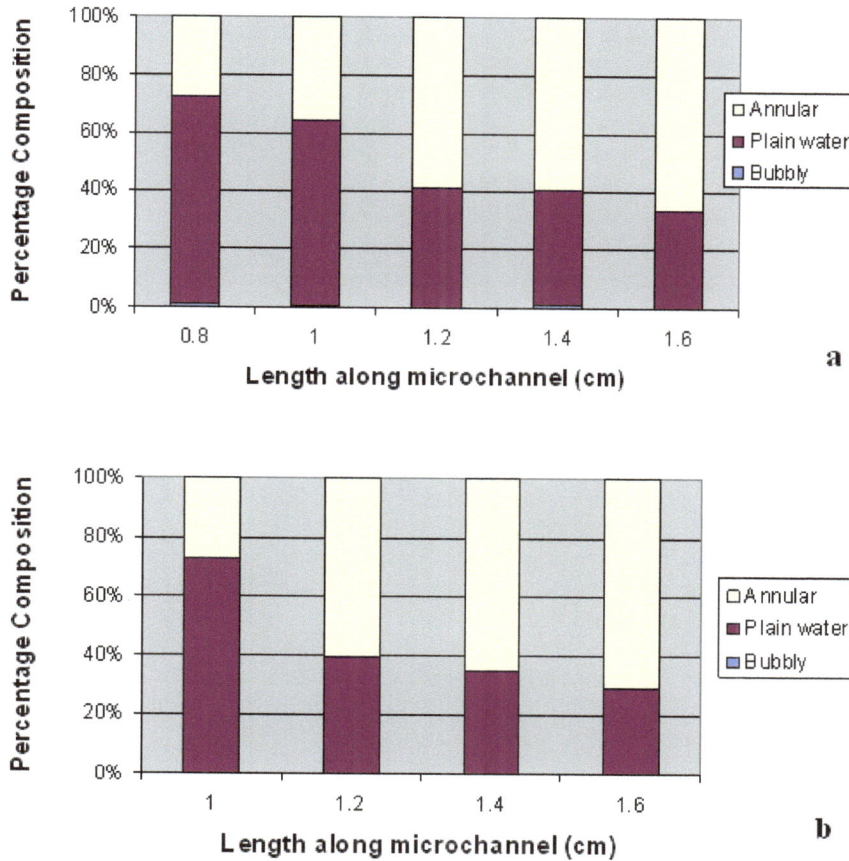

Figure 31: Flow composition versus along the length of microchannel for flow rates of 0.10 ml/min (79 kg/m^{2}s) and 0.15 ml/min (119 kg/m^{2}s) at 2.78 W (33.1 W/cm^2) of heating power.

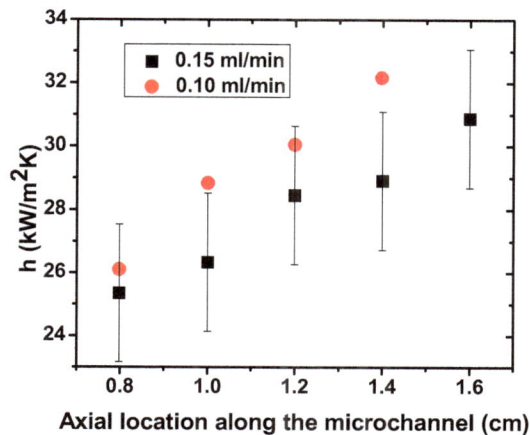

Figure 32: Heat transfer coefficient versus position for flow rates of 0.10 ml/min (79 kg/m^{2}s) and 0.15 ml/min (119 kg/m^{2}s) at 2.78 W (33.1 W/cm^2) of heating power.

As noted above, the flow oscillates between single phase liquid and annular regime (with negligible bubbly regime) for heat flux of 33.1 W/cm^2. The procedure outlined above is employed here, resulting in the heat transfer coefficient purely from the image analysis. Table **8** shows the value of heat transfer coefficient for the single phase and two phase along with their respective frequency of occurrence, from which the overall heat transfer coefficient can be obtained. Note that the average value (with respect to time) of δ has been employed for calculating the h_{tp}. From Table **8** it is observed that the heat transfer coefficient in two-phase is only 1.35 times larger than in the single phase. The factor is relatively small, due to the small hydraulic diameter of the channel (a small hydraulic diameter leads to a large single phase heat transfer coefficient, without a corresponding large increase in the heat transfer coefficient associated with two-phase regime). The heat transfer coefficient graph is plotted for both mass flow rates at a heat flux of 33.1 W/cm^2 (Fig. **32**). The heat transfer coefficient increases monotonically with length in microchannel, which is attributed to a monotonic increase in the frequency of annular regime. Further, the heat transfer coefficient decreases with an increase in the volumetric flow rate. Note that the magnitude of the heat transfer coefficient obtained here agrees with those of Qu and Mudawar [14] (who found h between 22-44 kW/m^2-K for mass flux 135–402 kg/m^2-s and exit quality between 0.01 to 0.17). However, Qu and Mudawar [14] report an initial decrease in the heat transfer coefficient before it increases. The initial decrease is not captured in the present experiments (probably due to the relatively large distance of the first recording station from the inlet). Qu and Mudawar [14] suggested that the decrease in heat transfer is due to substantial deposition of the liquid droplets from the core to the film on the walls of the microchannel. This deposition is again not sufficiently resolved in the present measurements.

The procedure discussed in the previous section can also be utilized in predominately bubbly regime, where h_{tp} can be determined using the homogeneous model. However, it cannot be used in the case of slug regime. Heat transfer coefficient in the other (bubbly/slug/ single phase liquid regime) is not estimated here owing to the non-negligible presence of slug flow.

Table 8: Heat transfer coefficient in the annular flow regime as a function of mass flow rate and position, for an input power of 2.78 W. The annular regime heat transfer coefficient, the percentage of time the flow is in single phase and annular, at different locations is also tabulated

Mass flow rate (ml/min)	Length from inlet (cm)	Percent time in single phase	Percent time in annular regime	h_{sp} (kW/m^{2}K)	h_{tp} (kW/m^{2}K)	h_{total} (kW/m^{2}K)
0.15	0.8	70.3	28.0	24.0	30.0	25.3
0.15	1	62.7	36.2	24.0	31.1	26.3
0.15	1.2	41.0	59.0	24.0	31.5	28.4
0.15	1.4	39.0	59.4	24.0	32.9	28.9
0.15	1.6	33.3	66.7	24.0	34.3	30.8
0.10	1.0	72.9	27.2	24.0	31.3	26.0
0.10	1.2	39.5	60.6	24.0	32.0	28.8
0.10	1.4	35.0	65.0	24.0	33.2	30.0
0.10	1.6	29.3	70.7	24.0	35.4	32.1

CONCLUSIONS

Microfluidics based remote cooling technology is being widely researched as a cooling technique for next generation of electronic devices. This chapter describes our effort in understanding the fundamental issues relevant to this futuristic technique. The variation of pressure drop, flow instability, void fraction, heat transfer coefficient and flow regimes, with variation in mass and heat flux over a sufficiently wide range of parameters has been documented. This exercise has led to a comprehensive understanding of single phase and two-phase fluid flow in microchannels. This work has several novel aspects. For example, to be the best of our knowledge this is the first instance that flow map with water boiling in microchannel has been generated. An image analysis based method has been developed to measure frequency of occurrence of different flow regimes, and determination of the void fraction and heat transfer coefficient. Channel engineering by optimizing the aspect ratio of microchannels is shown to result in a pressure drop minima in two-phase flow. Also we report minima in flow instability in two phase flow.

In summary, we have observed the following:

- The pressure drop across the channel initially increases with flow rate, reaches maximum and finally decreases at high flow rate.

- It is noted that up to points with equal pressure drop exists. Appropriate point can be employed in practice depending on that which yields the maximum heat transfer coefficients, leads to a minimum surface temperature, or based on some other criterion.

- Pressure drop characteristics for two-phase is found to match with annular flow model. However, our experimental data did not match with that predicted by the existing correlations.

- Channel engineering by optimizing aspect ratio (AR) of microchannel resulted in a pressure drop minima for two-phase flow. The optimum AR range is about 1.5. A small dependence on heat flux and flow rate is observed.

- Pressure instability is observed in 2-phase flow. A minima in pressure instability ($P_{RMS}/\Delta P$) with quality is noted and linked to the underlying flow regime.

- The flow is found to be a mixture of bubbly, slug and annular. The frequency of individual flow regime varies with heat and mass fluxes and also the location along the channel.

- It is demonstrated for the first time that the image analysis technique can be used to calculate the void fraction with water boiling in microchannel.

- Flow map for water flow in microchannel is generated for the first time.

- Heat transfer coefficient in the annular regime is calculated by using the image analysis technique.

These results have great practical significance as pressure drop penalty and instability have been considered to be the major limitations for use of flow boiling in electronic cooling. The above findings can help optimize the operating conditions for a microchannel based heat-sink for use in electronics cooling and other applications.

Future Scope of Work

Some suggestions for undertaking future work in this area are noted here:

1. The heat input to the microchannel can be made non-uniform (this can be achieved by varying the diameter of the heater filament to replicate the power map of a silicon die). The resultant effect of temperature on flow and heat transfer can be studied.

2. Explore further ways to reduce the instabilities during boiling.

3. Develop technique to accurately measure the temporal variation of heat transfer coefficient in microchannels.

NOMENCLATURE

A	Area	m^2
C_p	Specific heat of the fluid	J/kgK
d_h	Hydraulic Diameter	m
f	Friction Factor	-
H	Height of channel	m
h	Heat Transfer Coefficient	W/m^2K
h_{fg}	Latent heat of evaporation of fluid	J/kg
G	Mass Flux	kg/m^2s
L	Length in the channel	m
$\dot{m}$	Mass Flow Rate	kg/s
n	Number of readings	-
P	Power	W
ΔP	Pressure drop	mbar
q"	Heat Flux	W/m^2
Q	Volumetric Flow Rate	m^3/s
T	Temperature	^oC
u	Velocity	m/s
W	Width of the channel	m
x	Quality	-
z	Depth	m
ε	Surface roughness	m
ρ	Density	kg/m^3
ρ_H	Homogenous Density	kg/m^3
υ	Specific Volume	m^3/kg
μ	Viscosity	cP
σ	Surface Tension	N/m
Re	Reynolds Number	-
Nu	Nusselt Number	-
Fr_H	Froude Number	-
We_H	Weber Number	-

Subscripts

w	Wall
sat	Saturation
in	Inlet
exit	Exit
cross	Cross-Sectional
h	Heated
air	Ambient

supp	Supplied
res	Reservoir
chan	Channel
pred	Predicted
expt	Experimental
top	Top
bot	Bottom
sl	Slant
f	Fluid
g	Vapour
tp	Two Phase
sp	Single Phase
i,f	Frictional Pressure Drop
i,f	Acceleration Pressure Drop
avg	Average
i	i^{th} Interval
fo	Fluid only
go	Vapour only

REFERENCES

[1]   S. Oktay, R. Hanneman, and A. Bar-Cohen, "Heat transfer: High heat from a small package," *Mechanical Eng.*, vol. 109, pp. 36-42. 1986.

[2]   D.B. Tuckerman and R.F.W. Pease, "High Performance Heat sinking for VLS1," *IEEE Elecrron Dev. Lett.*, vol. 2, pp. 126, 1981.

[3]   S. Chakraborty and S.K. Som, "Heat transfer in an evaporating thin liquid film moving slowly along the walls of an inclined microchannel," *Int. J. Heat Mass Transfer*, vol. 48, pp. 2801-2805, 2005.

[4]   T.H. Yen, M. Shoji, F. Takemura, Y. Suzuki, and N. Kasagi, "Visualization of convective boiling heat transfer in single microchannels with different shaped cross-sections," *Int. J. Heat Mass Transfer*, vol. 49, pp. 3884-3894, 2006.

[5]   L. Jiang, M. Wong, and Y. Zohar, "Phase change in microchannel heat sinks with integrated temperature sensors," *J. Microelectromechanical System*, vol. 8, pp. 358–365, 1999.

[6]   L. Zhang, J.M. Koo, L. Jiang, M. Asheghi, K.E. Goodson, and J. G. Santiago "Measurements and modeling of two-phase flow in microchannels with nearly constant heat flux boundary conditions," *J. Microelectromechanical System*, vol. 11, pp. 12-19, 2001.

[7]   G. Hetsroni, A. Mosyak, Z. Segal, "Nonuniform temperature distribution in electronic devices cooled by flow in parallel microchannels," *IEEE Trans. Comp.Pack. Tech.*, vol. 24, pp. 16–23, 2001.

[8]   *http://news.bbc.co.uk/2/hi/technology/7439406.stm*

[9]   S.G. Singh and C. S. Tan "Impact of thermal through silicon *via* (TTSV) on the temperature profile of multi-layer 3-D device stack," *IEEE 3-D ICs conference, San-Francisco*, 2009.

[10]   J.R. Thome, "Boiling in microchannels: a review of experiment and theory," *Int. J. Heat Fluid Flow*, vol. 25, pp. 128-139, 2004.

[11]   X.F. Peng and B.X. Wang, "Forced convection and flow boiling heat transfer for liquid flowing through microchannels," *Int. J. Heat and Mass Transfer*, vol. 36, pp. 3421-3427, 1993.

[12]   J. Lee and I. Mudawar, "Two phase flow in high-heat-flux microchannel heat sink for refrigeration cooling applications: Part I- pressure drop characteristics," *Int. J. Heat Mass Transfer*, vol. 48, pp. 928-940, 2005.

[13]   T.N. Tran, M.W. Wambsganss, and D.M. France, "Small circular and rectangular channel boiling with two refrigerants," *Int. J. Multiphase Flow*, vol. 22, pp. 485–498, 1996.

[14]   R. Revellin and J.R.Thome, "A new type of adiabatic flow pattern map for boiling heat transfer in microchannels," *J. Micromechanics Microengineering*, vol. 17, pp. 788-796, 2007.

[15]   W. Qu and I. Mudawar, "Flow boiling heat transfer in two-phase micro-channel heat sinks—I. Experimental investigation and assessment of correlation methods," *Int. J. Heat and Mass Transfer*, vol. 46, pp. 2755-2771, 2003.

[16]   M. Lee, Y. Y. Wong, M. Wong, and Y. Zohar, "Size and shape effects on two phase flow patterns in microchannel forced convection boiling," *J. Micromechanics Microengineering*, vol. 13, pp. 155-164, 2003.

[17]   L. Jiang, M. Wong, and Y. Zohar, "Forced convection boiling in a microchannel heat sink," *J. Microelectromechanical System*, vol. 10, pp. 80-87, 2001.

[18]   L. Zhang, E.N. Wang, K.E. Goodson, and T.W. Kenny, "Phase change phenomena *in silico*n microchannels," *Int. J. Heat Mass Transfer*, vol. 48, pp. 1572-1582, 2005.

[19]   S.G. Kandlikar and P. Balasubramanian, "High speed photographic observation of flow patterns during flow boiling in a single rectangular minichannel," Paper No. HT2003-47175, *Proc. ASME Summer Heat Transfer Conf.*, 2003.

[20]   M. E. Steinke and S. G. Kandlikar, "Control and effect of dissolved air in water during flow boiling in microchannels," *Int. J. Heat Mass Transfer*, vol. 47, pp. 1925-1935, 2004.

[21]   S.G. Kandlikar, "Heat transfer mechanisms during flow boiling in microchannels," *J. Heat Transfer*, vol. 126, pp. 8-16, 2004.

[22]   A. Serizawa, Z. Feng, and Z. Kawara, "Two-phase flow in microchannels," *Exp. Therm. Fluid Sci.*, vol. 26, pp. 703, 2002.

[23]   W. Qu and I. Mudawar, "Flow boiling heat transfer in two-phase micro-channel heat sinks - 2. Annular two-phase flow model," *Int. J. Heat Mass Transfer*, vol. 46, pp.2773-2784, 2003.

[24]   G. Hetsroni, Z. Mosyak, and G. Ziskind, "A uniform temperature heat sink for cooling of electronics device," *Int. J. Heat Mass Transfer*, vol. 45, pp. 3275-3286, 2002.

[25]   R. Yun, Y. Kim, and M. Kim, "Convective boiling heat transfer characteristics of CO_2 in microchannels," *Int. J. Heat Mass Transfer*, vol. 48, pp. 235-242, 2005.

[26]   M.C. Diaz and J. Schmidt, "Experimental investigation of transient boiling heat transfer in microchannels," *Int. J. Heat Fluid Flow*, vol. 28, pp. 95-102, 2007.

[27]   P.C. Lee and C. Pan, "Boiling heat transfer and two-phase flow of water in a single shallow microchannel with a uniform or diverging cross section," *J. Micromechans Microengineering*, vol. 18, 025005, 2008.

[28]   G. Hetsroni, A. Mosyak, E. Pogrebnyak, and Z. Segal, "Periodic boiling in parallel microchannels at low vapor quality," *Int. J. Multiphase Flow*, vol. 32, pp. 1141-1159, 2006.

[29]   G. Ribatski, L. Wojtan, and J.R. Thome, "An analysis of experimental data and prediction methods for two-phase frictional pressure drop and flow boiling heat transfer in micro-scale channels," *Exp. Therm. Fluid Sci.*, vol. 31, pp. 1-19, 2006.

[30]   H. Muller-Steinhagen and K. Heck, "A simple friction pressure drop correlation for two-phase flow in pipes," *Chem. Eng. Proc.*, vol. 20, pp. 297-308, 1986.

[31]   L. Zhang, J.M. Koo, L. Jiang, M. Asheghi, K.E. Goodson, J.G. Santiago, and T.W. Kenny, "Measurements and modelling of two phase flow in microchannels," *J. Microelectromechanical Systems*, vol. 11, pp. 12-19, 2002.

[32]   W. Qu and I. Mudawar, "Measurement and prediction of pressure drop in two-phase microchannel heat sinks," *Int. J. Heat Mass Transfer*, vol. 46, pp. 2737-2753, 2003.

[33]   H.Y. Wu, P. Cheng, and H. Wang, "Pressure drop and flow boiling instabilities *in silico*n microchannel heat sinks," *J. Micromechanics Microengineering*, vol. 16, pp. 2138-2146, 2006.

[34]   M. B. Bowers and I. Mudawar, "High flux boiling in low flow rate, low pressure drop mini-channel and micro-channel heat sinks," *Int. J. Heat Mass Transfer*, vol. 37, pp. 321-332, 1994.

[35]   G. Hetsroni, A. Mosyak, E. Pogrebnyak, and Z. Segal, "Explosive boiling of water in parallel microchannels," *Int. J. Multiphase Flow*, vol. 31, pp. 371-392, 2005.

[36]   R. Muwanga, I. Hassan, and R. MacDonald, "Characteristics of flow boiling oscillations *in silico*n microchannel heat sinks," *J. Heat Transfer*, vol. 129, pp. 1341-1351, 2007.

[37]   H. Y. Wu and P. Cheng, "Boiling instability in parallel silicon micro-channels at different heat flux," *Int. J. Heat Mass Transfer*, vol. 47, pp. 3631-3641, 2004.

[38]   G. Wang, P. Cheng, and H. Wu, "Unstable and stable boiling in parallel microchannels and in a single microchannel," *Int. J. Heat Mass Transfer*, vol. 50, pp. 4297-4310, 2007.

[39]   M. Lee, Y.K. Lee, and Y. Zohar, "Two-phase flow oscillations in microchannel convective boiling," *MEMS2007*, Kobe, Japan, 21-25 (2007) pp. 635-638, 2007.

[40]   N.S. Chavan, A. Bhattacharya, and K. Iyer, "Modeling of Two Phase Flow Instabilities in Microchannels," *Proceedings of ICMM2005 3rd International Conference on Micro and Minichannels June 13-15*, 2005, Toronto, Ontario, Canada.

[41]   S. G. Kandlikar, D. A. Willistein, and J. Borrelli, "Experimental evaluation of pressure drop elements and fabricated nucleation sites for stabilizing flow boiling in minichannels and microchannels," *Proc. 3rd Int. Conf. on Microchannels and Minichannels*, ICMM2005-75197, Toronto, Ontario, Canada, 2005.

[42] R.R. Bhide, S.G. Singh, A. Sridharan, S. P. Duttagupta and A. Agrawal, "Pressure drop and heat transfer characteristics of boiling water in sub-hundred micron channel," *Exp. Therm. Fluid Sci.*, vol. 33, pp. 963-975. 2009.

[43] G. L. Morini, "Laminar liquid flow through silicon microchannel," *J. Fluids Engineering*, vol. 126, pp. 485 - 490, 2004.

[44] G. M. Mala, D. Q. Li, and C. Werner, "Flow characteristics of water through a microchannel between two parallel plates with electrokinetic effects," *Int. J. Heat Fluid Flow*, vol. 18, pp. 489-496. 1997.

[45] M. M. Rahman, "Measurements of heat transfer in microchannel heat sinks," *Heat Mass Transfer*, vol. 27, pp. 495–506, 2000.

[46] H. Y. Wu and P. Cheng, "Friction factors in smooth trapezoidal silicon microchannels with different aspect ratio," *Int. J. Heat Mass Transfer*, vol. 46, pp. 2519-2525, 2003.

[47] X. F. Peng and G. P. Peterson, "Convective heat transfer and flow friction for water flow in microchannel structures," *Int. J. Heat Mass Transfer*, vol. 39, pp. 2599–2608, 1996.

[48] R. K. Shah and A. L. London, "*Laminar Flow Forced Convection in Ducts: A Source Book for Compact Heat Exchanger Analytical Data*," New York: Academic, 1978.

[49] W. Qu and I. Mudawar, "Measurement and correlation of critical heat flux in two-phase micro-channel heat sinks," *Int. J. Heat Mass Transfer*, vol. 47, pp. 2045-2059, 2004.

[50] A.E. Bergles and S.G. Kandlikar, "On the nature of critical heat flux in microchannels," *J. Heat Transfer*, vol. 127, pp. 101-107, 2005.

[51] L. Zhang, J.M. Koo, L. Jiang, M. Asheghi, K.E. Goodson, and J.G. Santiago, "Measurements and modeling of two-phase flow in microchannels with nearly constant heat flux boundary conditions," *J. Microelectromechanical System*, vol. 11, pp. 12-17, 2002.

[52] K. Mishima and T. Hibiki, "Some characteristics of air water two-phase flow in small diameter vertical tubes," *Int. J. Multiphase Flow*, vol. 22, pp. 703-712, 1996.

[53] H. J. Lee and S. Y. Lee, "Heat transfer correlation for boiling flows in small rectangular horizontal channels with low aspect ratios," *Int. J. Multiphase Flow*, vol. 27, pp. 2043–2062, 2001.

[54] W. Yu, D. M. France, M. W. Wambsganss and J. R. Hull, "Two phase pressure drop, boiling heat transfer, and critical heat flux to water in a small-diameter horizontal tube," *Int. J. Multiphase Flow*, vol. 28, pp. 927–941, 2002.

[55] P. Patankar, "Fluid flow and heat transfer in microchannels", M.Tech. Thesis, Department of Mechanical Engineering, IIT Bombay, 2007.

[56] A. Jain, "Two phase flow pressure characteristics in microchannels," M.Tech. Thesis, Department of Mechanical Engineering, IIT Bombay, 2008.

[57] H.B. Ma and G.P. Peterson, "Laminar friction factor in microscale ducts of irregular cross-section microscale," *Thermophysical Eng.*, vol. 1, pp. 253-265, 1997.

[58] G.L. Morini and M. Spiga, "The rarefaction effect on the friction factor of gas flow in microchannels," *Superlattices & Microstructures*, vol. 35, pp. 587–599, 2004.

[59] H .Y. Wu and P. Cheng, "An experimental study of convective heat transfer *in silicon* microchannels with different surface conditions," *Int. J. Heat Mass Transfer*, vol. 46, pp. 2547-2556, 2003.

[60] S.G. Singh, S. P. Duttagupta and A. Agrawal, "Impact of aspect ratio effect on water boiling in rectangular microchannel," *Exp. Therm. Fluid Sci.*, vol. 33, pp. 153-160, 2008.

[61] R.J. Jones, D.T. Pate, S.H. "Control of instabilities in two-phase microchannel flow using artificial nucleation sites," Paper No. IPACK2007-33602, Proceedings of the ASME InterPACK 2007 Conference, Jul. 2007.

[62] S.G. Singh, R.R. Bhide, S.P. Duttagupta, B.P. Puranik, and A. Agrawal, "Two-phase flow pressure drop characteristics in trapezoidal silicon microchannels," *IEEE Trans. Comp. Pack. Tech.*, vol. 32, pp. 887-900, 2009.

[63] A. Agrawal, "Three-dimensional simulation of gaseous slip flow in different aspect ratio microducts," *Phys. Fluids*, vol. 18, pp. 103604, 2006.

[64] C. C. Hsieh, S. B. Wang, and C. Pan, "Dynamic visualization of two-phase flow patterns in a natural circulation loop," *Int. J. Multiphase flow*, vol. 23, pp. 147-154, 1997.

[65] T. J. Lin, K. Tsuchiya, and L. S. Fan, "Bubble flow characteristics in bubble columns at elevated pressure and temperature," *AIChE J.*, **vol. 44,** pp. 545-560, 1998.

[66] Y. Zhao, G. Chen, and Q. Yuan, "Liquid-liquid two-phase flow patterns in a rectangular microchannel", *AIChE J.*, vol. 52, pp. 4052 – 4060, 2006.

[67] S.G. Singh, A. Jain, A. Sridharan, S.P. Duttagupta, and A. Agrawal, "Flow map and measurement of void fraction and heat transfer coefficient using image analysis technique for flow boiling of water *in silicon* microchannel," *J. Micromechanics Microengineering*, vol. 19, 075004, 2009.

CHAPTER 7

Flow Pattern, Pressure Drop and Heat Transfer Coefficient for Evaporative Refrigerants in Horizontal Small Tubes

A.S. Pamitran[a], Kwang-Il Choi[a] and Jong-Taek Oh[b,*]

[a]Graduate School, Chonnam National University, San 96-1, Dunduk-dong, Yeosu, Chonnam 550-749, Republic of Korea and [b]Department of Refrigeration and Air Conditioning Engineering, Chonnam National University, San 96-1, Dunduk-dong, Yeosu, Chonnam 550-749, Republic of Korea

Abstract: An experimental investigation on two-phase flow boiling flow pattern, pressure drop and heat transfer of R-22, R-134a, R-410A and natural refrigerants of C_3H_8 and CO_2 in horizontal circular small tubes is presented in this study. The experimental data were obtained over a heat flux range of 5 to 40 kW/m^2, mass flux range of 50 to 600 $kg/(m^2 \cdot s)$, saturation temperature range of 0 to 15°C, and quality up to 1.0. The test section was made of stainless steel tubes with inner diameters of 0.5 mm, 1.5 mm and 3.0 mm, and lengths of 330, 1500, 2000 and 3000 mm. The effects of mass flux, heat flux, saturation temperature and inner tube diameter on the pressure drop and heat transfer coefficient are reported in this thesis. The experimental pressure drop and heat transfer coefficient were compared with the predictions given by some existing correlations. New correlations of pressure drop and heat transfer coefficient for small tubes that is based on the present experimental data are developed.

Keywords: Refrigerant, flow boiling, heat transfer coefficient, pressure drop, horizontal small tubes, Correlation.

INTRODUCTION

Background and Objective

Refrigeration and air-conditioning industries have become greatly aware of environmental protection efforts and process intensification recently. Such awareness has led to a demand for environmental friendly evaporation refrigerants in smaller evaporators, which are used in the refrigeration, air conditioning and process industries. However, the flow pattern, pressure drop and heat transfer of two-phase flows in small tubes cannot be properly predicted using existing procedures and correlations because they are more suitable for large tubes.

Hydrofluorocarbons (HFCs) and natural refrigerants, as long term alternative refrigerants, will be important in the future for compact heat exchanger applications due to their performance and their environmental friendliness with zero Ozone Depletion Potential (ODP) and low or zero Global Warming Potential (GWP). One important reason of the increasing attention to the HFC and natural refrigerants as working fluid in various refrigeration and air conditioning applications with various conditions is cause of a growing demand of new ecological refrigerants with low or zero ODP and GWP. Surrounding the earth at a height of about 25 km is the rich in ozone layer, stratosphere, which prevents the sun's harmful ultra-violet (UV-B) rays from reaching the earth. In 1974, Rowland and Molina hypothesized that chlorinated compounds were able to persist in the atmosphere where solar radiation would break up the molecules and release chlorine atoms that would destroy the ozone. Mounting evidence and discovery of the Antarctic ozone hole and the others recently lead to a global program to cut annual consumption of Ozone Depleting Substances (ODS). This program was signed on The Montreal Protocol on Substances that Deplete the Ozone Layer, in 1987. Two of the concern ODS are chlorofluorocarbon (CFC) and hydrochlorofluorocarbon (HCFC). The Montreal Protocol, an international environmental agreement, provisions of CFC consumption are to cease it at the end of 1995 in developed nations, and by 2010 in developing nations. These requirements were

*****Address correspondence to Jong-Taek Oh:** Department of Refrigeration and Air Conditioning Engineering, Chonnam National University, San 96-1, Dunduk-dong, Yeosu, Chonnam 550-749, Republic of Korea; Tel: +82-61-659-3273; Fax: +82-61-659-3279; E-mail: ohjt@chonnam.ac.kr

Lixin Cheng and Dieter Mewes (Eds)

later modified, leading to the phase-out in 1996 of CFC production in all developed nations. In addition, a 1992 amendment to the Montreal Protocol established a schedule for the phase-out of HCFCs. HCFCs are less damaging to the ozone layer than CFCs, but still contain ozone-destroying chlorine.

Although most HCFCs' chemicals break down before reaching the ozone layer, the chlorine produced reaches the stratosphere and depletes the ozone layer. R-22, a HCFC, is still widely used in the refrigeration and air-conditioning industry even though some countries have ceased using HCFCs. In an attempt to find a replacement for R-22 as an environmental conservation effort, HFC and natural refrigerants, such as R-134a, R-410A, CO_2 and C_3H_8 have been studied extensively because they do not contain chlorine, which can deplete the ozone layer. The zero or very low GWP is another advantage of these natural refrigerants. And recently, energy and material efficiencies have emerged as important topics in refrigeration and air conditioning. Recent awareness of the advantages of process intensification has led to the demand for smaller evaporators in refrigeration, air conditioning and processing due to their energy and material efficiencies. However, existing methods for predicting the pressure drop and heat transfer of two-phase flow in large tubes cannot properly predict same type of heat transfer in small tubes. Furthermore, compared with evaporation in conventional tubes, that in a small tube may yield a higher heat transfer coefficient due to the larger contact area per unit volume of fluid.

There are relatively few published works on two-phase flow pattern, pressure drop and heat transfer of refrigerants in small tubes, unlike the large amount of data for large tubes. Evaporation in small tubes has great advantages: high heat transfer coefficient, significant size reduction of compact heat exchangers, and lower required fluid mass. Despite these advantages, the pressure drop of the refrigerants higher in small tubes than in conventional tubes because of the increase of tube wall friction. Several studies have dealt with two-phase flow boiling in small tubes. However, the existing prediction methods of flow pattern, pressure drop and heat transfer coefficient could not predict well the current experimental data. Chisholm [10] proposed a theoretical basis for the Lockhart-Martinelli correlation for two-phase flow. The Friedel [20] correlation was obtained by optimizing an equation for the two-phase frictional multiplier based on data taken from a large measurement database. Many studies have developed pressure drop correlations based on the Chisholm [10] and Friedel [20] correlations. Kawahara *et al.* [30] and Yu *et al.* [57] developed pressure drop correlations based on the Chisholm [10] correlation. Zhang and Webb [60], Chang *et al.* [7], and Chen *et al.* [8] developed pressure drop correlations based on the Friedel [20] correlation. Tran *et al.* [48] measured two-phase flow pressure drops of refrigerants R-134a, R-12, and R-113 in small round and rectangular channels, and they modified Chisholm's [10] correlation and proposed a new correlation. In evaporation with small tubes, such as reported by Zhang *et al.* [61], Tran *et al.* [49], Pettersen [43] and Yun *et al.* [58], the contribution of nucleate boiling is predominant.

This chapter was undertaken to obtain experimental data for some working refrigerants and to determine the pressure drop and local heat transfer coefficient of these refrigerants during their evaporation in small tubes. The flow pattern of flow boiling was evaluated with existing flow pattern maps. The effects of mass flux, heat flux, saturation temperature and inner tube diameter on the pressure drop and heat transfer coefficient were presented. The pressure drop and heat transfer coefficient of the present working refrigerants were compared. Also, the experimental results were compared with the predictions from several existing pressure drop and heat transfer coefficient prediction methods. Due to the limitations of the correlation for forced convective boiling of refrigerants in small tubes, a new correlation for evaporative refrigerants in small tubes was developed in this paper based on superposition.

Experimental Working Refrigerants

Refrigeration and air-conditioning systems use working refrigerants as a means of transferring heat between heat sources and heat sinks. The working refrigerants must generally absorb heat at low temperatures, below 0 °C, without freezing, and reject heat at higher temperatures and pressures. Ammonia was one of the most common refrigerants until the 1930s. However, due to its toxicity and flammability, this natural refrigerant was gradually replaced by artificial ones, such as halocarbons, which were seen as a safer alternative to ammonia. The first halocarbon refrigerant, R-12, was introduced in the early 1930s. Twenty years later the new compounds had conquered the greater part of refrigeration applications the world over, starting with the smaller equipment and air conditioning, and gradually penetrating into even the large industrial area. However, we have heard a great deal lately of the harmful effects to the environment when halocarbon refrigerants are lost to the atmosphere. Research has recently been focused on development of new refrigerants to replace CFCs and HCFCs. These new working fluids are synthetic compounds-namely HFCs. Although the ODP of some HFCs is zero, their GWP can be large. The other alternative refrigerants to HCFCs and also HFCs are the use of natural refrigerants such as propane and carbon dioxide. The using of natural compounds, which are already circulating in quantity in the biosphere and are known to be harmless, is much more preferable. The comparisons of ODP and GWP for the current working refrigerants are illustrated in Fig. **1**.

R22 is a single HCFC compound. It has the formula $CHClF_2$, and a boiling point of -40.7 °C, in atmosphere pressure. R-22 has been the refrigerant of choice for refrigeration and air-conditioning systems for more than four decades. It is commonly used in air-conditioning applications, such as residential split systems, rooftop units and window air conditioners. Unfortunately for the environment, releases of R-22, such as those from leaks, contribute to ozone depletion. R-22 has low chlorine content and ozone depletion potential and only a modest global warming potential. R-22 has ODP = 0.05 and GWP = 1700. However, even this lower ozone depletion potential is no longer considered acceptable. It will be phased out soon under the Montreal Protocol, to be replaced by refrigerants with zero ODP such as propane, and other refrigerants (even though they don't have very similar properties): R-134a and R-410A. The possibility to find a substitute to R-22 has been discussed in numerous investigations. The general consensus today is that there is no single replacement available, pure or mixture. Different alternatives must be used for different applications and operating conditions with considering the performance enhancement. In order to replace R-22, refrigerant mixtures might be one alternative because of the possibility for manipulating their properties to find such desired requirements as thermo-physical, chemical, and safety.

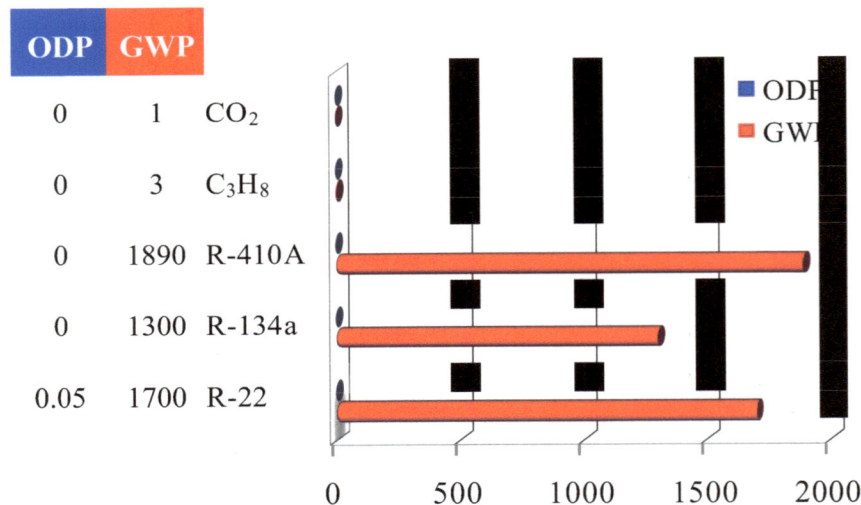

ODP	GWP		
0	1	CO_2	
0	3	C_3H_8	
0	1890	R-410A	
0	1300	R-134a	
0.05	1700	R-22	

Figure 1: Comparisons of ODP and GWP for the current working refrigerants.

R-134a, is a haloalkane refrigerant with thermodynamic properties similar to R-12, but without its ODP. It has the formula CH_2FCF_3, and a boiling point of -26.3 °C, in atmosphere pressure. R-134a first appeared in the early 1990s as a replacement for R-12. It is used in new medium- and high-temperature stationary commercial refrigeration, as well as chiller systems and home appliances. In addition, it can be used to retrofit existing R-12 refrigeration and air conditioning systems. It is also the global standard for new mobile air conditioning and can be used to retrofit existing R-12 mobile air conditioning systems. However, recently, R-134a has been subject to use restrictions due to its theorized contribution to climate change. R-134a has been atmospherically modeled for its impact on depleting ozone and as a contributor to global warming. R-134a has insignificant ODP, significant GWP of 1300. In the EU, it will be banned as from 2011 in all new cars. SAE (International, an auto engineers association) has proposed R-134a to be best replaced by a new fluorochemical refrigerant HFO-1234yf ($CF_3CF=CH_2$) in automobile air-conditioning systems.

It is known from many experiments that heat transfer coefficients during evaporation of mixtures are lower than those of the pure components of the mixture. In evaporation of refrigerant mixture, the volatile difference appears, further it results resistance on heat transfer. The heat transfer coefficient decreases in the region in which vapor and liquid compositions strongly different. For such non-azeotropic mixture as R-410A, however, due to mixtures' gliding temperature during evaporation, it is though the mixtures might lead to an improved thermodynamic performance of the cycle. It is possible to benefit from the gliding temperature to increase system energy efficiency; this is partly compensated due to the lower heat transfer of mixture compared with their single component. R-410A is a near-azeotropic mixture of R-32 and R-125 which is used as a refrigerant in air conditioning applications. It has the formula CH_2F_2/CHF_2CF_3, and a boiling point of -52 °C, in atmosphere pressure. Unlike many haloalkane refrigerants it does not contribute to ozone depletion, and is therefore becoming more widely used as ozone-depleting refrigerants like R-22 are phased out. However, R-410A has a high GWP of 1890.

Hydrocarbons have been used as refrigerants for many years in the petrochemical industry, where handling of flammable fluids is customary. Due to their flammability, hydrocarbons are mainly regarded as an alternative in systems with low working fluid charge. Hydrocarbons have been used in low charge systems in residential and commercial applications (water and space heating) and low-capacity air-conditioners for automotive/transport and space cooling. In recent years HCs have also been widely used in domestic refrigerators and heat pumps. In fact, research projects are being carried out in order to study the behavior and components of low charge vapor compression systems of relatively large capacities. These projects indicate that C_3H_8 systems have a heating capacity, when used as a heat pump, which is about 8% lower than R-22, while the COP is 5-7% higher. It was also shown that the refrigeration charge of a R-290 system is reduced by about 50%, when compared with R-22, for the same geometry. This is due to its smaller molar mass. C_3H_8 has a boiling point of -42 °C, in atmosphere pressure. When R-22 is to be substituted by C_3H_8, the costs of the necessary changes (expansion valve, lubricant) are negligible when compared with the costs of the substitution by HFCs. The very important disadvantage of C_3H_8 is that it is combustible, with a very low ignition concentration limit in air 2.2/9.5, and this drawback has been blown up to unreasonable proportions. As a fact they are popular fuels available everywhere and used with simple precautions even in private homes, caravans and small boats. With reasonably careful design it must be even simpler to ensure safety in a hermetic closed refrigeration circuit.

Carbon dioxide was a commonly used refrigerant from the late 1800s and well into our century. In the first decades of this century, CO_2 was extensively used as a working fluid in marine installations mainly due to its safety. However, since the 1950s CO_2 was replaced by halocarbons, in the sequence of the generalization of these refrigerants. Only recently has the interest in CO_2 increased, due to the development of transport air-conditioning systems. Unlike the other natural refrigerants, CO_2, is a non-flammable working fluid, odorless, with zero ODP and zero effective GWP, not toxic in practice. CO_2 is inexpensive when compared with halocarbons because it can be obtained as a waste product. Furthermore, its liquid density is lower and, as a consequence, the system charge and size will also be lower. Another advantage of CO_2, is its compatibility with normal lubricants and construction materials. Due to the thermodynamic properties of CO_2, the vapor compression cycle and the components of the system should differ from the ones with low-pressure refrigerants. In fact, for moderate ambient air temperatures, the pressure at which the fluid rejects heat must be supercritical, with variable fluid temperature. As pressure and temperature are independent properties in the supercritical region, the system must have a high side-pressure adjustment, since the COP is pressure dependent. The COP has a maximum value for a given high side pressure. The high pressure (> 100 bar), combined with the low molar mass of CO_2, reduces the volumetric flow and the dimensions of the system components (compressor, valves, piping). With regard to personal safety, CO_2 is at least as good as the best of the halocarbons with some advantages. One problem of this refrigerant, and in some applications an important asset, is its relatively low critical point of 31 °C. With condensation well below this temperature, in a cool climate or in a low stage of a cascade system, it works with condensation like any other refrigerant. As the critical temperature is approached or even exceeded, the losses by superheat and throttling increase. It turns out, however, that in some cases this can be compensated by much improved compressor performance as a result of very low pressure ratio and small volume requirement. This has been amply demonstrated by the motorcar air conditioning system.

Flow Boiling Pressure Drop in Small Tubes

Processes involving boiling of refrigerants in a compact heat exchanger are extensively encountered in industry. In order to enhance the compact heat exchanger performances, several investigations on boiling heat transfer in small tubes have been reported. Chisholm [12], by using steam-water flow in tube, introduced a procedure to evaluate the two-phase frictional multiplier based on pressure gradient for liquid alone flow. Chisholm and Sutherland [14] proposed a coefficient that is used in the two-phase frictional multiplier equation. One of the most accurate two-phase pressure drop correlations is that of Friedel [20]. The correlation was obtained by optimizing an equation for the two-phase frictional multiplier using a large data base of two-phase pressure drop measurement. The Friedel [20] correlation is valid for vertical and for horizontal flow. Chang and Ro [6] investigated the pressure drop of HFC refrigerants and their mixtures in capillary tube. They found that the friction factor of refrigerant flow in capillary tubes has to describe the friction factor as a function of Reynolds number and the roughness of the tube wall. Their study showed that Cicchitti et al. [16] was the most adequate for adiabatic capillary flow. Ould Didi et al. [39] presented an extensive two-phase pressure drop database for five refrigerants originating from the study of Kattan [29]. In irrespective of flow pattern, their study showed that the pressure drop method of Müller-Steinhagen and Heck [37] and the method of Grönnerud [22] consistently gave the best predictions. Kawahara et al. [30] investigated two-phase flow characteristics in a 100 µm diameter circular tube using de-ionized water and nitrogen gas. Single-phase friction factor and two-phase friction multiplier data were obtained from their pressure drop data. The single-phase friction factor was shown to be in good agreement with the conventional laminar correlation. The two-phase friction multiplier data were over-predicted by the homogeneous flow model, but correlated well, within ±10%, with the separated flow model of Lockhart and Martinelli [34]. Yu et al. [57] using water as working fluid studied two-phase pressure drop in a small horizontal tube of 2.98 mm inside diameter and 0.91 m heated length. The two-phase pressure drop data of the small tube of their study were consistently lower than would be expected in larger tubes at the same mass fluxes. A modification of the Chisholm correlation was developed to better predict activity in small-diameter tubes.

Flow Boiling Heat Transfer in Small Tubes

The greatest advantages of the small tubes are its high heat transfer coefficient and significant decreasing in size of compact heat exchangers. A higher heat transfer in small tubes is provided by an increased heat transfer surface area and a large ratio of heat transfer surface to fluid flow volume. As the tube diameter becomes smaller, the ratio of heat transfer surface to the fluid flow volume increases in inverse proportion to the tube diameter; hence small tube can support high heat flux with small temperature gradients, and furthermore the heat transfer coefficient increases for smaller tube diameters. The reducing size also allows heat exchangers to achieve a significant weight reduction, a lower fluid inventory, a low capital and installation cost and an energy savings.

Kew and Cornwell [31] studied boiling heat transfer of single substance R-141b in small-diameter tubes of 3.69 mm and 2.87 mm. The heat transfer coefficient increases with heat flux at low quality, while at higher qualities the heat transfer coefficient is a function of quality and is essentially independent of heat flux. These trends are less apparent in the smaller tube of 1.39 mm inner diameter, which is at high mass flux the heat transfer coefficient falls rapidly with increasing quality. The heat transfer coefficient comparison of measured and the predicted by six proposed correlation, in tube of 2.87 mm inner diameter, show that the Lazarek and Black [33] correlation gives a deviation of 19%, and the Cooper correlation gives a deviation of 21%. For the comparison with tube of 1.39 mm inner diameter, Lazarek and Black correlation gives a deviation of 69%, and Cooper correlation gives a deviation of 36%.

At two-phase flow in small tubes, the discrete bubbles, resulting from nucleation activity on the wall, are present in the subcooled boiling and low quality regions. Kandlikar [27] showed that, at the incipient of evaporation, the more rapid of bubble growing appears for the higher Re and wall temperature, and then the growing decreases by the time. Wambsganss *et al.* [50], by using R-113 in a small tubes with inner diameter of 2.92 mm, showed a significant effect of the inner tube diameter, that is, the slug flow pattern regime occurs over a much larger parameter range, and nucleate boiling mechanisms dominate the heat transfer in the small diameter tube. Peng and Peterson [41], by using a binary mixture and 0.133 mm to 0.367 mm hydraulic diameters, concluded that the transition Re (Reynolds number) and the transition range (Re$_{cr}$) decrease with decreasing small tube dimensions. Study of boiling heat transfer by using refrigerant mixtures has also been presented by Zhang *et al.* [59]. They showed that the mass transfer resistance near the interface significantly reduces the heat transfer coefficient of the mixture, especially in low quality and low mass flux regions. Jung *et al.* [25] concluded that mass transfer resistance was responsible for the heat transfer degradation from the ideal value, but for low vapor-liquid composition difference, mass transfer becomes negligible in the convective region. But these proposed studies are only valid for evaporated tube with moderate inner diameters, namely experimental data of the boiling characteristics for a small tube is limited. Hence in this paper, the effect of pressure drop and heat transfer coefficient of refrigerants boiling in horizontal small tube is disclosed; and new developed pressure drop and heat transfer coefficient correlations are proposed.

EXPERIMENTAL ASPECT

Experimental Apparatus and Method

The test photographs are presented in Figs. **3** and **4**, and the experimental facilities are schematically shown in Figs. **5** and **6**. The test facility consisted of a condenser, subcooler, receiver, refrigerant pump, mass flow meter, preheater, and test sections. For the test with the 3.0 and 1.5 mm inner diameter tubes, a variable A.C output motor controller was used to control the flow rate of the refrigerant. And for the test with the 0.5 mm inner diameter tube in an open-loop system, the flow rate was controlled with a needle valve. A Coriolis-type mass flow meter was installed in the horizontal layout for the test with the 1.5 and 3.0 mm inner diameter tubes, whereas a weighing balance was used for the test with the 0.5 mm inner diameter tube to measure the refrigerant flow rate. A preheater was installed to control the mass quality by heating the refrigerant before it entered the test section. For evaporation at the test section, a certain heat flux was conducted from a variable A.C voltage controller. The vapor refrigerant from the test section was then condensed in the condenser and subcooler, and then the condensate was supplied to the receiver.

The test section was made of stainless steel circular smooth tubes with inner tube diameters of 3.0, 1.5 and 0.5 mm. The rate of input electric potential and current were adjusted to control the input power and to determine the applied heat flux, which was measured by a standard multimeter. The test sections were uniformly and constantly heated by applying electric current directly to the tube walls, which were well insulated with foam and rubber. The outside tube wall temperatures at the top, both sides and bottom were measured at 100 mm (for the test section with inner tube diameters of 3.0 mm and 1.5 mm) and at 30 mm (for the test section with inner tube diameter of 0.5 mm) axial intervals from where heating was started by using T-type copper-constantan thermocouples at each site. The junctions of the copper-constantan thermocouples were attached to the lateral surface and were well electrically insulated. The local saturation pressure, which was used to determine the saturation temperature, was measured using bourdon tube type pressure gauges with 0.005 MPa scale at the inlet and the outlet of the test sections. The differential pressure was measured by the bourdon tube type pressure gauges and a differential pressure transducer. Sight glasses with the same inner tube diameter as the test section were installed to visualize the flow.

The experimental conditions used in this study were listed in Table **1**. The temperature and flow rate measurements were recorded using a DAQ logger software program and a Micro Motion ProLink Software package, respectively. The physical properties of the refrigerants were obtained from REFPROP 8.0. The experimental uncertainty associated with all the parameters is tabulated in Table **2**. The uncertainties were obtained using both random and systematic errors, and these values changed according to the flow conditions, so their minimum to maximum ranges were shown.

Figure 2: Test photograph for test sections with inner tube diameters of $D_i = 3.0$ mm and $D_i = 1.5$ mm.

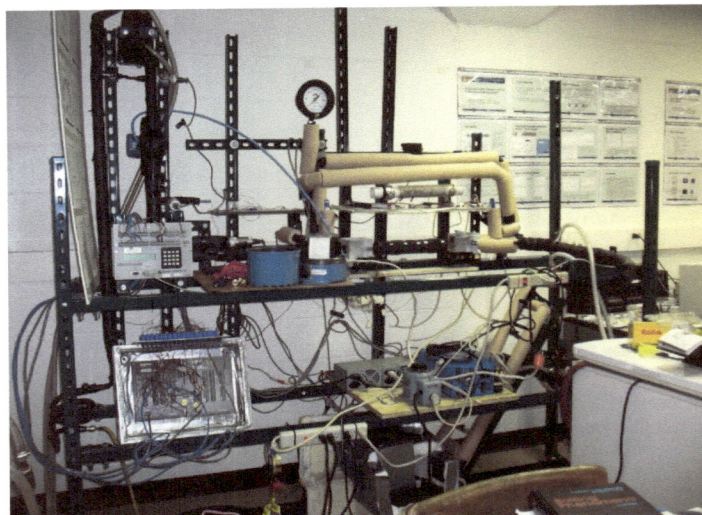

Figure 3: Test photograph for test section with inner tube diameter of $D_i = 0.5$ mm.

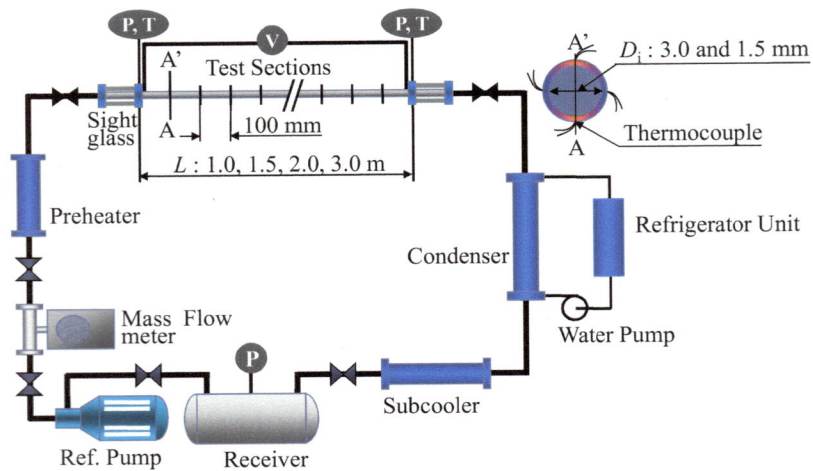

Figure 4: Experimental test facility for test section with inner tube diameters of $D_i = 3.0$ mm and $D_i = 1.5$ mm.

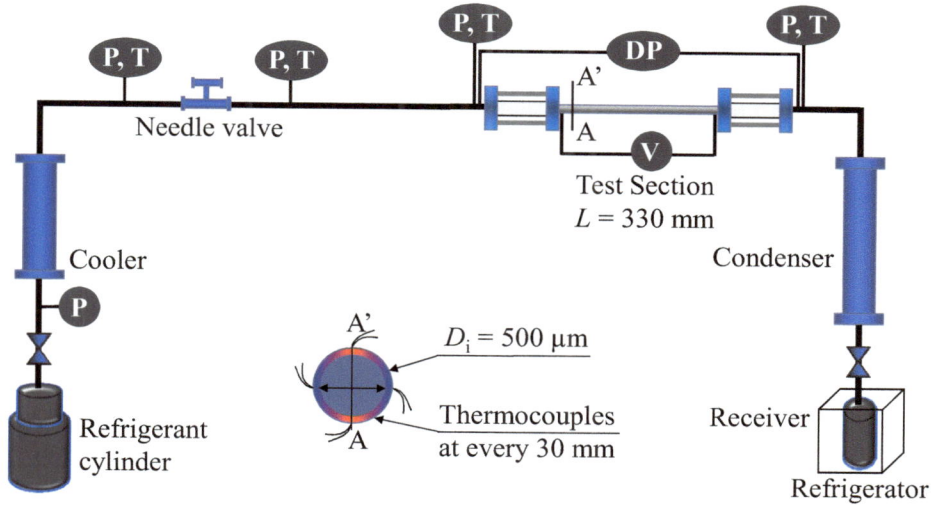

Figure 5: Experimental test facility for test section with inner tube diameter of $D_i = 0.5$ mm.

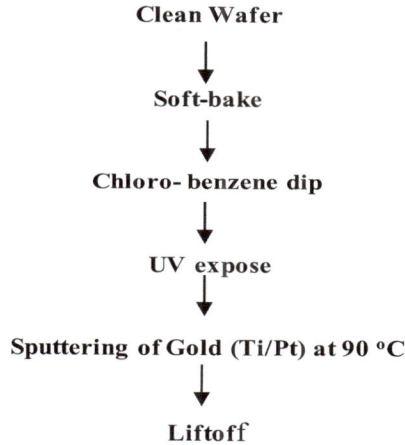

Figure 6: Void fraction comparison.

DATA REDUCTION

Pressure Drop

The saturation pressure at the initial point of saturation is determined by interpolating the measured pressure and the subcooled length; the subcooled length calculation is shown in Eq. (9). The experimental two-phase frictional pressure drop of flow boiling in a horizontal layout test section can be obtained by subtracting the calculated accelerational pressure drop from the measured pressure drop:

$$\left(\frac{dp}{dz}F\right) = \left(\frac{dp}{dz}\right) - \left(\frac{dp}{dz}a\right) = \left(\frac{dp}{dz}\right) - G^2 \frac{d}{dz}\left(\frac{x^2}{\alpha\rho_g} + \frac{(1-x)^2}{(1-\alpha)\rho_f}\right) \tag{1}$$

The results from some published void fraction prediction methods of Steiner [47], Chisholm [11], Premoli *et al.* [44], Lockhart-Martinelli [34], Baroczy *et al.* [3], Zivi *et al.* [63], Smith *et al.* [46] and Ali *et al.* [1] are compared with the current experimental data to reduce the deviation of the calculated pressure drop. Fig. **6** shows the comparison of selected predicted void fraction with those of Steiner [47], Chisholm [11] and Premoli *et al.* [44]. The absolute deviations from the homogenous void fractions for the Steiner [47], Chisholm [11] and Premoli *et al.* [44] void fractions are 6.83%, 11.17% and 4.98%, respectively. The predicted pressure drop based on the present experimental data using previous pressure drop correlations with the Steiner [47] void fraction was the best prediction among those obtained from the presented void fraction prediction methods. Therefore, the void fraction in the present paper is obtained from the Steiner [47] void fraction:

$$\alpha = \frac{x}{\rho_g}\left[\left(1+0.12(1-x)\right)\left(\frac{x}{\rho_g}+\frac{(1-x)}{\rho_f}\right)+\left(\frac{1.18}{G^2}\right)\left(\frac{(1-x)}{\rho_f^{0.5}}\right)\left(g\sigma\left(\rho_f+\rho_g\right)\right)^{0.25}\right]^{-1} \qquad (2)$$

Table 1: Experimental conditions

Test section	Horizontal circular smooth small tubes		
Quality	up to 1.0		
Working refrigerant	R-22		
Inner diameter (mm)	3.0	1.5	
Tube length (mm)	2000	2000	
Mass flux (kg/(m²·s))	400 – 600	300 – 600	
Heat flux (kW/m²)	20 – 40	10 – 20	
Inlet T_{sat} (°C)	10	10	
Working refrigerant	R-134a		
Inner diameter (mm)	3.0	1.5	0.5
Tube length (mm)	2000	2000	330
Mass flux (kg/(m²·s))	200 – 600	200 – 400	100
Heat flux (kW/m²)	10 – 40	10	5 – 20
Inlet T_{sat} (°C)	10	10	6 – 10
Working refrigerant	R-410A		
Inner diameter (mm)	3.0	1.5	0.5
Tube length (mm)	3000	1500	330
Mass flux (kg/(m²·s))	300 – 600	300 – 600	70 – 400
Heat flux (kW/m²)	10 – 40	10 – 30	5 – 20
Inlet T_{sat} (°C)	10 – 15	10 – 15	1 – 11
Working refrigerant	C_3H_8		
Inner diameter (mm)	3.0	1.5	
Tube length (mm)	2000	2000	
Mass flux (kg/(m²·s))	50 – 240	100 – 400	
Heat flux (kW/m²)	5 – 25	5 – 20	
Inlet T_{sat} (°C)	0 – 11	0 – 12	
Working refrigerant	CO_2		
Inner diameter (mm)	3.0	1.5	
Tube length (mm)	2000	2000	
Mass flux (kg/(m²·s))	200 – 600	300 – 600	
Heat flux (kW/m²)	20 – 30	10 – 30	
Inlet T_{sat} (°C)	1 – 10	0 – 11	

Table 2: Summary of the estimated uncertainty

Parameters	Uncertainty (%)				
	R-22	R-134a	R-410A	C_3H_8	CO_2
T_{wi}	±0.44 to ±3.90	±0.29 to ±3.92	±0.23 to ±6.53	±0.18 to ±5.58	±2.10 to ±4.56
P	±2.5 kPa	±2.5 kPa	±2.5 kPa	±2.5 kPa	±2.5 kPa
G	±1.84 to ±3.16	±1.85 to ±3.80	±1.84 to ±9.48	±3.24 to ±9.78	±1.85 to ±9.48
q	±1.78 to ±2.26	±1.79 to ±2.59	±1.67 to ±3.20	±2.07 to ±3.58	±1.67 to ±2.70
x	±1.88 to ±3.39	±2.23 to ±4.19	±1.78 to ±9.85	±4.27 to ±9.82	±1.79 to ±9.71
h	±2.74 to ±9.07	±2.75 to ±9.24	±2.59 to ±10.33	±1.78 to ±9.89	±4.46 to ±8.23

Heat Transfer Coefficient

The local heat transfer coefficients at position z along the length of the test section were defined as follow:

$$h = \frac{q}{T_{wi} - T_{sat}}$$

(3)

The inside tube wall temperature, T_{wi} was the average temperature of the top, both right and left sides, and the bottom wall temperatures, and was determined based on the steady-state one-dimensional radial conduction heat transfer through the wall with internal heat generation. The rate of input electric power Q at which energy is generated by passing a current I through a medium of electrical resistance R is calculated as follow:

$$Q = I^2 R$$

(4)

Power generation is then obtained from the rate of input electric power divided by volume of the current-carrying medium.

$$\dot{q} = \frac{Q}{V} = \frac{Q}{\frac{1}{4}\pi L \left(D_o^2 - D_i^2\right)}$$

(5)

The heat flux that was used in Eq. (3) is then derived from Eq. (6).

$$q = \frac{\dot{q}\left(D_o^2 - D_i^2\right)}{4D_i}$$

(6)

The vapor quality, x, at the measurement locations, z, were determined based on the thermodynamic properties.

$$x = \frac{i - i_f}{i_{fg}}$$

(7)

The outlet mass quality was then determined using the following equation:

$$x_o = \frac{\Delta i + i_{fi} - i_f}{i_{fg}}$$

(8)

The refrigerant flow at the inlet of the test section was not completely saturated. Even though it was almost completely saturated, it was necessary to determine the subcooled length to ensure reduction data accuracy. The subcooled length was calculated using the following equation to determine the initial point of saturation.

$$z_{sc} = L\frac{i_f - i_{fi}}{\Delta i} = L\frac{i_f - i_{fi}}{(Q/W)} \tag{9}$$

EXPERIMENTAL RESULTS AND DISCUSSION

The flow with heat addition, or diabatic flow, is a coupled thermodynamic problem. On the one hand, heat transfer leads to a phase change, which leads to the change of phase distribution and flow pattern; on the other hand, it causes a change in the hydrodynamics, such as the pressure drop along the flow path, which affects the heat transfer characteristics. Furthermore, a single component, two-phase flow in a tube can hardly become fully developed at low pressure because of the shape change in large bubbles and the inherent pressure change along the tube, which continually change the state of the fluid and thereby change the phase distribution and flow pattern. The current study presents the characteristics of a two-phase flow pattern and pressure drop in small tubes for some refrigerants.

Flow Pattern

The hydrodynamic characteristic of two-phase flows, such as the pressure drop, void fraction, or velocity distribution, varies systematically with the observed flow pattern, just as in the case of a single-phase flow, whose behavior depends on whether the flow is in the laminar or turbulent regime. However, in contrast to a single-phase flow, liquid-vapor flows are difficult to describe by general principles, which could serve as a framework for solving practical work. The identification of a flow regime provides a picture of the phase boundaries.

The present experimental results were mapped on Wang *et al.* [51] and Wojtan *et al.* [52] flow pattern maps, which were developed for diabatic two-phase flows. The Wang *et al.* [51] flow pattern map is a modified Baker [2] map, developed using R-22, R-134a, and R-407C inside a 6.5 mm horizontal smooth tube. The Wang *et al.* [51] study showed that the flow transition for a mixture refrigerant showed a considerable delay compared with that of pure refrigerants. For R-410A, at the initial stage of evaporation, R-32 evaporated faster than R-125. Therefore, R-32 increased the concentration of the vapor phase, and R-125 increased the concentration of the liquid phase throughout the evaporation process at the liquid–vapor interface. This resulted in a higher mean vapor velocity and a lower mean liquid velocity during evaporation. For the other working refrigerants, the physical properties of the refrigerant such as density, viscosity and surface tension have a strong effect on the flow pattern.

The predicted flow pattern for the selected current experimental data by the existing flow pattern maps of Wang *et al.* [51] and Wojtan *et al.* [52] can be seen in Figs. **7** to **20**, respectively. The Wang *et al.* [51] map showed a better prediction of the flow pattern of the current experimental data for the beginning of annular flow than the Wojtan *et al.* [52] map; however, the Wang *et al.* [51] map could not show the prediction for dry-out condition. The flow pattern prediction of the current test results for the all test conditions with the Wang *et al.* [51] flow pattern map was for the intermittent, stratified wavy, and annular flow. The present experimental results flow patterns with Wang *et al.* [51] map showed that the mass flux, heat flux and the inner diameter have an effect on the flow pattern. The selected flow pattern maps to show the effect of mass flux, heat flux and inner tube diameter are illustrated in Figs. **9** to **14**. The figures show that annular flow and dry-out occur at lower vapor quality for the conditions of higher heat flux and lower inner tube diameter. The figures also depicts that the stratified wavy flow appeared earlier for higher mass fluxes and its regime was longer for the low mass flux condition. The annular flow appeared earlier for higher mass fluxes.

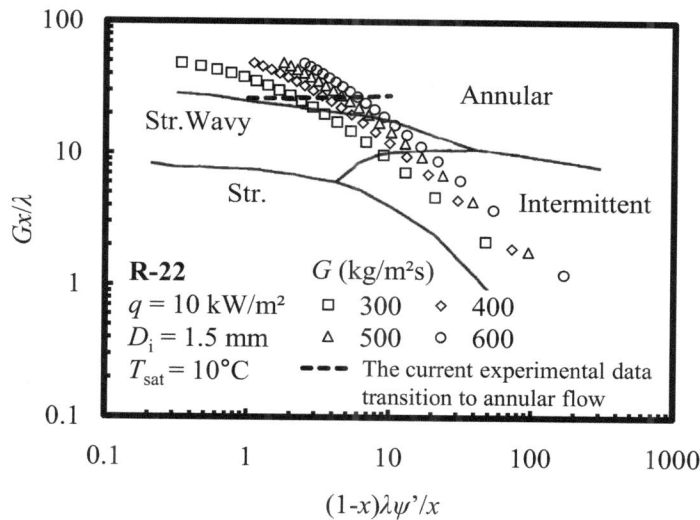

Figure 7: Present experimental results mapped on Wang *et al.* (1997) flow pattern map for R-22 at $q = 10$ kW/m^2, $D_i = 1.5$ mm, $T_{sat} = 10$ °C.

The Wojtan *et al.* [52] flow pattern map is a modified Kattan *et al.* [29] map, developed using R-22 and R-410A inside a 13.6 mm horizontal smooth tube. Kattan *et al.* [29] used five refrigerants R-134a, R-123, R-402A, R-404A, and R-502 inside 12 mm and 10.92 mm (only for R-134a) tubes, which were heated by hot water flowing counter-currently to develop their flow pattern maps on the basis of the Steiner [47] flow pattern map. The flow pattern prediction of the experimental results for all the test conditions with the Wojtan *et al.* [52] flow pattern map was on intermittent, annular, dry-out, and mist flows. Because the Wojtan *et al.* [52] map is developed using a conventional tube, the flow pattern transition of the present experimental data showed a delay on this map. However, as shown in Figs. **7** to **20**, the Wojtan *et al.* (2005) map could predict the dry-out condition. The Wojtan *et al.* [52] map predict the dry-out condition of the R-410A experimental data better than those of the other working refrigerants. Overall, Wang *et al.* [51] flow pattern map provides a better flow pattern prediction for the current experimental results than the Wojtan *et al.* [52] flow pattern map.

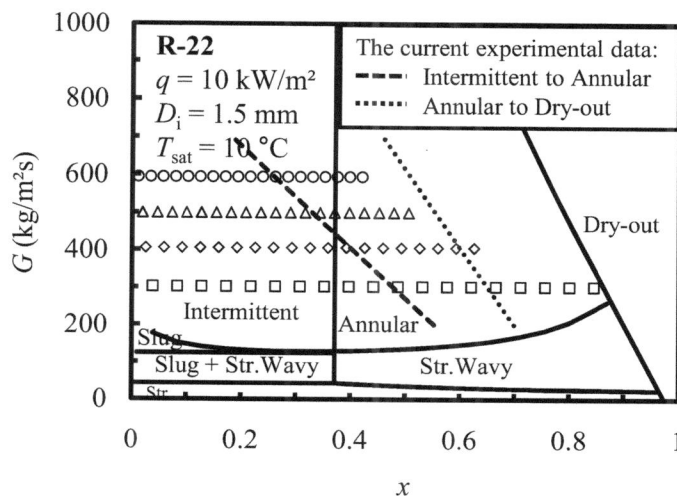

Figure 8: Present experimental results mapped on Wojtan *et al.* [52] flow pattern map for R-22 at $q = 10$ kW/m^2, $D_i = 1.5$ mm, $T_{sat} = 10$ °C.

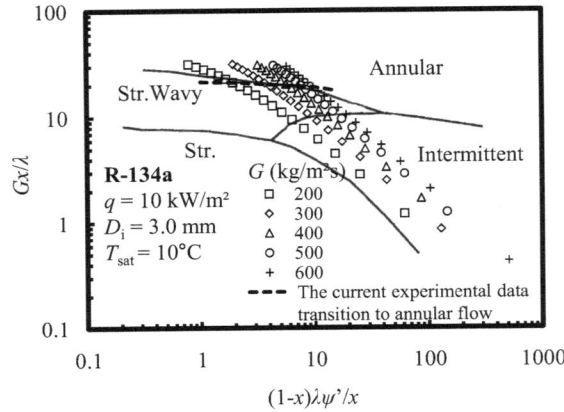

Figure 9: Present experimental results mapped on Wang *et al.* [51] flow pattern map for R-134a at $q = 10$ kW/m^2, $D_i = 3.0$ mm, $T_{sat} = 10$ °C.

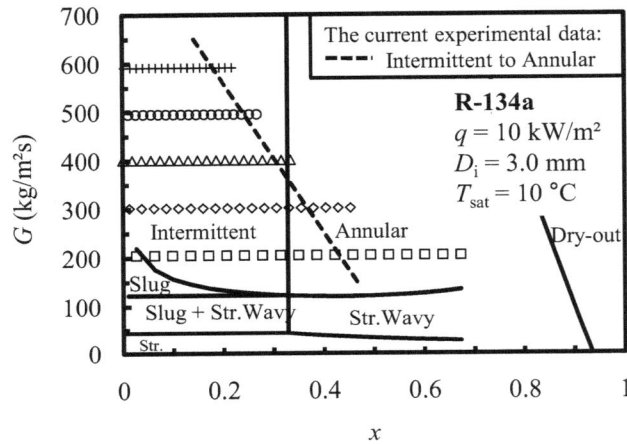

Figure 10: Present experimental results mapped on Wojtan *et al.* [52] flow pattern map for R-134a at $q = 10$ kW/m^2, $D_i = 3.0$ mm, $T_{sat} = 10$ °C.

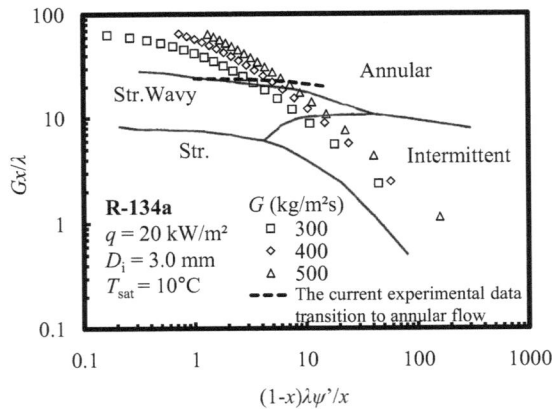

Figure 11: Present experimental results mapped on Wang *et al.* [51] flow pattern map for R-134a at $q = 20$ kW/m^2, $D_i = 3.0$ mm, $T_{sat} = 10$ °C.

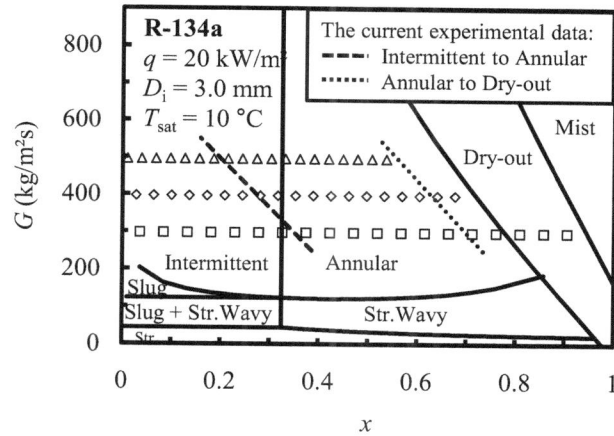

Figure 12: Present experimental results mapped on Wojtan *et al.* [52] flow pattern map for R-134a at $q = 20$ kW/m^2, D_i = 3.0 mm, $T_{sat} = 10$ °C.

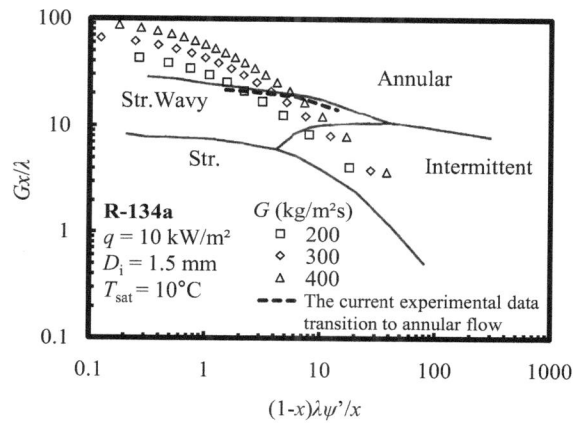

Figure 13: Present experimental results mapped on Wang *et al.* [51] flow pattern map for R-134a at $q = 10$ kW/m^2, D_i = 1.5 mm, $T_{sat} = 10$ °C.

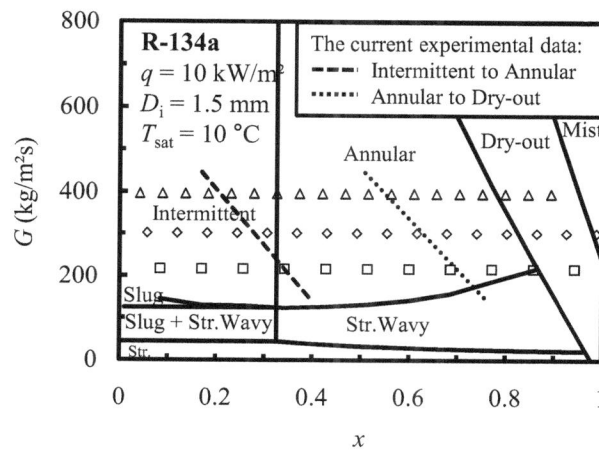

Figure 14: Present experimental results mapped on Wojtan *et al.* [52] flow pattern map for R-134a at $q = 10$ kW/m^2, D_i = 1.5 mm, $T_{sat} = 10$ °C.

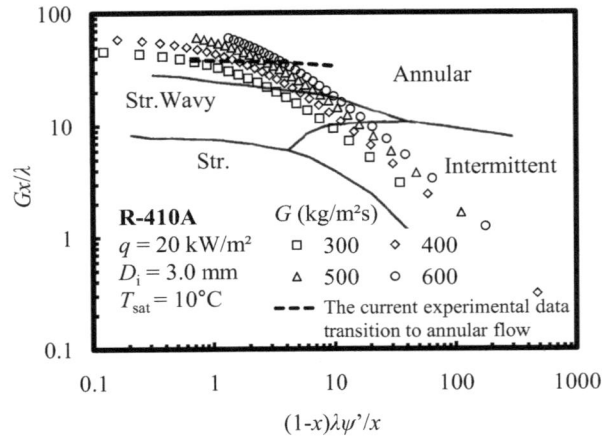

Figure 15: Present experimental results mapped on Wang *et al.* [51] flow pattern map for R-410A at $q = 20$ kW/m^2, D_i = 3.0 mm, $T_{sat} = 10$ °C.

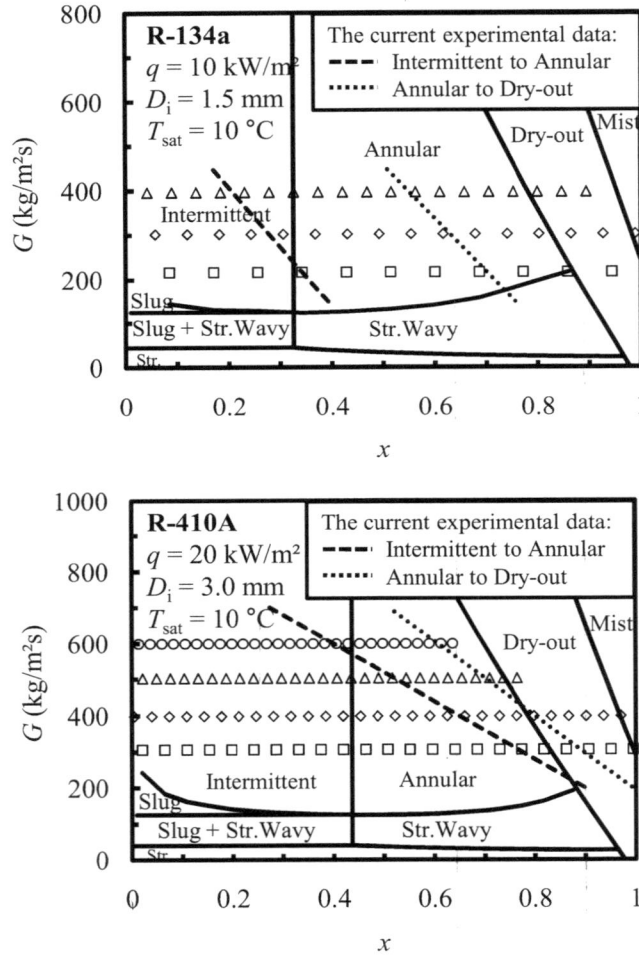

Figure 16: Present experimental results mapped on Wojtan *et al.* [52] flow pattern map for R-410A at $q = 20$ kW/m^2, D_i = 3.0 mm, $T_{sat} = 10$ °C.

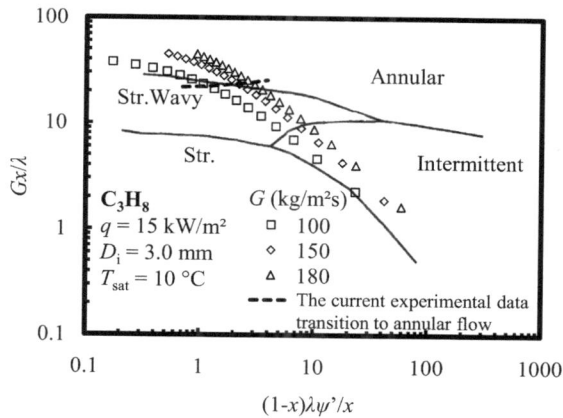

Figure 17: Present experimental results mapped on Wang *et al.* [51] flow pattern map for C_3H_8 at $q = 15$ kW/m², $D_i = 3.0$ mm, $T_{sat} = 10$ °C.

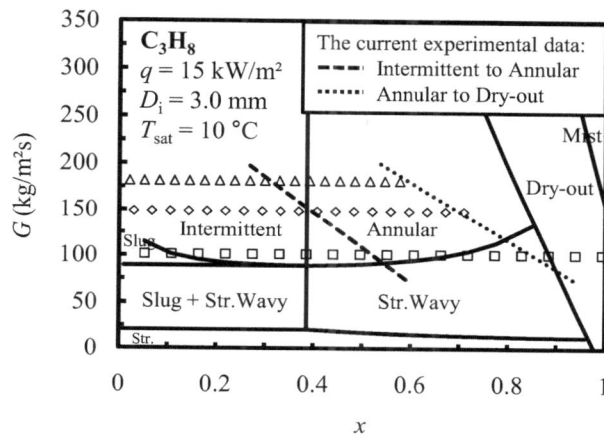

Figure 18: Present experimental results mapped on Wojtan *et al.* [52] flow pattern map for C_3H_8 at $q = 15$ kW/m², $D_i = 3.0$ mm, $T_{sat} = 10$ °C.

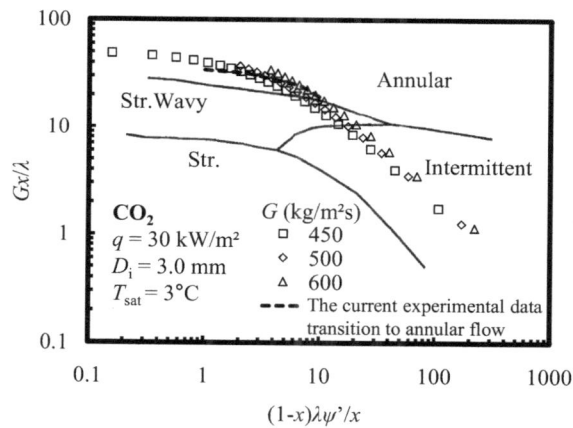

Figure 19: Present experimental results mapped on Wang *et al.* [51] flow pattern map for CO_2 at $q = 30$ kW/m², $D_i = 3.0$ mm, $T_{sat} = 3$ °C.

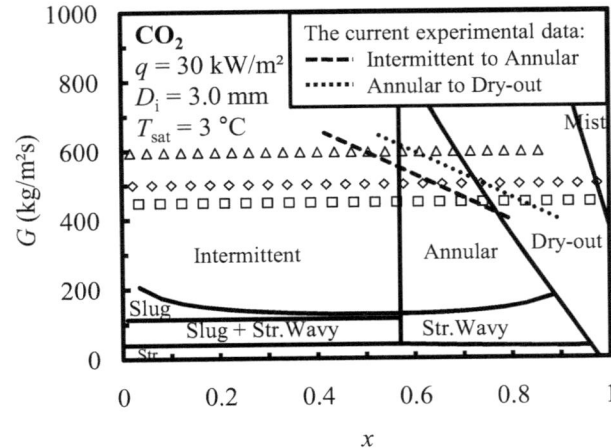

Figure 20: Present experimental results mapped on Wojtan *et al.* [52] flow pattern map for CO_2 at $q = 30$ kW/m^2, $D_i = 3.0$ mm, $T_{sat} = 3$ °C.

Effect on Pressure Drop

Figs. **21** to **25** show that mass flux has a strong effect on the pressure drop. An increase in the mass flux results in a higher flow velocity, which increases the frictional and accelerational pressure drops. The potential for turbulent flow and (stratified) wavy flow is also higher in the flow of a higher mass flux; this condition results in a higher pressure drop. This similar trend is shown by some previous studies such as Zhao *et al.* [62], Yoon *et al.* [56], Park and Hrnjak [40], Oh *et al.* [38] and Cho and Kim [15].

Figs. **21** to **25** also illustrate that the pressure drop increases as the heat flux increases. It is presumed that the increasing heat flux results in more vaporization, which increases the average fluid vapor quality and flow velocity. In diabatic two-phase flow, the increasing heat transfer to the flow path increases bubble and/or slug formations in the boiling working fluid, which then lead to a higher pressure drop. The addition of heat in the case of single-component flow causes a phase change along the channel; consequently, the amount of vapor void increases and the phase (also velocity) distribution as well as the momentum of the flow vary, accordingly. The trend of the current experimental results is similar to that shown by Zhao *et al.* [62].

The effect of the inner tube diameter on the pressure drop is also illustrated in Figs. **21** to **25**. The pressure drop in the smaller diameter tubes is higher than that in the bigger ones. This is due to the fact that the smaller inner tube diameter results in a higher wall shear stress, wherein for a given temperature condition, it results in a higher friction factor, which then results in higher frictional pressure drops. A smaller tube has less potential for turbulent flow, thus, a less pressure drop. However, the current experimental results show that the effect of wall shear stress on the pressure drop is greater than that on the laminar-turbulent flow.

Figs. **26** to **28** depict the effect of saturation temperature on pressure drop, where a lower saturation temperature results in a higher pressure drop. This result can be explained by the effect of the physical properties of density and viscosity on the pressure drop at different temperatures. The selected physical properties for the present working refrigerants are shown in Table **3** The liquid density, ρ_f, and liquid viscosity, μ_f, increase as the temperature decreases, whereas the vapor density, ρ_g, and vapor viscosity, μ_g, decrease as the temperature decreases. As the saturation temperature decreases for a constant mass flux condition, the increasing liquid density and liquid viscosity result in a lower liquid velocity, whereas the decreasing vapor density and vapor viscosity result in a higher vapor velocity. It is clear that the pressure drop increases during evaporation, and this increase rate of the pressure drop becomes higher as the saturation temperature decreases.

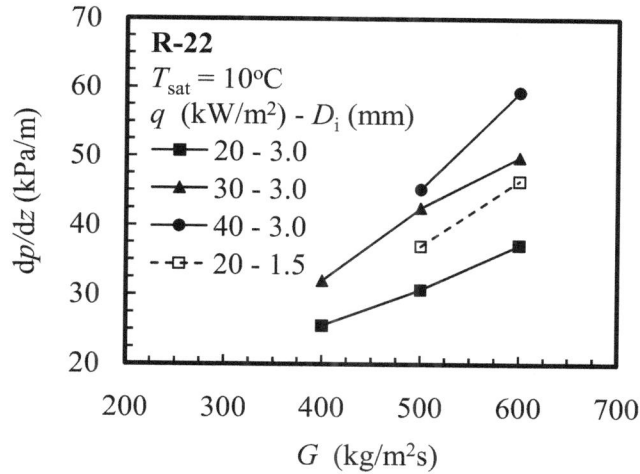

Figure 21: Effect of mass flux, heat flux and inner tube diameter on pressure drop for R-22 at T_{sat} = 10 °C.

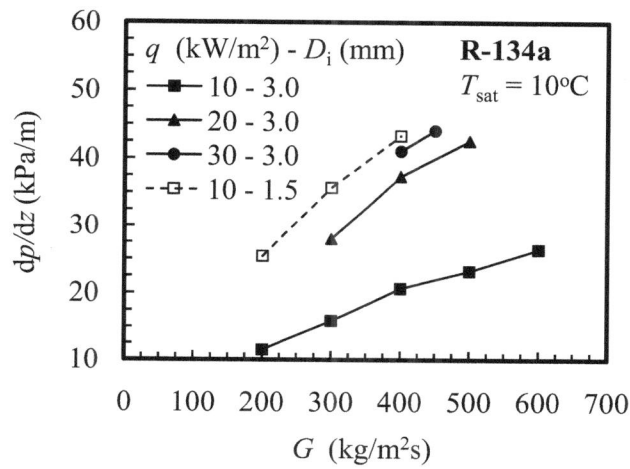

Figure 22: Effect of mass flux, heat flux and inner tube diameter on pressure drop for R-134a at T_{sat} = 10 °C.

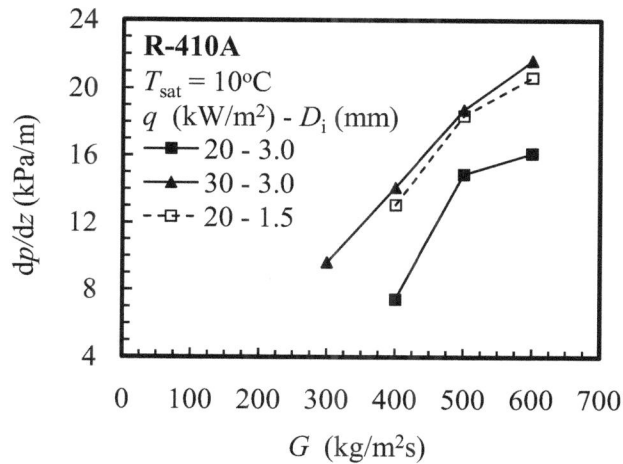

Figure 23: Effect of mass flux, heat flux and inner tube diameter on pressure drop for R-410A at T_{sat} = 10 °C.

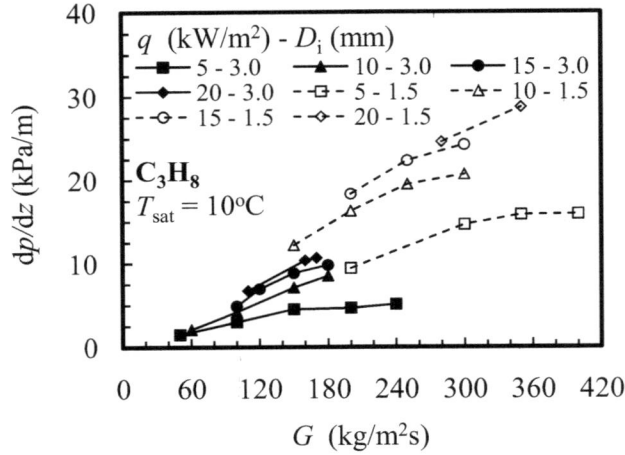

Figure 24: Effect of mass flux, heat flux and inner tube diameter on pressure drop for C_3H_8 at $T_{sat} = 10$ °C.

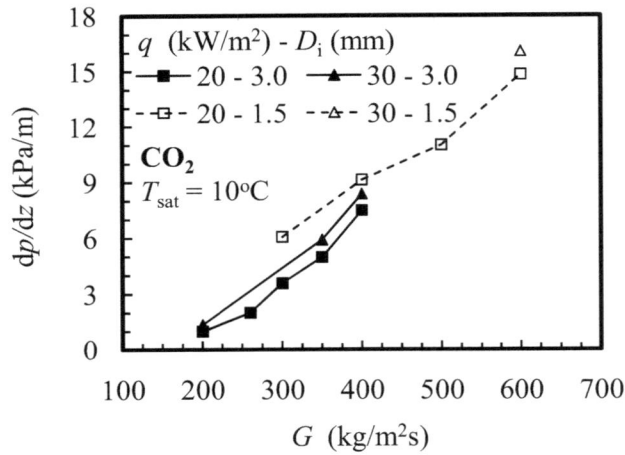

Figure 25: Effect of mass flux, heat flux and inner tube diameter on pressure drop for CO_2 at $T_{sat} = 10$ °C.

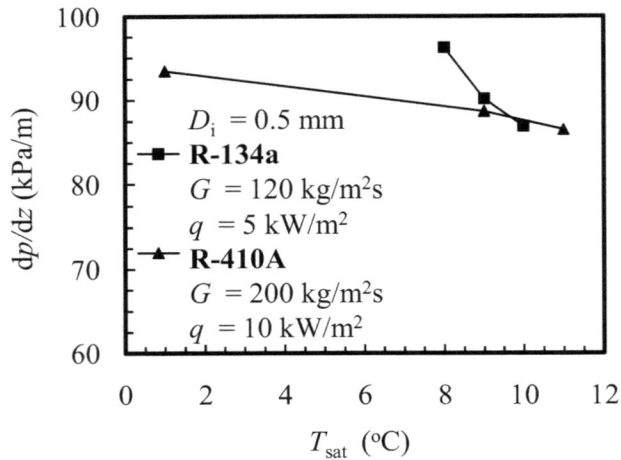

Figure 26: Effect of saturation temperature on pressure drop for R-134a (at $G = 120$ kg/(m^2·s), $q = 5$ kW/m^2, $D_i = 0.5$ mm) and R-410A (at $G = 200$ kg/(m^2·s), $q = 10$ kW/m^2, $D_i = 0.5$ mm).

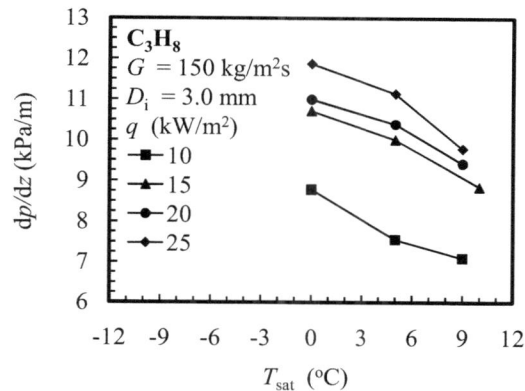

Figure 27: Effect of saturation temperature on pressure drop for C_3H_8 at $G = 150$ kg//(m^2·s), $D_i = 3.0$ mm.

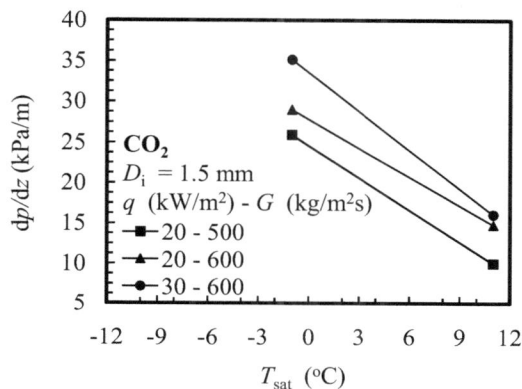

Figure 28: Effect of saturation temperature on pressure drop for CO_2 at $D_i = 1.5$ mm.

Comparison of Pressure Drop

Comparison of the Current Experimental Pressure Drop

The pressure drop of the present working refrigerants is compared at several same experimental conditions. The pressure drop comparisons are illustrated in Figs. **29** to **33**. The order of the pressure drop from the highest to the lowest is R-134a, R-22, C_3H_8, R-410A, and CO_2. The pressure drop is strongly affected by the physical properties of the working fluid such as density, viscosity, surface tension and pressure. As shown in Table **3**, the working refrigerant with a higher pressure drop has a higher density ratio ρ_f/ρ_g, viscosity ratio μ_f/μ_g, surface tension, and a lower pressure, generally. Therefore, it is clear that in the present comparison CO_2 has the lowest pressure drop, making CO_2 an effective future environmental friendly refrigerant.

Pressure Drop Comparison with Some Existing Correlations

The current experimental frictional two-phase pressure drop data were compared with the predictions from eleven existing correlations, as is shown in Figs. **34** to **44**. The pressure drop models of Friedel [20] provided the best prediction among the other methods, yielding a mean deviation of around 33%. Friedel's [20] correlation was developed using a large database; it was valid for horizontal flow and vertical upward flow. The prediction by Cicchitti *et al.* homogeneous model [16] and Chisholm [13] methods yielded a mean deviation of lower than 40%. The homogenous model assumes that vapor and liquid velocities are equal. The remaining predictions of Zhang-Webb [60], Beattie-Whalley homogeneous model [5], McAdams homogeneous model [35], Chang-Ro [6], Dukler *et al.* homogeneous model [19], Tran *et al.* [48], Mishima and Hibiki [36], and Lockhart-Martinelli [34] provided a large mean deviation of more than 40%; they failed to predict the present experimental data.

Table 3: Physical properties of R-22, R-134a, R-410A, C_3H_8 and CO_2 at 10 and 5 °C

T (°C)	Refrigerant	P (MPa)	ρ_f (kg/m³)	ρ_g (kg/m³)	ρ_f/ρ_g	μ_f (10^{-6} Pa s)	μ_g (10^{-6} Pa s)	μ_f/μ_g	σ (10^{-3} N/m)	h_{fg} (kJ/kg)
10	R-22	0.681	1247	28.82	43.27	195.7	11.96	16.36	10.22	196.7
	R-134a	0.415	1261	20.23	62.33	238.8	11.15	21.42	10.14	190.7
	R-410A	1.085	12.91	146.6	27.07	41.74	1130	11.36	7.16	208.6
	C_3H_8	0.636	515	13.8	37.32	113.8	8.15	13.96	8.85	360.3
	CO_2	4.497	861.7	134.4	6.41	86.37	15.46	5.59	2.77	197.2
5	R-22	0.584	1264	24.79	50.99	206.7	11.73	17.62	10.95	192.9
	R-134a	0.350	1278	17.13	74.61	254.4	10.94	23.25	10.84	194.7
	R-410A	0.934	1151	35.73	32.21	156.0	12.60	12.38	7.90	215.2
	C_3H_8	0.551	522	11.98	43.57	119.8	7.97	15.03	9.48	367.7
	CO_2	3.965	896.7	114.1	7.85	95.84	14.83	6.46	3.64	215.0

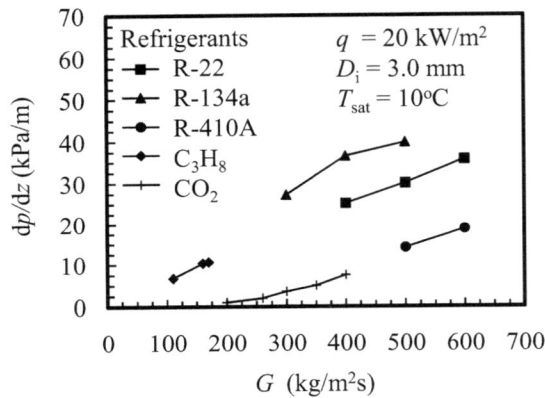

Figure 29: Pressure drop comparison of the present working refrigerants at q = 20 kW/m², D_i = 3.0 mm, T_{sat} = 10 °C.

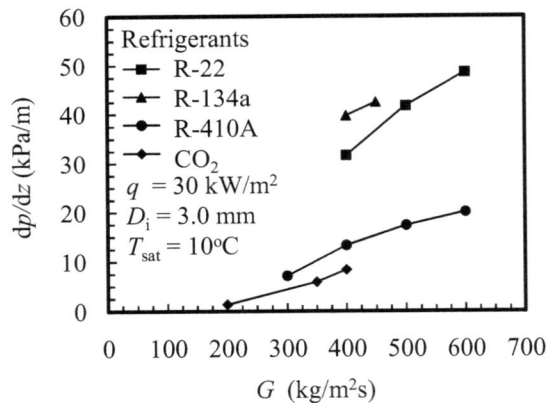

Figure 30: Pressure drop comparison of the present working refrigerants at q = 30 kW/m², D_i = 3.0 mm, T_{sat} = 10 °C.

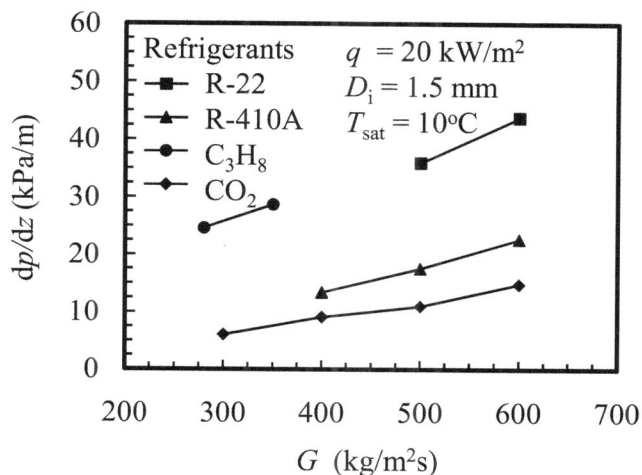

Figure 31: Pressure drop comparison of the present working refrigerants at $q = 20$ kW/m^2, $D_i = 1.5$ mm, $T_{sat} = 10$ °C.

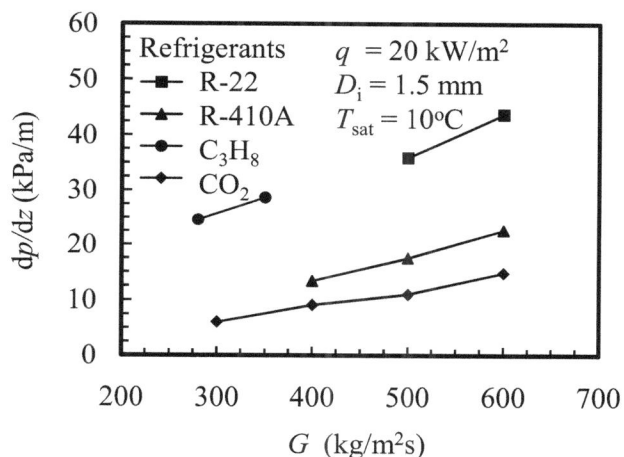

Figure 32: Pressure drop comparison of the present working refrigerants at $q = 20$ kW/m^2, $D_i = 1.5$ mm, $T_{sat} = 10$ °C.

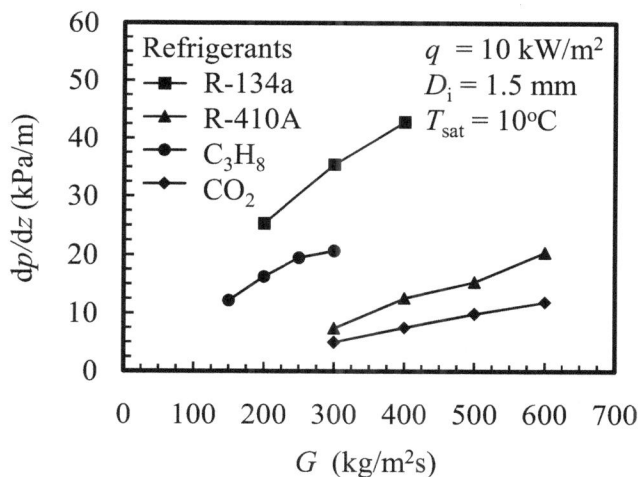

Figure 33: Pressure drop comparison of the present working refrigerants at $q = 10$ kW/m^2, $D_i = 1.5$ mm, $T_{sat} = 10$ °C.

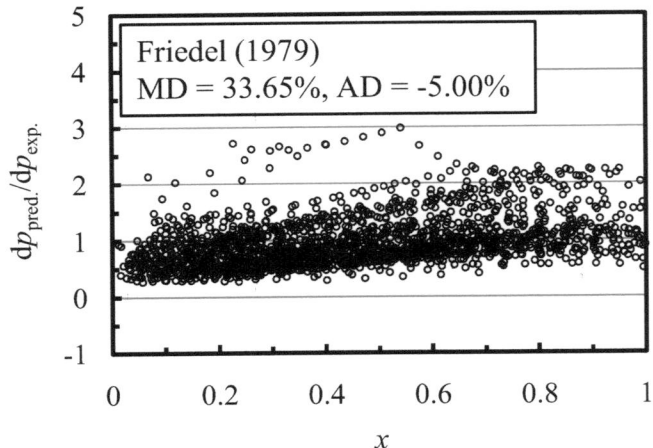

Figure 34: Comparison between the experimental and the prediction pressure drop with Friedel [20] correlation.

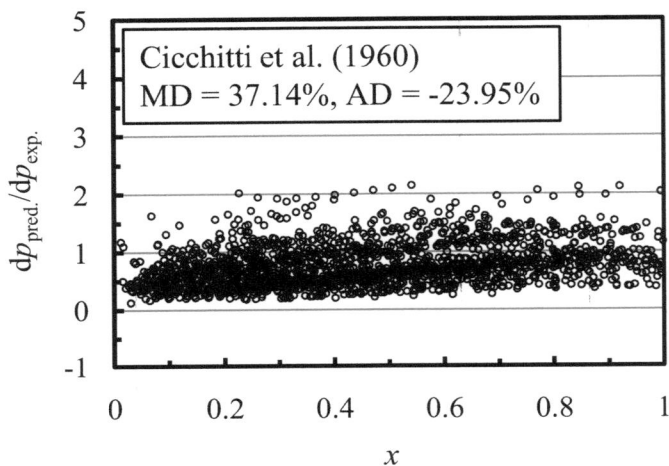

Figure 35: Comparison between the experimental and the prediction pressure drop with Cicchitti *et al.* [16] correlation.

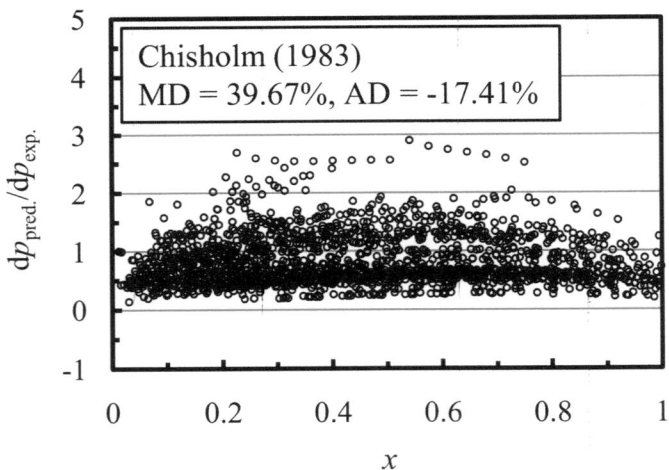

Figure 36: Comparison between the experimental and the prediction pressure drop with Chisholm [13] correlation.

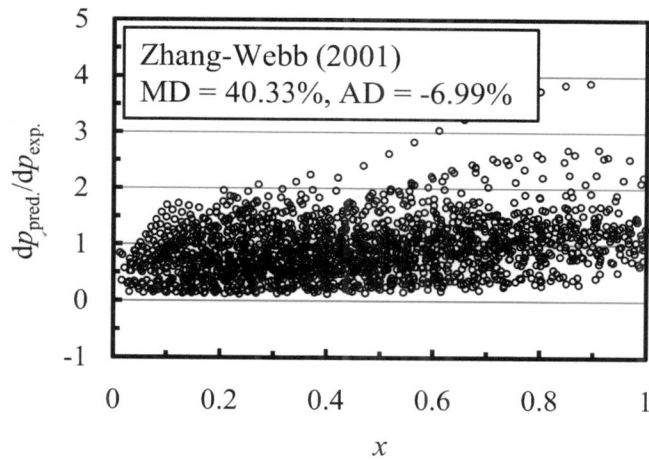

Figure 37: Comparison between the experimental and the prediction pressure drop with Zhang and Webb [60] correlation.

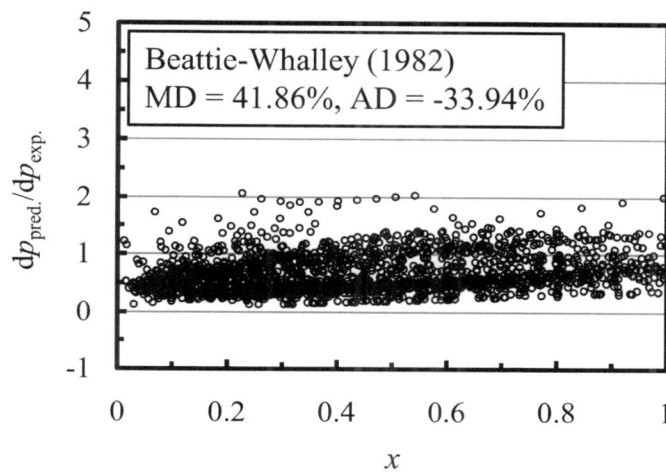

Figure 38: Comparison between the experimental and the prediction pressure drop with Beattie-Whalley [6] correlation.

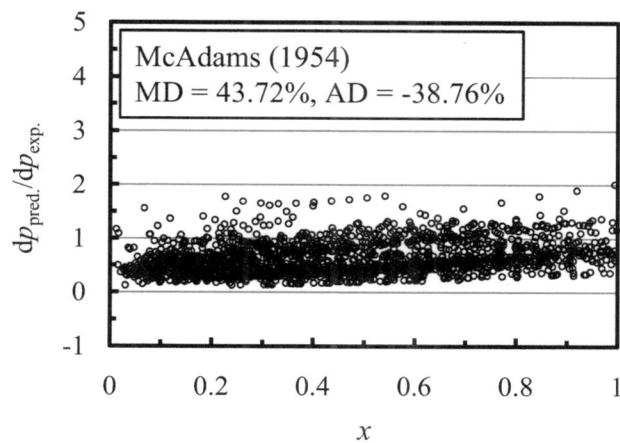

Figure 39: Comparison between the experimental and the prediction pressure drop with McAdams [35] correlation.

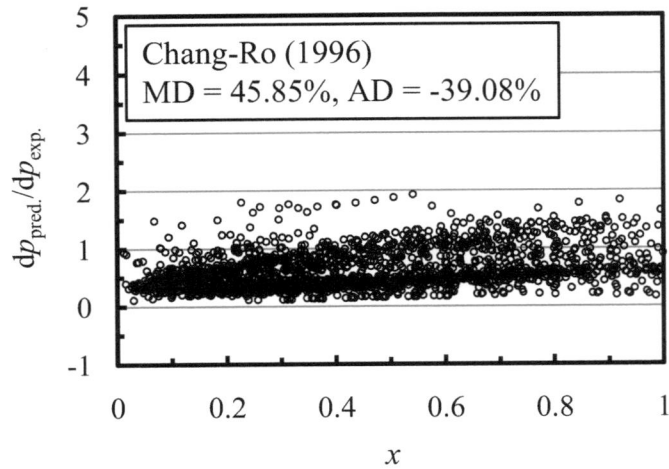

Figure 40: Comparison between the experimental and the prediction pressure drop with Chang-Ro [6] correlation.

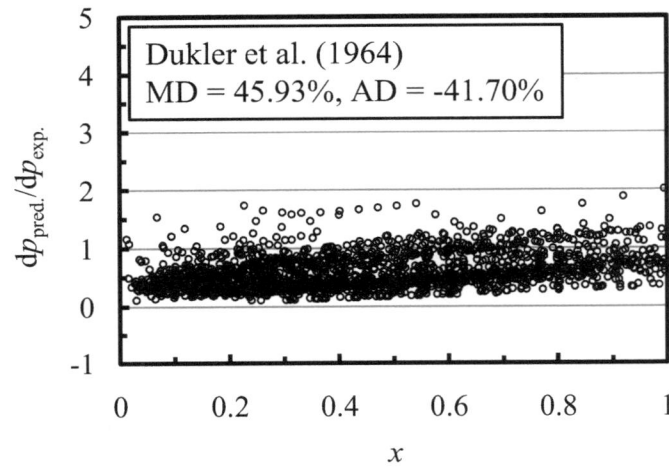

Figure 41: Comparison between the experimental and the prediction pressure drop with Dukler *et al.* [19] correlation.

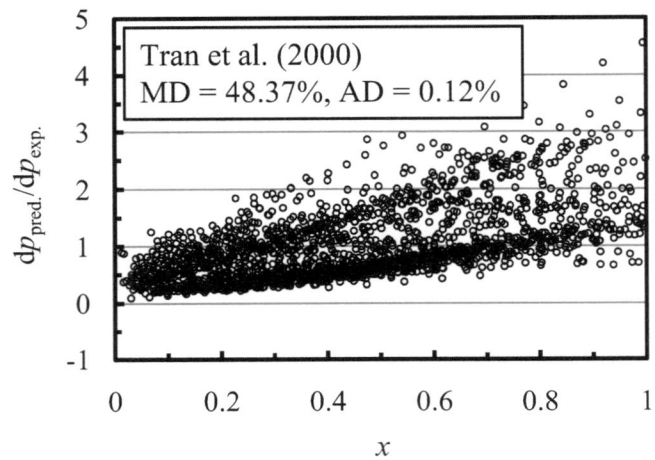

Figure 42: Comparison between the experimental and the prediction pressure drop with Tran *et al.* [48] correlation.

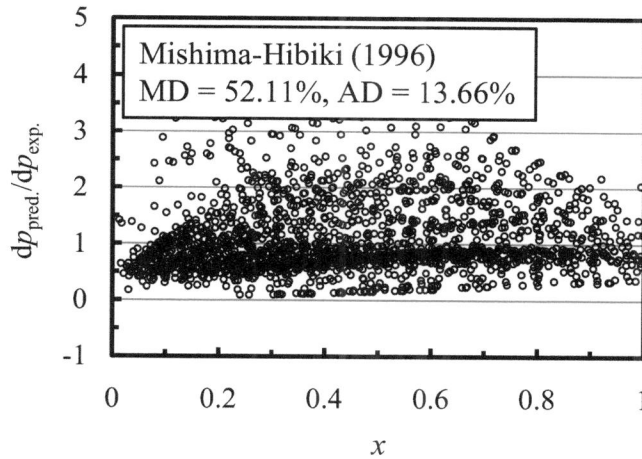

Figure 43: Comparison between the experimental and the prediction pressure drop with Mishima-Hibiki [36] correlation.

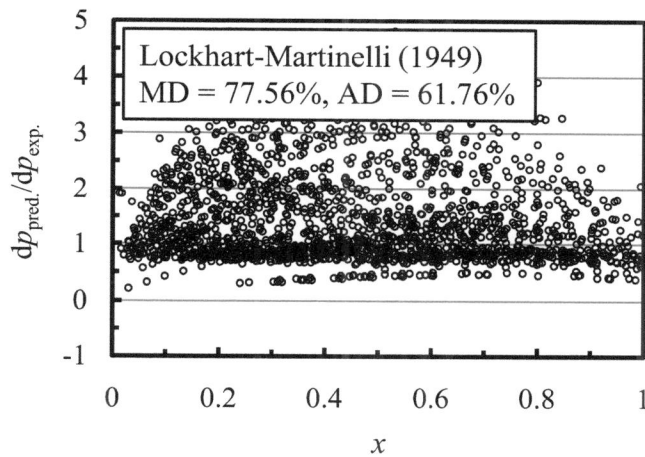

Figure 44: Comparison between the experimental and the prediction pressure drop with Lockhart-Martinelli [34] correlation.

Heat Transfer Coefficient

Effect on Heat Transfer Coefficient

Effect of Mass Flux on Heat Transfer Coefficient

Figs. **45** to **49** show the effect of mass flux on the heat transfer coefficient. Mass flux had an insignificant effect on the heat transfer coefficient in the low quality region. This indicates that nucleate boiling heat transfer was predominant. The high nucleate boiling heat transfer occurred because of the physical properties of the refrigerants, namely their surface tension and pressure, and the geometric effect of the small tubes. A higher mass flux corresponded to a higher heat transfer coefficient at the moderate-high vapor quality region, due to an increase of the convective boiling heat transfer contribution; this result is similar to the results reported by Kuo-Wang [32], who used R-22 in a 9.52 mm tube. In the high quality region, the heat transfer coefficient dropped at lower qualities for a relatively higher mass flux; this trend agrees with the results of Pettersen [43] and Yun *et al.* [58]. For the higher mass flux condition in the convective evaporation region, the increase in the heat transfer coefficient appeared at a lower quality, which can be explained by the annular flow becoming dominant. Nucleate boiling suppression appeared earlier for the higher mass flux, and this means that convective heat transfer appeared earlier under the higher mass flux condition. The lower mass flux condition results show smaller increases in the heat transfer coefficient in the convective region. The heat transfer coefficients suddenly increase in the annular flow region before the initial dry-out, and this can be explained by the fact that as quality was increased in the annular flow, the effective wall superheat decreased due to the thinner liquid film or less thermal resistance. The steep decrease of the heat transfer coefficient at high qualities was due to the effect of the small diameter on the boiling flow pattern; dry-patch occurs easier in smaller diameter tubes and at higher mass fluxes. Several previous studies such as those of Kew and Cornwell [31], Lazarek and Black [33], Wambsganss et al. [50], Tran et al. [49] and Bao et al. [4] used small tubes and showed that nucleate boiling in small tubes tends to be predominant.

Effect of Heat Flux on Heat Transfer Coefficient

Figs. **50** to **54** the dependence of heat flux on heat transfer coefficients in the low-moderate quality region. Nucleate boiling is known to be dominant in the initial stage of evaporation, particularly under high heat flux conditions. The large effect of heat flux on the heat transfer coefficient shows the dominance of nucleate boiling heat transfer contribution. At the higher quality region, nucleate boiling was suppressed, or convective heat transfer contribution was predominant; this is indicated by the low effect of heat flux on the heat transfer coefficient. As the heat flux increased at high qualities, the evaporation was more active and the dry-out quality reduced. The trend illustrated in Figs. **50** to **54** agree with previous studies, e.g. Kew and Cornwell [31], Yan and Lin [54], and Kuo-Wang [32].

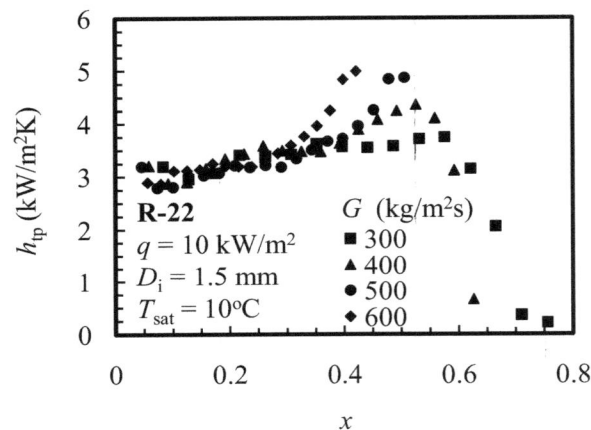

Figure 45: Effect of mass flux on heat transfer coefficient for R-22 at $q = 10$ kW/m^2, $D_i = 1.5$ mm, $T_{sat} = 10$ °C.

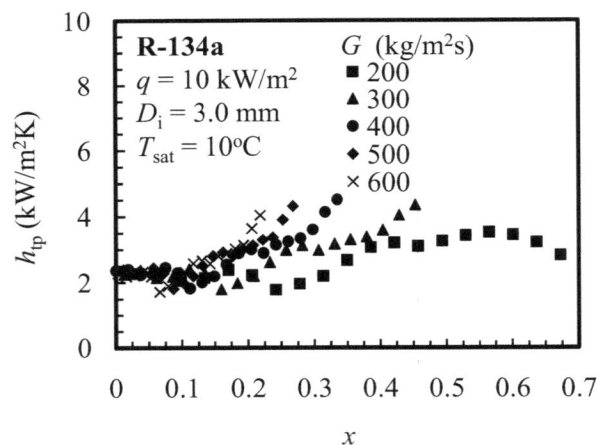

Figure 46: Effect of mass flux on heat transfer coefficient for R-134a at $q = 10$ kW/m^2, $D_i = 3.0$ mm, $T_{sat} = 10$ °C.

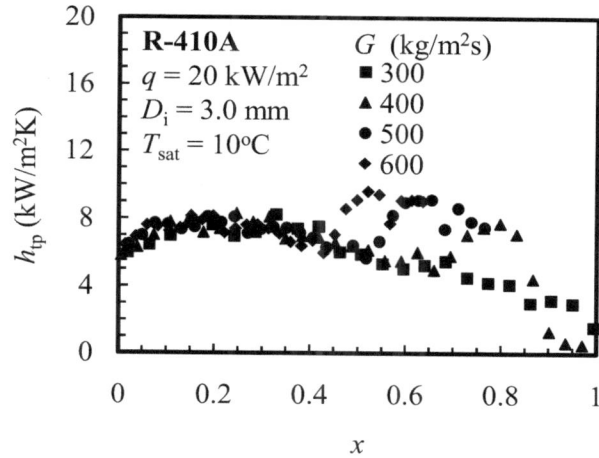

Figure 47: Effect of mass flux on heat transfer coefficient for R-410A at $q = 20$ kW/m^2, $D_i = 3.0$ mm, $T_{sat} = 10$ °C.

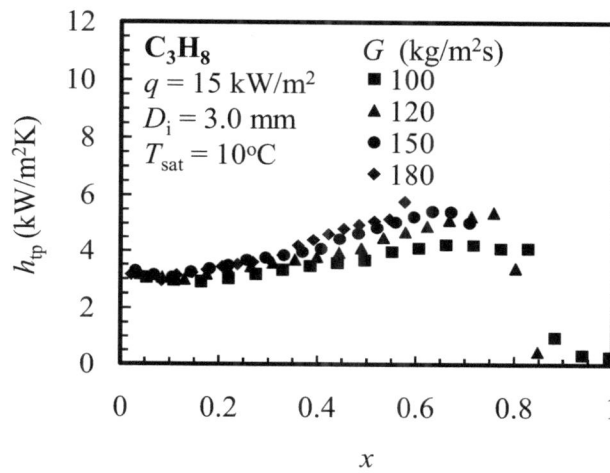

Figure 48: Effect of mass flux on heat transfer coefficient for C_3H_8 at $q = 15$ kW/m^2, $D_i = 3.0$ mm, $T_{sat} = 10$ °C.

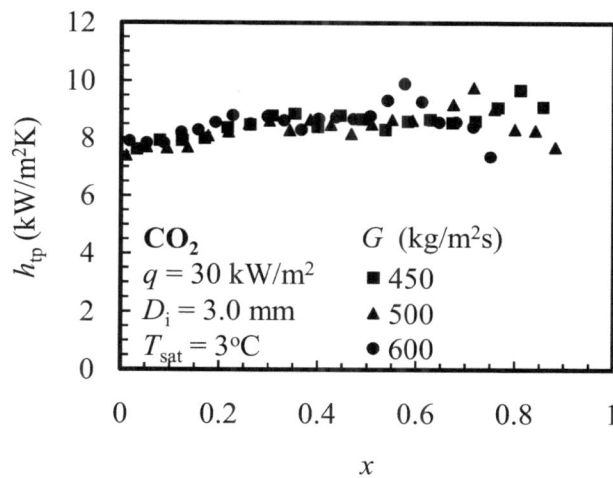

Figure 49: Effect of mass flux on heat transfer coefficient for CO_2 at $q = 30$ kW/m^2, $D_i = 3.0$ mm, $T_{sat} = 3$ °C.

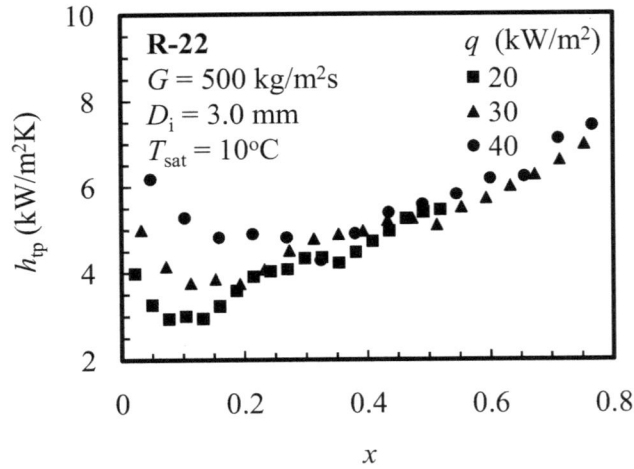

Figure 50: Effect of heat flux on heat transfer coefficient for R-22 at G = 500 kg//(m^2·s), D_i = 3.0 mm, T_{sat} = 10 °C.

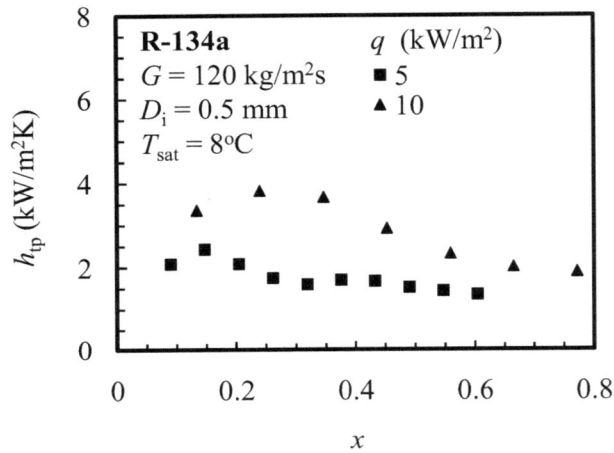

Figure 51: Effect of heat flux on heat transfer coefficient for R-134a at G = 120 kg//(m^2·s), D_i = 0.5 mm, T_{sat} = 8 °C.

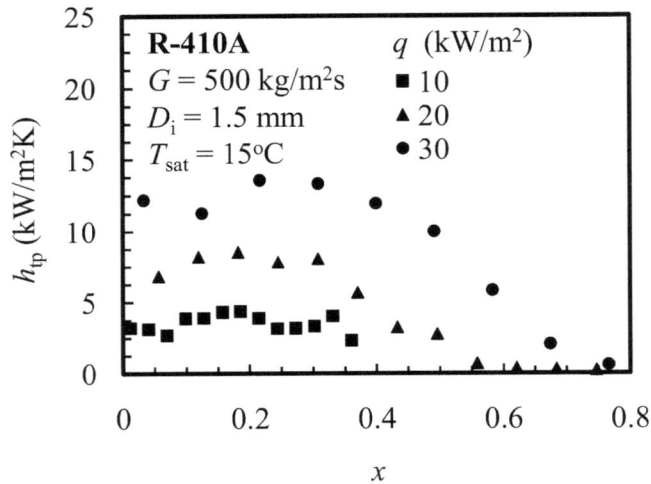

Figure 52: Effect of heat flux on heat transfer coefficient for R-410A at G = 500 kg//(m^2·s), D_i = 1.5 mm, T_{sat} = 15 °C

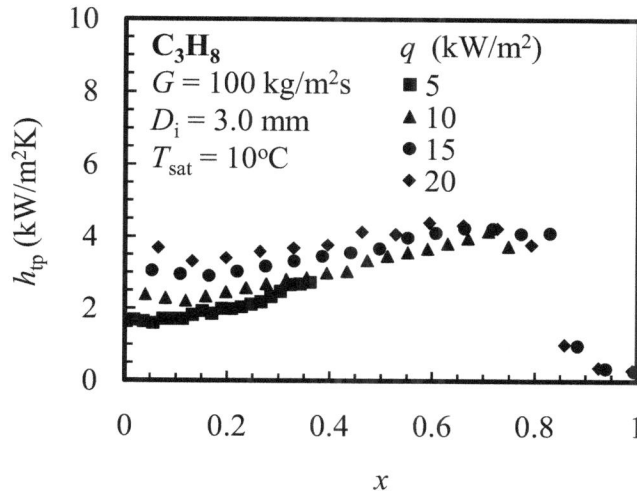

Figure 53: Effect of heat flux on heat transfer coefficient for C_3H_8 at $G = 100$ kg//(m^2·s), $D_i = 3.0$ mm, $T_{sat} = 10$ °C.

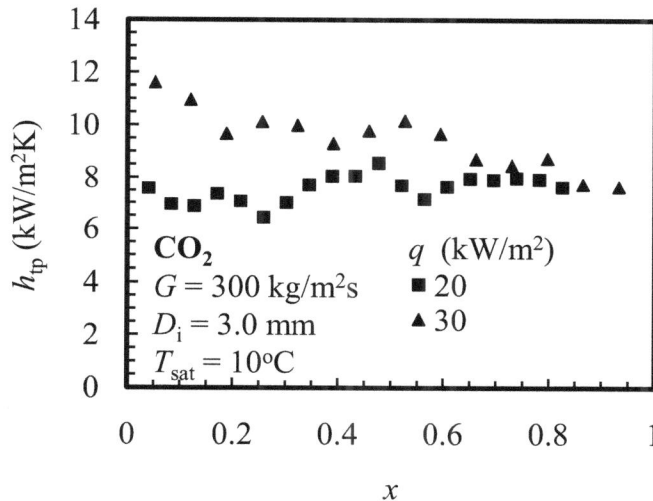

Figure 54: Effect of heat flux on heat transfer coefficient for CO_2 at $G = 300$ kg//(m^2·s), $D_i = 3.0$ mm, $T_{sat} = 10$ °C.

Effect of Saturation Temperature on Heat Transfer Coefficient

The effect of saturation temperature on heat transfer coefficient is depicted in Figs. **55** to **58**. The heat transfer coefficient increases with the increase in saturation temperature because of the larger effect of nucleate boiling. A higher saturation temperature leads to lower surface tension and higher pressure, as shown in Table **3**. The vapor formation in the boiling process indicates that a lower surface tension and higher pressure provide a higher heat transfer coefficient.

Effect of Inner Tube Diameter on Heat Transfer Coefficient

Figs. **59** to **62** show that a smaller inner tube diameter has a higher heat transfer coefficient at low quality regions. This is due to the more active nucleate boiling in a smaller diameter tube. As the tube diameter becomes smaller, the contact surface area for heat transfer increases. More active nucleate boiling causes dry-patches to appear earlier. The quality for a rapid decrease in the heat transfer coefficient can be lowered for a smaller tube. It is supposed that annular flow would appears at a lower quality in a smaller tube, and therefore, the dry-out quality would be relatively lower for a smaller tube.

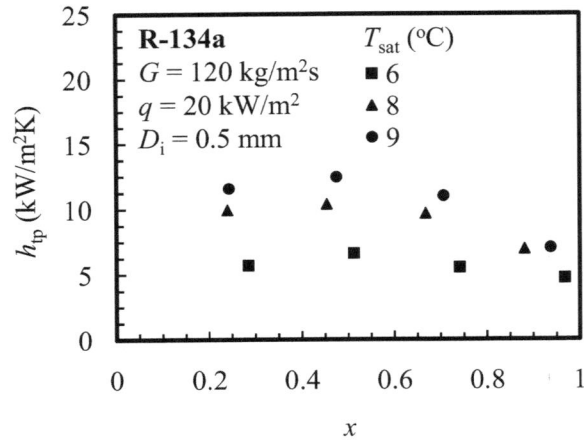

Figure 55: Effect of saturation temperature on heat transfer coefficient for R-134a at $G = 120$ kg/(m^2·s), $q = 20$ kW/m^2, $D_i = 0.5$ mm.

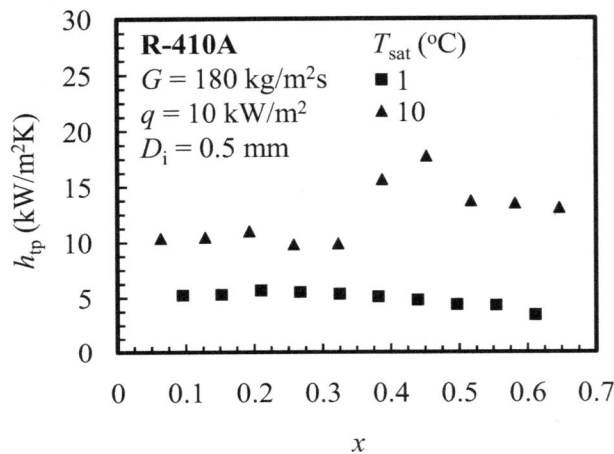

Figure 56: Effect of saturation temperature on heat transfer coefficient for R-410A at $G = 180$ kg/(m^2·s), $q = 10$ kW/m^2, $D_i = 0.5$ mm.

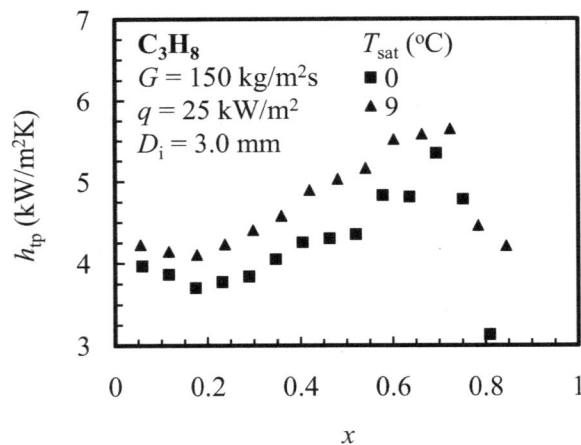

Figure 57: Effect of saturation temperature on heat transfer coefficient for C$_3$H$_8$ at $G = 150$ kg/(m^2·s), $q = 25$ kW/m^2, $D_i = 3.0$ mm.

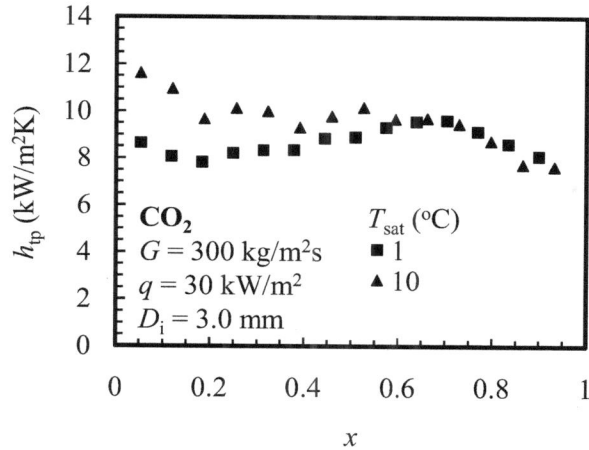

Figure 58: Effect of saturation temperature on heat transfer coefficient for CO_2 at $G = 300$ kg//(m²·s), $q = 30$ kW/m², $D_i = 3.0$ mm.

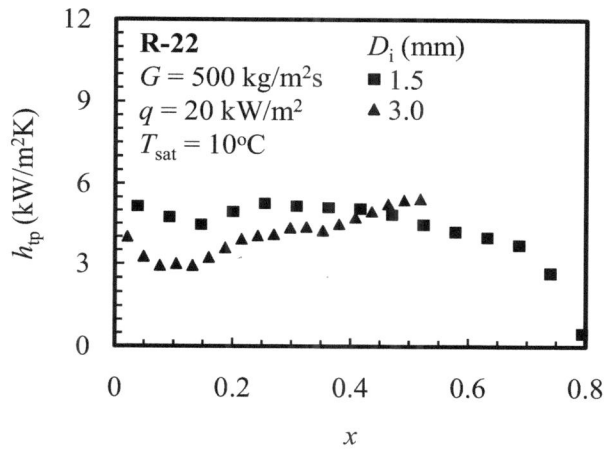

Figure 59: Effect of inner tube diameter on heat transfer coefficient for R-22 at $G = 500$ kg/(m²·s), $q = 20$ kW/m², $T_{sat} = 10$ °C.

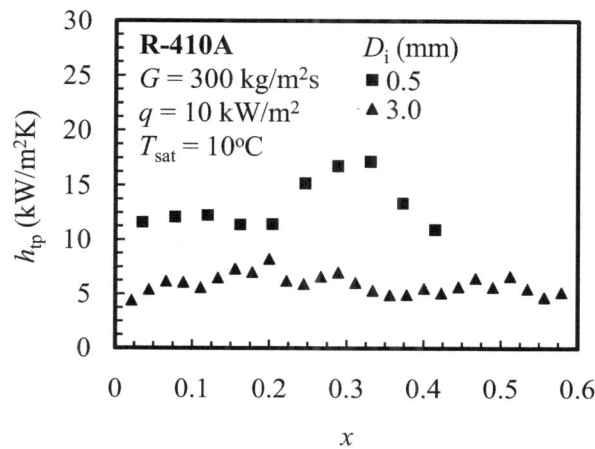

Figure 60: Effect of inner tube diameter on heat transfer coefficient for R-410A at $G = 300$ kg/(m²·s), $q = 10$ kW/m², $T_{sat} = 10$ °C.

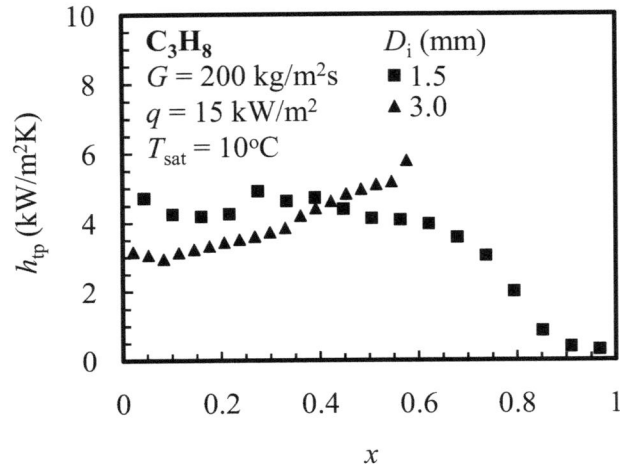

Figure 61: Effect of inner tube diameter on heat transfer coefficient for C_3H_8 at $G = 200$ kg//(m^2·s), $q = 15$ kW/m^2, $T_{sat} = 10$ °C.

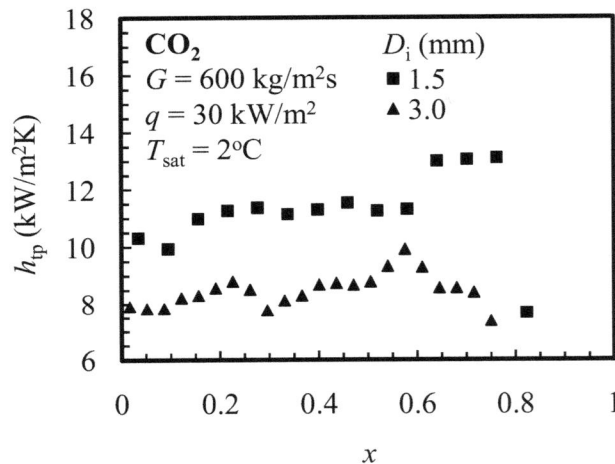

Figure 62: Effect of inner tube diameter on heat transfer coefficient for CO_2 at $G = 600$ kg//(m^2·s), $q = 30$ kW/m^2, $T_{sat} = 2$ °C.

Comparison of Heat Transfer Coefficient

Comparison of the Current Experimental Heat Transfer Coefficient

Figs. **63** to **65** show the comparisons of the heat transfer coefficients of R-22, R-134a, R-410A, C_3H_8 and CO_2 at some experimental conditions. The mean heat transfer coefficient ratio of R-22, R-134a, R-410A, C_3H_8 and CO_2 was approximately 1.0 : 0.8 : 1.8 : 0.7 : 2.0. The heat transfer coefficient of CO_2 was higher than that of the other working refrigerants during evaporation under all test conditions. The higher heat transfer coefficient of CO_2 is believed to be due to its high boiling nucleation. CO_2 has much lower surface tension and applies much higher pressure than the other working refrigerants. The heat transfer coefficients of R-22, R-134a and C_3H_8 are similar due to their similar physical properties. The comparisons of the physical properties of the present working refrigerants are given in Table **3**. CO_2 has a much lower viscosity ratio μ_f/μ_g than the other working refrigerants, which means that the liquid film of CO_2 can break easier than those of the other refrigerants. CO_2 has also a much lower density ratio ρ_f/ρ_g than the other working refrigerants, which leads to a lower vapor velocity, which in turn causes less suppression of nucleate boiling.

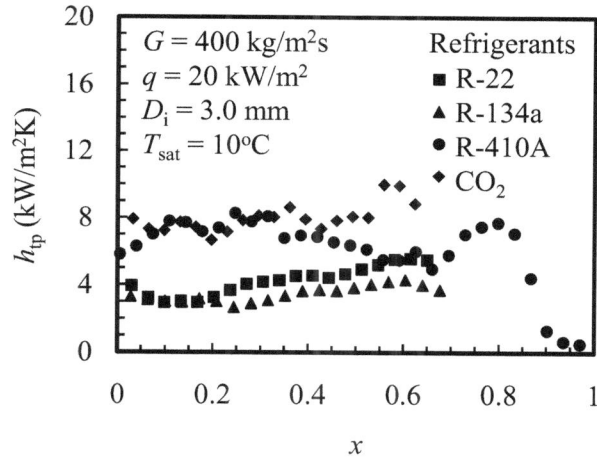

Figure 63: Heat transfer coefficient comparison of the present working refrigerants at $G = 400$ kg//(m^2·s), $q = 20$ kW/m^2, $D_i = 3.0$ mm, $T_{sat} = 10$ °C.

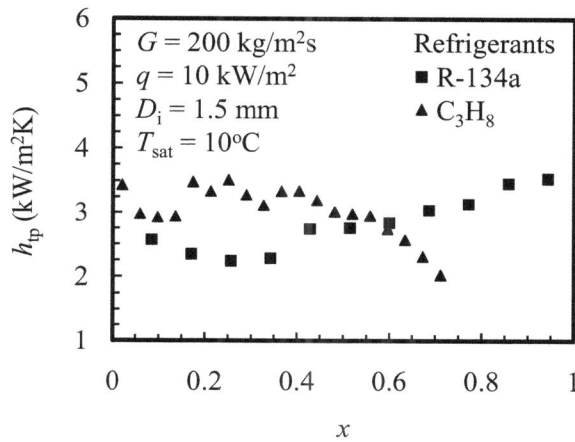

Figure 64: Heat transfer coefficient comparison of the present working refrigerants at $G = 200$ kg//(m^2·s), $q = 10$ kW/m^2, $D_i = 1.5$ mm, $T_{sat} = 10$ °C.

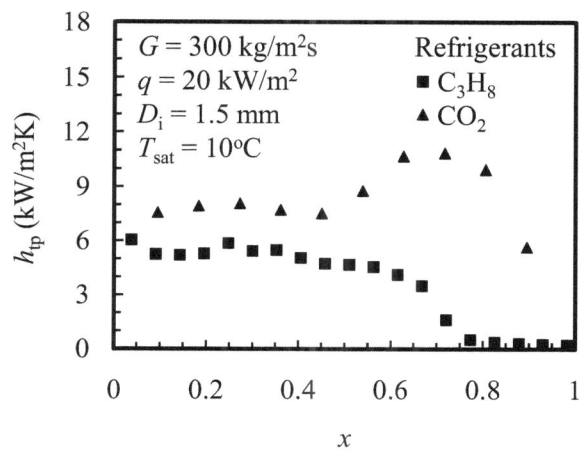

Figure 65: Heat transfer coefficient comparison of the present working refrigerants at $G = 300$ kg//(m^2·s), $q = 20$ kW/m^2, $D_i = 1.5$ mm, $T_{sat} = 10$ °C.

Heat Transfer Coefficient Comparison with Some Existing Correlations

The heat transfer coefficients of the present study are compared with the results given by seven correlations for boiling heat transfer coefficient as shown in Figs. **66** to **73**. The Gungor-Winterton [23], Jung *et al.* [25], Shah [45] and Tran *et al.* [49] correlations provided better predictions, with mean deviations of lower than 30%, than the other correlations. The Gungor-Winterton [23] correlation was a modification of the superposition model; it was developed using fluids in several small and conventional tubes under various test conditions. The Jung *et al.* [25] correlation was developed with pure and mixture refrigerants in conventional channels; its F factor contributed a big calculation deviation with the current experimental data. The Shah [45] correlation was developed using a large data set for conventional tubes. The prediction with Shah [45] correlation was fair under prediction at low quality region. The Tran *et al.* [49] correlation was developed for R-12 and R-113 in small tubes. The correlations of Chen [9] and Wattelet *et al.* [53], which were developed for large tubes, have a high prediction deviation. The correlations of Kandlikar [26] and Zhang *et al.* [61] were developed for small tubes; however, the correlations could not predict well the present experimental data. The correlations of Wattelet *et al.* [53], Kandlikar [26] and Zhang *et al.* [61] showed a large deviation in the prediction of the CO_2 data. The Kandlikar [26] correlation failed to predict the heat transfer coefficient at the high quality region.

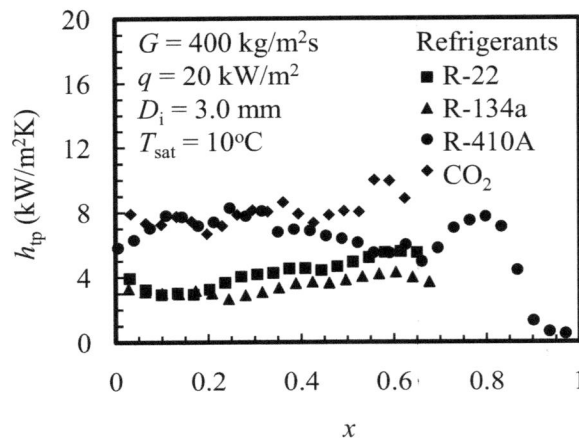

Figure 66: Heat transfer coefficient comparison of the present working refrigerants at $G = 400$ kg//(m^2·s), $q = 20$ kW/m^2, $D_i = 3.0$ mm, $T_{sat} = 10$ °C.

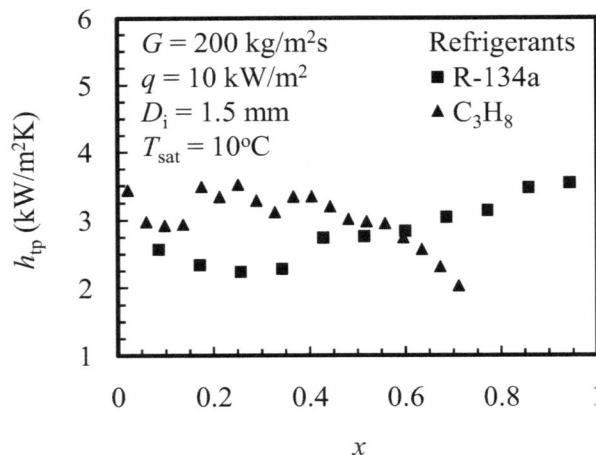

Figure 67: Heat transfer coefficient comparison of the present working refrigerants at $G = 200$ kg//(m^2·s), $q = 10$ kW/m^2, $D_i = 1.5$ mm, $T_{sat} = 10$ °C.

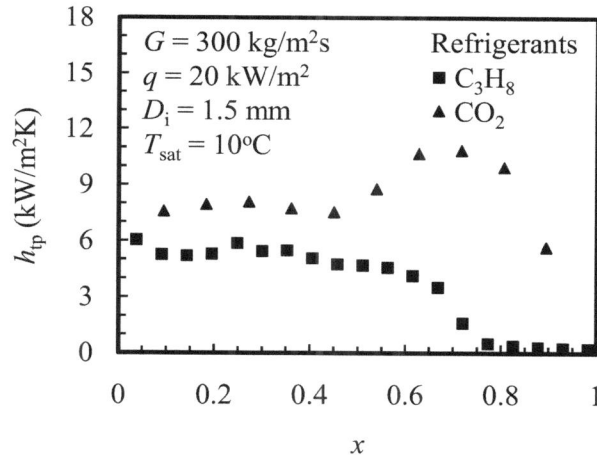

Figure 68: Heat transfer coefficient comparison of the present working refrigerants at $G = 300$ kg//(m²·s), $q = 20$ kW/m², $D_i = 1.5$ mm, $T_{sat} = 10$ °C.

Heat Transfer Coefficient Comparison with some Existing Correlations

The heat transfer coefficients of the present study are compared with the results given by seven correlations for boiling heat transfer coefficient as shown in Figs. **66** to **73**. The Gungor-Winterton [23], Jung *et al.* [25], Shah [45] and Tran *et al.* [49] correlations provided better predictions, with mean deviations of lower than 30%, than the other correlations. The Gungor-Winterton [23] correlation was a modification of the superposition model; it was developed using fluids in several small and conventional tubes under various test conditions. The Jung *et al.* [25] correlation was developed with pure and mixture refrigerants in conventional channels; its F factor contributed a big calculation deviation with the current experimental data. The Shah [45] correlation was developed using a large data set for conventional tubes. The prediction with Shah [45] correlation was fair under prediction at low quality region. The Tran *et al.* [49] correlation was developed for R-12 and R-113 in small tubes. The correlations of Chen [9] and Wattelet *et al.* [53], which were developed for large tubes, have a high prediction deviation. The correlations of Kandlikar [26] and Zhang *et al.* [61] were developed for small tubes; however, the correlations could not predict well the present experimental data. The correlations of Wattelet *et al.* [53], Kandlikar [26] and Zhang *et al.* [61] showed a large deviation in the prediction of the CO_2 data. The Kandlikar [26] correlation failed to predict the heat transfer coefficient at the high quality region.

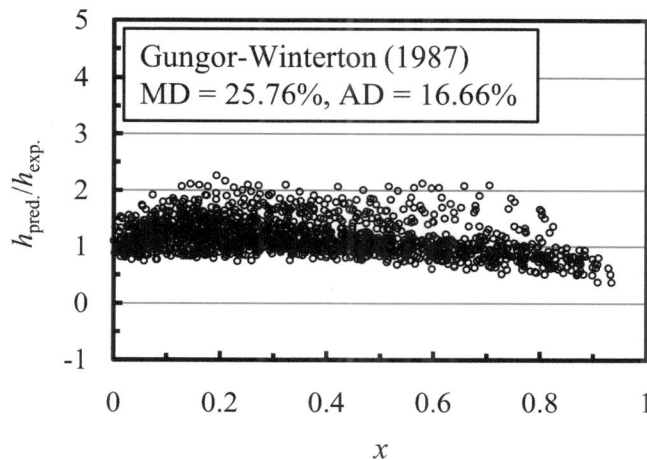

Figure 69: Comparison between the experimental and the prediction heat transfer coefficient with Gungor-Winterton [23] correlation.

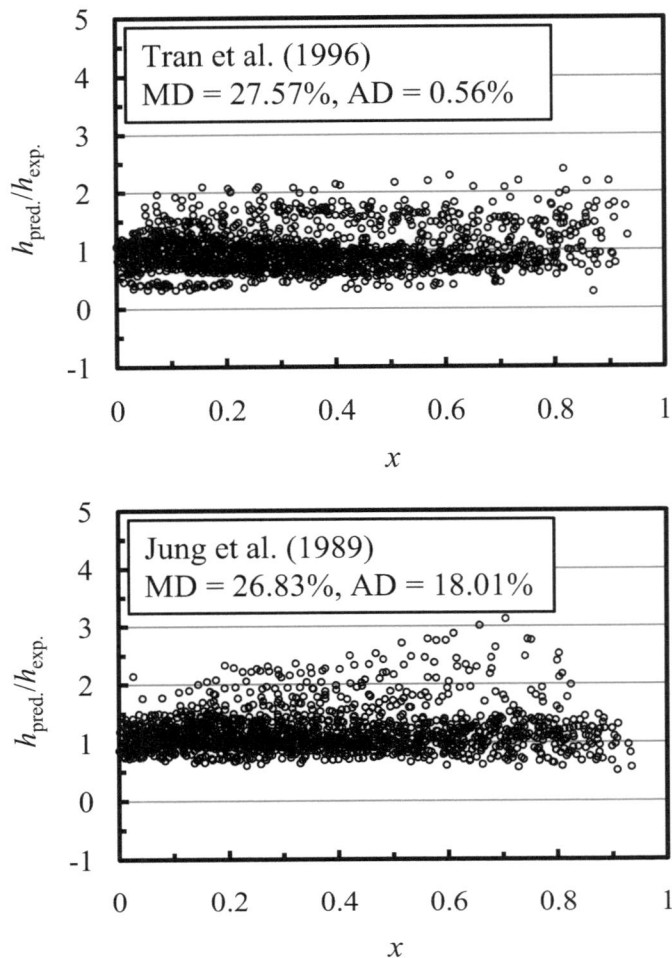

Figure 70: Comparison between the experimental and the prediction heat transfer coefficient with Jung *et al.* [25] correlation.

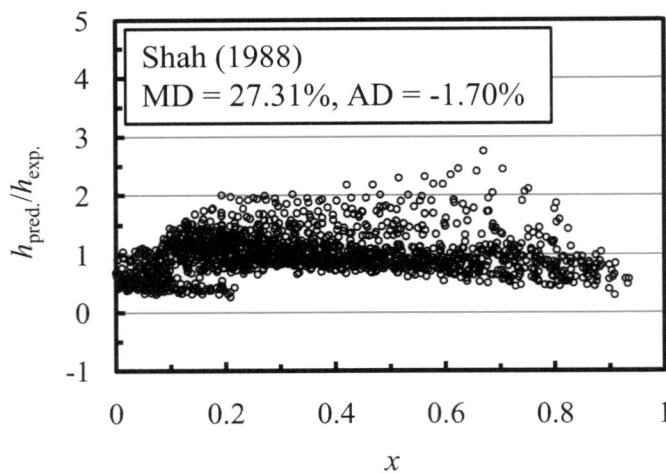

Figure 71: Comparison between the experimental and the prediction heat transfer coefficient with Shah [45] correlation.

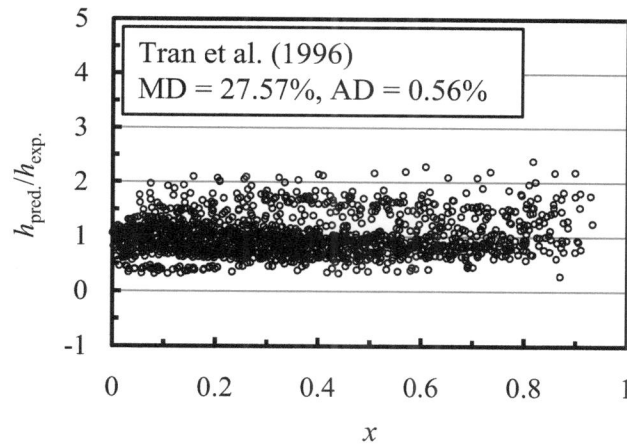

Figure 72: Comparison between the experimental and the prediction heat transfer coefficient with Tran *et al.* [49] correlation.

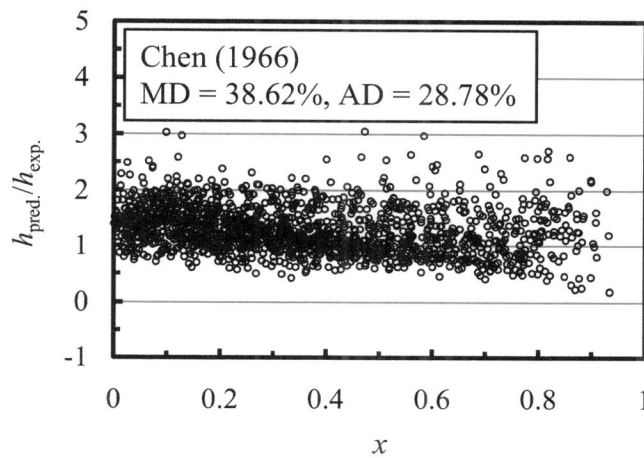

Figure 73: Comparison between the experimental and the prediction heat transfer coefficient with Chen [9] correlation.

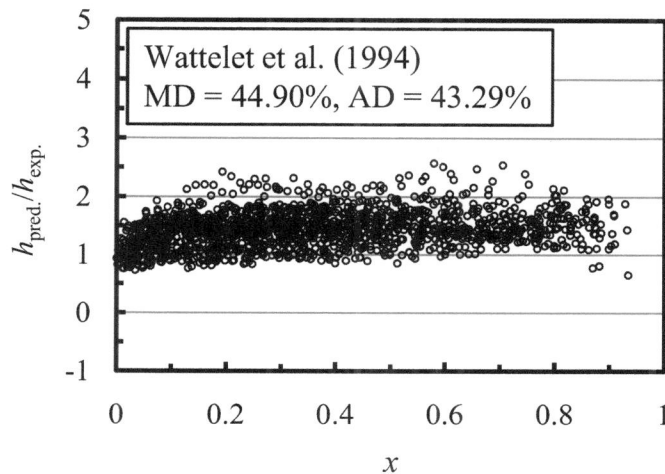

Figure 74: Comparison between the experimental and the prediction heat transfer coefficient with Wattelet *et al.* [53] correlation.

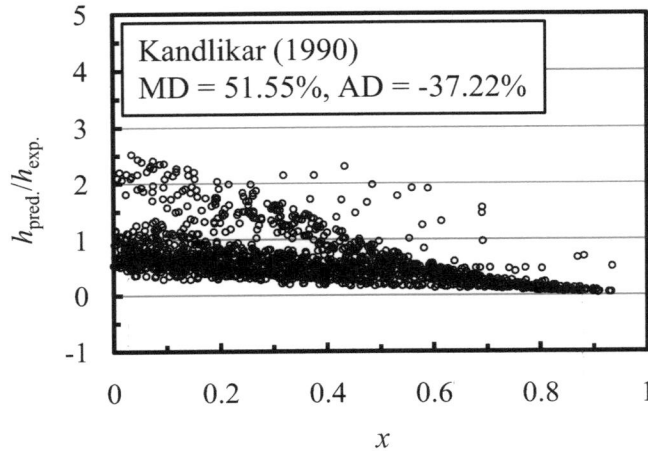

Figure 75: Comparison between the experimental and the prediction heat transfer coefficient with Kandlikar [26] correlation.

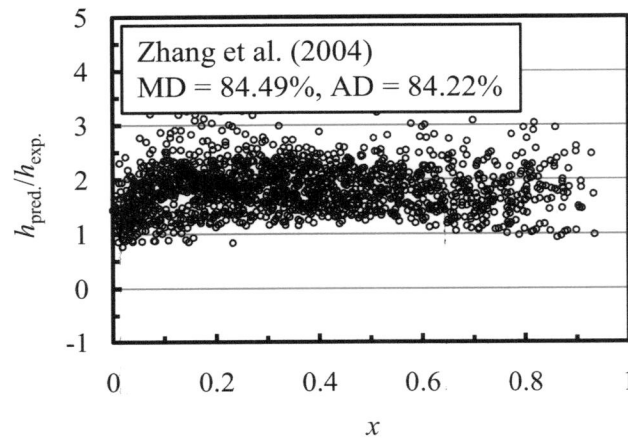

Figure 76: Comparison between the experimental and the prediction heat transfer coefficient with Zhang *et al.* [61] correlation.

PRESSURE DROP AND BOILING HEAT TRANSFER COEFFICIENT CORRELATIONS

Development of Pressure Drop Correlation

The total pressure drop consists of friction, acceleration and static head components as illustrated in the Eq. (10) below:

$$\left(\frac{dp}{dz}\right) = \left(\frac{dp}{dz}F\right) + \left(\frac{dp}{dz}a\right) + \left(\frac{dp}{dz}z\right) \tag{10}$$

For flow boiling in a horizontal layout test section, the static head pressure drop is excluded. Therefore, the experimental two-phase frictional pressure drop can be obtained by subtracting the calculated accelerational pressure drop from the measured pressure drop as shown in Eq. (11).

$$\left(\frac{dp}{dz}F\right) = \left(\frac{dp}{dz}\right) - \left(\frac{dp}{dz}a\right) = \left(\frac{dp}{dz}\right) - G^2\frac{d}{dz}\left(\frac{x^2}{\alpha\rho_g} + \frac{(1-x)^2}{(1-\alpha)\rho_f}\right) \tag{11}$$

Eq. (10) may then be expressed in terms of the single-phase pressure drop for the liquid phase, considered to exist in the tube, as below

$$\left(\frac{dp}{dz}F\right) = \left(\frac{dp}{dz}F\right)_f \varphi_f^2 \tag{12}$$

where φ_f^2 is known as the two-phase frictional multiplier, and

$$\left(\frac{dp}{dz}F\right)_f = \frac{2f_f G^2 (1-x)^2}{\rho_f D} \tag{13}$$

is the liquid frictional pressure drop calculated from the Fanning equation. The frictional pressure drop can then be rewritten as Eq. (14) below.

$$\left(\frac{dp}{dz}F\right) = \left[\frac{2f_f G^2 (1-x)^2}{\rho_f D}\right] \varphi_f^2 \tag{14}$$

The friction factor in Eq. (14) was obtained by considering the flow conditions of laminar-turbulent flows, with the following equation,

$$f = 16\,\mathrm{Re}^{-1}, \text{ for } \mathrm{Re} < 2300 \tag{15}$$

and turbulent flow, with the Blasius equation

$$f = 0.079\,\mathrm{Re}^{-0.25}, \text{ for } \mathrm{Re} > 3000 \tag{16}$$

A new modified pressure drop correlation was developed on the basis of the Lockhart-Martinelli method. The two-phase pressure drop of Lockhart-Martinelli consists of the following three terms: the liquid phase pressure drop, the interaction between the liquid phase and the vapor phase, and the vapor phase pressure drop. The relationship among these terms is expressed in Eq. (17).

$$\left(-\frac{dp}{dz}F\right)_{tp} = \left(-\frac{dp}{dz}F\right)_f + C\left[\left(-\frac{dp}{dz}F\right)_f \left(-\frac{dp}{dz}F\right)_g\right]^{1/2} + \left(-\frac{dp}{dz}F\right)_g \tag{17}$$

The two-phase frictional multiplier based on the pressure drop for liquid flow, φ_f^2, is calculated by dividing Eq. (17) by the liquid phase pressure drop as shown in Eq. (18).

$$\varphi_f^2 = \frac{\left(-\frac{dp}{dz}F\right)_{tp}}{\left(-\frac{dp}{dz}F\right)_f} = 1 + C\left[\frac{\left(-\frac{dp}{dz}F\right)_g}{\left(-\frac{dp}{dz}F\right)_f}\right]^{1/2} + \frac{\left(-\frac{dp}{dz}F\right)_g}{\left(-\frac{dp}{dz}F\right)_f} = 1 + \frac{C}{X} + \frac{1}{X^2} \tag{18}$$

The Martinelli parameter, X, is defined by the following equation:

$$X = \left[\frac{\left(-\frac{dp}{dz}F\right)_f}{\left(-\frac{dp}{dz}F\right)_g}\right]^{1/2} = \left[\frac{2f_f G^2 (1-x)^2 \rho_g / D}{2f_g G^2 x^2 \rho_f / D}\right]^{1/2} = \left(\frac{f_f}{f_g}\right)^{1/2}\left(\frac{1-x}{x}\right)\left(\frac{\rho_g}{\rho_f}\right)^{1/2} \tag{19}$$

The friction factor in Eq. (19) was obtained by considering the flow conditions of laminar and turbulent, as cited in Eqs. (7) and (8). The pressure drop prediction method can be summarized as follow

$$
\left(\frac{\mathrm{d}p}{\mathrm{d}z}\right) = \left(\frac{\mathrm{d}p}{\mathrm{d}z}F\right) + \left(\frac{\mathrm{d}p}{\mathrm{d}z}a\right)
$$

$$
= \left(\frac{2f_\mathrm{f}G^2(1-x)^2}{\rho_\mathrm{f}D}\right)\varphi_\mathrm{f}^2 + G^2\frac{\mathrm{d}}{\mathrm{d}z}\left(\frac{x^2}{\alpha\rho_\mathrm{g}} + \frac{(1-x)^2}{(1-\alpha)\rho_\mathrm{f}}\right)
$$

$$
= \left(\frac{2f_\mathrm{f}G^2(1-x)^2}{\rho_\mathrm{f}D}\right)\left(1+\frac{C}{X}+\frac{1}{X^2}\right) + G^2\frac{\mathrm{d}}{\mathrm{d}z}\left(\frac{x^2}{\alpha\rho_\mathrm{g}} + \frac{(1-x)^2}{(1-\alpha)\rho_\mathrm{f}}\right)
$$

$$
= \left(\frac{2f_\mathrm{f}G^2(1-x)^2}{\rho_\mathrm{f}D}\right)\left(1+\frac{C}{\left(\frac{f_\mathrm{f}}{f_\mathrm{g}}\right)^{1/2}\left(\frac{1-x}{x}\right)\left(\frac{\rho_\mathrm{g}}{\rho_\mathrm{f}}\right)^{1/2}}+\frac{1}{\left(\frac{f_\mathrm{f}}{f_\mathrm{g}}\right)\left(\frac{1-x}{x}\right)^2\left(\frac{\rho_\mathrm{g}}{\rho_\mathrm{f}}\right)}\right)
$$

$$
+ G^2\frac{\mathrm{d}}{\mathrm{d}z}\left(\frac{x^2}{\alpha\rho_\mathrm{g}} + \frac{(1-x)^2}{(1-\alpha)\rho_\mathrm{f}}\right)
\tag{20}
$$

The friction factor is found by considering the flow conditions of laminar and turbulent. The void fraction is calculated with the Steiner [47] void fraction prediction method.

The calculated factor *C* is obtained from Chisholm [10]. For the liquid-vapor flow condition of turbulent-turbulent (tt), laminar-turbulent (vt), turbulent-laminar (tv) and laminar-laminar (vv), the values of the Chisholm [10] parameter, *C*, are 20, 12, 10, and 5, respectively. The value of *C* in this thesis is obtained by considering the flow conditions of laminar and turbulent with thresholds of Re=2300 and Re=3000 for the laminar and turbulent flows, respectively. The laminar-turbulent transition Reynolds number was referred from Yang and Lin [55]. Fig. **74** shows a comparison of the two-phase frictional multiplier data with the values predicted by the Lockhart-Martinelli correlation with *C*=5 and *C*=20. The figure shows that the present data are located at mostly between the baseline of *C*=5 and *C*=20, which means that laminar, turbulent and the co-current laminar-turbulent flows exist in the present data. For liquid-vapor flow condition, the present experimental data shows 5.43% laminar-laminar, 28.46% laminar-turbulent, 3.78% turbulent-laminar, and 62.33% turbulent-turbulent.

Fig. **75** illustrates the two-phase pressure drop comparison between the present experimental data and the prediction with the new modified pressure drop correlation. The comparison shows a smaller deviation than the pressure drop prediction with the existing pressure drop correlations. The mean deviation and average deviation of the comparison are 33.56% and -2.67%, respectively.

Development of Heat Transfer Coefficient Correlation

It is well known that the flow boiling heat transfer is mainly governed by two important mechanisms, namely nucleate boiling and forced convective heat transfer.

$$
h_\mathrm{tp} = h_\mathrm{nb} + h_\mathrm{c}
\tag{21}
$$

In two-phase flow boiling heat transfer, the nucleate boiling heat transfer contribution is suppressed by the two-phase flow. Therefore, the nucleate boiling heat transfer contribution may be correlated with a nucleate boiling suppression factor, *S*. Another contribution of convective heat transfer may be correlated with a liquid single phase heat transfer. The F factor is introduced as a convective two-phase multiplier to account for enhanced convective due to co-current flow of liquid and vapor. A superposition model of heat transfer coefficient may be written as follow,

$$h_{tp} = Sh_{nbc} + Fh_f \qquad (22)$$

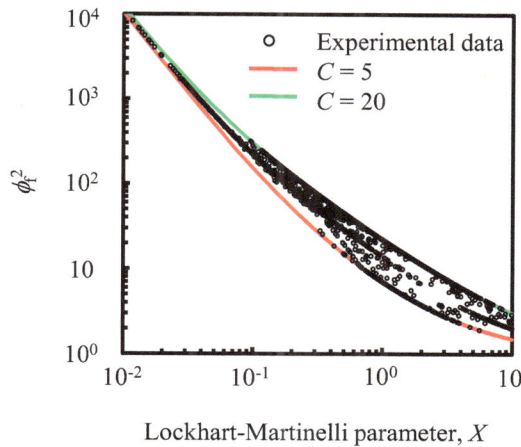

Figure 77: Variation of the two-phase frictional multiplier data with the Lockhart-Martinelli parameter.

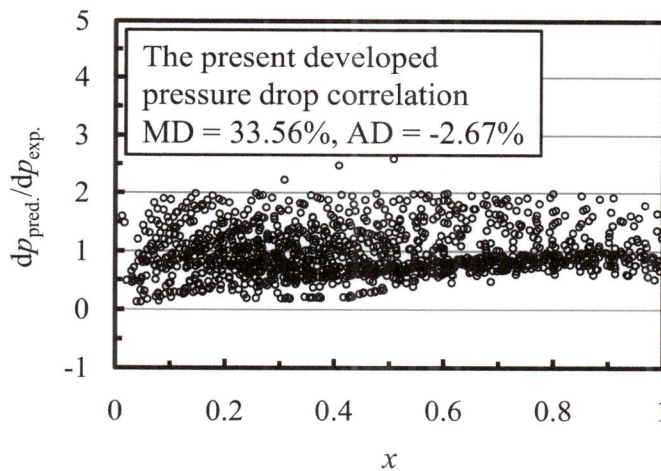

Figure 78: Comparison between the experimental and the prediction pressure drop with the new modified pressure drop correlation.

The appearance of convective heat transfer for boiling in small tubes occurs later than it does in larger tubes because of its high boiling nucleation. Chen [9] introduced a multiplier factor, $F=\text{fn}(X_{tt})$, to account for the increase in the convective turbulence that is due to the presence of the vapor phase. The function should be physically evaluated again for flow boiling heat transfer in a small tube that has a laminar flow condition, which is due to the small diameter effect. By considering the flow conditions (laminar or turbulent) in the Reynolds number factor, F, Zhang *et al.* [61] introduced a relationship between the factor F and the two-phase frictional multiplier that is based on the pressure gradient for liquid alone flow, φ_f^2. This relationship is $F=\text{fn}(\varphi_f^2)$, where φ_f^2 is a general form for four conditions according to Chisholm [10]. For the liquid-vapor flow condition of turbulent-turbulent (tt), laminar-turbulent (vt), turbulent-laminar (tv) and laminar-laminar (vv), the values of the Chisholm parameter, C, are 20, 12, 10, and 5, respectively. The value of C is obtained by the modified pressure drop prediction method as explained in the foregoing.

The F factor in this chapter is developed as a function of φ_f^2, F=fn(φ_f^2), where φ_f^2 is obtained from Eq. (10). The liquid heat transfer is defined by existing liquid heat transfer coefficient correlations by considering flow conditions of laminar and turbulent. For laminar flow, $Re_f < 2300$, where

$$Re_f = \frac{G(1-x)D}{\mu_f} \tag{23}$$

the liquid heat transfer coefficient is obtained from the following correlation:

$$h_f = 4.36\frac{k_f}{D} \tag{24}$$

For flow with $3000 \le Re_f \le 10^4$, the liquid heat transfer coefficient is obtained from Gnielinski [21] correlation:

$$h_f = \frac{\left(Re_f - 1000\right)Pr_f\left(\dfrac{f_f}{2}\right)\left(\dfrac{k_f}{D}\right)}{1 + 12.7\left(Pr_f^{2/3} - 1\right)\left(\dfrac{f_f}{2}\right)^{0.5}} \tag{25}$$

where the friction factor is calculated from Eqs. (7) and (8). For flow with $2300 \le Re_f \le 3000$, the liquid heat transfer coefficient is calculated by interpolation. For turbulent flow with $10^4 \le Re_f \le 5\times10^6$, the liquid heat transfer coefficient is obtained from Petukhov and Popov (1963) correlation:

$$h_f = \frac{Re_f\, Pr_f\left(\dfrac{f_f}{2}\right)\left(\dfrac{k_f}{D}\right)}{1 + 12.7\left(Pr_f^{2/3} - 1\right)\left(\dfrac{f_f}{2}\right)^{0.5}} \tag{26}$$

Dittus Boelter [18] correlation is used for turbulent flow with $Re_f \ge 5\times10^6$.

$$h_f = 0.023\,Re_f^{0.8}\,Pr_f^{0.4}\frac{k_f}{D} \tag{27}$$

and then a new factor F, as is shown in Fig. **76**, is developed using a regression method.

$$F = MAX\left[\left(0.0.23\varphi_f^{2.2} + 0.76\right),\ 1\right] \tag{28}$$

where φ_f^2 is obtained from Eq. (10).

The prediction of the nucleate boiling heat transfer for the present experimental data used Cooper [17], which is a pool boiling correlation developed based on an extensive study.

$$h = 55P_{red}^{0.12}\left(-0.4343\ln P_{red}\right)^{-0.55} M^{-0.5}q^{0.67} \tag{29}$$

where the heat flux, q, is in W m^{-2}. Kew and Cornwell [31] and Jung et al. [24] showed that the Cooper [17] pool boiling correlation best predicted their experimental data.

Chen [9] defined the nucleate boiling suppression factor, S, as a ratio of the mean superheat, ΔT_e, to the wall superheat, ΔT_{sat}. Jung et al. [25] proposed a convective boiling heat transfer multiplier factor, N, as a function of quality, heat flux and mass flow rate (represented by employing X_{tt} and Bo) to represent the strong effect of nucleate boiling in flow boiling as it is compared with that in nucleate pool boiling, h_{nbc}/h_{nb}. The Martinelli parameter, X_{tt}, is replaced by a two-phase frictional multiplier, φ_f^2, in order to consider laminar flow in small tubes. By using the experimental data of this study, a new nucleate boiling suppression factor, as a ratio of h_{nbc}/h_{nb}, is proposed as follows:

$$S = 0.279 \left(\varphi_f^2 \right)^{-0.029} Bo^{-0.098} \tag{30}$$

The new heat transfer coefficient correlation, as summarized in Table **4**, is developed using a regression method with 1588 data points. The comparison of the experimental heat transfer coefficient, $h_{tp,\,exp}$, and the predicted heat transfer coefficient, $h_{tp,\,pred}$, is illustrated in Fig. **77**. The new correlation agrees closely for the comparison with a mean deviation of 15.28% and an average deviation of -0.48%.

Table 4: Summary of the new heat transfer coefficient correlation

$$h_{tp} = Sh_{nbc} + Fh_f$$

$$S = 0.279 \left(\varphi_f^2 \right)^{-0.029} Bo^{-0.098}$$

$$F = \mathrm{MAX}\left[\left(0.023\varphi_f^{2.2} + 0.76 \right),\ 1 \right]$$

$$h_{nbc} = 55 P_r^{0.12} \left(-0.4343 \ln P_r \right)^{-0.55} M^{-0.5} q^{0.67}, \quad \text{where } q \text{ is in Wm}^{-2}$$

$$h_f \begin{cases} = 4.36 \dfrac{k_f}{D} \quad \text{if Re}_f < 2300 \\[3mm] = \dfrac{\left(\text{Re}_f - 1000 \right) \text{Pr}_f \left(\dfrac{f_f}{2} \right) \left(\dfrac{k_f}{D} \right)}{1 + 12.7 \left(\text{Pr}_f^{2/3} - 1 \right) \left(\dfrac{f_f}{2} \right)^{0.5}} \quad \text{if } 3000 \le \text{Re}_f \le 10^4 \\[5mm] = \dfrac{\text{Re}_f \, \text{Pr}_f \left(\dfrac{f_f}{2} \right) \left(\dfrac{k_f}{D} \right)}{1 + 12.7 \left(\text{Pr}_f^{2/3} - 1 \right) \left(\dfrac{f_f}{2} \right)^{0.5}} \quad \text{if } 10^4 \le \text{Re}_f \le 5 \times 10^6 \\[5mm] = 0.023 \dfrac{k_f}{D} \left[\dfrac{G(1-x)D}{\mu_f} \right]^{0.8} \left(\dfrac{C_{pf}\mu_f}{k_f} \right)^{0.4} \quad \text{if Re}_f \ge 5 \times 10^6 \end{cases}$$

$$\varphi_f^2 = 1 + \dfrac{C}{X} + \dfrac{1}{X^2}$$

$$X = \left(\dfrac{f_f}{f_g} \right)^{1/2} \left(\dfrac{1-x}{x} \right) \left(\dfrac{\rho_g}{\rho_f} \right)^{1/2}$$

$$C \begin{cases} = 5 \text{ for } \text{Re}_f < 2300 \text{ and } \text{Re}_g < 2300 \\ = 10 \text{ for } \text{Re}_f > 3000 \text{ and } \text{Re}_g < 2300 \\ = 12 \text{ for } \text{Re}_f < 2300 \text{ and } \text{Re}_g > 3000 \\ = 20 \text{ for } \text{Re}_f > 3000 \text{ and } \text{Re}_g > 3000 \end{cases}$$

$$f \begin{cases} = 16 \, \text{Re}^{-1} \text{ for } \text{Re} < 2300 \\ = 0.079 \, \text{Re}^{-0.25} \text{ for } \text{Re} > 3000 \end{cases}$$

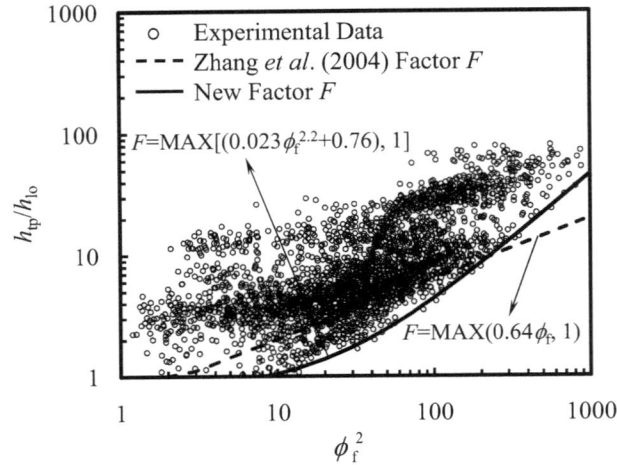

Figure 79: Convective heat transfer multiplier as a function of φ_f^2.

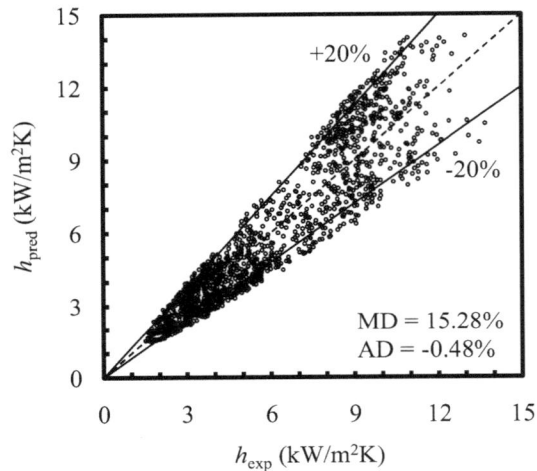

Figure 80: Diagram of the experimental heat transfer coefficient, h_{exp}, vs prediction heat transfer coefficient, h_{pred}, with the new developed heat transfer coefficient correlation.

CONCLUSIONS

Convective boiling pressure drop and heat transfer experiments were performed in horizontal small tubes with R-22, R-134a, R-410A, C_3H_8 and CO_2. The current experimental data are mapped on Wang et al. [51] and Wojtan et al. [52] flow pattern maps. Wang et al. [51] predicted the current experimental results flow pattern better than Wojtan et al. [52]. Wang et al. [51] map showed a better flow pattern prediction for the beginning of annular flow. Wojtan et al. [52] map showed a better flow pattern prediction for the dry-out with R-410A experimental data than with the other working refrigerants. In comparison with the flow pattern predicted by Wang et al. [51] and Wojtan et al. [52] flow pattern maps, the current experimental data shows that annular flow and dry-out occur at a lower vapor quality for the evaporation with a higher heat flux, higher mass flux and a lower inner tube diameter. Stratified wavy flow appears earlier for higher mass flux condition, and the regime is longer for a lower mass flux condition.

The pressure drop is higher for the conditions of higher mass and heat fluxes, and for the conditions of lower saturation temperature and smaller inner tube diameter. The order of the pressure drop from the highest to the lowest is R-134a, R-22, C_3H_8, R-410A, and CO_2. The pressure drop is strongly affected by physical properties of the working fluid such as density, viscosity, surface tension and pressure. The Friedel [20] and Cicchitti *et al.* [16] models predicted the current experimental pressure drop relatively well. CO_2 has a lowest pressure drop among the present working refrigerants. The pressure drop is higher for the conditions of higher mass and heat fluxes, and for the conditions of smaller inner tube diameter and lower saturation temperature. A modified pressure drop correlation was proposed on the basis of the Lockhart-Martinelli method by considering the laminar-turbulent flow conditions.

Mass flux has insignificant effect on heat transfer coefficient at low quality region, but it has significant effect on heat transfer coefficient at high quality region. Heat flux has significant effect on heat transfer coefficient at low quality region, but it has insignificant effect on heat transfer coefficient at high quality region. heat transfer coefficient increases with an increased saturation temperature. Heat transfer coefficient is higher in the smaller inner diameter tube, especially at low quality region. The heat transfer coefficient in evaporation with small tubes is dominated by nucleate boiling heat transfer contribution. CO_2 has a highest heat transfer coefficient among the present working refrigerants. The mean heat transfer coefficient ratio of R-22, R-134a, R-410A, C_3H_8 and CO_2 was approximately 1.0 : 0.8 : 1.8 : 0.7 : 2.0. The average heat transfer coefficient increases with a smaller inner tube diameter and with an increased saturation temperature. The Gungor-Winterton [23], Jung *et al.* [25], Shah [45] and Tran *et al.* [49] models predicted the current experimental heat transfer coefficient relatively well.

The geometric effect of the small tube must be considered to develop a new heat transfer coefficient correlation. Laminar flow appears for flow boiling in small tubes, so the modified correlation of the multiplier factor for the convective boiling contribution, F, and the nucleate boiling suppression factor, S, is developed in this study using laminar and turbulent flows consideration. A new boiling heat transfer coefficient correlation that is based on a superposition model for refrigerants in small tubes was presented with 15.28% mean deviation and -0.48% average deviation.

NOMENCLATURE

AD	Average Deviation, $AD = \left(\dfrac{1}{n}\right)\sum_1^n\left(\left(dp_{\text{pred}} - dp_{\text{exp}}\right)\times 100 \big/ dp_{\text{exp}}\right)$ for pressure drop or
	$AD = \left(\dfrac{1}{n}\right)\sum_1^n\left(\left(h_{\text{pred}} - h_{\text{exp}}\right)\times 100 \big/ h_{\text{exp}}\right)$ for heat transfer coefficient
Bo	Boiling number, $Bo = q\big/Gi_{\text{fg}}$
C	Correction factor for two-phase pressure drop
c_p	Specific heat capacity at constant pressure (kJ/(kg·K))
D	Diameter (m)
E	Electric potential (V)
F	Multiplier factor for convective heat transfer contribution
f	Friction factor
G	Mass flux (kg/(m²·s))
g	Acceleration due to gravity (m/s²)
h	Heat transfer coefficient (kW/(m²·K))
I	Electric current (A)
i	Enthalpy (kJ/(kg·K))
k	Thermal conductivity (kW/(m·K))
L	Tube length (m)
M	Molecular weight of the liquid
MD	Mean Deviation, $MD = \left(\dfrac{1}{n}\right)\sum_1^n\left|\left(\left(dp_{\text{pred}} - dp_{\text{exp}}\right)\times 100 \big/ dp_{\text{exp}}\right)\right|$ for pressure drop or
	$MD = \left(\dfrac{1}{n}\right)\sum_1^n\left|\left(\left(h_{\text{pred}} - h_{\text{exp}}\right)\times 100 \big/ h_{\text{exp}}\right)\right|$ for heat transfer coefficient
n	Number of data

P	Pressure (N/m^2)
Pr	Prandtl number, $\mathrm{Pr} = c_p \mu / k$
Q	Electric power (kW)
q	Heat flux (kW/m^2)
$\dot{q}$	Heat generation (kW/m^3)
R	Electrical resistance (ohm)
Re	Reynolds number, $\mathrm{Re} = GD/\mu$
R_p	Surface roughness parameter (μm)
S	Suppression factor
T	Temperature (K)
V	Volume (m^3)
W	Mass flow rate (kg/s)
X	Lockhart-Martinelli parameter
x	Mass quality
z	Length (m)

Greek symbols

λ	Correction factor on Baker (1954) flow pattern map, $\lambda = \left[\left(\rho_g / \rho_A \right) \left(\rho_f / \rho_w \right) \right]^{1/2}$
α	Void fraction
μ	Viscosity (Pa·s)
ρ	Density (kg/m^3)
σ	Surface tension (N/m)
ϕ^2	Two-phase frictional multiplier
ψ	Correction factor on Wang *et al.* (1997) flow pattern map,

$$\psi = \left(\sigma_w / \sigma \right)^{1/4} \left[\left(\mu_f / \mu_w \right) \left(\rho_w / \rho_f \right)^2 \right]^{1/3}$$

Gradients and differences

(dp/dz)	Pressure gradient (N/m^{2}m)
$(dp/dz\ F)$	Pressure gradient due to friction (N/m^{2}m)
$(dp/dz\ a)$	Pressure gradient due to acceleration (N/m^{2}m)
$(dp/dz\ z)$	Pressure gradient due to static head (N/m^{2}m)

Subscripts

A	Air
c	Convective
exp	Experimental value
f	Saturated liquid
g	Saturated vapor
i	Inner tube
fo	Liquid only
nb	Nucleate boiling
nbc	Nucleate boiling contribution
o	Outer tube
pb	Pool boiling
pred	Predicted value
red	Reduced
sat	Saturation
sc	Subcooled
t	Turbulent flow
tp	Two-phase

v   Laminar flow
W   Water
w   Wall

REFERENCES

[1]   M. I. Ali, M. Sadatomi, and M. Kawaji, "Two-phase flow in narrow channels between two flat plates," *Can. J. Chem. Eng.*, vol. 71, pp. 657-666, 1993.

[2]   O. Baker, "Design of pipe lines for simultaneous flow of oil and gas", *Oil & Gas J.*, July, 26. 1954.

[3]   C. J. Baroczy, "Correlation of liquid fraction in two-phase flow with application to liquid metals", *NAA-SR-8171*, 1963.

[4]   Z. Y. Bao, D. F. Fletcher, and B. S. Haynes, "Flow boiling heat transfer of Freon R11 and HCFC123 in narrow passages," *Int. J. Heat and Mass Transfer*, vol. 43, pp. 3347-3358, 2000.

[5]   D. R. H. Beattie and P. B. Whalley, "A simple two-phase flow frictional pressure drop calculation method," *Int. J. Multiphase Flow*, vol. 8, pp. 83-87, 1982.

[6]   S. D. Chang and S. T. Ro, "Pressure drop of pure HFC refrigerants and their mixtures flowing in capillary tubes," *Int. J. Multiphase Flow*, vol. 22, pp. 551-561, 1996.

[7]   Y. J., Chang, S. K. Chiang, T. W. Chung, and C. C. Wang, "Two-phase frictional characteristics of R-410A and air-water in a 5 mm smooth tube," *ASHRAE Trans*, DA-00-11-3, pp. 792-797, 2000.

[8]   I. Y., Chen, K. S. Yang, Y. J. Chang, and C. C. Wang, "Two-phase pressure drop of air-water and R-410A in small horizontal tubes," *Int. J. Multiphase Flow*, vol. 27, pp. 1293-1299, 2001.

[9]   J. C. Chen, "A correlation for boiling heat transfer to saturated fluids in convective flow," *Ind. Eng. Chem., Process Des. Dev.*, vol. 5, pp. 322-329, 1966.

[10]   D. Chisholm, "A theoretical basis for the Lockhart-Martinelli correlation for two-phase flow," *Int. J. Heat Mass Transfer*, vol. 10, pp. 1767-1778, 1967.

[11]   D. Chisholm, "An equation for velocity ratio in two-phase flow," *NEL Report*, 535, 1972.

[12]   D. Chisholm, "The influence of mass velocity on friction pressure gradients during steam-water flow," Paper 35 presented at *Thermodynamics & Fluid Mechanics Convention*, Institutes of Mechanical Engineers (Bristol), March, 1968.

[13]   D. Chisholm, "Two-phase flow in pipelines and heat exchangers," Longman, New York, 1983.

[14]   D. Chisholm and L. A. Sutherland, "Predicted of pressure gradients in pipeline systems during two-phase flow," Paper 4 presented at *Sym. Fluid Mechanics & Measurements in Two-phase Flow Systems*, Leeds, 24-25, September, 1969.

[15]   J. M. Cho and M. S. Kim, "Experimental studies on the evaporative heat transfer and pressure drop of CO_2 in smooth and micro-fin tubes of the diameters of 5 and 9.52 mm," *Int. J. Refrig.*, vol. 30, pp. 986-994, 2007.

[16]   A. Cicchitti, C. Lombardi, M. Silvestri, G. Solddaini, and R., Zavalluilli, "Two-phase cooling experiments—Pressure drop, heat transfer and burnout measurement," *Energia Nucl.*, vol. 7, pp. 407-425, 1960.

[17]   M. G. Cooper, "Heat flow rates in saturated nucleate pool boiling–a wide-ranging examination using reduced properties," *Advs. Heat Transfer*, Academic Press, vol. 16, pp. 157-239, 1984.

[18]   F. W. Dittus and L. M. K. Boelter, "Heat transfer in automobile radiators of the tubular type," *University of California Publication in Engineering*, vol. 2, pp. 443-461, 1930.

[19]   A. E. Dukler, I. I. M. Wicks, and R. G. Cleveland, "Pressure drop and hold-up in two-phase flow," *AIChE J.*, vol. 10, pp. 38-51, 1964.

[20]   L. Friedel, "Improved friction pressure drop correlations for horizontal and vertical two-phase pipe flow," Presented at the *European Two-phase Flow Group Meeting*, Ispra, Italy, Paper E2, June, 1979.

[21]   V. Gnielinski, "New equations for heat and mass transfer in turbulent pipe and channel flow", *Int. Chem. Eng.*, vol. 16, pp. 359-368, 1976.

[22]   R. Grönnerud, "Investigation of Liquid Hold-Up, Flow-Resistance and Heat Transfer in Circulation Type Evaporators," Part IV: Two-phase flow resistance in boiling refrigerants, Annexe 1972-1, *Bull. de l'Inst. du Froid*, International Inst. of Refrigeration, Paris, 1979.

[23]   K. E., Gungor and H. S. Winterton, "Simplified general correlation for saturated flow boiling and comparisons of correlations with data," *Chem. Eng. Res.*, vol. 65, pp. 148-156, 1987.

[24]   D. Jung, Y. Kim, Y., Ko, and K. Song, "Nucleate boiling heat transfer coefficients of pure halogenated refrigerants," *Int. J. Refrig.*, vol. 26, pp. 240-248, 2003.

[25] D. S. Jung, M. McLinden, R. Radermacher, and D. Didion, "A Study of Flow Boiling Heat Transfer with Refrigerant Mixtures," *Int. J. Mass Transfer*, vol. 32, pp. 1751-1764, 1989.

[26] S. G. Kandlikar, "A general correlation for saturated two-phase flow boiling heat transfer inside horizontal and vertical tubes", *J. Heat Transfer*, vol. 112, pp. 219-228, 1990.

[27] S. G. Kandlikar, "Fundamental issues related to flow boiling in minichannels and microchannels," *Exp. Therm. Fluid Sci.*, vol. 26, pp. 389-407, 2002.

[28] N. Kattan, "Contribution to the heat transfer analysis of substitute refrigerants in evaporator tubes with smooth or enhanced tube surfaces," PhD thesis No 1498, Swiss Federal Institute of Technology, Lausanne, Switzerland, 1996.

[29] N. Kattan, J. R. Thome, and D. Favrat, "Flow boiling in horizontal tubes: part 1 – development of a diabatic two-phase flow pattern map," *J. Heat Transfer*, vol. 120, pp. 140-147, 1998.

[30] A. Kawahara, P. M. Y. Chung, and M. Kawaji, "Investigation of two-phase flow pattern, void fraction and pressure drop in a microchannel," *Int. J. Multiphase Flow*, vol. 28, pp. 1411-1435, 2002.

[31] P. A. Kew and K. Cornwell, "Correlations for the Prediction of Boiling Heat Transfer in Small-Diameter Channels," *Appl. Therm. Eng.* Vol. 17, pp. 705-715, 1997.

[32] C. S. Kuo and C. C. Wang, "In-tube evaporation of HCFC-22 in a 9.52 mm micro-fin/smooth tube," *Int. J. Heat Mass Transfer*, vol. 39, pp. 2559-2569, 1996.

[33] G. M. Lazarek and S. H. Black, "Evaporative heat transfer, pressure drop and critical heat flux in a small diameter vertical tube with R-113," *Int. J. Heat Mass Transfer*, vol. 25, pp. 945-960, 1982.

[34] R. W. Lockhart and R. C. Martinelli, "Proposed correlation of data for isothermal two-phase, two-component flow in pipes," *Chem. Eng. Prog.*, vol. 45, pp. 39-48, 1949.

[35] W. H. McAdams, *Heat Transmission*, Third ed. New York: McGraw-Hill, 1954.

[36] K. Mishima and T. Hibiki, "Some characteristics of air-water two-phase flow in small diameter vertical tubes," *Int. J. Multiphase Flow*, vol. 22, pp. 703-712, 1996.

[37] H. Müller-Steinhagen and K. Heck, "A simple friction pressure drop correlation for two-phase flow in pipes," *Chem. Eng. Processing*, vol. 20, pp. 297-308, 1986.

[38] H. K. Oh, H. G. Ku, G. S. Roh, C. H. Son, and S. J. Park, "Flow boiling heat transfer characteristics of carbon dioxide in a horizontal tube," *Appl. Therm. Eng.*, vol. 28, pp. 1022-1030, 2008.

[39] M. B. Ould Didi, N. Kattan, and J. R. Thome, "Prediction of Two-phase Pressure Gradients of Refrigerants in Horizontal Tubes," *Int. J. Refrig.*, vol. 25, pp. 935-947, 2002.

[40] C. Y. Park and P. S. Hrnjak, "CO_2 and R410A flow boiling heat transfer, pressure drop, and flow pattern at low temperatures in a horizontal smooth tube," *Int. J. Refrig.*, vol. 30, pp. 166-178, 2007.

[41] X. F. Peng and G. P. Peterson, "Forced convective heat transfer of single-phase binary mixtures through microchannels," *Exp. Therm. Fluid Sci.*, 12, 98-104, 1996.

[42] B. S. Petukhov and V. N. Popov, "Theoretical calculation of heat exchanger in turbulent flow in tubes of an incompressible fluid with variable physical properties," *High Temp.*, vol. 1(1), 69-83, 1963.

[43] J. Pettersen, "Flow vaporization of CO_2 in microchannels tubes," *Exp. Therm. Fluid Sci.*, vol. 28, pp. 111-121, 2004.

[44] A. Premoli, D., Francesco, and A., Prina, "A dimensionless correlation for determining the density of two-phase mixtures," *Lo Termotecnica*, vol. 25, pp. 17-26, 1971.

[45] M. M. Shah, "Chart correlation for saturated boiling heat transfer: equations and further study," *ASHRAE Trans.*, 2673, pp. 185-196, 1988.

[46] S. L. Smith, "Void fractions in two phase flow: a correlation based upon an equal velocity head model," *Proc. Inst. Mech. Engrs*, London, vol. 184, pp. 647–657 Part 1, 1969.

[47] D. Steiner, "Heat transfer to boiling saturated liquids," *VDI-Wärmeatlas* (*VDI Heat Atlas*), Verein Deutcher Ingenieure, ed., VDI-Gessellschaft Verfahrenstechnik und Chemieinge- nieurwesen (GCV), Düsseldorf, Germany, (J.W. Fullarton, translator), 1993.

[48] T. N. Tran, M.C. Chyu, M. W. Wambsganss, and D. M. France, "Two-phase pressure drop of refrigerants during flow boiling in small channels: An experimental investigation and correlation development," *Int. J. Multiphase Flow*, vol. 26, pp. 1739-1754, 2000.

[49] T. N. Tran, M. W. Wambsganss, and D. M. France, "Small circular- and rectangular-channel boiling with two refrigerants," *Int. J. Multiphase Flow*, vol. 22, pp. 485-498, 1996.

[50] M. W. Wambsganss, D. M. France, J. A. Jendrzejczyk, and T. N. Tran, "Boiling heat transfer in a horizontal small-diameter tube," *AMSE Trans.*, vol. 115, pp. 963-975, 1993.

[51]   C. C. Wang, C. S. Chiang, and D. C. Lu, "Visual observation of two-phase flow pattern of R-22, R-134a, and R-407C in a 6.5-mm smooth tube," *Exp. Therm. Fluid Sci.*, vol. 15, pp. 395-405, 1997.

[52]   L. Wojtan, T. Ursenbacher, and J. R. Thome, "Investigation of flow boiling in horizontal tubes: part I – a new diabatic two-phase flow pattern map," *Int. J. Heat Mass Transfer*, vol. 48, pp. 2955-2969, 2005.

[53]   J. P. Wattelat, J. C. Chato, A. L. Souza, and B. R. Christoffersen, "Evaporative characteristics of R-12, R-134a, and a mixture at low mass fluxes," *ASHRAE Trans.*, 94-2-1, pp. 603-615, 1994.

[54]   Y. Y. Yan and T. F. Lin, "Evaporation heat transfer and pressure drop of refrigerant R-134a in a small pipe," *Int. J. Heat Mass Transfer*, vol. 41, pp. 4183-4194, 1998.

[55]   C. Y. Yang and T. Y. Lin, "Heat transfer characteristics of water flow in microtubes," *Exp. Therm. Fluid Sci.*, vol. 32, pp. 432-439, 2007.

[56]   S. H. Yoon, E. S. Cho, Y. W. Hwang, M. S. Kim, K. Min, and Y. Kim, "Characteristics of evaporative heat transfer and pressure drop of carbon dioxide and correlation development," *Int. J. Refrig.*, vol. 27, pp. 111-119, 2004.

[57]   W. Yu, D. M. France, M. W. Wambsganss, and J. R. Hull, "Two-phase pressure drop, boiling heat transfer, and critical heat flux to water in a small-diameter horizontal tube," *Int. J. Multiphase Flow*, vol. 28, pp. 927-941, 2002.

[58]   R. Yun, Y. Kim and M. S. Kim, "Convective boiling heat transfer characteristics of CO_2 in microchannels," *Int. J. Heat. Mass Transfer*, vol. 48, pp. 235-242, 2005.

[59]   L. Zhang, E. Hihara, T. Saito, and J. T. Oh, "Boiling heat transfer of a ternary refrigerant mixture inside a horizontal smooth tube," *Int. J. Mass Transfer*, vol. 40, pp. 2009-2017, 1997.

[60]   M. Zhang and R. L. Webb, "Correlation of two-phase friction for refrigerants in small-diameter tubes," *Exp. Therm. Fluid Sci.*, vol. 25, pp. 131-139, 2001.

[61]   W. Zhang, T. Hibiki, and K. Mishima, "Correlation for flow boiling heat transfer in mini-channels," *Int. J. Heat and Mass Transfer*, vol. 47, pp. 5749-5763, 2004.

[62]   Y. Zhao, M. Molki, M. M. Ohadi, and S. V., Dessiatoun, "Flow boiling of CO_2 in microchannels," *ASHRAE Trans.*, DA-00-2-1, pp. 437-445, 2000.

[63]   S. M. Zivi, "Estimation of steady-state steam void-fraction by means of the principle of minimum entropy generation," *J. Heat Transfer*, vol. 86, pp. 247-252, 1964.

CHAPTER 8

Flashing-Induced Density Wave Oscillations in a Boiling Natural Circulation System

Masahiro Furuya[*]

Nuclear Energy System Department, Central Research Institute of Electric Power Industry, 2-11-1 Iwado-kita, Komae, Tokyo 201-8511, Japan

Abstract: This chapter addresses characteristics of flashing-induced density wave oscillations on the basis of the experimental results in a boiling natural circulation system with an adiabatic chimney. Flashing is caused by the sudden increase of vapor generation due to the reduction in hydrostatic head, since saturation enthalpy changes with pressure. Flashing-induced density wave oscillations may, therefore, occur at low pressure. The oscillation period correlates well with the passing time of bubbles in the chimney section regardless of the system pressure, the heat flux, and the inlet subcooling. According to the stability map, the flow became stable below a certain heat flux regardless of the channel inlet subcooling. The stable region enlarged with increasing system pressure. Therefore, the stability margin becomes larger by pressurizing the loop sufficiently before heating.

Keyword: Natural circulation, boiling two-phase flow, flashing, stability, BWR.

INTRODUCTION

Natural circulation is a key issue in design of chemical plants and nuclear power stations for simplicity, inherent safety, and maintenance reduction features. Those natural circulation systems are susceptible to the flow instability, science the driving force of circulation is sensitive to momentum and energy balances.

Instability Classification

In general, several types of thermal-hydraulic instabilities may occur in a boiling two-phase system [1, 6, 20, 13, 9]. Some of these instabilities arise from the steady state characteristics of the system such as flow excursion (or Ledinegg instability) and relaxation instability (flow pattern transition, bumping, geysering, and chugging, *etc.*). The remaining types of instabilities are due to the dynamic nature of the system such as density wave oscillations, pressure drop oscillations, acoustic oscillations and thermal oscillations. The most commonly encountered instabilities are so-called density wave oscillations, which may occur in the operation range from start-up to normal operating conditions.

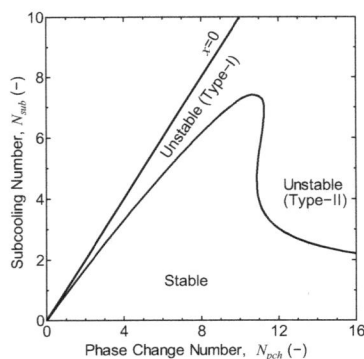

Figure 1: Stability boundaries are shown for Type-I and Type-II instabilities in nondimensional plane of the phase change number and the subcooling number. The Type-I instability region becomes smaller as the chimney length is decreased.

*Address correspondence to Masahiro Furuya: Nuclear Energy System Department, Central Research Institute of Electric Power Industry, 2-11-1 Iwado-kita, Komae, Tokyo 201-8511, Japan; E-mail: furuya@criepi.denken.or.jp

Type-I and Type-II Instabilities

Within the class of density wave oscillations, two distinct types of instabilities can be found. Fukuda and Kobori [3] classified them as the Type-I instability and the Type-II instability. The Type-I instability is dominant when the flow quality is low, while the Type-II instability is dominant when the flow quality is high. The Type-II instability is the most commonly observed form of density wave oscillations.

Numerous analyses were performed on the sensitivities of Type-I and Type-II instabilities to system parameters such as heat flux, subcooling and pressure. Ishii and Zuber [8], Nayak *et al.* [15], van Bragt and van der Hagen [17], [18] have conducted a linear stability analysis by creating a set of linear equations to describe the boiling two-phase flow system. They determined the stability boundaries of Type-I and Type-II instabilities, which can be represented as a function of the dimensionless phase change number (also known as the Zuber number) and the subcooling number. The phase change number, N_{pch}, is proportional to the heating power and inversely proportional to the coolant flow rate, while the subcooling number, N_{sub}, is proportional to the channel inlet subcooling as defined by the following equations:

$$N_{pch} \equiv \frac{P_e q'' L_c}{u_{in} A_c h_{gl}} \left(\frac{\rho_l}{\rho_g} - 1 \right) \tag{1}$$

$$N_{sub} \equiv \frac{Cp_l \Delta T_{sub}}{h_{gl}} \left(\frac{\rho_l}{\rho_g} - 1 \right). \tag{2}$$

P_e denotes the perimeter; q'', the heat flux; L_c, the heated length; u_{in}, the inlet flow velocity; A_c, the flow area; ρ_l and ρ_g, the liquid density and gas phase density, respectively, Cp_l, the liquid phase specific heat capacity at constant pressure; ΔT_{sub}, the degree of subcooling; and h_{gl}, the latent heat. N_{sub} yields a negative value when the channel inlet temperature is higher than the saturation temperature at the system pressure.

Fig. **1** shows a typical stability map in the dimensionless plane of the phase change number and the subcooling number for a boiling two-phase flow loop consisting of a heated core, adiabatic chimney and a downcomer [21]. The straight line in the figure indicates the operational points where the flow quality at the core exit is zero. Therefore, this line divides the stability map into two regions: a single phase flow region ($N_{sub} \geq N_{pch}$) and a boiling two-phase flow region ($N_{sub} < N_{pch}$).

The stability boundary is drawn in the two-phase flow region as a curve in the figure. The flow is stable below the curve. Type-I instability is dominant when the flow quality is small (in between the zero quality line and the stability boundary curve), while Type-II instability is dominant at the large flow quality. The stability boundary is independent of the pressure, if one applies the homogeneous equilibrium mixture (HEM) model and neglects changes in thermo-dynamic properties such as densities and liquid saturation enthalpy. The HEM model assumes a perfect mixture of liquid and vapor. The liquid density differences (liquid single-phase natural circulation) have been, therefore neglected. Refer to the reference [21] for more detail and the effect of liquid single-phase natural circulation effect.

Flashing-Induced Instability

Wissler and Amudson [19] were the first to report that flashing causes instability at low pressure. Flashing is caused by the sudden increase of vapor generation due to the reduction in hydrostatic head, since saturation enthalpy changes with pressure. Natural circulation system is susceptible to the flashing-induced instability, since the recirculation flow rate is significantly dependent of vapor amounts in the flow loop. Flashing-induced instability is not clearly determined and often confused as it is in a category of geysering instability.

This chapter is devoted to the flashing-induced density wave oscillations in a boiling natural circulation system with an adiabatic chimney. The following sections will describe experimental results with the test facility, which simulates natural circulation nuclear reactor depicted in Fig. **2** as an example of a boiling natural circulation system. The characteristic of the flashing-induced density wave oscillations will be discussed by comparing with other instabilities.

EXPERIMENTAL FACILITY, SIRIUS-N

Thermal-Hydraulic Loop

Fig. **3** shows a schematic of the test facility SIRIUS-N. The thermal-hydraulic loop consists of two channels, a chimney, an upper-plenum (separator), a downcomer, a subcooler, and a preheater. The channel length, l_c, is 1.7 m and the chimney length, l_r, is 5.7 m, being around 70% of the actual values used in a prototypical natural circulation BWR (Boiling Water Reactor, as shown in the Fig. **2**). A heater pin is installed concentrically in each channel. The test fluid is water that has passed through an ion exchange resin.

Measurement regions of the differential pressure sensors (R1-R8) and locations for temperature measurements (T) are shown in Fig. **3**. In addition to these temperature measurements, type-K thermocouples were embedded in the surface of the sheath heater. The recirculation flow rate was measured by an orifice flowmeter attached to the downcomer. The system pressure refers to the pressure in the separator dome.

The nondimensional parameters used for scaling of the SIRIUS-N facility are listed in Table **1**. The Froude number, N_{Fr}, and flashing parameter, N_f are key nondimensional parameters that determine the magnitude of flashing. These are defined as follows:

$$N_{Fr} \equiv \frac{u_{in}^2}{gL_c} \tag{3}$$

$$N_f \equiv \frac{h_{l,ch,in} - h_{l,r,ex}}{h_{gl}} \left(\frac{\rho_l}{\rho_g} - 1 \right) \tag{4}$$

g denotes the gravitational acceleration; $h_{l,ch,in}$ and $h_{l,r,ex}$, the saturation enthalpy at channel inlet and chimney exit, respectively. These two values in the SIRIUS-N facility are approximately 70% of the values found in a prototypical natural circulation BWR. The facility height is limited within the availability of vertical space of the existing facility building. As a result, the height of the facility had to be less than 13 m. The value of the nondimensional drift velocity is also approximately 70% of the value common to prototypical natural circulation BWRs. However, the difference in the drift velocity may not affect the stability behavior, since the drift velocity is sufficiently smaller than the single-phase natural circulation flow rate in the range of experiments. The other parameter values agree well with the actual reactor specifications. Detailed descriptions of the nondimensional parameters are summarized in reference [5].

Experimental Procedure

The experimental parameters are the system pressure, heat flux and channel inlet temperature. The natural circulation flow rate is dependent on these experimental parameters. The experimental system pressures ranges from 0.1 MPa to 7.2 MPa. The experiments were carried out with an open valve, attached at the upper separator, because condenser performance was insufficient to maintain the system pressure at 0.1 MPa. For other system pressures, the system pressure was maintained at a specific level by controlling the performance of the condenser.

Subsequently, the channel inlet temperature was maintained at the desired value by controlling the performance of the subcooler. The system allows to perform measurements for a range of the channel inlet temperature at one- or two-degree Kelvin intervals. Experiments were repeated to verify reproducibility. The time-average and standard deviation of the acquired data were correlated over four oscillation periods.

Table 1: Comparison of the SIRIUS-N facility with a prototypical natural circulation BWR.

System Pressure, P_s	0.1 MPa		7.2 MPa	
Signalling Target	**Reactor**	**SIRIUS**	**Reactor**	**SIRIUS**
Flashing Parameter, N_f	67	46	0.057	0.036
Froude Number, N_{Fr}	10.5×10^{-4}	7.6×10^{-4}	0.058	0.053
Phase Change Number, N_{pch}	11.6	13.1	3.7	3.7
Subcooling Number, N_{sub}	9.0	9.0	0.58	0.58
Nondimensional Drift Velocity, $\widehat{v}_{gj}$	1.32	1.97	0.138	0.183
Ratio of Vapor to Liquid Densities, R_{gl}	6.2×10^{-4}	6.2×10^{-4}	0.052	0.052
Ratio of Channel Inlet to Chimney Exit Vapor Densities, $\rho_{g,ch,in}/\rho_{g,r,ex}$	2.01	1.63	1.01	1.01
Friction Coefficient in the Channel, f_{ch}	6.9	5.7	3.4	2.7
Orifice Coefficient at the Channel Inlet, κ_i	10 - 50	30	10 - 50	30
Orifice Coefficient at the Chimney Exit, $\kappa_{r,ex}$	20 - 40	21	20 - 40	21
Nondimensional Downcomer Cross-Sectional Area, $\widehat{A}_d$	1.05	1.11	1.05	1.11
Nondimensional Chimney Cross-Sectional Area, $\widehat{A}_r$	2.59	2.47	2.59	2.47
Non dimensional Chimney Length, $\widehat{L}_r$	3.34	3.38	3.34	3.38

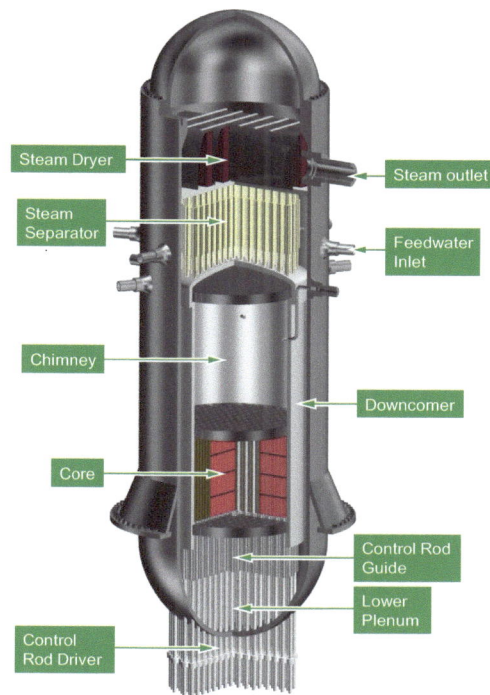

Figure 2: Schematic of a Natural Circulation BWR.

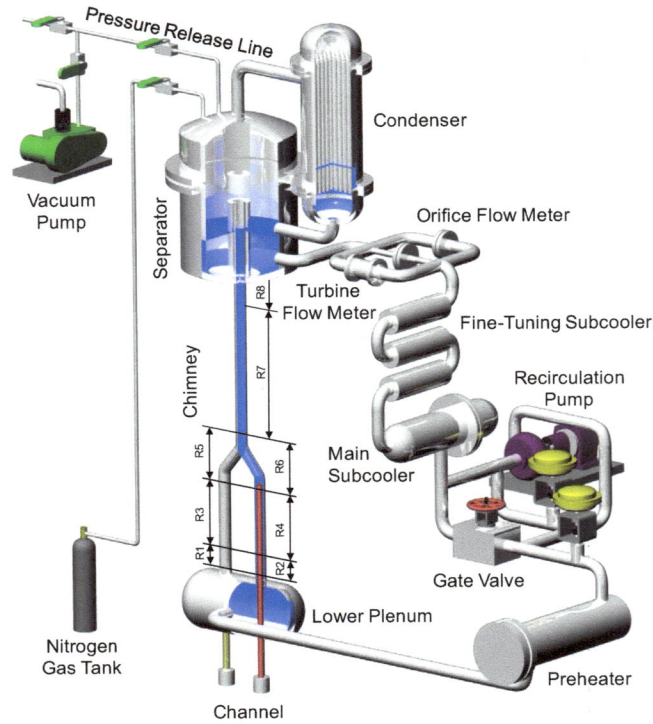

Figure 3: Schematic of the SIRIUS-N Facility.

Void Fraction Estimation

In order to investigate the thermal-hydraulic stability of a boiling two-phase flow driven by natural circulation, the void fraction must be estimated accurately, since that determines the flow rate. In this study, the void fraction was calculated in each measured section using the measured differential pressure, inlet velocity, and inlet temperature. The analysis in estimating the void fraction is based on:

[1] The flow is in a one-dimensional steady-state condition. This assumption is valid only if the integration time in the acquisition system (10 ms) is sufficiently shorter than the flow oscillation period ($\tau_{f0} > 10$ s).

[2] The following pressure losses are taken into account: acceleration, gravitation, wall friction, and local pressure losses.

[3] Each measurement region (denoted as R1 through R8 in Fig. **3**) was divided into a liquid single-phase flow part and a boiling two-phase flow part, separated by a boiling boundary. The effect of subcooled boiling is neglected, since flashing mainly takes place in the adiabatic chimney.

[4] Thermo-physical properties refer to those under saturated conditions for the system pressure, except that the liquid density in the connecting pipes of the differential pressure measurement system (DPMS) is determined by an experimentally obtained correlation.

The momentum equation, thereby using the above assumptions, yields:

$$\frac{d}{d_z}\left[\frac{1}{2}\alpha\rho_g u_g^2 + \frac{1}{2}(1-\alpha)\rho_l u_l^2\right] + \frac{d}{d_z}P + M_{wall} + M_{local} + \left[\alpha\rho_g + (1-\alpha)\rho_l\right]g = 0. \tag{5}$$

In this equation, α is the void fraction, ρ the density, P the pressure and g the gravitational acceleration. The subscripts g and l denote vapor and liquid phases respectively. The terms on the left-hand side of the equation correspond to acceleration, total pressure, wall friction, local (concentrated) friction, and gravitational losses, respectively. The wall friction loss, M_{wall}, is given by:

$$M_{wall} = \begin{cases} \dfrac{f_{1\varphi}}{2d}\rho_l u_{in}^2 & (\text{single} - \text{phase flow}) \\ \dfrac{f_{2\varphi,g}}{2d}\alpha\rho_g u_g^2 + \dfrac{f_{2\varphi,l}}{2d}(1-\alpha)\rho_l u_l^2 & (\text{two} - \text{phase flow}) \end{cases} \tag{6}$$

The loss coefficient was estimated on the basis of Blasius's correlation for liquid single-phase flow ($f_{1\varphi}$), and Martinelli-Nelson correlation with Jones's fitting for two-phase flow ($f_{2\varphi}$). The local pressure loss, M_{local}, is expressed as follows:

$$M_{local} = \begin{cases} \displaystyle\sum_i \dfrac{\kappa_{1\varphi,i}}{2}\rho_l u_{in}^2 \delta(z-z_i) & (\text{single} - \text{phase flow}) \\ \displaystyle\sum_i \left[\dfrac{\kappa_{2\varphi,g,i}}{2}\alpha\rho_g u_g^2 + \dfrac{\kappa_{2\varphi,l,i}}{2}(1-\alpha)\rho_l u_l^2 \right] \times \delta(z-z_i) & (\text{two} - \text{phase flow}) \end{cases} \tag{7}$$

A single-phase flow experiment was conducted to determine the local friction coefficients of the orifices and changes in the flow area. The measured values of these coefficients are summarized in Table **1**. Detailed experimental procedures are summarized in reference [4]. Assuming thermal equilibrium conditions, the non-boiling length (distance between the inlet of the measurement region and the boiling boundary), $Z_{1\varphi}$, and the boiling length, $Z_{2\varphi}$, is expressed on the basis of energy equation as:

$$Z_{1\varphi} = \frac{\rho_l Cp_l \Delta T_{sub,in} u_{in} A_c}{\pi D q''} \tag{8}$$

$$Z_{2\varphi} = l - Z_{1\varphi} \tag{9}$$

The u_{in} denotes the velocity at the channel inlet, A_c is the flow area of the channel, D denotes the heater diameter and q'' refers to the heat flux, l is length of measurement region. The liquid and vapor velocities are related by a drift flux model [7]. The differential pressure of the channel region, $P_{ex} - P_{in}$, is expressed as the measured value of the differential pressure measurement system (DPMS), ΔP_{DPMS}:

$$\Delta P_{DPMS} = P_{ex} - P_{in} + \rho_{DPMS} g \left(Z_{1\varphi} + Z_{2\varphi} \right) \tag{10}$$

where ρ_{DPMS} is the liquid density in the connecting pipes of the DPMS, which is higher than that under saturated conditions because the liquid flow is stagnant and at a lower temperature (due to heat loss). Neglecting this effect would result in a large error in the estimated void fraction, especially when the void fraction is low and the system pressure is high. Therefore, ρ_{DPMS} was determined experimentally as a function of ambient temperature and the local saturation temperature.

The void fraction in the two-phase region, $\alpha_{2\varphi}$, is obtained by integrating Eq. (2) over the liquid single-phase and two-phase regions with equation (10) as a constraint. This procedure requires an iterative method, since coefficients in the momentum equation are functions of the void fraction. The total void fraction in each measurement region of the DPMS, α_{total}, is then calculated as follows:

$$\alpha_{total} = \frac{z_{2\varphi}}{z_{1\varphi} + z_{2\varphi}} \alpha_{2\varphi} \tag{11}$$

EXPERIMENTAL RESULTS

Signal Time Traces

In this section, the measured time trace signals for different operational conditions are discussed. More specifically, the phenomenon of flashing is investigated.

Figs. **4a-d** show time traces of the signals acquired when P_s = 0.2 MPa and q'' = 53 kW/m^2. The graphs (from top to bottom) show the void fractions in the chimney, the temperature, and the channel inlet velocity in time. Each figure was obtained at a different channel inlet subcooling condition: (a) stable condition, (b) intermittent but periodic oscillation, (c) sinusoidal oscillations, and (d) stable condition. Similar waveforms were observed at other system pressures and heat fluxes.

As for the stable condition at the higher subcooling which is shown in Fig. **4a**, temperatures at the chimney inlet (= the channel exit) and center may coincide with the local saturation temperatures. When the flow rate is small, the temperature at the chimney exit indicates the temperature in the separator dome, which is, in fact, somewhat lower than the fluid temperature at the chimney exit. Although void fractions of 0.6% and 2.7% were generated by flashing in the chimney center (corresponding to region R7, see Fig. **3**) and exit (R8), respectively, both the void fractions and the temperatures remained constant. Therefore, no instabilities were found.

When the channel inlet temperature exceeds a certain value, the channel inlet flow rate and the void fractions pulsate intermittently as shown in Fig. **4b**. This phenomenon will be discussed later. Although the channel inlet temperature was kept constant, temperatures pulsated because of flow oscillations induced by flashing. Note that the time scale in Fig. **4b** is three times larger than the time scale in the other figures.

A further increase in inlet temperature decreases the oscillation period and amplitude, resulting in a sinusoidal curve as shown in Fig. **4c**. In both oscillatory conditions (shown in Fig. **4b** and **c**), a phase lag occurs between the void fractions at different locations in the chimney. Flashing bubbles are initiated in the chimney inlet for N_{sub} = 23.6 and in the chimney center for N_{sub} = −0.080. The bubble volume increases and travels through the chimney toward its exit. Moreover, an increase in the inlet temperature reduces the flow rate amplitude, void fraction, and temperature fluctuations. Finally, the flow stabilizes as the temperature approaches its local saturation condition at the chimney inlet as shown in Fig. **4d**.

As mentioned in the previous section (experimental procedure), the experiments were performed by increasing the inlet temperature. In order to investigate the influence of the experimental procedure on the stability boundary, the experiments were also carried out by decreasing inlet temperature. The stability boundaries obtained for increasing and decreasing inlet temperature agreed well. In other words, the stability boundary remained the same whether it was determined from the stable or from the unstable side of the boundary. This fact indicates that the nonlinearity of the instability phenomenon caused by flashing is small when the experimental parameter is close to the stability boundary.

The interaction between two parallel heated channels was investigated to determine the instability mode (in phase or out of phase). The oscillation mode was found to be in phase; no significant phase lags and amplitude differences were observed between the channels for all tested conditions. As discussed later, this lack of phase lag or amplitude difference is directly related to bubbles generated by flashing in the chimney and not to phenomena occurring in the two heated sections.

Stability Maps for Different Pressures

In order to investigate the effects of the experimental parameters (system pressure, heat flux, and inlet subcooling), stability maps were obtained. Figs. **5a-d** show the stability maps for nondimensional inlet subcooling and heat flux at system pressures of 0.1, 0.2, 0.35, and 0.5 MPa. In the figures, the horizontal axis refers to the heat flux instead of the phase change number in order to avoid a busy graph all points are situated near the zero-quality line).

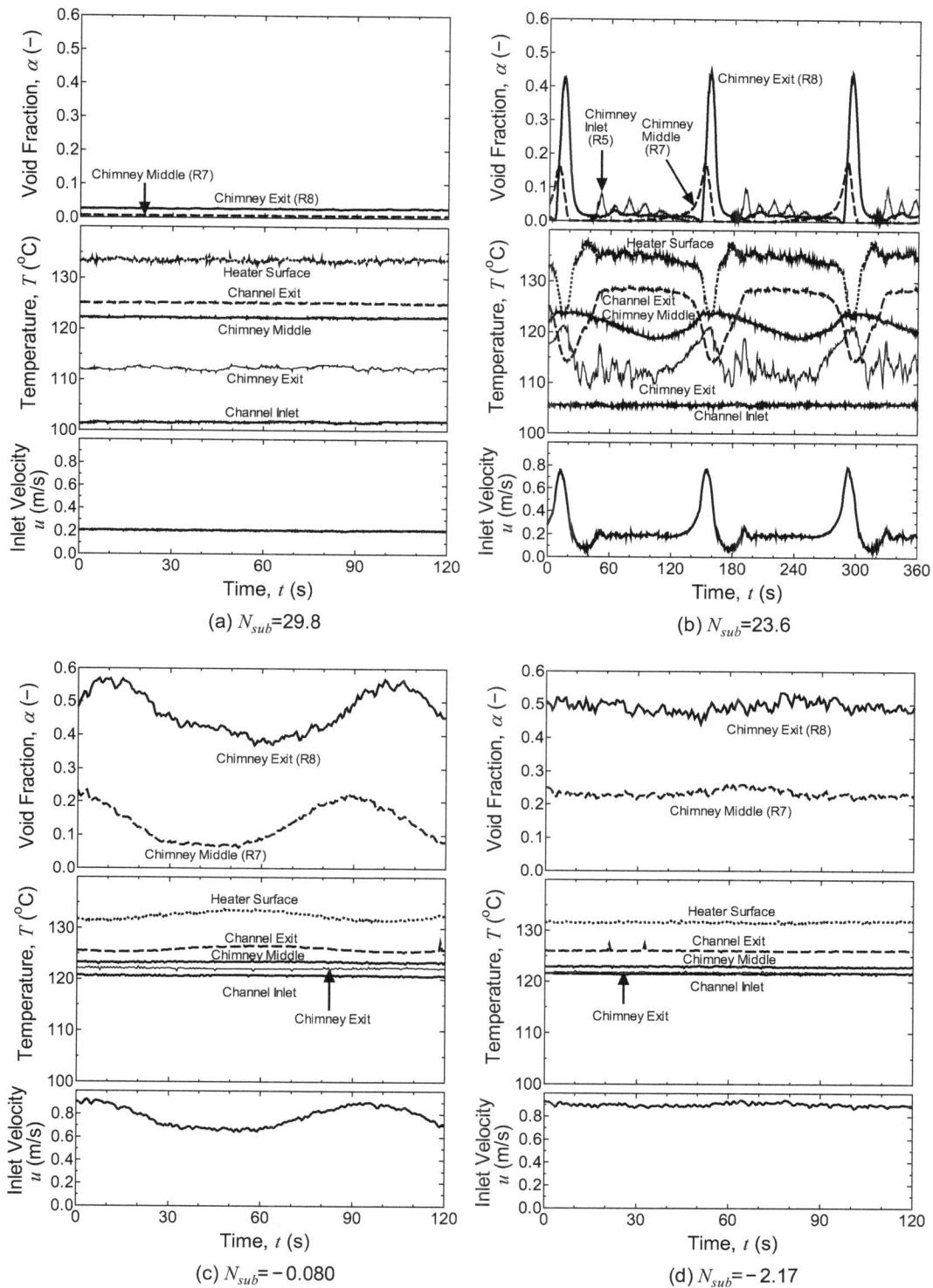

Figure 4: Time traces of signals measured when P_s = 0.2 MPa and q'' = 53 kW/m^2. (a) shows stable flow condition. Void fractions, temperatures, and inlet velocity are constant in time. Void fractions of 0.6% and 2.7% appear in the chimney, generated by flashing in the chimney middle (R7) and exit (R8). (b) shows intermittent oscillations when the channel inlet temperature exceeds a certain value. Temperatures at the chimney inlet and exit are out of phase.

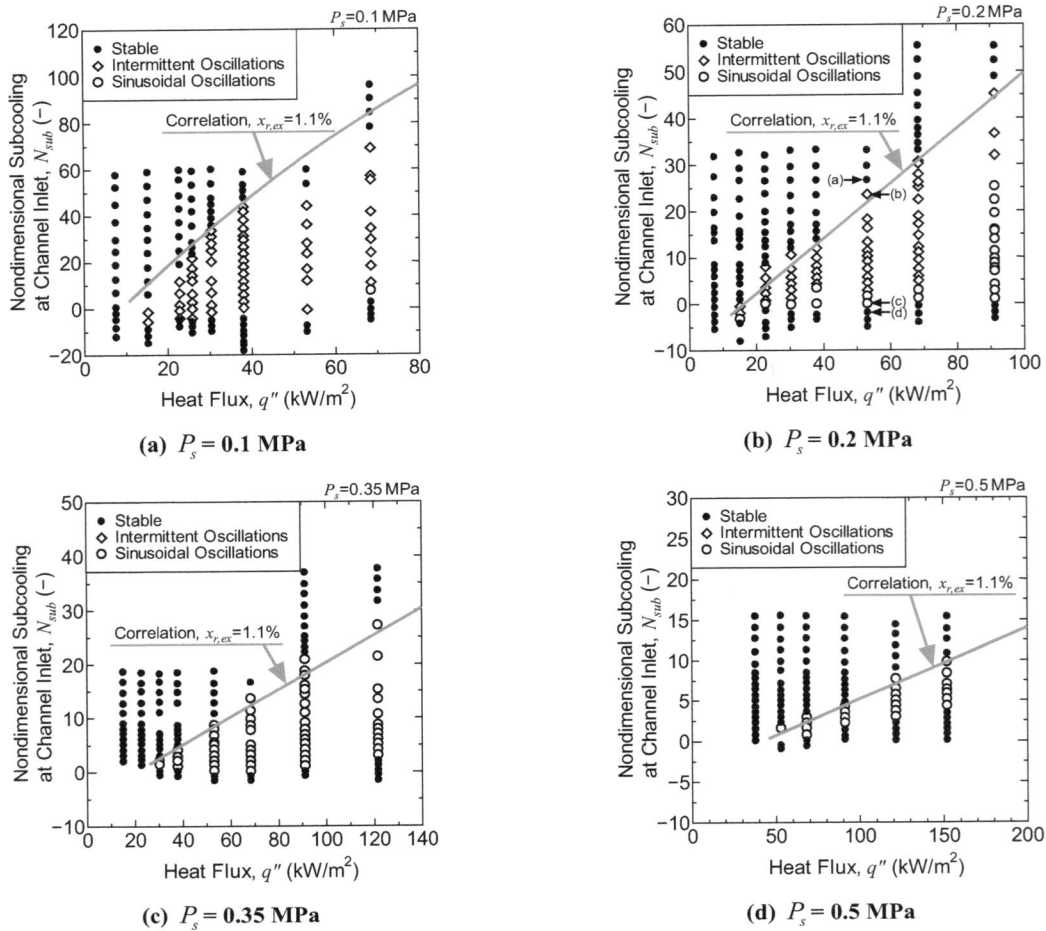

(a) $P_s = 0.1$ MPa   (b) $P_s = 0.2$ MPa

(c) $P_s = 0.35$ MPa   (d) $P_s = 0.5$ MPa

Figure 5: Stability maps for a range of heat fluxes and inlet subcooling. For every system pressure used in the study (Figs. **5a-d**), instability occurred within a certain range of the inlet subcooling. The unstable range grows as the heat flux was increased. The stability boundary at higher subcooling agrees well with the correlation ($x_{r,ex} = 1.1$ %), as is discussed later.

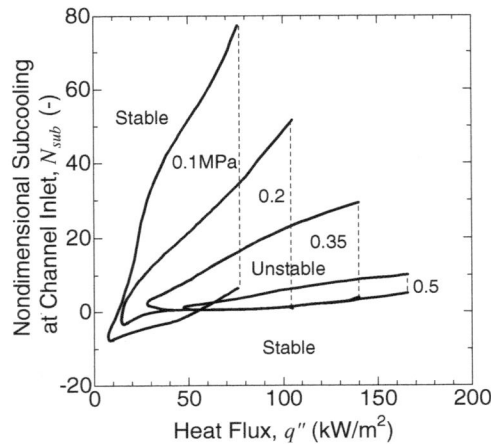

Figure 6: Stability maps for $P_s = 0.1, 0.2, 0.35,$ and 0.5 MPa. Instability was observed within the region surrounded by solid lines. The region of stability increased in size as the system pressure was increased. No instabilities were observed below a certain heat flux. This region is of practical importance, because one can start up a reactor, from cold state to high power conditions without encountering a region of instability.

The flow conditions were classified and plotted using different symbols, based on the acquired signal time traces. The symbol ' • ' means stable condition; ' ◊ ', means intermittent but periodic oscillations, as shown in Fig. **4b**; and ' ○ ', means sinusoidal oscillations, as shown in Fig. **4c**. In this study, the flow condition was classified as stable when the root mean square (r.m.s.) of the inlet velocity was less than 10% of the average inlet velocity. This threshold value is often applied in thermal-hydraulic stability estimations.

Fig. **5b** shows the stability map at 0.2 MPa. Symbols (a)-(d) in Fig. **5b** correspond to the conditions of the representative signal time traces in Fig. **4a-d**. The solid line in the figure represents a correlation indicating conditions with an exit quality of 1.1%, which will be described in Section 6.

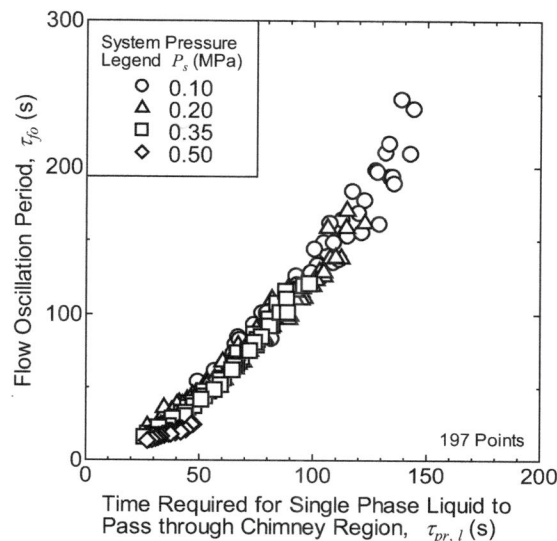

Figure 7: The relationship of the enthalpy wave propagation time and the oscillation period. All unstable data correlate well in a single curve regardless of system pressure, heat flux, and inlet subcooling. The oscillation period is nearly one to two times the time required for the fluid to travel through the chimney section, which is typical for density-wave oscillations.

As mentioned above, instability occurs within a certain range of the inlet subcooling. Fig. **5** shows that the unstable range enlarges as the heat flux increases. It is interesting to note that there were no instabilities observed below 8 kW/m^2. This region is of practical importance: one can start up a reactor without encountering an area of instability.

Intermittent oscillations occurred at higher subcooling numbers, whereas sinusoidal oscillations were observed in the lower subcooling region. The border between intermittent and sinusoidal oscillations remained unclear, as the oscillation period and amplitude continuously changed with inlet subcooling.

When the stability map at 0.2 MPa is compared with those at other system pressures, intermittent oscillations are observed mainly under unstable conditions at 0.1 MPa, as shown in Fig. **5a**. However, only sinusoidal oscillations were observed at 0.35 and 0.5 MPa, as shown in Figs. **5c** and **d**. This occurs because flashing becomes more dominant as the pressure decreases. Note that both scales are different in Figs. **5a-d**.

Fig. **6** shows the stability boundaries at system pressures of 0.1, 0.2, 0.35, and 0.5 MPa, as determined from the data in Fig. **5**. Instability is shown within the internal region surrounded by solid lines. It is obvious from this figure that the stable region increases in size as the system pressure goes up. Other studies reported the same tendency [11, 10]. Furthermore, as the system pressure increases, the region of instability shifts to a higher heat flux. According to the stability map, one can conclude that stability can be improved during the cold startup process of a reactor by pressurizing the reactor sufficiently before increasing reactor power.

INSTABILITY MECHANISM

The types of the instabilities which should be considered in this facility configuration are: flow pattern transition instability, geysering, natural circulation oscillations, and density-wave oscillations [1]. Acoustic oscillation is excluded because its mechanism is related to pressure-wave resonance with a period much shorter than that of the oscillations observed in this experiment (ranging from 13 to 250 seconds).

Pressure drop oscillations is the dynamic instability caused by a dynamic interaction between a heated channel and a compressible volume. Pressure perturbations in the separator dome are, however, relatively small ($\leq 0.5\%$ of P_s). Pressure drop oscillations are excluded for this reason. Hereafter, the discussions on possible instability mechanisms is therefore restricted to density-wave oscillations, geysering, flow pattern transition instability, and natural circulation oscillations.

Density-wave oscillations can be induced by multiple regenerative feedback among the flow rate, vapor generation rate, and pressure drops. Experiments show that the oscillation period is approximately one-and-a-half to two times the period required for the fluid to travel through the pipe [1].

Fig. **7** illustrates the relationship between the flow oscillation period, τ_{f0}, and the time required for the fluid to pass through the chimney region, $\tau_{pr,l}$, which was calculated as the volume, V_r, divided by the average volumetric flow rate of the liquid, $\langle Q_r \rangle$:

$$\tau_{pr,l} = \frac{V_r}{\langle Q_r \rangle} \tag{12}$$

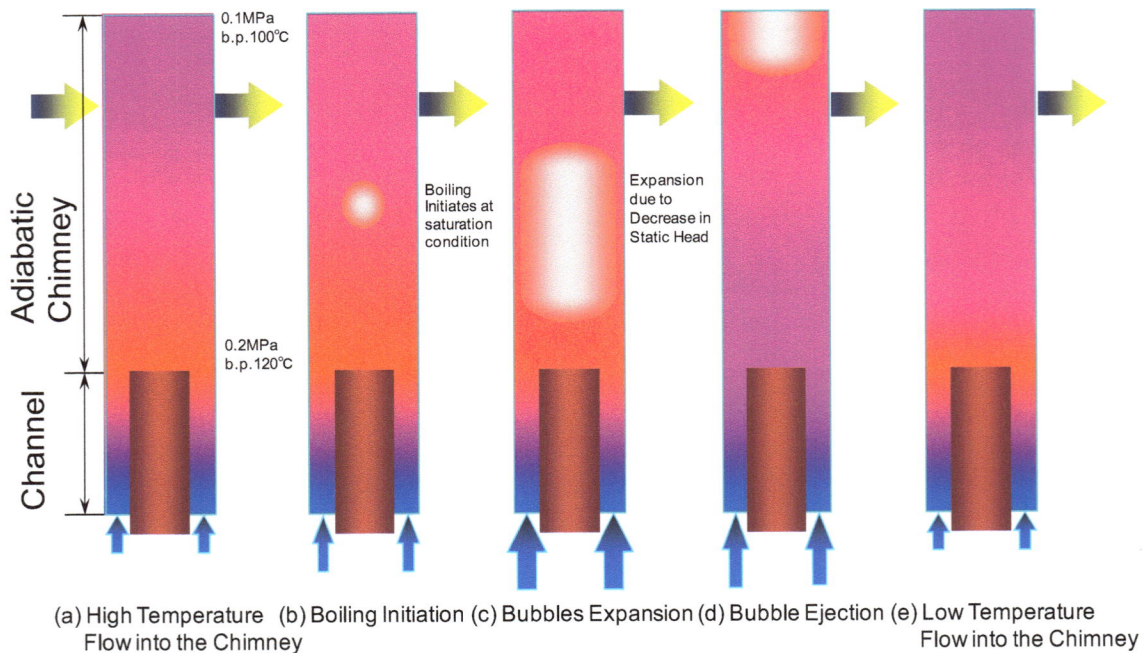

(a) High Temperature (b) Boiling Initiation (c) Bubbles Expansion (d) Bubble Ejection (e) Low Temperature
Flow into the Chimney   Flow into the Chimney

Figure 8: Proposed process for flashing-induced density-wave oscillations. Consider an adiabatic vertical pipe with a heater rod inserted from the bottom. The colors in the figure indicate the water temperature from cold (blue) to hot (red). The magnitudes of the arrows at the bottom indicate the flow rate. Flashing-induced density-wave oscillation is believed to occur through the following steps: (a) Water heated by the heater flows into the chimney. (b) Boiling occurs where the water temperature exceeds the local saturation temperature. (c) A decrease in the static head of water immediately promotes further evaporation (*i.e.*, the flashing phenomenon). (d) The natural circulation flow rate increases as a result of the enlarged vapor volume, resulting in an outflow of steam bubbles. (e) After the chimney is filled with cold water, the flow rate decreases. Then the process starts all over again from step (a).

Time $\tau_{pr,l}$ accounts only for the chimney region, since the density-waves (known as the void-sweep phenomenon) propagate mainly in the chimney region. All unstable data (197 points) correlate closely in a single curve, regardless of system pressure, heat flux, and inlet subcooling. In addition, the oscillation period is roughly one-and-a-half to two times the period required for the fluid to travel through the chimney region. These characteristics are typical for density-wave oscillations. Intermittent (but periodic) and sinusoidal oscillations in this experiment are both included in the density-wave oscillation category, because data for both oscillations correlate well on the same curve.

Fig. **8** illustrates the proposed process for flashing-induced density-wave oscillations. Imagine an adiabatic vertical pipe with a heater rod inserted from the bottom. Suppose that the pressure at the top of the chimney is 0.1 MPa (corresponding to a saturation temperature of approximately 100 °C) and that the length of the chimney is 10 m. The pressure at the bottom of the chimney will then be approximately 0.2 MPa because of the static head of water (with a saturation temperature of approximately 120 °C). The colors in Fig. **8** indicate water temperature from cold (blue) to hot (red). The magnitudes of the arrows at the bottom indicate the flow rate. Flashing-induced density-wave oscillations is believed to occur as a result of the steps outlined in Fig. **8**.

a)   Water heated by the heater (at 110 °C for instance) flows into the chimney.

b)   Boiling is initiated at a location where the water temperature exceeds the local saturation temperature.

c)   A decrease in the static head of water immediately promotes further evaporation, a phenomenon known as flashing.

d)   The natural circulation flow rate increases due to an enlarged vapor volume, resulting in an outflow of steam bubbles. The temperature at the chimney inlet becomes relatively low in turn because of the short dwell time in the heated region.

e)   After the chimney is filled with cold water, the flow rate decreases and the temperature at the chimney inlet becomes relatively high because of the long dwell time in the heated region caused by flow stagnation. The process then repeats itself beginning with process (a).

The periods of flashing and bubble outflow, corresponding to processes (b) through (d), are relatively short compared to the time scale of the entire loop. It is found that the oscillation period indeed almost agrees with the time required for the single-phase liquid to pass through the chimney region, as shown in Fig. 7.

COMPARISON WITH OTHER KINDS OF INSTABILITIES

In the previous section, the observed oscillations were shown to be caused by the flashing-induced density-wave oscillations, and a process for flashing-induced density-wave oscillations (DWO) was proposed. This section elaborates on the characteristics of flashing-induced density-wave oscillations by comparing them with other kinds of instabilities.

Geysering

Geysering has been observed in a variety of closed end (or forced flow at low flow rate) vertical columns of liquid which are heated at the base. The process can be broken down into three sub-processes: boiling delay, condensation (or expulsion of vapor) and liquid returning to the main column of liquid [14]. Studies have also shown that the period of flow oscillation, τ_{fo}, is nearly equal to the boiling delay time, τ_{bd}, because the boiling delay time is much longer than that of the other two sub-processes. Since geysering has not yet been clearly defined for natural circulation flow in relation to density-wave oscillations, geysering characteristics in a natural circulation system have been assumed to be the same as those in a closed end or forced circulation system. The boiling delay time is defined as the time required for fluid having some degree of subcooling $\Delta T_{sub,in}$ to be heated to saturation temperature, based on the pressure at the channel inlet. It is expressed by the following equation [14]:

$$\tau_{bd} = \frac{\rho_{l,in} C p_{l,in} \Delta T_{sub,in} A_c l_c}{q''} \qquad (13)$$

Fig. **9** depicts the relationship between boiling delay time and the flow oscillation period. The density, $\rho_{l,in}$, and the specific heat capacity at constant pressure, $Cp_{l,in}$, both depend on the channel inlet temperature. A_c and l_c are the flow area and length of the heated section, respectively. The oscillation period was found to be one order of magnitude longer than the boiling delay time and increased in proportion to the boiling delay time. Based on these results, we can conclude that the flow oscillations are not caused by geysering. Two additional facts also support the conclusion that geysering does not cause the instability: (1) the flow rate is large enough to prevent formation of a superheated water layer; and (2) there is not enough subcooled water to condense large bubbles, as shown in Figs. **4b** and **c**.

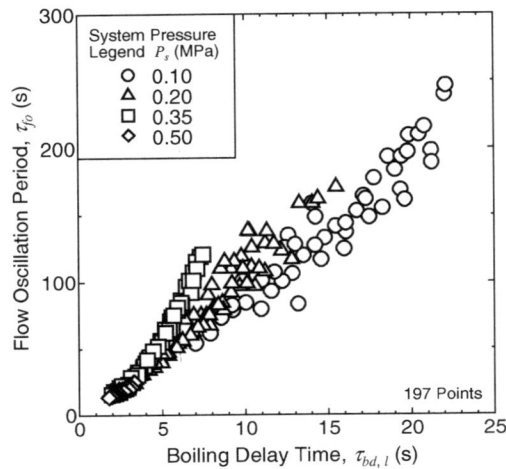

Figure 9: The relationship between boiling delay time and the oscillation period. The oscillation period was one order of magnitude longer than the boiling delay time. This shows that the flow oscillations were not caused by geysering.

Flow Pattern Transition Instability

Flow pattern transition instabilities have been postulated to occur when the flow conditions are close to the point of transition between bubbly flow and annular flow pattern. Nayak *et al.* [16] performed numerical analyses and took into account flow pattern transition criteria for a natural circulation heavy-water-moderated boiling light water cooled reactor. Analytical results show that increasing the pressure and decreasing the inlet subcooling has a stabilizing effect on the flow pattern transition instability. Although the facility used in this study was roughly as high as a standard natural circulation BWR, the chimney diameter of the facility was made relatively small for accurate demonstration of high quality flow patterns, such as churning and annular flow. The transient flow pattern was examined to clarify its relationship with flow pattern transition instability.

Fig. **10** shows experimentally obtained transient flow patterns drawn according to the method used in Mishima-Ishii's diagram [12]. The experimental data are plotted for approximately three oscillation cycles at the chimney exit region, R8 (see Fig. **3**), when $P_s = 0.2$ MPa and $q'' = 53$ kW/m². Labels (b), (c), and (d) correspond to those in Figs. **4b, c,** and **d**. Because of the low void fraction, no plot is shown for a highly subcooled stable condition as shown in Fig. **4(a)**.

The flow pattern changed from liquid single-phase flow to slug flow during the intermittent oscillations. Boiling always took place in the chimney during the sinusoidal oscillations, so the flow regime was slug flow and the locus resembles an ellipse. Decreasing inlet subcooling reduces the area of the ellipse, and the flow then stabilizes. Consequently, the flow pattern never changed to a churn or annular flow pattern, hence flow pattern transition instabilities were not found in our experiments.

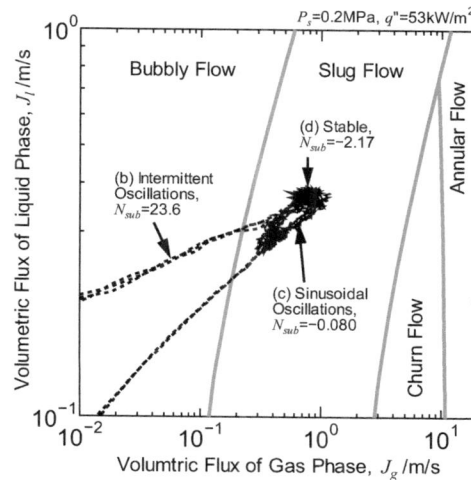

Figure 10: Transient flow patterns plotted for three cycles show a change from a bubbly flow pattern to a sluggish one. The flow pattern never changed to a churn or annular flow pattern.

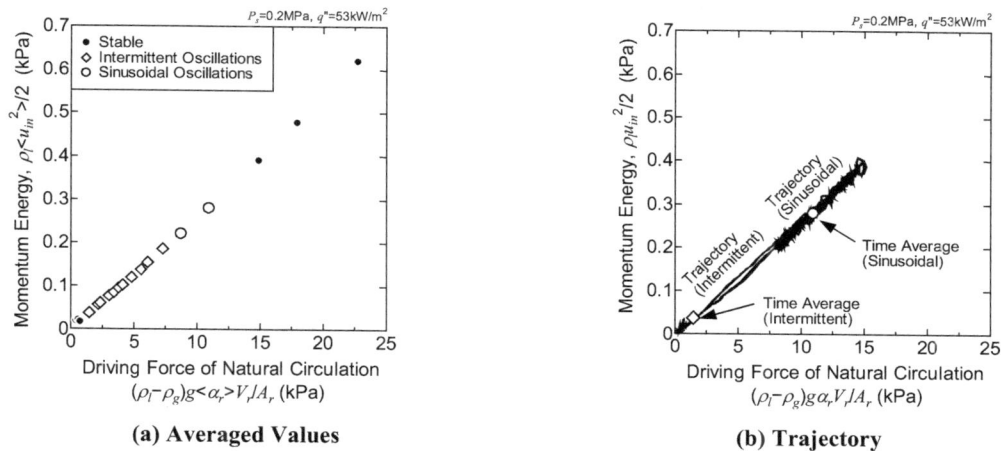

Figure 11: Momentum response to a natural circulation driving force. **(a)** shows that the average values form a straight line that passes through the origin in such a way that the static characteristics are well correlated whether the flow condition is stable or unstable. **(b)** shows that the trajectory follows the straight line of the static characteristics. Therefore, the resulting oscillations are not caused by natural circulation oscillations for which the trajectory would be ellipsoidal.

Natural Circulation Oscillation

Chiang *et al.* [2] observed an interesting flow oscillation phenomenon in a loop having a section in which bubbles could accumulate. They called this phenomenon "natural circulation oscillation," and postulated the following mechanism for it:

a)   Vapor accumulates in an unheated pipe (connecting the outlet plenum to the separator tank, which consists of vertical and horizontal pipes), when the evaporation rate is insufficient.

b)   Bubbles coalesce with the incoming vapor, and then the hydrostatic head decreases gradually.

c)   An increase in the flow rate flushes the accumulated vapor and water fills in.

d)   The hydrostatic head increases and therefore the circulation rate decreases (resulting in a return to process (a)) and then the process repeats itself.

This phenomenon repeats periodically. Although bubbles cannot accumulate in the present SIRIUS-N facility, a method to investigate the natural circulation oscillation was proposed. Fig. **11** illustrates the response of the momentum energy to the natural circulation driving force at $P_s = 0.2$ MPa and $q'' = 53$ kW/m^2. Natural circulation was induced by a decrease in apparent density as a result of vapor generation. Therefore the driving force can be expressed as $\left(\rho_l - \rho_g \right) g \alpha_r V_r / A_r$. The subscript ' r ' denotes the value defined at the chimney. The momentum energy is expressed as $\rho_l u_{in}^2 / 2$.

The average values are plotted in Fig. **11a**. The values form a straight line that passes through the origin. This means that static characteristics are well correlated under both stable and unstable conditions.

The trajectory is shown in Fig. **11b** for representative intermittent and sinusoidal oscillations, as shown in Figs. **4b** and **c**. The trajectory follows the straight line of the static characteristics, as illustrated in Fig. **11a**. Apparently, the response of the momentum energy to the natural circulation driving force was so fast that vapor did not accumulate in the two-phase section. Since the trajectory is not ellipsoidal, we conclude that natural circulation oscillations are absent here.

Type-1 Instability

The importance of flashing is quantified by the flashing number. The flashing number is a nondimensional form of saturated enthalpy difference, which is defined as follows :

$$N_f \equiv \frac{h_{l,in}\left(P_{in}\right) - h_l}{h_{gl}} \left(\frac{\rho_l}{\rho_g} - 1 \right) \tag{14}$$

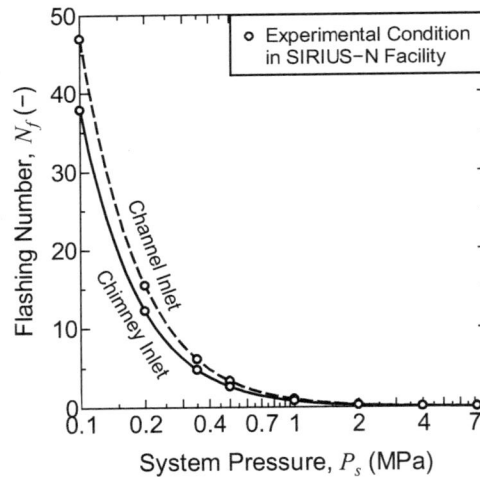

Figure 12: Effect of flashing in terms of pressure. The pressure conditions tested in the SIRIUS-N facility are plotted as open circles. Flashing becomes less dominant as pressure increases.

Fig. **12** shows the effect of flashing in terms of the system pressure. The pressure conditions tested in the SIRIUS-N facility are included in the figure (open circle). As shown in the figure, flashing becomes less dominant above roughly 1 MPa.

In contrast to Type-I density wave oscillations at high pressure, the oscillation period correlates well with single-phase liquid travel time for flashing induced density wave oscillations at low pressure as shown in Fig. **14**.

The horizontal axis denotes the time required for single-phase liquid to pass through the chimney region, $\tau_{pr,l}$, which is equal to the volume, V_r, divided by the average volumetric flow rate of the liquid, $\langle Q_{l,r} \rangle$. The obtained high pressure data are also plotted in the same figure. Two distinct branches in Fig. **14** clearly indicate that the driving mechanisms of the instabilities are different for low and high pressures. It can be seen in the figure that some plots of 0.5 MPa are laid in between the two branches, which implies the transition region of the two mechanisms. Therefore, the demarcation lies somewhere in between 0.5 MPa and 1.0 MPa.

The difference of the two mechanisms at high and low pressures can be explained by the role of flashing. As shown previously in Fig. **12**, flashing becomes less dominant above roughly 1 MPa at which the two mechanism are divided. At high pressures, the oscillation period correlates well with the bubble passing time in the chimney, since the instability is caused by a delay of the bubbles created in the channels flowing into the chimney to dominate natural circulation flow. On the other hand, the oscillation period at low pressures correlates well with the single-phase liquid passing time in the chimney, since the instability is caused by a delay of single phase liquid flowing into the chimney to dominate natural circulation flow by flashing.

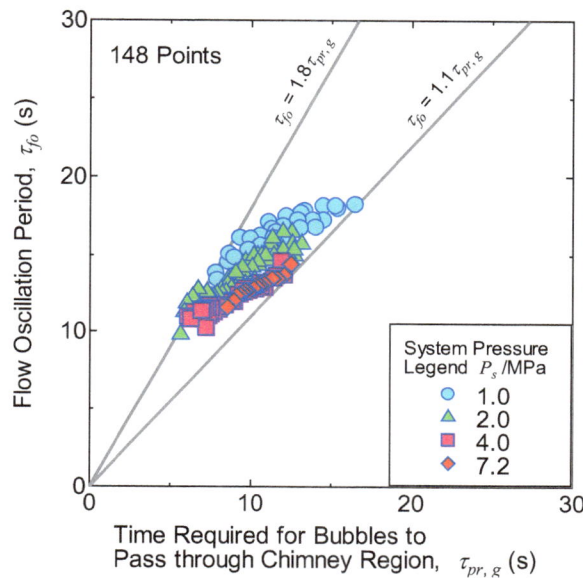

Figure 13: All unstable data correlate well in the single clusters regardless of the system pressure, heat flux, and inlet subcooling. The oscillation period was 1.1 to 1.8 times the time required for bubbles to travel through the chimney region, exhibiting practical characteristics of density wave oscillations.

STABILITY BOUNDARY FOR HIGHER SUBCOOLING

The unstable region shown in Fig. **6** can be considered to be composed of two lines at higher and lower subcooling. Those two lines may be considered to merge near the origin. In this section, this linear approximation of the stability boundary at higher subcooling is investigated.

Fig. **15** shows the exit quality at the stability boundary for the upper line, $x_{r,ex}$, versus the heat flux. $x_{r,ex}$ is estimated as follows:

$$x_{r,ex} = \frac{h_{ch,ex} - h_{l,r,ex}}{h_{g,r,ex} - h_{l,r,ex}} \tag{15}$$

where $h_{g,r,ex}$ and $h_{l,r,ex}$ are the vapor and liquid saturation enthalpy at the chimney exit respectively (based on the system pressure). $h_{ch,ex}$ is the enthalpy estimated by the pressure and temperature at the channel exit.

The enthalpy defined at the channel exit is used in place of that defined at the chimney exit, because these two values are almost the same under the adiabatic condition. Remarkably, variations in $x_{ch,ex}$ are very small, ranging from 0.9% to 1.3% regardless of heat flux and system pressure. Flashing occurred when the exit quality exceeded zero at the chimney exit. For flow oscillations to be caused by flashing, the exit quality should probably be higher than approximately 1.1%. For a reference, a constant exit-quality line ($x_{r,ex} = 1.1\%$) is drawn by the homogeneous equilibrium mixture (HEM) analysis in the stability maps in Fig. **5**. Flashing induces flow oscillations below this line in the stability map.

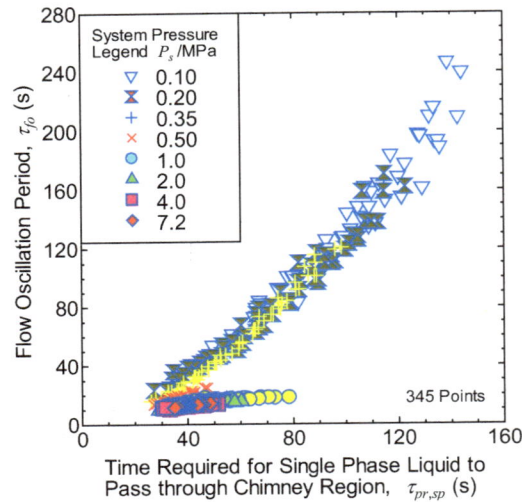

Figure 14: Two distinct branches are found in the relation between the oscillation period and liquid passing time in the chimney. This indicates that the driving mechanisms of the instabilities are different for low and high pressures. The demarcation lies somewhere in between 0.5 MPa and 1.0 MPa.

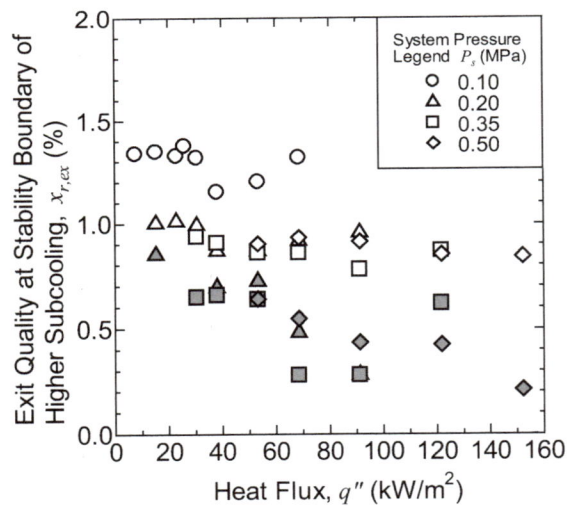

Figure 15: Exit quality at the stability boundary for higher subcooling. Variations of $x_{r,ex}$ are very small, ranging from 0.9% to 1.3% regardless of the heat flux and system pressure. Flashing occurs when the exit quality exceeds zero at the chimney exit. Flow oscillations caused by flashing occur for the exit qualities higher than approximately 1.1%.

CONCLUDING REMARKS

Characteristics of flashing-induced density wave oscillations are addressed on the basis of the experimental results in a boiling natural circulation system with an adiabatic chimney. Two distinct branches are found in the relation of the oscillation period and the liquid passing time in the chimney, indicating that the driving mechanisms of the instabilities are different for low and high pressures. The demarcation lies between 0.5 MPa and 1.0 MPa as implied by flashing parameter.

At lower pressure, the flashing-induced density wave oscillations can occur, since the oscillation period correlates well with the passing time of bubbles in the chimney section regardless of the system pressure, the heat flux, and the inlet subcooling. In contrast, type-I instability can occur, since the oscillation period correlates well with the passing time of bubbles in the chimney section as well.

According to the stability map, the flow became stable below a certain heat flux regardless of the channel inlet subcooling. The stable region enlarged with increasing system pressure. Therefore, the stability margin becomes larger by pressurizing the loop sufficiently before heating.

ACKNOWLEDGEMENTS

The Authors would like to thank Dr. Y. K. Cheung of GE Nuclear Energy for providing the design parameters of the natural circulation BWR. Thanks are extended to Mr. Y. Shiratori of Electric Power Engineering Systems Co. for conducting the experiments.

NOMENCLATURES

A	flow area
A_r	ratio of chimney to channel flow area
Cp	heat capacity
D	heater rod diameter
f	wall friction coefficient
g	gravitational acceleration
h	enthalpy
h_{gl}	latent heat
L_r	nondimensional chimney length
l	length of DPMS measurement region
P_e	perimeter
P_s	system pressure (separator dome pressure)
q''	heat flux
T	temperature
t	time
M_{local}	local pressure loss
M_{wall}	wall friction pressure loss
N_{Fr}	flashing parameter
N_f	Froude number
N_{pch}	phase change number
N_{sub}	subcooling number
R_{gl}	ratio of vapor to liquid densities
u	velocity
v_{gj}	nondimensional drift velocity

Z	length of a region
z	vertical axis

Greek Symbols

α	void fraction
ΔT_{sub}	degree of subcooling
ΔP_{DPMS}	measured value of DPMS
κ	local pressure loss coefficient
ρ	fluid density or reactivity
ρ_{g2}/ρ_{g1}	ratio of chimney exit to channel inlet vapor densities
τ_{f0}	flow oscillation period
$\tau_{bd,l}$	boiling delay time
$\tau_{pr,g}$	time required for bubbles to pass through the chimney region
$\tau_{pr,sp}$	time required for single-phase liquid to pass through the chimney region

Subscript and Superscript

1φ	liquid single-phase
2φ	two-phase
c	connecting pipes of DPMS
ch	heated channel
ex	exit
f	fuel rod
g	vapor phase
in	channel inlet
l	liquid phase
s	saturation condition

REFERENCES

[1]   J. A. Bouré, A. E. Bergles, and L. S. Tong, "Review of two-phase flow instability," *Nucl. Engng. Des.*, vol. 25, pp. 165-192, 1973.

[2]   M. Ishii, "Study of flow instabilities in two-phase mixtures," *Tech. Rep. ANL76-23*, ANL, 1976.

[3]   G. Yadigaroglu, "Two-phase Flow Instabilities and Propagation Phenomena," Hemisphere Publishing Corporation, Ch. 17, pp. 353-403, 1981.

[4]   S. Nakanishi, "Recent Japanese research on two-phase flow instabilities," In: *Proceedings of Japan-US Seminar on Two-phase Flow Dynamics*, Hemisphere Publishing Corporation, 1979.

[5]   S. Kakac and H. T. Liu, "Two-phase flow dynamics instabilities in boiling systems," In: *Multiphase Flow and Heat Transfer*, Second International Symposium. Hemisphere, Washington, DC, pp. 403-444, 1991.

[6]   K. Fukuda and T. Kobori, "Classification of two-phase flow instability by density wave oscillation model," *J. Nucl. Sci. Technol.*, vol. 16 (2), pp. 95-108, 1979.

[7]   M. Ishii and N. Zuber, Thermally induced flow instabilities in two-phase mixtures", In: *Proc. 4th International Heat Transfer Conference*, vol. B5.11. Paris, 1970.

[8]   A. Nayak, P. K. Vijayan, D. Saha, R. V. Venkat, and M. Aritomi, "Linear analysis of thermohydraulic instabilities of the advanced heavy water reactor (AHWR)," *J. Nucl. Sci. Technol.*, vol. 35, pp. 768-778, 1998.

[9]   D. Van Bragt and T. Van der Hagen, "Stability of natural circulation boiling water reactors: Part I: Description stability model and theoretical analysis in terms of dimensionless groups," *Nucl. Technol.*, vol. 121, pp. 40-51, 1998.

[10]   D. Van Bragt and T Van der Hagen, T., "Stability of natural circulation boiling water reactors: Part II: Parametric study of coupled neutronic-thermalhydraulic stability," *Nucl. Technol.*, vol. 121, pp. 52-61, 1998.

[11]   M. Furuya, "Experimental and analytical modeling of natural circulation and forced circulation BWRs - Thermal-hydraulic, core-wide, and regional stability phenomena," Chapter 3, PhD Dissertation, Delft University of Technology, ISBN1-58603-605-X, IOS Press Co.

[12]   E. Wissler, H. S. ISBIN, and N. R. AMUDSON, "Oscillatory behavior of a two-phase natural-circulation loop," *AIChE J.*, vol. 2, pp. 157-162, 1956.

[13]   F. Inada and M. Furuya, "Thermo-hydraulic instability of natural circulation BWRs at low pressure start-up: experimental estimation of instability region with test facility considering scaling law," *Proc. 3rd JSME/ASME Joint Int. Conf. Nucl. Engng.*, vol. 1, pp. 173-178, 1995.

[14]   M. Furuya, F. Inada, and A. Yasuo, "Density wave oscillations of a boiling natural circulation loop induced by flashing," *Proc. 7th Int. Meeting on Nucl. Reactor Thermal-Hydraulics*, NUREG/CP-0142 2, pp. 923-932, 1995.

[15]   M. Ishii, "One-dimensional drift-flux model and constitutive equations for relative motion between phases in various two-phase flow regimes," *Tech. rep., ANL77-47*, 1977.

[16]   Y. Masuhara, *et al.*, "Research on geysering phenomena in the natural circulation BWR". *Proc. 2nd JSME/ASME Joint Int. Conf. Nucl. Engng.*, vol. 1, pp. 135-141, 1993.

[17]   A. Manera and T. H. J. J. van der Hagen, "Stability of natural- circulation-cooled boiling water reactors during startup: experimental results," *J. Nuclear Technology*, vol. 143 (1), pp. 77-88, 2003.

[18]   S. Nakanishi, S. Ishigai, M. Ozawa, Y. Mizuta, and H. Tarui, "Flow instability in boiling channels: 2. Geysering," *Trans. JSME J. Heat Transfer* (in Japanese), vol. 44 (388), pp. 4252-4262, 1978.

[19]   A. K. Nayak, P. K. Vijayan, V. Jain, D. Saha, and R. K. Sinha, "Study on the flow-pattern-transition instability in a natural circulation heavy water moderated boiling light water cooled reactor," *Nucl. Engng. Design.*, vol. 225 (2-3), pp. 159-172, 2003.

[20]   K. Mishima and M. Ishii, "Flow regime transition criteria for upward two-phase flow in vertical tubes," *Int. J. Heat Mass Transfer*, vol. 27, pp. 723-737, 1984.

[21]   J. H., Chiang, M. Aritomi, and M. Mori, "Fundamental study on thermo-hydraulics during start-up in natural circulation boiling water reactors, (II) natural circulation oscillation induced by hydrostatic head fluctuation," *J. Nucl. Sci. Technol.*, vol. 30 (3), pp. 203-211, 1993.

CHAPTER 9

Nonlinear Dynamic Characteristics of Bubbling Fluidization

Reza Zarghami[1], Navid Mostoufi[1], Rahmat Sotudeh-Gharebagh[1] and Jamal Chaouki[2,*]

[1]*School of Chemical Engineering, College of Engineering, University of Tehran, Tehran, Iran and* [2]*Department of Chemical Engineering, Ecole Polytechnique de Montreal, Montreal, Canada*

Abstract: Nonlinear time series analysis techniques were applied to characterise bubbling fluidization. The experiments were carried out in a laboratory scale fluidized bed, operated under ambient conditions and various sizes of particles, settled bed heights, measurement heights and superficial gas velocities. It was found that a minimum in average cycle frequency, wide band energy and entropy with an increase in the velocity corresponds to the transition between macro structures and finer structures of the fluidization system. This minimum was mostly found in the macro structures of the bubbling fluidization system. Hurst exponent of the pressure fluctuations showed that the fluidized bed has a bifractal behaviour. The reciprocal of the break point in the Hurst profile is similar to the domain frequency of the bed. The method of delays was used to reconstruct the state space attractor to carry out analysis in the reconstructed state space. The state space reconstruction parameters, *i.e.*, time delay and embedding dimension, were determined and the results showed that their values are different for various types of methods introduced in literature. Chaotic behaviour of fluidized system was determined by introducing two nonlinear dynamic invariants, correlation dimension and entropy in different ways. The state-space analysis reflected that a low dimension behaviour of the bubbling fluidization system. The nonlinearity test showed that nonlinearity cannot be concluded at all gas velocities.

Keywords: Nonlinear, pressure fluctuation, hurst exponent, chaotic attractor, entropy, correlation dimension.

INTRODUCTION

Fluidization is a process in which solid particles are suspended in a gas or liquid and become fluidized, similar to state of a liquid. Today, the fluidization phenomenon has increasingly developed and is considered as a suitable alternative for many old industrial processes. The most significant benefits of this process are the high rate of heat and mass transfer in compare to the similar processes. Fluidized beds are important for their effective contact between solid and fluid in many chemical industries (*e.g.*, oil, petrochemical, chemical, mineral, biochemical, pharmaceutical, food, *etc.*).

Despite numerous benefits of fluidized beds, fluidization is one of the most complicated unit operations in practice. Its typical particle-fluid two-phase flow patterns exhibit nonlinear dynamic characteristics with heterogeneous flow structure. Therefore, its quantitative understanding and practical application present a challenge in science and engineering which makes its application to be limited in industry. An unwanted change in the fluidized bed hydrodynamic due to sudden change of the excess gas velocity (the difference between superficial gas velocity and minimum fluidization velocity) is disadvantageous. The quality of mixing of solid particles is affected by this phenomenon and may result in total or partial defluidization of the bed, blocking parts of gas distributors, creating hot spots or increasing the local gas velocity. Therefore, permanent monitoring of hydrodynamic conditions of the fluidized bed is very important and further investigations are still needed on this issue.

The hydrodynamics of gas-solids fluidized bed is governed by a complex nonlinear dynamic relationship

*Address correspondence to Jamal Chaouki:** Department of Chemical Engineering, Ecole Polytechnique de Montreal, Montreal, Canada; E-mail: jamal.chaouki@polymtl.ca

which is mainly controlled by different dynamic phenomena that occur in the bed. Examples of these phenomena are bubble formation, bubble coalescence and splitting, bubble passage as well as particles behaviours. If the hydrodynamics of the fluidized system is modeled with a set of nonlinear governing equations, then a proper understanding of the state of the fluidized bed at a certain time can be determined. However, the governing equations of these systems are complex and unknown [1]. In this case, a quantitative interpretation of the hydrodynamics of the fluidized bed can be achieved through time series evaluation of the measured signals, such as pressure or local void fraction in the bed.

A great advantage of the pressure fluctuations is that they are easy to measure and include the effect of many different dynamic phenomena taking place in the bed, such as gas turbulence, bubbles hydrodynamics and the bed operating conditions [2, 3]. On the other hand, most of other laboratory measurement techniques are not applicable in industrial processes [4].

The most usual methods to characterize time dependent signals (time series analysis) from fluidized beds (which describe the properties of the fluidized system) are time, frequency domain and nonlinear state space analyses. Time domain approaches typically include observation of the time sequence of the measured signal [5, 6], standard deviation analysis [5-8], analysis of other statistical moments like skewness, kurtosis and flatness [5-9], auto and cross correlation functions [6, 9], mutual information function [9], and less frequently used rescaled range (known also as R/S or Hurst exponent) analysis [9-11].

The standard deviation defines the deviation of the distribution from a normal distribution and is proportional to size and number of bubbles [12], whereas the higher order moments give information about intermittency in the time series [6]. The flatness in intermittent systems also can be considered as the ratio of the time spent under inactive conditions to the time spent under active conditions. Skewness express lack of symmetry (0 for a Gaussian distribution) and flatness presents sharpness in a probability distribution (3 for a Gaussian distribution).

Most of time domain analyses have been used for determination of regime transitions [12-14] rather than the fluidization quality [15]. These researchers believe that the maximum value of standard deviation versus superficial gas velocity is corresponding to transition point from bubbling to turbulent regime. van Ommen *et al.* [16] stated that since standard deviation changes with superficial gas velocity, it is not appropriate for monitoring of the status of industrial fluidized beds in which the superficial gas velocity is frequently changing. Lee and Kim [5] showed a shift from negative to positive skewness and a maximum in flatness by increasing velocity and considered the zero point in skewness and the flatness-maximum to be corresponding to the transition velocity. However, Bi and Grace [17] found that this transition depends on the type of measurement as well as if skewness or standard deviation is used. Zarghami [9] found that a shift of skewness from negative to positive and a minimum in flatness against velocity corresponds to shift from macro structures and finer structures of the flow rather than transition.

Autocorrelation and mutual information functions are more frequently used in nonlinear state space analysis to determine time delay of reconstructed attractor [18-20]. Hurst exponent analysis was developed for the first time by Hurst [10] to distinguish completely random time series from correlated time series. Hurst exponent was used by Fan *et al.* [21], Franca *et al.* [22], Drahos *et al.* [23], Cabrejos *et al.* [24], Briens *et al.* [25], Karamavruc and Clark [26] to assess the hydrodynamic status of the fluidized bed. Hurst analysis was used by Zhao *et al.* [11] to analyze the multifractal characteristics of different level of decomposed signals with wavelet transform.

The average cycle frequency has been used by many researchers [27-29] in the frequency domain due to its simple calculation for obtaining embedding parameters of the fluidized reconstructed attractor. The average cycle frequency has not been considered to investigate the transition point and/or conditional monitoring of the fluidized bed.

Fast Fourier Transform (FFT) and Wavelet Transform (WT) are also the mathematical tools used to analyse the pressure fluctuations in fluidized bed, which express the behaviour of a time series in the frequency domain. Analysis of the pressure fluctuations in the frequency domain was first adopted to find the dominant frequencies of the bed phenomena. The peak dominant frequency (frequency ranges 0.5 to 2.5 Hz) in a power spectrum in bubbling and slugging beds is corresponding to the bubbles/slugs pass through the bed [3, 30-32]. Frequency domain analysis was also reported to quantify fluidization regimes [33-34]. It was also applied to prove the existence of the same regime in different scaling systems by comparing dominant frequencies of a scaled model with those of the full-scale unit [35] or by comparing the frequency spectra in a certain range of frequencies [36]. This was done since wavelet analysis also has great potential in signal processing and has been widely studied in the fluidization field [11, 37-44] for studying of transition velocity from bubbling to turbulent fluidization [44], dynamic behavior [41-42] and multi-scale nature of gas-solid fluidization [11, 37-39, 43]. Evaluation of time series in frequency domain is rarely used for fluidization monitoring. van Ommen *et al.* [16] showed that the signal power spectrum density of the pressure fluctuations is not sensitive to small changes in particle size distribution and cannot be used for permanent monitoring to detect changes in particle size.

In the recent years, frequency domain analysis has been proposed to study the complex dynamics of a fluidization system through identifying different scale of structures in the fluidized bed. Three different structures, *i.e.*, macro, mezzo, and micro, have been identified in literature from both FFT [6] and WT methods [11, 37-39, 43]. However, Zarghami [9] showed that the range of frequencies identified for each structure is not the same in these methods and should be used cautiously.

Various nonlinear analysis methods have been used for analyzing the dynamic changes in fluidization hydrodynamics such as time delay embedding theory [45-56]. However, all methods of nonlinear time series analysis are based on the construction of an attractor of the dynamic evolution of the system in the state space. The state of a fluidized bed at a certain time can be determined by projecting all governing variables of the system into a multidimensional space, *i.e.*, the state space. However, it is not possible to determine all governing variables of a fluidized bed. Takens [57] proved that the dynamic state of a system can be reconstructed from the time series of only one characteristic variable, such as the local pressure, in a fluidized bed.

Several researchers have found characteristics of low dimensional (typically less than five) deterministic chaos from time series of the pressure fluctuations in fluidized beds [6, 28, 46, 51-52]. Nonlinearity tests applied to bubbling fluidized bed data have confirmed that the data are nonlinear [29]. However, Johnssona *et al.* [6] and Zarghami [9] found that nonlinearity can be only addressed at low correlation dimensions. Some researchers have attributed the fluctuations of the pressure signal in the fluidized beds to the noise in the measurements rather than the chaotic behaviour of a low dimension system [58-59]. Nonlinear analysis of the measured time series in a fluidized bed, usually containing noise, is not straightforward and still needs more research. Zarghami *et al.* [60] found a kind of uncertainty in reconstruction of fluidized bed attractor from its pressure fluctuations. They reported that different methods of reconstruction lead to different embedding parameters.

All methods of characterizing the attractor, in principal, are based on evaluation of the correlation dimension and the (Kolmogorov) entropy [6, 61] and/or determination of the Lyapunov exponents [62-63]. These parameters are statistics in the sense that they are averaged values and are discriminant because they clearly evaluate different values for regular (non-chaotic) and chaotic deterministic data. Entropy and Lyapunov exponents identify the deterministic behaviour of a system and the sensitivity to initial conditions, whereas, the correlation dimension expresses the number of degrees of freedom of the system. While the correlation dimension and the (Kolmogorov) entropy have been determined in fluidized beds by many researchers [46, 50, 58], estimation of Lyapunov exponents is not straightforward and the analysis methods are still under investigation [29].

While only time and frequency domain analyses of dynamical variables of the fluidized bed are not capable of determining the whole complexity of the fluidized bed system, the study of nonlinear behavior of fluidization still is a relatively new subject. Nonlinear analysis of experimental time series can be highlighted in different steps of preliminary analysis in time and frequency domains, attractor reconstruction (estimation of embedding parameters), characterization of reconstructed attractor through evaluation of geometric structure of the attractor (estimation of entropy and correlation dimension and stationarity), supporting side calculation (*e.g.*, the separation of high dimensional and stochastic dynamics from low dimensional deterministic signals, nonlinearity detection, *etc.*) and finally, conditional monitoring, prediction, modeling, scaling and control. A schematic representation of these different steps is given in Fig. **1**.

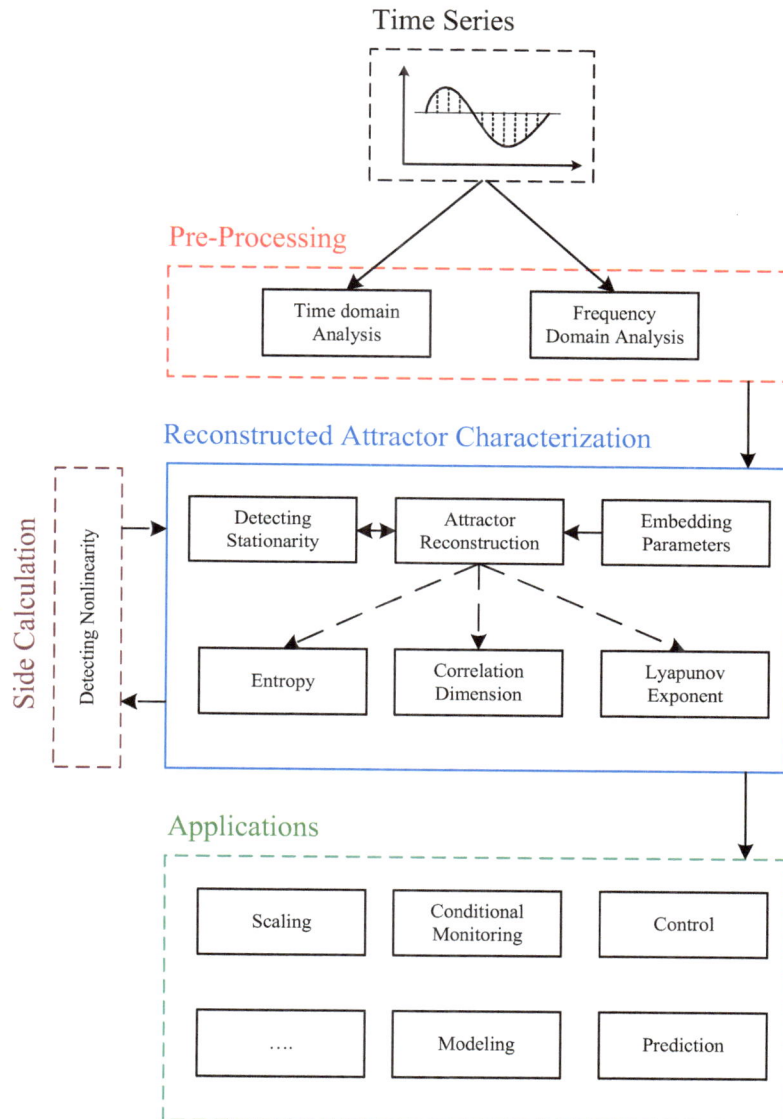

Figure 1: Schematic representation of time series analysis.

The present work intends to address some nonlinear characteristics of fluidized bed hydrodynamics through the evaluation of the measured local pressure fluctuations within the bed. First, after a brief review of sources of the pressure fluctuations, a preliminary study is carried out with the aim of characterizing time series in terms of long term memory (R/S analysis) and average cycle frequency. In the second step, stationarity behaviour and state space reconstruction parameters, *i.e.*, time delay and embedding dimension are obtained for different hydrodynamic state and then the correlation dimension and the (Kolmogorov) entropy have been analysed in the reconstructed state space. In order to ensure the correct interpretations of these discriminant statistics, the nonlinearity of the pressure fluctuations were investigated.

EXPERIMENTAL DATA

The experimental setup is schematically shown in Fig. **2**. Experiments were carried out in a gas-solid fluidized bed made of a Plexiglas-pipe of 15 cm inner diameter (D) and 2 m height (L). Air at ambient temperature entered the column through a perforated plate distributor with 435 holes of 7 mm arranged in a triangular pitch. A cyclone was used to separate particles from air at high superficial gas velocities.

The pressure fluctuations were recorded as absolute pressure through a probe of 50 mm length and diameter of 4 mm with a fine mesh net at the side facing of the fluidized bed. The used Piezoresistive transducer (Kobold, SEN-3248 B075) has a response time less than 1 millisecond. van Ommen *et al.* [64] showed that the model of Bergh and Tijdeman [65] gives reliable predictions of the frequency response characteristics of probe-transducer systems over a wide range of probe lengths and diameters. The model for transducer with assumed inner space volume of 1500 mm^3 predicts the first resonance frequency of 679 Hz and the amplitude ratio of 1.1 at 200 Hz (Nyquist frequency) and lower at lower frequencies. This shows that the measuring technique is suitable for gathering dynamic information from pressure signals in the expected range of frequencies, typically less than 20 Hz in a fluidized bed [6, 9].

The measured signals were band-pass (hardware) filtered at lower cut-off frequency of 0.1 Hz to remove the bias value of the pressure fluctuations and upper cut-off Nyquist frequency (200 Hz). The filtered signals then were amplified with a gain of 100. The pressure transducer was connected to a 16 bit data acquisition board (Advantech 1712L). The sampling frequency for pressure fluctuation signals was 400 Hz, satisfying the Nyquist criterion. This sampling frequency is also in according with criterion of 50 to 100 times of the average cycle frequency (typically between 100 to 600 Hz) which is required for nonlinear evaluation of the pressure fluctuations in bubbling fluidized bed [6, 27, 29]. Total number of data points in each sample was 65535, corresponding to about 164 seconds of sampling time which is according to the sampling time criterion, 100 times of average cycle time, addressed by Johnssona *et al.* [6].

Sand particles (Geldart B) with three different mean sizes of 150, 280 and 490 μm and a particle density of 2640 kg/m^3 were used in the experiments. The bed was operated with different loaded sand heights (L/D of 0.5 to 2) and with gas velocities ranging from 0.1 to 1.2 m/s.

Samples of time sequences of the pressure fluctuations measured at 20 cm above the distributor at superficial gas velocities of 0.1, 0.5, and 0.9 m/s, particles size 150 μm, and L/D=1.5 are shown in Fig. **3**. There are differences both in the amplitude (standard deviation, σ) and in the time behaviour of these three signals. The standard deviation at gas velocities of 0.1, 0.5 and 0.9 m/s is 82, 349 and 424 Pa, respectively. The higher superficial gas velocity provides greater amplitude due to the larger bubbles.

SOURCES OF PRESSURE FLUCTUATIONS

Before evaluating the bed pressure fluctuations in the concerned domains, sources of the pressure fluctuations should be determined. The pressure fluctuations within a fluidized bed have been widely studied in literature, including modelling of the pressure fluctuation [66-68], cause of the pressure fluctuations [3], nature of the pressure fluctuations [56], signal composition of the pressure fluctuations [38, 69] and origin, propagation and attenuations of the pressure waves [2]. Fig. **4** shows a general view of sources of the pressure fluctuations within the bed. As can be seen in this figure, there is not a single source of the pressure fluctuations, the fluctuations are resulting from a number of phenomena within the bed such as bubble generation [69], bubble passage [66-68], bubbles interaction [3], bubble eruption [69-70], gas flow fluctuations and self-excited oscillations of fluidized particles as well as by the influence of solids behaviours [2, 71, 72]. It should be noted that the bed mass oscillations cannot be considered as an independent sources of the pressure fluctuation. The bed mass oscillation is a result of formation of bubbles at the gas distributor and eruption of bubbles at the bed surface.

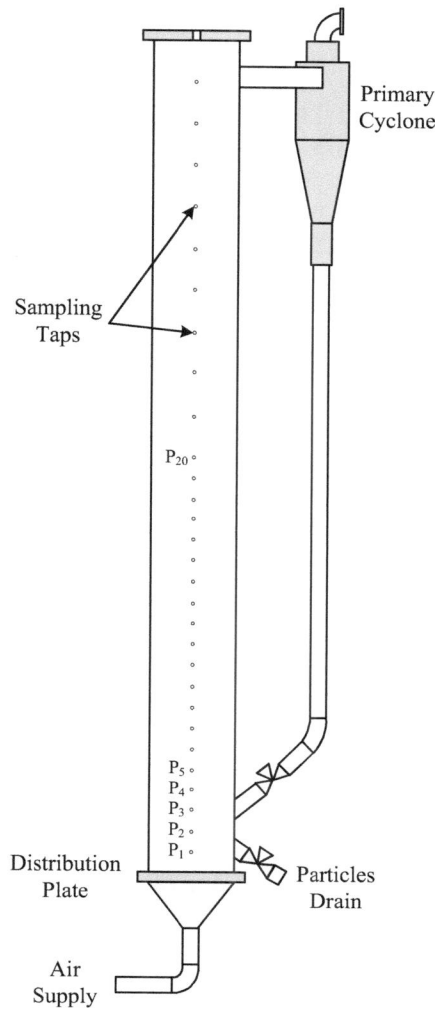

Figure 2: Schematic of the experimental setup. The sampling taps are indicated by P_1, P_2.

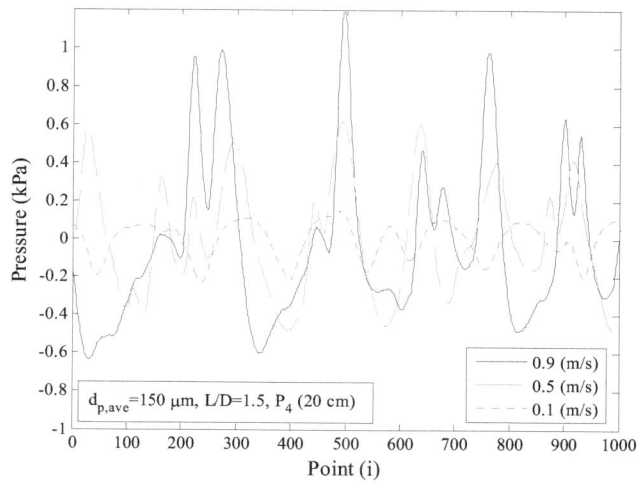

Figure 3: Time sequences of the pressure fluctuations measured in tap height 20 cm above the distributor, particles size 150 µm, superficial gas velocities of 0.1, 0.5 and 0.9 m/s and L/D=1.5.

It is known that the pressure fluctuations are a result of kinematic (or continuity) waves corresponding to slow propagating pressure waves and dynamic waves with high propagating pressure waves. Pressure waves with high propagation velocities (> 10 m/s), as a result of more global phenomena which may be recorded far away from the source, are identified as compression waves which move upward and downward. While the amplitude of downward propagating compression waves stays constant, the amplitude of an upward traveling compression wave decreases linearly to zero at the bed surface (attenuation) [2]. Alternatively, pressure waves with propagation velocities of the order of the minimum fluidization velocity (less than 2 m/s) are identified as a result of more local phenomena which propagate only upwards. Consequently, the measured pressure fluctuation signal anywhere in the fluidized bed is a combination of fluctuations due to phenomena close to the tip of the probe and of waves propagated from other locations [2, 73]. To consider these effects, van Ommen *et al.* [74] suggested a method for optimal placement of the pressure probes in relatively large units.

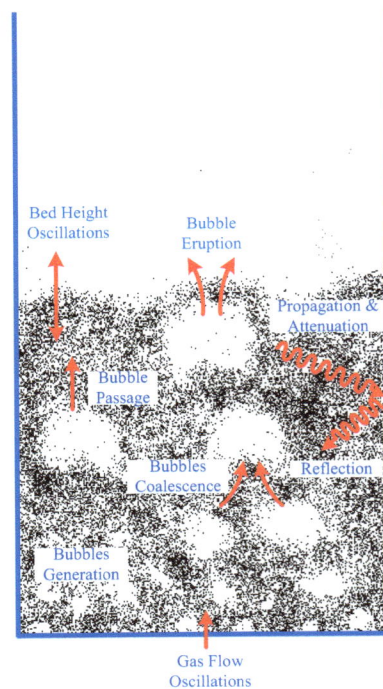

Figure 4: Origins of fluidized bed pressure fluctuations.

van der Schaaf *et al.* [2] claimed that upward moving compression waves originate from bubble formation and bubble coalescence while downward moving compression waves are caused by bubble eruptions at the fluidized bed surface, bubble coalescence and by changes in bed voidage. They also mentioned that upward slow propagation waves are caused by rising bubbles. However, gas voidage, local gas bubble dynamics and particle clusters were considered as sources of slow propagation waves by Musmarra *et al.* [73] and van Ommen *et al.* [74]. They also argued that gas flow fluctuations are also a source of high propagation compression waves. On the other hand, several researchers [9, 75-77] have stated that the main sources of pressure fluctuations are from events closer to the top surface of the fluidized bed.

Zarghami [9] showed that pressure fluctuations represent interaction of three different structures in a fluidized bed: large pressure fluctuations of low frequency correspond to the macro structure (a size comparable to the physical dimension of the bed diameter such as large bubble eruptions and movement of large bubbles), mezzo structure of higher frequencies (reflecting dynamic feature of clusters of dense phase and small bubbles or voidage changes close to the tip of the probe) and micro-structure of very high frequencies which represents behaviour of the interaction among single particles and between particles and fluid as well as noise effects.

As it is seen, there still are conflicting conclusions with respect to the sources of the pressure fluctuations and more researches are needed on this issue. Although, modeling of the fluid dynamics of fluidized beds is challenging, pressure fluctuations in fluidized beds also can be described by models of the two-phase medium [1] and may resolve these conflicts.

PRELIMINARY ANALYSIS

R/S Analysis

Rescaled range analysis (R/S analysis) was first introduced by Hurst [10] for studying long-term memory and fractality of a time series hydrology. Mandelbrot [78] showed that the R/S analysis is a more powerful tool in detecting long range dependence compared to more conventional analysis like autocorrelation analysis and spectral analysis. In this method, first cumulative deviation from the mean of the time series $x(i)$ in time window n is calculated:

$$x(i,n) = \sum_{j=1}^{i}\left(x(j) - \overline{x}_n\right)$$

(1)

where

$$\overline{x}_n = \frac{1}{n}\sum_{i=1}^{n} x(i)$$

(2)

Then, the range function $R(n)$ is determined as maximum and minimum difference of time series $x(i)$ in each time interval n:

$$R(n) = \max\left(x(i,n)\right) - \min\left(x(i,n)\right) \quad 1 \le i < n$$

(3)

Rescaled range function R/S is obtained by dividing $R(n)$ by the standard deviation $S(n)$:

$$\frac{R}{S} = \frac{R(n)}{S(n)}$$

(4)

where the standard deviation $S(n)$ is:

$$S(n) = \sqrt{\frac{1}{n}\sum_{i=1}^{n}\left(x(i) - \overline{x}_n\right)^2}$$

(5)

It has been found that, for some time series, the dependence of R/S on the number of data points (or time) follows an empirical power law described as [10]:

$$\left(\frac{R}{S}\right)_n \propto n^H$$

(6)

where H is the Hurst exponent and varies between 0 and 1. The Hurst exponent is equal to 0.5 for stochastic (*e.g.*, white noise) series, less than 0.5 for rough anticorrelated series and greater than 0.5 for positively correlated series known as persistence.

For the persistent data set, if the trend or behaviour in the data set is increasing or decreasing over a certain unit interval of time, it would have a tendency to persist in increase or decrease over such an interval. For this time series data, H is related to its fractal dimension, D_F, by the relation $D_F = 2-H$, which represents the self-similarity of the studied curves. Therefore, a higher fractal dimension corresponds to a lower Hurst exponent [19, 20].

Hurst exponent can be estimated by linear regression of $ln(R/S)$ versus $ln(n)$. Fig. **5** shows the Hurst diagram of the pressure fluctuations, measured 20 cm above the distributor, at gas velocity of 0.6 m/s, particles size of 150 μm and initial aspect ratio L/D of 1.5. As can be seen in this figure, Hurst exponent of the pressure fluctuations shows that the fluidized bed has a bifractal behaviour with two different Hurst exponents. Hurst exponent 1.0 at smaller n is much larger than 0.5, indicating a highly persistent dynamic feature of the fluidized bed. Hurst exponent 0.11 at larger n is much less than 0.5, which indicates a highly antipersistent dynamic feature of the fluidized system. Corresponding fractal dimensions are about 1.0 and 1.9 for higher and lower Hurst exponents, respectively. In general, as stated by Karamavruc and Clark [26], bubble motions correspond to higher Hurst exponents than do particle motions. It can be concluded that while Hurst exponent at smaller n or smaller fractal dimension represents a dynamic feature of macro structures, Hurst exponent at larger n or larger fractal dimension represents dynamic feature of finer structures.

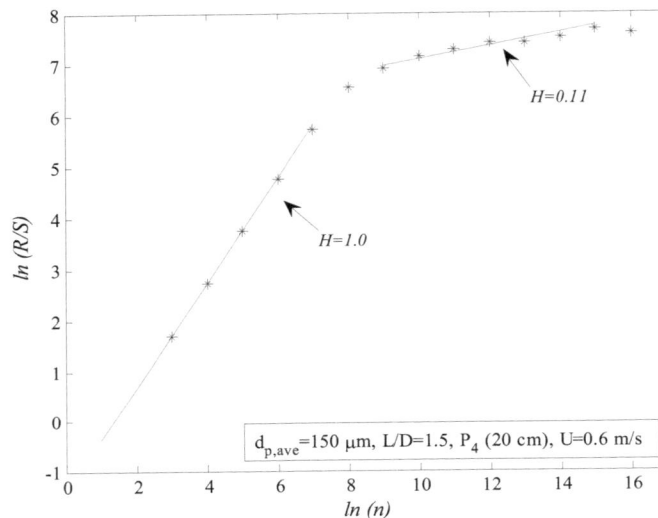

Figure 5: Hurst exponent diagram of the pressure fluctuations signal measuring in tap height 20 cm above the distributor at superficial gas velocity of 0.6 m/s, particles size 150 μm, and L/D=1.5.

Fig. **6** shows Hurst diagram of the pressure fluctuations measured 20 cm above the distributor at various gas velocities, particles size of 150 μm and initial aspect ratio L/D equal to 1.5. As can be seen in this figure, this bifractal behaviour of the bubbling fluidization does not change by changing gas velocity or particles size. On the other hand, as pointed out by Fan *et al.* [48], the reciprocal of the break point in the Hurst profile is similar to the domain frequency of the bed. As shown in Fig. **6**, the break point occurs at n equal to 256 points which is equal to a time interval of about 0.64 s. With sample frequency of 400 Hz, this presents an equivalent domain frequency of about 1.56 Hz which is close to the values estimated from the power spectrum analysis of the bed pressure fluctuations at the same condition [9].

Average Cycle Frequency

Average cycle frequency is the number of cycles per signal total time. The number of cycles is the number of times that the time series crosses its average value divided by two. Fig. **7** shows the average cycle frequency of the measured pressure fluctuations 20 cm above the distributor with gas velocity for different sizes of sands and initial bed height L/D equal to 1.5. As it can be observed, average cycle frequency of all three different sizes of particles initially decreases and approaches to the peak dominant frequency of the bed and then increases with an increase in velocity.

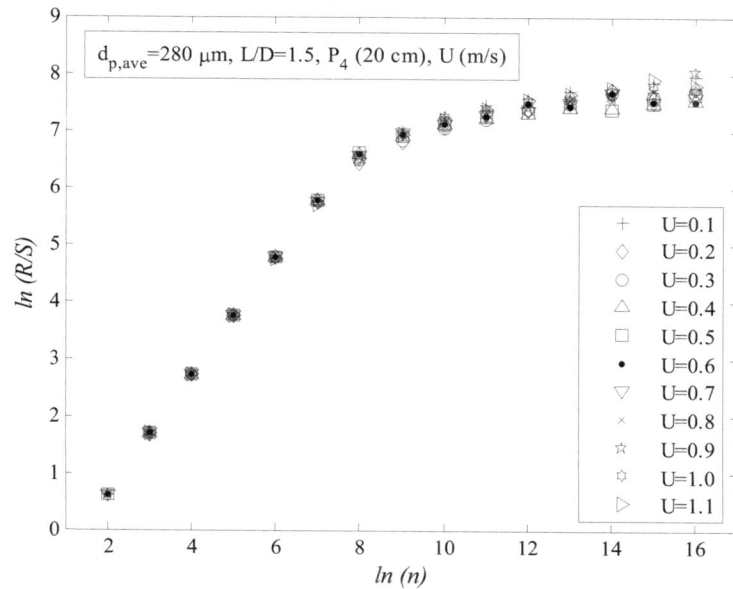

Figure 6: Hurst exponent diagram of the pressure fluctuations signal measuring in tap height 20 cm above the distributor at different superficial gas velocities, particles size 150 μm, and L/D=1.5.

While average cycle frequency of a perfect periodic time series is identical to the peak dominant frequency of its power spectrum. A difference between these two frequencies indicates deviations from a strictly periodicity of the system. The main frequencies of a bubbling fluidized bed (between 1.5 to 2.5 Hz for this work) correspond to the macro structures of the bed [6, 9]. The greater the difference between average cycle frequency and the main frequencies of the bed, the more the deviation from the larger structures of the bed and the more dominant the finer structures. Thus, a minimum in average cycle frequency of the pressure fluctuation signal indicates a minimum deviation from periodicity of the bed. This minimum average cycle frequency occurs in gas velocities of 0.4, 0.5, and 0.7 m/s for sands with mean diameters of 150, 280, and 490 μm, respectively. In addition, as it can be seen in Fig. 7, average cycle frequency is higher for smaller particles which shows that finer structures become more important in a bed of smaller particles.

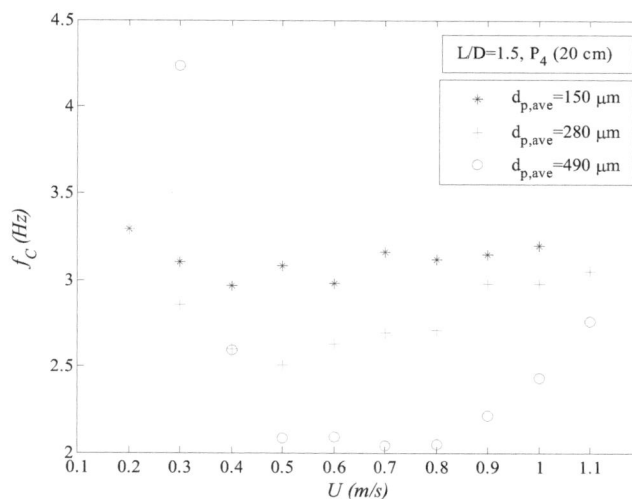

Figure 7: Average cycle frequency versus superficial gas velocity for different sizes of particles, at measuring tap height 20 cm above the distributor and L/D=1.5.

Fourier Analysis

Fourier analysis is an extremely useful tool for data analysis in frequency domain. Fourier analysis decomposes a signal into constituent sinusoids of different frequencies and is performed using the discrete Fourier transform (DFT). Estimated Fourier transform, $X(f)$, of the measured pressure time series $x(i)$ consists of N point is equal to [19]:

$$X(f) = \sum_{i=1}^{N} x(i)\exp(-j2\pi i f) \tag{7}$$

in which f and j are frequency and the complex number. If N is a power of 2, above relation becomes the fast Fourier algorithm which is efficient an algorithm for computing the DFT of a sequence.

The power spectrum of a signal which represents the contribution of every frequency of the spectrum to the power of the overall signal can be estimated from the magnitude of $X(f)$ squared. The variance of such estimation of the power spectrum does not decrease with an increase in N. In order to decrease the variance, the signal is repeatedly divided into windows and an average of the power spectrum within the windows is used to obtain an estimate of the power spectrum (the Welsh method of power spectrum estimation). However, decrease of the samples within the windows gives poor frequency resolution. Hence, an appropriate window width should be chosen to obtain a satisfactory trade off between frequency resolution and variance [6].

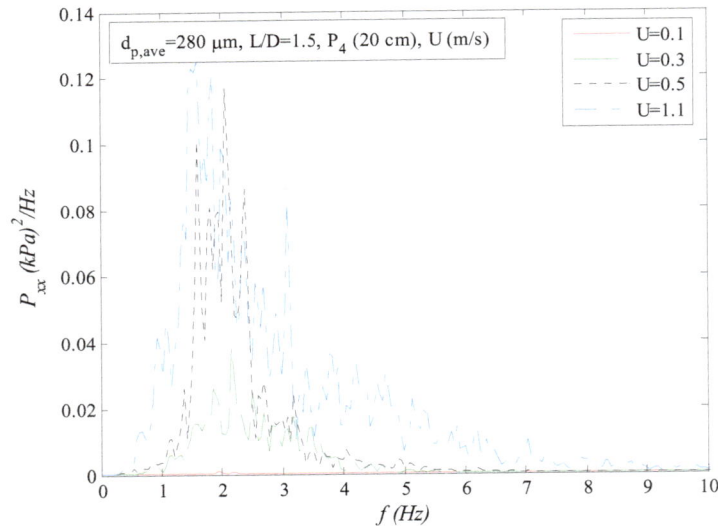

Figure 8: The power spectrum estimation of the measured pressure fluctuations at tap height 20 cm above the distributor with different superficial gas velocities, particles size 280 μm and L/D=1.5.

Using the Hanning window and without any overlap between the windows, the averaged power spectrum becomes [6]:

$$P_{xx}(f) = \frac{1}{L}\sum_{n=1}^{L} P_{xx}^{n}(f) \tag{8}$$

where L is number of windows and $P_{xx}^{n}(f)$ is the power spectrum estimate of each window.

In the present work, the measured pressure signal was divided into 8 non-overlapping windows of 8192 samples which give a frequency resolution of 0.048 Hz with the 400 Hz sampling frequency. Fig. **8** shows the power spectrum estimation of the pressure fluctuations measured 20 cm above the distributor at different gas velocities, particles size of 280 μm and L/D=1.5. As can be seen in this figure, the main frequency of the pressure fluctuations is less than 10 Hz. On the other hand, Fig. **8** shows that the dominant frequency (peak) of the pressure fluctuations is about 1.5 to 2.5 Hz which corresponds to the macro structures of the bed [6, 9]. Johnsson *et al.* [6] showed that the low frequency region (up to 3 Hz) in the power spectrum diagram is related to the macro structures, frequency range of 3 to 20 Hz corresponds to the mezzo structures and high frequency (20 to 200 Hz) is related to the micro structures of the bed.

Energy of the signal (*i.e.*, squared sum of amplitudes) can be defined in the frequency domain with Parseval's theorem:

$$E = \sum_{i=1}^{N} |x(i)|^2 \approx \sum_{k=1}^{N_f} P_{xx}(f) \tag{9}$$

where N_f is the number of points in the frequency domain. The energy of the signal in a given frequency range is calculated from summation of the power spectrum over its frequency range. The energy of the finer structures (frequency range of 3 to 200 Hz) can be expressed in relation to the total energy of the power spectrum as wide band energy, E_{WB}. Fig. **9** shows wide band energy of the pressure fluctuations measured 20 cm above the distributor against gas velocity for various sizes of sands and initial aspect ratio L/D equal to 1.5. As can be seen in this figure, the wide band energy of the finer structures initially decrease, then increase with increasing gas velocity for different sizes of particles. Minimum energy of the wide band energy (1-E_{WB}) indicates a maximum contribution of the macro structures of the bed. Figs. 7 and 9 show similar trends of average cycle frequency and wide band energy against gas velocity. The minimum energy of the wide band energy approximately occurs at the minimum deviation from a strictly periodicity of the system. Fig. **9** shows that this minimum point occurs at gas velocities of 0.3, 0.4 and 0.6 m/s for sands with mean diameters of 150, 280 and 490 μm, respectively. Moreover, the wide band energy of the finer structures is higher for smaller particles which show that the finer structures become more important in a bed of smaller particles.

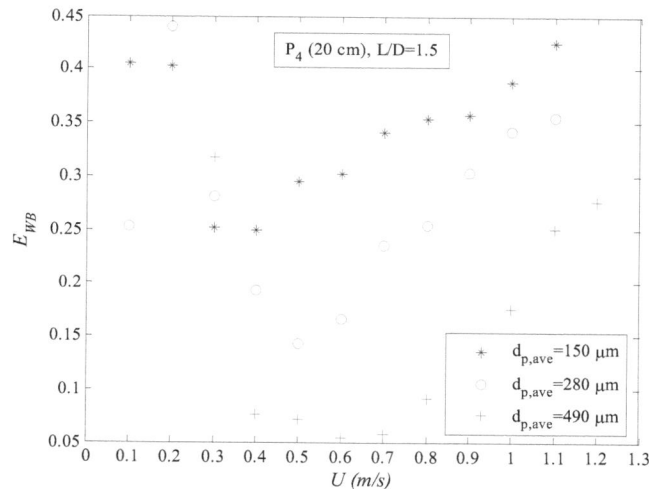

Figure 9: The wide band energy of the measured pressure fluctuations 20 cm above the distributor with gas velocity for different sizes of sands and initial bed height L/D equal to 1.5.

DETECTING STATIONARITY

A time series is said to be strictly stationary if its statistical distribution does not change across time which implies that the parameters of the system which has generated the time series remain constant. Since all nonlinear methods assume that the analyzing time series is stationary, the stationarity of the fluidized bed pressure fluctuations should be checked before any evaluation. There are different deterministic and probabilistic methods for detecting non-stationary from a time series [19]. However, in this work a relatively simple stationarity test, called space time separation plot, is addressed [79].

In the presence of temporal correlation, the probability that a given pair of points on the reconstructed attractor are closer than ε (geometry distance on the attractor) depends not only on the position of the points but also on the time that has elapsed between them, Δt. This dependence can be detected by plotting the number of neighbor points as a function of two variables, the time separation $n\Delta t$ and the spatial distance ε. The space time separation of the time series of the pressure fluctuations measured 20 cm above distributor at gas velocity of 1.0 m/s, particles size of 280 μm, and initial aspect ratio L/D equal to 1.5 is shown in Fig. **10**. The contour lines on this figure correspond to the percent of neighbouring points at different spatial distances and time separation. As it can be seen in this figure, the curves reach to a steady state at n equal to about 200 points which shows stationarity of the pressure fluctuations of the bubbling fluidized bed.

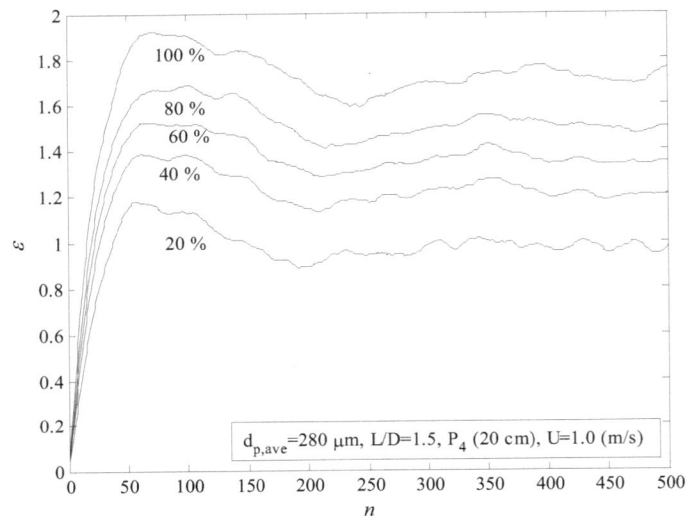

Figure 10: Space time separation plot for the pressure fluctuations measuring in tap height 20 cm above the distributor at superficial gas velocity of 1.0 m/s, particles size 280 μm, and L/D=1.5.

State Space and Embedding Theory

The state of a dynamical system can be defined in a state, such that a point in this state space specifies the state of the system. For a dissipative dynamical system such as fluidized bed [28-29], a set of initial conditions, after some transient time, is attracted to a sub-set of state spaces which is called attractor of the system [19-20]. While attractors of non-chaotic systems are simple, attractors of chaotic systems such as fluidized beds have complex geometrical structures and are called strange (or fractal) attractors [19]. To characterize the structure of the underlying attractor from an observed time series, it is necessary to reconstruct a state space from the time series. The process of reconstructing the attractor of the system is commonly referred to as embedding [19].

There are several methods of reconstructing the state space from the observed quantity which may differ in the quality of the resulting coordinates [57, 80]. A number of these methods are used to reconstruct the attractor of the fluidized bed system from pressure or void fluctuations [6, 28-29, 49-50, 58]. It is still not clear; however, which method is the most efficient. The lack of a unique solution is a source of uncertainty in nonlinear study of chaotic behaviour of fluidized bed systems [58]. The simplest method to embed scalar data, such as pressure fluctuations, is the method of delays which was introduced by Takens [57] and used in this work. In this method, the state space is reconstructed from a scalar time series, by using delayed copies of the original time series as components of the reconstructed state space (RSS). The RSS is sometimes referred to as the reconstructed state vector. Letting $x(i)$, with $i = 1, 2,..., N$, represent the measured pressure signal at equidistant time intervals, Δt (*i.e.*, with a sampling frequency of $f_s=1/\Delta t$) and N be the total number of samples, the RSS could be represented as:

$$s(i) = \left(x(i), x(i+\tau), \cdots, x\left(i+(d-1)\tau\right)\right) \tag{10}$$

where d is the number of elements of the reconstructed state vector which is equal to the number of coordinates in the reconstructed state space and is called the embedding dimension and τ is the embedding delay. It should be noted that the total number of state vectors (points) is equal to $N-(d-1)\tau$. A sample of reconstructed state space method is illustrated in Fig. **11**. In this figure, $d = 3$ and the time delay is equal to the time between the samples, *i.e.*, $\tau = 1$ (corresponding to $1/f_s$).

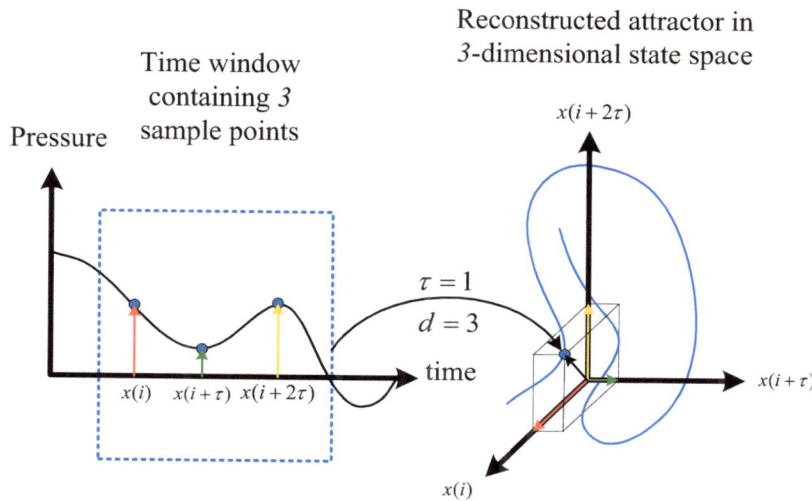

Figure 11: Attractor reconstruction using the method of time delays.

Embedding Parameters

Determining the embedding parameters, time delay and embedding dimension, are the most important steps in the nonlinear time series modelling. Several methods have been developed for determining the time delay and the embedding dimension. However, no rigorous method exists for determining the optimal value of the time delay and embedding dimension. A number of these methods for selection of these parameters are described below and applied to the pressure fluctuations time series of fluidized bed.

Time Delay

The first step in the state space reconstruction is to choose the optimum delay parameter, τ. Generally, time delay is not the subject of the embedding theorem. Therefore, from a mathematical point of view, its determination is arbitrary [19, 20, 29]. If it is taken too small, there is almost no difference between different elements of the delay vectors, such that all points are accumulated around the bisector of the embedding space. If τ is very large, different coordinates may be uncorrelated. In application, however, the proper choice of the delay time is quite important and is made based on geometrical arguments the attractor should be unfolded in the optimum delay time.

Different prescriptions have appeared in literature to choose τ [18-20, 81]. Nevertheless, since τ has no relevance in the mathematical framework, they are all empirical and do not necessarily provide appropriate unique estimate. Autocorrelation function [19-20] and mutual information function [18-20] have been used to determine the optimum value of the time delay for the state space reconstruction.

Autocorrelation Function

From the basic mathematics, the autocorrelation function, *ACF*, compares linear dependence of two time series separated by delay and is defined as [19-20]:

$$ACF = \frac{\sum_{i=1}^{N-\tau}\left[x(i)-\bar{x}\right]\left[x(i+\tau)-\bar{x}\right]}{\sum_{i=1}^{N-\tau}\left[x(i)-\bar{x}\right]^2} \tag{11}$$

Where

$$\bar{x} = \frac{1}{N}\sum_{i=1}^{N}x(i) \tag{12}$$

The delay for the attractor reconstruction, τ, is then taken at a specific threshold value of *ACF* where a linear independence is found. The recommendation for this threshold is the first value of τ which *ACF* is equalled to one half or zero or the first inflection point of *ACF*.

Mutual Information

While the autocorrelation function measures the linear dependence of two variables, Fraser and Swinney [18] suggested using the mutual information $I(\tau)$ function as a kind of nonlinear correlation function to determine when the values of $x(i)$ and $x(i+\tau)$ are independent enough of each other to be useful as coordinates in a time delay vector, but not so independent as to have no connection which each other at all. The mutual information of the attractor reconstruction co-ordinates is defined as:

$$I(\tau) = \sum_{i=1}^{N-(d-1)\tau} P\big(x(i),x(i+\tau),\cdots,x(i+(d-1)\tau)\big)\log\left[\frac{P\big(x(i),x(i+\tau),\cdots,x(i+(d-1)\tau)\big)}{P\big(x(i)\big)P\big(x(i+\tau)\big)...P\big(x(i+(d-1)\tau)\big)}\right] \tag{13}$$

where $P(x(i))$ refers to the individual probability of the time series variable $x(i)$ and $P(x(i), x(i+\tau), \ldots, x(i+(d-1)\tau))$ is the joint probability density of attractor coordinate $s(i) = (x(i), x(i+\tau), \ldots, x(i+(d-1)\tau))^T$. A suitable choice of time delay requires the mutual information to be minimum [19-20, 49-50]. In this case, the attractor is unfolded and stretched out. This condition for the choice of the delay time is known as the minimum mutual information criterion.

In general, the time delay provided by the $I(\tau)$ criteria is normally smaller than what calculated by the $ACF(\tau)$ and provides appropriate characteristic time scales for the motion. As mentioned above, $I(\tau)$ presents a kind of nonlinear correlation concept, while the $ACF(\tau)$ provides an optimum linear correlation criterion.

The autocorrelation function $ACF(\tau)$ and mutual information profile $I(\tau)$ of the pressure fluctuations, measured 20 cm above distributor at gas velocity of 1.0 m/s, particles size of 280 μm and initial aspect ratio L/D equal to 1.5 is illustrated in Fig. **12**. The first pass of the autocorrelation function from one half and the time delay at which the *ACF* becomes zero occur at 23 and 41, respectively. The first minimum of the mutual information occurs at a delay time of 41. While the autocorrelation function compares linear dependence, this value of mutual information, provide a measure of nonlinear dependence and a better estimate of time delay.

Embedding Dimension

As mentioned earlier, the dimension of a reconstructed state space is called the embedding dimension, *d*. Working in a dimension larger than the minimum required by the data will lead to excessive requirements in terms of the number of data points and computation times when investigating different issues such as nonlinear dynamic invariants calculation, prediction, *etc.* According to Whitney's embedding theorem [82], every m dimensional manifold can be embedded within an embedding space of dimension $d = 2m + 1$. Therefore, an attractor may, in general, be reconstructed within an embedding space of dimension $d = 2m + 1$ where m is the dimension of the state space of the underlying real attractor. Alternatively, Sauer *et al.* [83] generalized Whitney's embedding theorem with connection to the correlation dimension. If the correlation dimension, D_C, of the attractor is known, then a closer estimate of the minimum embedding would be $d > 2D_C$.

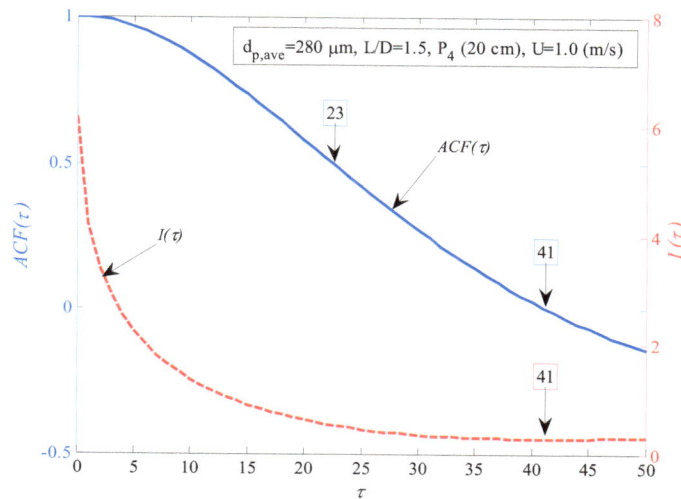

Figure 12: The autocorrelation function and mutual information profile against delay time for pressure fluctuations measuring in tap height 20 cm above the distributor at superficial gas velocity of 1.0 m/s, particles size 280 μm and L/D=1.5.

Several practical methods have been addressed in literature for determining the embedding dimension from an observed time series. However, different methods lead to different values for the embedding dimension [60]. The usual method for choosing the minimum embedding dimension is to calculate some invariants (correlation dimension, entropy, *etc.*) of the reconstructed attractor. Since these invariants are geometric properties of the attractor, they become independent of the dimension after the geometry is unfolded [19-20]. Embedding dimension can be then determined when the value of the invariant stops changing and reaches its steady state status. The false nearest neighbors (FNN) addressed by Kennel *et al.* [84], use of correlation dimension [85, 86] and principal components analysis [87] are the methods of choosing the minimum embedding dimension of a time series. An alternative method has been proposed to select the optimum value of the time window (multiple values of embedding dimension and time delay) [28-29].

Correlation Dimension

Practical method to determine a minimum dimension of the embedding space is to calculate the correlation dimension for the reconstructed attractor. The correlation dimension can be calculated from the power-law relation between the correlation integral, C, of an attractor and a neighborhood radius, ε. Generally, correlation dimension, which is the slope of the log plot of C versus ε, initially increases with the embedding dimension and reaches a limiting value in which further increase in the embedding dimension would not increase the correlation dimension [19-20, 85-86]. The embedding dimension can be specified when the correlation dimension reaches its saturation state.

The power-law relation between the correlation integral of a reconstructed attractor and the neighborhood radius, ε, of the assumed hyper-sphere, $C(\varepsilon) \propto \varepsilon^{D_c}$, can be used to provide an estimate of the correlation dimension [19-20, 85-86]:

$$D_C = \lim_{M \to \infty} \lim_{\varepsilon \to 0} \frac{\partial \ln C(\varepsilon, M)}{\partial \ln \varepsilon} \tag{14}$$

where $C(\varepsilon)$ is the correlation integral defined as:

$$C(\varepsilon, M) = \frac{2}{M(M-1)} \sum_{i=1}^{M} \sum_{j=i+1}^{M} \Theta\left(\varepsilon - \|s(i) - s(j)\|\right) \tag{15}$$

in which s is a point on the attractor which has M such points; $M = N-(d-1)\tau$. Θ is the Heaviside function which is equal to unity/zero if the value inside the parentheses is positive/negative. $s(i)$ are the points on the reference trajectory and $s(j)$ are other points on the attractor in the vicinity of $s(i)$. The correlation integral is essentially a measure of the number of points within a neighborhood of radius ε, averaged over the entire attractor.

Correlation integral of the pressure fluctuations measured at 20 cm above distributor at gas velocity of 1.0 m/s, particles size of 280 μm, and initial aspect ratio L/D equal to 1.5 is plotted against hyper-sphere radius for a range of embedding dimensions in Fig. **13**. The time delay τ was assumed to be 41. Different curves in these figure are shown with embedding dimensions $d=1$ to $d=20$. Fig. **13** reveals that the log-log plot of the correlation sum, $C(\varepsilon)$, versus ε is not quite a straight line but more or less S-shaped, specially in large embedding dimensions. This trend is due to the temporal correlation problem in the definition of correlation sum, Eq. (15). According to Theiler *et al.* [87], the pairs of points in the time series which are measured within a short Δt lead to close points in the state space, thus, introduce a bias in estimation of the correlation sum. In fact, these points are close not because of the attractor geometry, which should be considered in definition of correlation sum, but because they are correlated. Temporal correlations are seen in almost every data set. However, their effect on the correlation sum is more severe for high resolution time series.

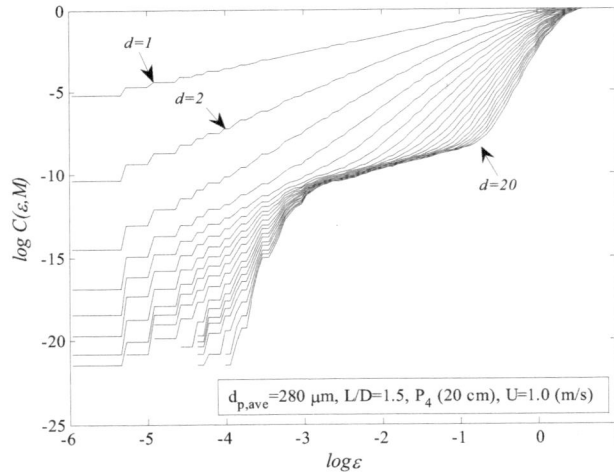

Figure 13: Correlation integral versus ε for pressure fluctuations measuring in tap height 20 cm above the distributor at superficial gas velocity of 1.0 m/s, particles size 280 μm, and L/D=1.5.

To solve this problem, the pairs of points which are close, not due to the attractor geometry, should be excluded. Therefore, the second summation in Eq. (15) should be started after a typical minimum correlation time, $t_{min} = n_{min}\Delta t$ has passed:

$$C(\varepsilon, M) = \frac{2}{(M - n_{min})(M - n_{min})} \sum_{i=1}^{M} \sum_{j=i+n_{min}}^{M} U\left(\varepsilon - |s(i) - s(j)|\right) \tag{16}$$

This correlation time, sometimes referred as Theiler window, can be estimated by introducing time separation plot as addressed by Provenzale *et al.* [79] and Kantz and Schreiber [19]. They explained that the pairs which are close in time should be excluded only for values of the time separation at saturation condition. The space time separation plot, Fig. **10**, shows that for the pressure fluctuations of the bubbling fluidized bed with corresponding condition, this saturation occurs at $n > 200$ points. Therefore, to be in the safe margin, a minimum correlation time of 250 points would be enough for calculating the correlation sum.

Fig. **14** illustrate the correlation integral of the pressure fluctuations measured at 20 cm above distributor at gas velocity of 1.0 m/s, particles size of 280 µm, and initial aspect ratio L/D equal to 1.5 when all pair points closer than 250 time steps are discarded. This figure shows straight lines as an indicator of self-similar geometry. These curves show a rigid saturation for higher embedding dimensions of about 20 corresponding to correlation dimension of D_C=10.11. Zarghami [9] showed that this method is time consuming and, in addition, the obtained dimension continues to increase in the presence of noise in measurement of the pressure fluctuations.

False Nearest Neighbors

Different false nearest neighbors (FNN) methods have been introduced in literature [84, 88] in order to determine embedding dimension. However, in this work, the method developed by Kennel *et al.* [84] is addressed. Based on this method, if the dynamics of a system is to be reconstructed successfully in embedding dimension d, the entire neighbour points in d would be also neighbours in embedding dimension $d+1$. This method, as shown in Fig. **15**, checks the neighbours in successively higher embedding dimensions until only a negligible number of false neighbors are found when increasing the dimension from d to $d+1$. This dimension is then chosen as the embedding dimension.

In Fig. **15**, a number of the neighbours of solid red circle are traced in different dimensions (from small to large). While only blue circle is a real neighborhood of the red circle, a number of incorrect neighbours of solid red circles (small hollow circles) decrease with increasing embedding dimension until the number of neighbours do not change when increasing the dimension from d to $d+1$ which is chosen as the embedding dimension.

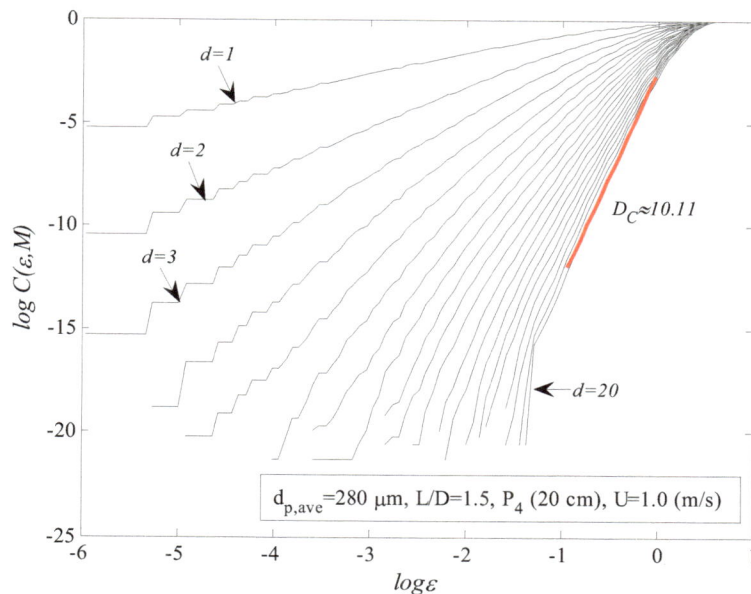

Figure 14: Correlation integral versus ε with discarding points which are less than 250 time steps apart, for pressure fluctuations measuring in tap height 20 cm above the distributor at superficial gas velocity of 1.0 m/s, particles size 280 µm and L/D=1.5.

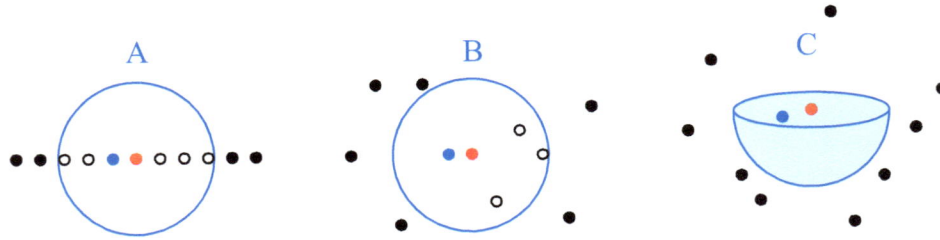

Figure 15: The number of false neighbors in different embedding dimension, from small (A) to large (C).

Percentage of false nearest neighbours for τ=41 is shown versus embedding dimension in Fig. **16** for pressure fluctuations measured 20 cm above distributor at gas velocity of 1.0 m/s, particles size of 280 μm and aspect ratio L/D equal to 1.5. It can be seen in this figure that there is a fast decay in the percentage of false nearest neighbours with embedding dimension. The minimum percentage of the false neighbour occurs in the embedding dimension of 4. This shows that the embedding dimension of the bubbling fluidization system by the false nearest neighbour method is about 4 which is totally different from the estimated dimension by the correlation method.

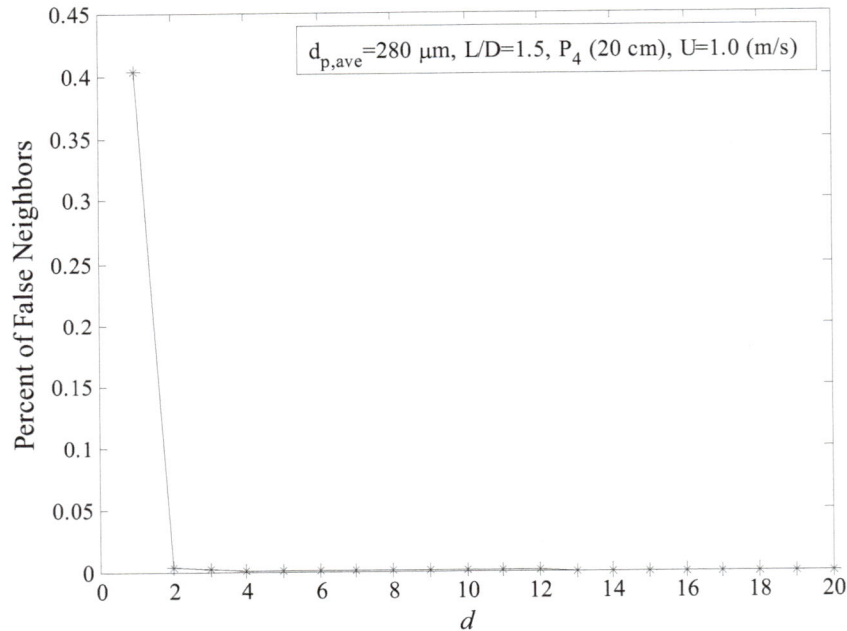

Figure 16: Percent of false neighbor versus embedding dimension for pressure fluctuations measuring in tap height 20 cm above the distributor at superficial gas velocity of 1.0 m/s, particles size 280 μm and L/D=1.5.

Principal Components Analysis

Principal components analysis is a multivariate statistical technique that can be used to remove the correlation between input variables. Points on the attractor can be projected to the new points of independent variables components. With the principal components analysis, each component of the new point is a linear combination of all components of the initial points on attractor [89]. In this method, initial normalized attractor matrix is determined first:

$$S = \frac{1}{(M-1)^{1/2}} \left\{ s(i) \big| i = 1, 2, ..., M \right\} \tag{17}$$

where M is the number of points on the attractor and coefficient $(M-1)^{1/2}$ is used to normalize the attractor matrix. By calculating the covariance of the normalized matrix, the eigenvectors and corresponding eigenvalues of the covariance matrix can be obtained. Eigenvectors which have higher eigenvalues, in fact, form the main components of the system while other components are less important. In this case, the less important components can be ignored and the attractor can be reconstructed into the principal components embedding dimension.

The idea of using the principal components analysis is to select the embedding dimension of the reconstructed attractor was first proposed by Broomhead and King [90]. In their method, the attractor is initially constructed with an assumed large embedding dimension and then by neglecting of less important components, new embedding dimension is found. While several researchers have proposed this method to select the embedding dimension [19], van der Stappen [29] reported that the method is not useful to determine the optimum reconstruction parameters in fluidization system, especially for small window lengths.

For $\tau=41$, the percentage of eigenvalues, λ_d, is shown versus embedding dimension in Fig. **17** for pressure fluctuations measured 20 cm above distributor at gas velocity of 1.0 m/s, particles size of 280 µm and initial aspect ratio L/D equal to 1.5. This figure shows that there is a fast decay in the percentage of eigenvalues with embedding dimension and dimensions higher than 4 can be ignored. This embedding dimension is equal to the estimated dimension of the false nearest neighbour method.

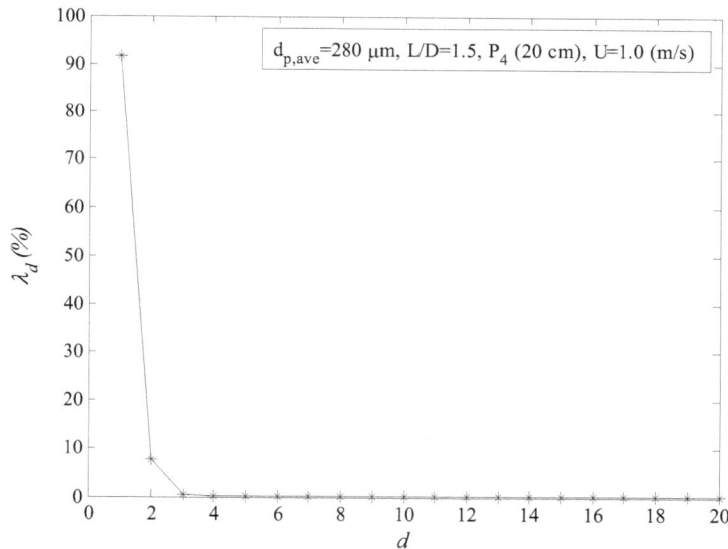

Figure 17: Percent of eigenvalues versus embedding dimension for pressure fluctuations measuring in tap height 20 cm above the distributor at superficial gas velocity of 1.0 m/s, particles size 280 µm and L/D=1.5.

Time Window

Another method to select embedding parameters is to find an optimum value for the time window $T_w = d\tau.\Delta t$. In this case, $d\tau$ can be determined for a specified sampling frequency. For the attractor systems with a dominant periodic characteristic, the time window is a segment $[t, t+T_w]$ of the time series and its optimum value can be chosen as one or one quarter of the dominant period or average cycle time [27-29]. Average cycle time is defined as the length of the time series (in time units) divided by the number of cycles.

The dominant period of a dynamic system is identified from the spectral analysis of the system in the frequency domain. However, average cycle time is calculated directly in the time domain and can be used in this work. In addition, the average cycle time was shown to be a good choice for the size of the time window for pressure fluctuation of fluidized beds [27]. This is due to the fact that one cycle approximately corresponds to the main physical characteristics of the bed like bubble passage or bubble coalescence [91].

In the present work, the time windows of 0.335 and 0.1 s were obtained for pressure fluctuations measured 20 cm above distributor at gas velocity of 1.0 m/s, particles size of 280 μm and initial aspect ratio L/D equal to 1.5, corresponding to 1 and 1/3 of average cycle times. These time windows contain 134 and 40 points per cycle, respectively, for the sampling frequency of 400 Hz. Therefore, the embedding dimensions of the reconstructed attractor for $\tau=1$ would be 134 and 40, respectively, which are far from those estimated by previous methods. However, it seems that the values of embedding dimension and time delay are interchangeable, thus, the optimum value for the time window is a more important parameter. For example, $d\tau=134$ of this method can be compared with $d\tau=164$ of false nearest neighbor and principal components analysis methods.

Nonlinear Dynamical Invariants

After reconstructing the attractor with proper selection of the embedding parameters, the most common characteristic of the reconstructed attractors and compare their geometric structure is evaluation of their nonlinear dynamical invariant such as the correlation dimension and entropy. The correlation dimension expresses the spatial complexity of the attractor in the state space, whereas the entropy is a measure of the predictability of the system. As mentioned before, the correlation dimension is also used for selection of the embedding dimension.

Correlation Dimension

Strange attractors of chaotic systems are objects with complex geometrical structure which are self-similar at various resolutions. Self-similarity in a geometrical structure of a system is a strong signature of the chaotic behaviour of the underlying system. There are several methods to quantify the self-similarity of a geometrical object by its dimension. Grassberger and Procaccia [85, 86] introduced the correlation dimension, D, for practical applications where the geometrical object has to be reconstructed from a finite sample of data points. Non-integer value of the correlation dimension shows self-similarity in geometrical structure of a state space attractor. In fact, when the correlation dimension is non-integer, the corresponding system has chaotic (nonlinear) behaviour. The correlation dimension for a simple system (linear) is an integer.

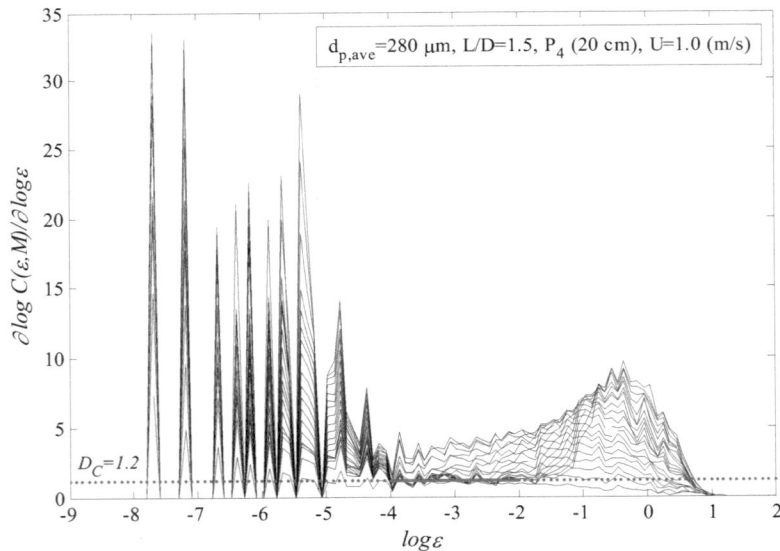

Figrue 18: Local slopes of correlation integral versus ε for pressure fluctuations measuring in tap height 20 cm above the distributor at superficial gas velocity of 1.0 m/s, particles size 280 μm and L/D=1.5.

Although no specific and clear embedding dimension could be obtained from the correlation dimension method, local slopes, $\partial lnC(\varepsilon, M)/\partial ln(\varepsilon)$, of the correlation integral versus ε for the pressure fluctuations were used to study the chaotic behaviour of the corresponding time series. These slopes are shown in Fig. **18**. Different curves in this figure represent different embedding dimensions d. The scaling behaviour, which shows self-similarity, is clearly visible in this figure. For large scales of ε the correlation integral does not obey the power law. On the smaller length scales of ε the curves are relatively flat and suggest a correlation dimension of 1.2 which confirms a low and chaotic dimensional behaviour for the fluidized bed. At very smaller scales, curves are separated again and self-similarity is obviously destroyed.

The direct method of using local slopes of the correlation integral versus ε is time consuming and sometime is not applicable for the experimental pressure fluctuations signal containing noise. There is another method of estimating the correlation dimension from the correlation integral, which is simple in use for the experimental time series analysis of data such as pressure fluctuations of fluidized beds [6, 27]. In this method, which was first introduced by Takens [92], a maximum likelihood estimation of the correlation dimension, D_{ML}, is used to characterize the geometrical structure of the attractor in the state space and the correlation dimension would be estimated from the distribution of distances between points on the reconstructed attractor [27]. It is assumed that the correlation integral can be estimated by $C(\varepsilon) \propto \varepsilon^D$ or using a cut-off length ε_0, $C(r) \propto r^{D_{ML}}$, with $r = (\varepsilon / \varepsilon_0)$ and $C(r)$ is scaled on the interval [0, 1]. A maximum likelihood estimation of the correlation dimension is [27, 29]:

$$D_{ML} = -\frac{M}{\ln(r_1 r_2 ... r_M)} \tag{18}$$

The series $r_1, r_2, ..., r_M$ are random sample of M observations from a probability function whose cumulative distribution is given by $C(r) \propto r^{D_{ML}}$. The cut-off distance, ε_0, is taken as the average absolute deviation (*AAD*) of the time series:

$$AAD = \frac{1}{N}\sum_{i=1}^{N}|x(i) - \bar{x}| \tag{19}$$

It should be noted that this dimension is different from that obtained by the direct method of using local slope of the correlation integral versus ε. This difference may be due to the noise effect on the experimental time series. The correlation dimension calculated by the maximum likelihood method gives an overestimation of the correct value for noisy data [27].

The maximum likelihood estimation of the correlation dimension, D_{ML}, for pressure fluctuations measured 20 cm above distributor at gas velocity of 1.0 m/s, particles size of 280 μm and aspect ratio L/D=1.5 was found to be equal to 3.574 (with 0.9 % relative error). Fig. **19** shows the maximum likelihood estimation of the correlation dimension of the pressure fluctuations measured 20 cm above distributor against gas velocity for various sizes of sands and initial aspect ratio L/D equal to 1.5. According to this figure, estimated correlation dimension does not show any regular trend with increasing gas velocity for three different sizes of particles. However, the maximum likelihood estimation of the pressure fluctuations in the fluidized bed shows a low dimensional (less than five) chaotic system which is in agreement with the values reported in literature [6, 28, 29, 38, 46, 51, 52].

Entropy

Entropy is a well known measure used to quantify the amount of disorder in a system. It has also been associated with the loss of information along the attractor. In fact, two very close points on separate trajectories of the attractor (*i.e.*, different initial conditions) are evolving into two different trajectories, thus, two distinguished states due to the divergence of the nearby trajectories. Therefore, the initial information would be lost after a certain time. When expressed in bits/s, entropy indicates the amount of information lost in the time unit [19]. Based on the information theory, Grassberger and Procaccia [85-86] stated that the required information to predict a system during the time interval $[t_1, t_2]$ is [6, 29]:

$$I_{[t_1,t_2]} = I_{[t_1]} + H(t_2 - t_1) \tag{20}$$

where $I_{[t_1]}$ is information in bits at time t_1 and H is the entropy. The value of H is zero for a predictable system (*e.g.*, fully periodic), infinity for a stochastic system and finite and positive for a chaotic system.

The entropy can be estimated as the second order Renyi entropy (H_2) and can be related to the correlation integral of the reconstructed attractor [19]:

$$H_2 = \frac{1}{\tau} \lim_{\substack{\varepsilon \to 0 \\ d \to \infty}} \ln \left(\frac{C_d(\varepsilon)}{C_{d+1}(\varepsilon)} \right) \tag{21}$$

where d is the embedding dimension and τ is the time delay used for the attractor reconstruction. In a practical situation, values of ε and d are restricted by the resolution of the attractor and the length of the time series.

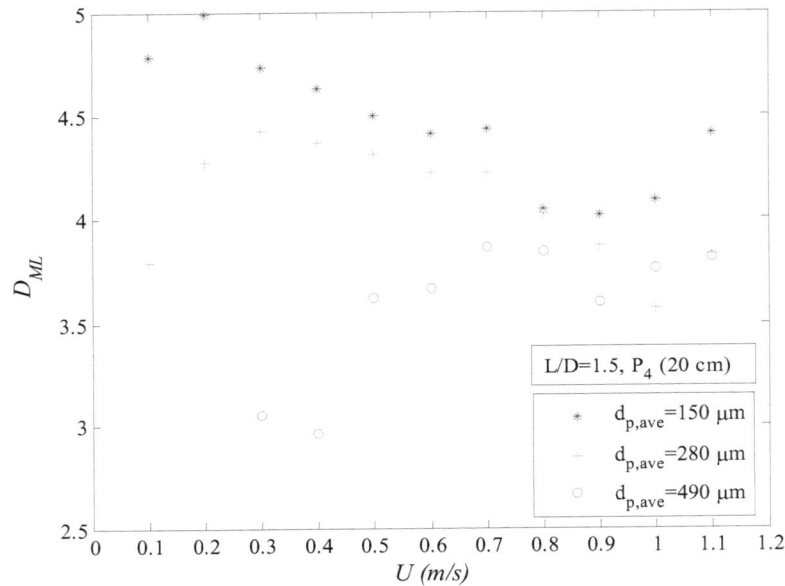

Figure 19: The maximum likelihood estimation of the correlation dimension versus superficial gas velocity for different sizes of particles, at measuring tap height 20 cm above the distributor and L/D=1.5.

The numerical values of the Renyi entropy for the pressure fluctuations measured 20 cm above distributor at gas velocity of 1.0 m/s, particles size of 280 μm and initial aspect ratio L/D=1.5 is plotted in Fig. **20**. It can be seen in this figure that a clear plateau is not visible at any embedding dimension, ranging from 1 to 10, and curves do not approach a specific limit. This might be attributed to the noise of the measurements.

Another method for estimation the entropy of fluidized beds has been proposed by Schouten *et al.* [93]. This method has a simple algorithm and easy to apply for pressure fluctuations signals. Schouten *et al.* [93] explained that entropy, as Kolmogorov entropy, can be calculated by the maximum likelihood method. It has been shown that the nearby trajectories diverge exponentially on the reconstructed attractors [29, 93]. This exponential divergence can be characterized by the following cumulative distribution:

$$C(b) = e^{\left(-\frac{Hb}{f_s} \right)} \tag{22}$$

where b is called the escape time and defined as the number of time steps that the trajectories starting from an arbitrary pairs of initial points on the attractor, remain close within a specified cut-off length ε_0. Probability to observe a pair of trajectories of the same escape time is equal to $C(b-1)-C(b)$. For this probability distribution, the maximum likelihood estimation for the entropy, H_{ML}, corresponding to M observation of a random sequence values of b, $\{b_1, b_2, \ldots, b_M\}$, is derived as:

$$H_{ML} = -f_s \ln\left(1 - \frac{1}{\bar{b}}\right) \tag{23}$$

Where

$$\bar{b} = \frac{1}{M}\sum_{i=1}^{M} b_i \tag{24}$$

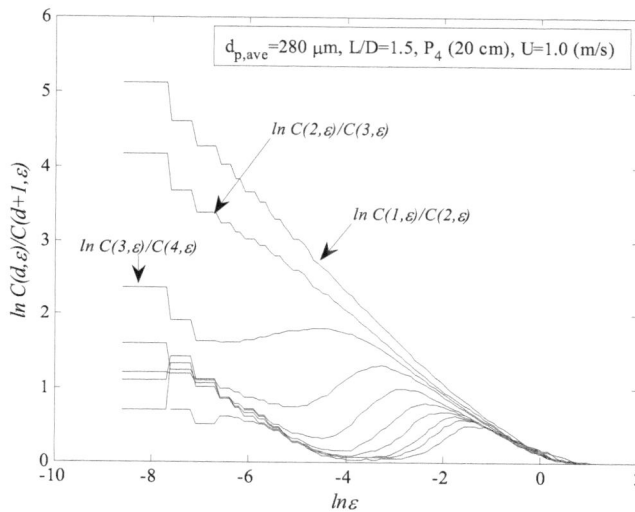

Figure 20: The estimated the Renyi entropy (H_2) of the measured pressure fluctuations 20 cm above distributors at gas velocity of 1.0 m/s, particles size 280 μm and L/D=1.5.

The maximum likelihood estimation of the Kolmogorov entropy, H_{ML}, for pressure fluctuation measured 20 cm above distributor at gas velocity of 1.0 m/s, particles size of 280 μm and initial aspect ratio L/D=1.5 was evaluated to be equal to 14.02 which shows an overall deterministic or predictable behaviour rather than the stochastic behaviour. However, it is worth mentioning that due to the large approximate entropy, its predictability is poor.

Fig. **21** shows the maximum likelihood estimation of the Kolmogorov entropy, H_{ML}, of the pressure fluctuations measured 20 cm above distributor as a function of gas velocity for different sizes of sand and initial aspect ratio L/D equal to 1.5. As shown in this figure, the maximum likelihood estimation of the Kolmogorov entropy for all three sizes of particles initially decreases and then increases with an increase in gas velocity. The minimum Kolmogorov entropy occurred in gas velocities of 0.4, 0.5, and 0.7 m/s for sand particles of mean diameter 150, 280 and 490 μm, respectively. Comparing Figs. **7**, **9**, and **21** reveals that the trends of average cycle frequency, wide band energy, and the Kolmogorov entropy against gas velocity are similar. It can be concluded that when there is a minimum deviation from periodicity of the bed, wide band energy and entropy are minimum. Since the periodicity of the bed corresponds more to the contributions of the larger structures of the bed, a minimum in the Kolmogorov entropy, wide band energy and average cycle frequency of the pressure fluctuations signal indicates a minimum deviation from the macro structures of the bed. This minimum in the Kolmogorov entropy, wide band energy, and average cycle frequency with an increase in velocity can be considered as the transition point between macro structures and finer structures in the bed. This shows that the finer structures of the bed first lose their contribution by increasing gas velocity when compared to the macro structures of the bed. After passing a transition velocity in the minimum entropy, their contribution increases with increasing gas velocity. In addition, as it can be seen in Fig. **21**, the Kolmogorov entropy is higher in smaller sizes of particles which shows that finer structures become more important in smaller particles.

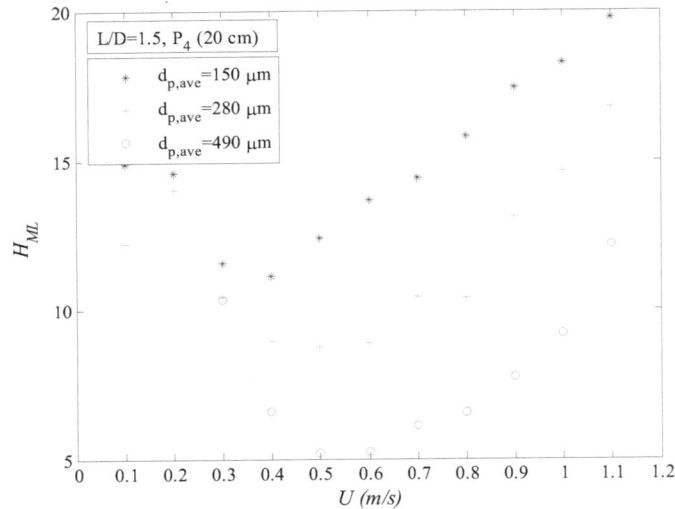

Figure 21: The maximum likelihood estimation of the Kolmogorov entropy versus superficial gas velocity for different sizes of particles, at measuring tap height 20 cm above the distributor and L/D=1.5.

Detecting Nonlinearity

Takens [94] showed that the stochastic time series with long time correlation and power law relation of power spectra, *e.g.*, fractal Brownian motion, may also have finite a dimension. Thus, it is important to investigate whether this finite dimension is generated from a low dimension nonlinear chaotic system or a stochastic linear system. Different tests have been proposed to investigate whether the irregularity of the experimental data is most likely due to nonlinear deterministic structure or due to stochastic inputs to the system or random fluctuations in the parameters [94-96]. These tests are based on creating a surrogate data series from the original signal. Different methods proposed in literature [19, 94-97] for generating a surrogate data series from the original series, *e.g.*, phase randomized surrogates with the same power spectrum as the original signal but with no phase correlation, shuffled surrogates with the same distribution as the original signal, *etc.* Normally, the original time series and the surrogate time series are compared by introducing the null hypothesis as well as an appropriate discriminating statistics. The discriminating statistics is calculated for both the surrogate and the original data. If the calculated statistics of the surrogate and the original data are significantly different, the null hypothesis is rejected.

A proper discriminating statistics has been proposed by Schouten and van den Bleek [98] based on combination of the methods given by Kennel and Isabelle [95] and Takens [96]. The test compares the original and surrogate time series with respect to their short-time predictability in state space, Z_{avg}. Schouten and van den Bleek [98] showed that with a Z_{avg} value less than -3 the null hypothesis is rejected at 99% confidence level. Hence, the lower the value of Z_{avg}, the higher the significance of rejecting the time series as being generated by a linear stochastic process would be [6]. In this method, by random choosing of N_A pairs on the reconstructed attractor of the original time series which are in the neighborhood radius ε_0 and trace their distance on the attractor during a subsequent period of 10 % of the average cycle time, set A with N_A new distances is formed. With similar method for surrogate time series, set B with N_B new distances is created. With the set A of distances of the original time series and the set B of distances of the successive time series, the Mann-Whitney rank-sum (sometime referred to Wilcoxon) statistics is formed as [98]:

$$W = \sum_{i=1}^{N_A} \sum_{j=1}^{N_B} U(A_i - B_i) \tag{25}$$

where U is the Heaviside step function. If N_A and N_B are large enough (at least equal to 8 [16]), W has a symmetrical distribution with following mean:

$$\mu_W = \frac{N_A N_B}{2} \tag{26}$$

and variance:

$$\sigma_W^2 = \frac{N_A N_B (N_A + N_B + 1)}{12} \tag{27}$$

Consequently, the quantity:

$$Z = \frac{W - \mu_W}{\sigma_W} \tag{28}$$

is normally distributed with zero mean and unit variance, under the null hypothesis that the two observed samples A and B came from the same distribution. It can be concluded that as the Z_{avg} value becomes less than -3, it can be stated with more than 99% confidence that the pressure fluctuations in the original and surrogate time series did not come from the same underlying distribution function.

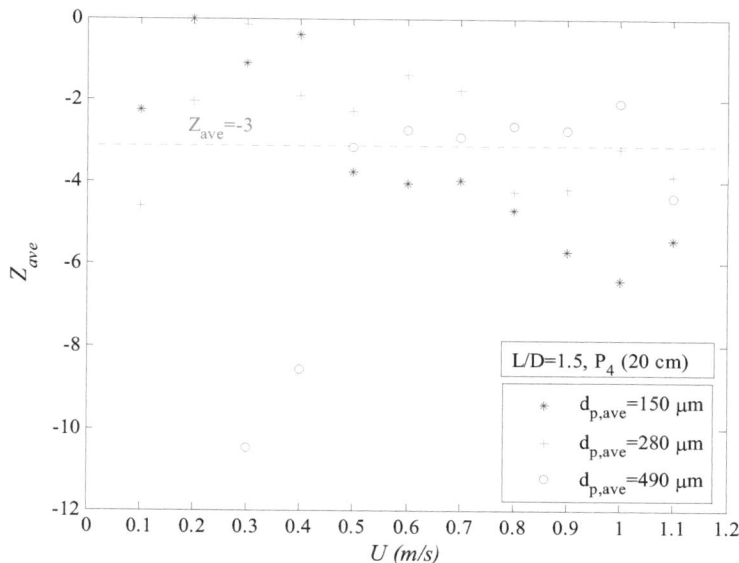

Figure 22: The Z_{avg} estimation of the measured pressure fluctuation versus superficial gas velocity for different sizes of particles, at measuring tap height 20 cm above the distributor and L/D=1.5.

Fig. **22** shows the Z_{avg} estimation of the pressure fluctuations measured 20 cm above distributor with gas velocity for different sizes of sands and initial aspect ratio L/D equal to 1.5. The nonlinearity test shows that for all three different sizes of particles, the Z_{avg} value initially increases then decreases with further increase in gas velocity. In addition, it can be found that the Z_{avg} value is not always less than -3 and nonlinearity cannot be claimed at all gas velocities. However, this does not mean that corresponding points necessarily are linear stochastic.

CONCLUSIONS

Average cycle frequency and Hurst exponent of the measured pressure fluctuations were calculated. A minimum in average cycle frequency of the pressure fluctuations indicated a minimum deviation from periodicity or, in other words, a minimum deviation from the larger structures, of the bed. It was concluded that while Hurst exponent at smaller fractal dimension represents a dynamic feature of macro structures, Hurst exponent at larger fractal dimension represents dynamic feature of finer structures of fluidization. The reciprocal of the break point in Hurst profile is similar to the main frequency of the bed.

The stationarity of the fluidized bed pressure fluctuations was proved by introducing the space time separation method. The method of reconstruction in the state space has applied theoretically based on the fact that the dynamic state of a system can be reconstructed from the time series of only one characteristic variable such as the local pressure in a fluidized bed. For practical applications, it also depends on the measurement techniques, which is considered properly.

First, the state space reconstruction parameters, *i.e.*, time delay and embedding dimension were determined for two different hydrodynamic states. Autocorrelation function and mutual information function have been used to determine the optimum value of the time delay. The false nearest neighbors and use of correlation dimension and time window were considered for choosing the minimum embedding dimension. It was shown that the values of the time delay and embedding dimensions were different for various methods.

Next, the nonlinear dynamical invariants, such as the correlation dimension and entropy, were evaluated by different proposed methods. It was shown that the results were different for different methods. The correlation dimensions results showed a low dimensional chaotic behaviour of fluidization systems. The estimated Kolmogorov entropy showed a similar trend of average cycle frequency and wide band energy with increasing gas velocity. As expected, the results showed a minimum entropy of the fluidization system when a minimum deviation from the macro structures (deviation from periodicity) of the bed existed. It was found that a minimum in average cycle frequency and wide band energy with an increase in velocity corresponds to the transition velocity between macro structures and finer structures of the bubbling fluidization system. It is found that the Kolmogorov entropy, wide band energy, and average cycle frequency are higher in smaller sizes of particles and this shows that contribution of the finer structures become more important in smaller particles. Method of surrogate was used to investigate nonlinearity of the bubbling fluidized bed pressure fluctuations. The results showed that nonlinearity cannot be distinguished for all superficial gas velocities.

NOMENCLATURE

AAD	average absolute deviation
ACF	autocorrelation function
b	escape time
C	correlation integral
d	embedding dimension
D	Bed diameter
DFT	discrete Fourier transform
D_C	correlation dimension
D_F	fractal dimension
D_{ML}	maximum likelihood estimation of the correlation dimension
E	energy of the signal
E_{WB}	wide band energy

FFT	fast Fourier transform
FNN	false nearest neighbours
f_s	sampling frequency, Hz
f_C	average cycle frequency, Hz
H	Horst exponent
H	Kolmogorov entropy, bits/s
H_2	Renyi entropy
H_{ML}	maximum likelihood estimation of the Kolmogorov entropy
I	mutual information
$I_{[t_1]}$	information in bits at time t_1
L	bed height; number of windows
M	number of points on attractor
m	manifold dimension
N	total number samples
N_f	number of points in the frequency domain
P	individual probability in Eq. 13
P_{xx}	power spectrum, kPa²/Hz
R	Rescaled Range function
r_i	random sample of M observation of C
S	standard deviation in Eq. 5
s	state vector, point on state space attractor
t	time, s
t_{min}	minimum correlation time, s
T_w	time window
WT	wavelet transform
X	Estimated Fourier transform of the measured pressure time series x
x	(pressure) signal (pa)

Greek Symbols

σ	standard deviation
Δt	time resolution, s
Θ	Heaviside function
ε	neighborhood radius around a point on attractor
τ	embedding time delay
ε_0	cutoff length
λ_d	eigenvalues

REFERENCES

[1]   S. Sasic, B. Leckner, and F. Johnsson, "Characterization of fluid dynamics of fluidized beds by analysis of pressure fluctuations", *Progr. Energy Combustion Sci.*, vol. 33, pp. 453-496, 2007.

[2]   J. van der Schaaf, J. C. Schouten, and C. M. van den Bleek, "Origin, propagation and attenuation of pressure waves in gas-solid fluidized beds," *Powder Tech.*, vol. 95, pp. 220-233, 1998,

[3]   L.T. Fan, T. C. Ho, S. Hiraoka, and W. P. Walawender, "Pressure fluctuations in a fluidized bed," *AIChE J.*, vol. 27, pp. 388-396, 1981.

[4]   J. Werther, "Measurement techniques in fluidized beds," *Powder Tech.*, vol. 102, pp. 15-36, 1999.

[5]   G. S. Lee and S. D. Kim, "Pressure fluctuations in turbulent fluidized beds," *J. Chem. Eng. Japan,* vol. 21, pp. 515-521, 1988.

[6]   F. Johnsson, R. C. Zijerveldb, J. C. Schoutenb, C. M. van den Bleek, and B. Lecknera, "Characterization of fluidization regimes by time-series analysis of pressure fluctuations", *Int. J. Multiphase Flow*, vol. 26, pp. 663-715, 2000.

[7]   H. T. Bi, J. R. Grace, and K. S. Lim, "Transition from bubbling to turbulent fluidization", *Ind. Chem. Res. Dev.*, vol. 34, pp. 4003-4008, 1995.

[8]   D. Bai, E. Shibuya, N. Nakagawa, and K. Kato, "Fractal Characteristics of gas-solids flow in a circulating fluidized bed," *Powder Tech.*, vol. 90, pp. 205-212, 1997.

[9]   R. Zarghami, "Conditional monitoring of fluidization quality in fluidized beds," Ph.D. Dissertation, University of Tehran, 2009.

[10]   H. E. Hurst, "Long-term storage capacity of reservoirs," *Trans. Am. Soc. Civ. Eng.*, vol. 116, pp. 770-799, 1951.

[11]   G. B. Zhao and Y. R. Yang, "Multiscale resolution of fluidized-bed pressure fluctuations," *AIChE J.*, vol. 49, pp. 869-882, 2003.

[12]   A. Chen, and H. T. Bi, "Pressure fluctuations and transition from bubbling to turbulent fluidization", *Powder Tech.*, vol. 133, pp. 237-246, 2003.

[13]   S. Alberto, C. Felipe, and S. C. S. Rocha, "Time series analysis of pressure fluctuation in gas-solid fluidized beds," *Braz. J. Chem. Eng.*, vol. 21, pp. 497-507, 2004.

[14]   H. T. Bi, I. A. Abba, N. Ellis, and J. R. Grace, "A state-of-the-art review of gas-solids turbulent fluidization," *Chem. Eng. Sci.*, vol. 55, pp. 4789-4825, 2000

[15]   Y. O. Chong, D. P. O'Dea, E. T. White, P. L. Lee, and L. S. Leung, "Control of quality of fluidization in a tall bed using the variance of pressure fluctuations," *Powder Tech.*, vol. 53, pp. 237-246, 1987.

[16]   J. R. van Ommen, J. C. Schouten, and C. M. van den Bleek, "An early-warning-method for detecting bed agglomeration in fluidized bed combustors," Paper No. FBC99-0150, *Proc. 15th Int. Conf. Fluidized Bed Combustion*, R. B. Reuther, ed., ASME, New York, 1999

[17]   H. T. Bi and J. R. Grace, "Effect of measurement method on the velocities used to demarcate the onset of turbulent fluidization," *Chem. Eng. J.*, vol. 57, pp. 261-271, 1995.

[18]   A. Fraser and H. Swinney, "Independent coordinates for strange attractors from mutual information," *Phys. Rev. A*, vol. 33, pp. 1134-1140, 1986.

[19]   H. Kantz, T. Schreiber, Ed., *Nonlinear time series analysis*. Cambridge University Press, 2002.

[20]   P. S. Addison, *Fractals and Chaos: An Illustrated Course*. IOP Publishing Ltd., 2005.

[21]   L. T. Fan, D. Neogi, M. Yashima, and R. Nassar, "Stochastic analysis of a three-phase fluidized bed: Fractal Approach," *AIChE J.*, vol. 36, pp. 1529-1535, 1990.

[22]   F. Franca, M. Acikgoz, R. T. Lahey, and A. Clausse, "The use of fractal techniques for flow regime identification," *Int. J. Multiphase Flow*, vol. 17, pp. 545-552, 1991.

[23]   J. Drahos, F. Bradka, M. Puncochar, "Fractal behaviour of pressure fluctuations in a bubble column," *Chem. Eng. Sci.*, vol. 47, pp. 4069-4075, 1992.

[24]   F. J. Cabrejos and G. E. Klinzing, "Characterization of dilute flows using the rescaled range analysis," *Powder Tech.*, vol. 84, pp. 139-156, 1995.

[25]   C. L. Briens, L. A. Briens, J. Hay, C. Hudson, and A. Margaritis, "Hurst's analysis to detect minimum fluidization and gas maldistribution in fluidized beds," *AIChE J.*, vol. 43, pp. 1904-1908, 1997.

[26]   A. I. Karamavruc and N. N. Clark, "A fractal approach for interpretation of local instantaneous temperature signals around a horizontal heat transfer tube in a bubbling fluidized bed," *Powder Tech.*, vol. 90, pp. 235-244, 1997.

[27]   J. C. Schouten, F. Takens, and C. M. van den Bleek, "Estimation of the dimension of a noisy attractor," *Phys. Rev. E*, vol. 50, pp. 1851-1861, 1994.

[28]   M. L. M. van der Stappen, J. C. Schouten, and C. M. van den Bleek, "Application of deterministic chaos theory in understanding the fluid dynamic behavior of gas-solids fluidization," *AIChE Symp. Ser.*, vol. 89, pp. 91-102, 1993.

[29]   M. L. M. van der Stappen, "Chaotic hydrodynamics of fluidized beds," Ph.D. Thesis, Delft University Press, Delft, 1996.

[30]   J. Verloop, P. M. Heertjes, "Periodic pressure fluctuations in fluidized beds," *Chem. Eng. Sci.*, vol. 29, pp. 1035-1042, 1974.

[31]   T. E. Broadhurst and H. A. Becker, "Measurement and spectral analysis of pressure fluctuations in slugging beds," In: Keairns, D.L. (Ed.), *Fluidization Tech.*, *Vol. 1, Engineering Foundation*, New York, pp. 63-85, 1976.

[32]   S. Satija and L. S. Fan, "Characteristics of slugging regime and transition to turbulent regime for fluidized beds of large coarse particles", *AIChE J.*, vol. 31, pp. 1554-1562, 1985.

[33]   F. Johnsson, A. Svensson, S. Andersson, and B. Leckner, "Fluidization regimes in boilers," *Fluidization VIII*, Tours, 129, 1995.

[34]   A. Svensson, F. Johnsson, and B. Leckner, "Fluidization regimes in non-slugging fluidized beds: the influence of pressure drop across the air distributor," *Powder Tech.*, vol. 86, pp. 299-312, 1996.

[35]   R. A. Newby and D. L. Keairns, "Test of the scaling relationships for fluid-bed dynamics," In: K. Ostergaard and A. Sorensen, A. (Eds.), *Fluidization V. Engineering Foundation*, New York, pp. 31-38, 1986.

[36]   L. R. Glicksman, M. Hyre, and K. Woloshun, "Simplified scaling relationships for fluidized beds," *Powder Tech.*, vol. 77, pp. 177-199, 1993.

[37]   B. R. Bakshi, H. Zhong, P. Jiang, and L. S. Fan, "Analysis of flow in gas-liquid bubble columns using multi-resolution methods," *Trans. Inst. Chem. Eng.*, vol. 73, Part A, pp. 608-614, 1995.

[38]   Z. He, W. Zhang, K. He, and B. Chen, "Modeling pressure fluctuations *via* correlation structure in a gas-solids fluidized bed," *AIChE J.*, vol. 43, pp. 1914-1920, 1997.

[39]   X. Lu and H. Li, "Wavelet analysis of pressure fluctuation signals in a bubbling fluidized bed," *Chem. Eng. J.*, vol. 75, pp. 113-119, 1999.

[40]   J. Li, "Compromise and resolution-exploring the multi-scale nature of gas-solid fluidization," *Powder Tech.*, vol. 111, pp. 50-59, 2000.

[41]   J. Ren, Q. Mao, J. Li, and W. Lin, "Wavelet Analysis of Dynamic Behavior in Fluidized Beds," *Chem. Eng. Sci.*, vol. 56, pp. 981-988, 2001.

[42]   Q. Guo, G. Yue, and J. Werther, "Dynamics of pressure fluctuation in a bubbling fluidized bed at high temperature," *Ind. Eng. Chem. Res.*, vol. 41, pp. 3482-3488, 2002.

[43]   H. Li, "Application of wavelet multi-resolution analysis to pressure fluctuations of gas-solid two-phase flow in a horizontal pipe," *Powder Tech.*, vol. 125, pp. 61-73, 2002.

[44]   T. Y. Yang and L. Leu, "Study of transition velocities from bubbling to turbulent fluidization by statistic and wavelet multi-resolution analysis on absolute pressure fluctuations," *Chem. Eng. Sci.*, vol. 63, pp. 1950-1970, 2008.

[45]   C. S. Daw, W. F. Lawkins, D. J. Downing, and N. E. Clapp, Jr., "Chaotic characteristics of a complex gas-solid flow," *Phys. Re. A*, vol. 41, pp. 1179-1181, 1990.

[46]   C. S. Daw and J. S. Halow, "Characterization of voidage and pressure signals from fluidized beds using deterministic chaos theory," In: E. J. Anthony, (Ed.), *Proc. Eleventh Int. Conf. Fluidized Bed Combustion*, ASME, New York, vol. 1, pp. 777-786, 1991.

[47]   C. S. Daw and E. J. Kostelich, "Self-Organization and chaos in a fluidized bed," *Phys. Rev. Lett.*, vol. 75, pp. 2308-2311, 1995.

[48]   L. T. Fan, Y. Kang, D. Neogi, and M. Yashima, "Fractal analysis of fluidized particle behavior in liquid-solid fluidized beds," *AIChE J.*, vol. 39, pp. 513-517, 1993.

[49]   C. M. van den Bleek and J. C. Schouten, "Can deterministic chaos create order in fluidized bed scale-up?", *Chem. Eng. Sci.*, vol. 48, pp. 2367-2373, 1993.

[50]   C. M. van den Bleek and J. C. Schouten, "Deterministic chaos: a new tool in fluidized bed design and operation," *Chem. Eng. J.*, vol. 53, pp. 75-87, 1993.

[51]   D. P. Skrzycke, K. Nguyen, and C. S. Daw, "Characterization of the fluidization behavior of different solid types based on chaotic time-series analysis of pressure signals," In: L. Rubow and G. Commonwealth (Eds.), *Proc. Twelfth Int. Conf. Fluidized Bed Combustion*, ASME Book No. I0344B, New York, pp. 155, 1993.

[52]   J. M. Hay, B. H. Nelson, C. L. Briens, and M. A. Bergougnou, "The calculation of the characteristics of a chaotic attractor in a gas-solid fluidized bed," *Chem. Eng. Sci.*, vol. 50, pp. 373-380, 1995.

[53]   D. Bai, E. Shibuya, Y. Masuda, N. Nakagawa, and K. Kato, "Flow structure in a fast fluidized bed," *Chem. Eng. Sci.*, vol. 51, pp. 957-966, 1996.

[54]   D. Bai, E. Shibuya, N. Nakagawa, and Kato, K., "Fractal Characteristics of gas-solids flow in a circulating fluidized bed," *Powder Tech.*, vol. 90, pp. 205, 1997.

[55]   D. Bai, H. T. Bi, and J. R. Grace, "Chaotic behavior of fluidized beds based on pressure and voidage fluctuations," *AIChE J.*, vol. 43, pp. 1357-1361, 1997.

[56]   G. B. Zhao, J. Z. Chen, and Y. R. Yang, "Predictive model and deterministic mechanism in a bubbling fluidized bed," *AIChE J.*, vol. 47, pp. 1524-1532, 2001.

[57]   F. Takens, "Detecting strange attractors in turbulence," *Lecture Notes in Mathematics, Dynamical Systems and Turbulence,* D. A. Rand and L. S. Young, (Eds.), Springer Verlag, Berlin, Germany, vol. 898, pp. 366-381, 1981.

[58]   S. W. Tam and M. K. Devine, "Is there a strange attractor in a fluidized bed?" In: N. B. Abraham, A. M. Albano, A. Passamante, P. E. Rapp, (Eds.). Measures of Complexity and Chaos, Plenum, New York, pp. 193-197, 1989.

[59]   N. Letaief, C. Roze, and G. Gouesbet, "Noise/chaos distinction applied to the study of a fluidized bed," *J. Physique II France*, vol. 5, pp. 1883-1899, 1995.

[60] R. Zarghami, N. Mostoufi, and R. Sotudeh-Gharebagh, "Nonlinear characterization of pressure fluctuations in fluidized beds", *Ind. Eng. Chem. Res.*, vol. 47, pp. 9497-9507, 2008.

[61] P. Grassberger and I. Procaccia, "Dimensions and entropies of strange attractors from a fluctuating dynamics approach," *Physica D: Nonlinear Phenomena*, vol. 13, pp. 34-54, 1984.

[62] J. P. Eckmann, S. Oliffson-Kamphorst, D. Ruelle, and S. Ciliberto, "Lyapunov exponents from time series", *Phys. Rev. A*, vol. 34, pp. 4971-4979, 1986.

[63] H. D. I. Abarbanel, R. Brown, J. B. Kadtke, "Prediction in chaotic nonlinear systems: methods for time series with broadband Fourier spectra," *Phys. Rev. A*, vol. 41, pp. 1742-1807, 1990.

[64] J. R. van Ommen, J. C. Schouten, M. L. M. van der Stappen, and C. M. van den Bleek, "Response characteristics of probe-transducer systems for pressure measurements in gas-solid fluidized beds: how to prevent pitfalls in dynamic pressure measurements", *Powder Tech.*, vol. 106, pp. 199-218, 1999.

[65] H. Bergh and H. Tijdeman, "Theoretical and experimental results for the dynamic response of pressure measuring systems", Report NLR-TRF.238, National Aero- and Astronautical Research Institute, Amesterdam, The Netherlands, 1965.

[66] J. F. Davidson, Symposium on Fluidization Discussion, *Trans. Instn. Chem. Engrs.*, vol. 39, pp. 230-232, 1961.

[67] Collins, R., "An extension of Davidson's theory of bubbles in fluidized beds," *Chem. Eng. Sci.*, vol. 20, pp. 747-755, 1965.

[68] P. S. B. Stewart, "Isolated bubbles in fluidized beds-theory and experiment," *Trans. Inst. Chem. Eng.*, vol. 46, pp. T60, 1968.

[69] H. Kage, N. Iwasaki, H. Yamaguchi, and Y. Matsuna, "Frequency analysis of pressure fluctuation in fluidized bed plenum," *J. Chem. Eng. Japan*, vol. 24, pp. 76-81, 1991.

[70] A. P. Baskakov, V. G. Tuponogov, and N. F. Filippovsky, "A study of pressure fluctuations in a bubbling fluidized bed," *Powder Tech.*, vol. 45, pp. 113-117, 1986.

[71] H. T. Bi, J. R. Grace, and J. Zhu, "Propagation of pressure waves and forced oscillations in gas-solid fluidized beds and their influence on diagnostics of local hydrodynamics," *Powder Tech.*, vol. 82, pp. 239-253, 1995.

[72] R. Roy, "Pressure fluctuations and scal-up," Thesis, University of Cambridge, 1989.

[73] D. Musmarra, M. Poletto, S. Vaccaro, R. Clift, "Dynamic waves in fluidized beds", *Powder Tech.*, vol. 82, pp. 255-268, 1995.

[74] J. R. van Ommen, J. van der Schaaf, J. C. Schouten, B. G. M. van Wachem, M. O. Coppens, and C. M. van den Bleek, "Optimal placement of probes for dynamic pressure measurements in large-scale fluidized beds," *Powder Tech.*, vol. 139, pp. 264-276, 2004.

[75] L. P. Leu and C. W. Lan, "Measurements of pressure fluctuations in two-dimensional gas solid fluidized beds at elevated temperatures," *J. Chem. Eng. Jpn.*, vol. 23, pp. 555-562, 1990.

[76] J. F. Davidson, "The two-phase theory of fluidization: successes and opportunities," *AIChE Symp. Ser.*, vol. 87, No. 281, pp. 1-12, 1991.

[77] A. P. Baskakov, A. V. Mudrichenko, N. F. Filipovskii, "On the mechanism of pressure pulsation in a bubbling fluidized bed," *J. Eng. Phys. Thermophys.*, vol. 66 (1), pp. 30-33, 1994.

[78] B. Mandelbrot, *The Fractal Geometry of Nature*, Freeman, San Francisco, 1982.

[79] A. Provenzale, L. A. Smith, R. Vio, and G. Murante, "Distinguishing between low dimensional dynamics and randomness in measured time series," *Physica D: Nonlinear Phenomena*, vol. 58, 31-49, 1992.

[80] N. Packard, J. Crutchfield, D. Farmer, and R. Shaw, "Geometry from a time series," *Phys. Rev. Lett.*, vol. 45, pp. 712-716, 1980.

[81] Th. Buzug, G. Pfister, "Comparison of algorithms calculating optimal parameters for delay time coordinate," *Physica D: Nonlinear Phenomena*, vol. 58, pp. 127-137, 1992.

[82] H. Whitney, "Differentiable manifolds," *Ann. Math.*, vol. 37, pp. 645-680, 1936.

[83] T. Sauer, J. A. Yorke, and M. Casdagli, "Embedology", *J. Stat. Phys.*, vol. 65, pp. 579-616, 1991.

[84] M. B. Kennel, R. Brown, and H. D. I. Abarbanel, "Determining embedding dimension for phase-space reconstruction using a geometrical construction," *Phys. Rev. A*, vol. 45, pp. 3403-3411, 1992.

[85] P. Grassberger and I. Procaccia, "Characterization of strange attractors," *Phys. Rev. Lett.*, vol. 50, pp. 346-349, 1983

[86] P. Grassberger and I. Procaccia, "Measuring the strangeness of strange attractors," *Physica D: Nonlinear Phenomena*, vol. 9, pp. 189-208, 1983.

[87] J. Theiler, "Spurious dimension from correlation algorithms applied to limited time series data," *Phys. Rev. A*, vol. 34, pp. 2427-2432, 1986.

[88] L. Cao, "Practical method for determining the minimum embedding dimension of a scalar time series," *Physica D: Nonlinear Phenomena*, vol. 110, pp. 43-50, 1997.

[89] J. E. Jackson, A user guide to principal components, John Wiley & Sons, 1991.

[90] D. S. Broomhead and G. P. King, "Extracting qualitative dynamics from experimental data," *Physica D,* vol. 20, pp. 217-236, 1986.

[91] J. R. van Ommen, M. O. Coppens, C. M. van den Bleek, and J. C. Schouten, "Early warning of agglomeration in fluidized beds by Attractor Comparison," *AIChE J.*, vol. 46, pp. 2183-2197, 2000.

[92] F. Takens, "On the numerical determination of the dimension of an attractor," In: B. L. J. Braaksma, H. W. Broer, F. Takens, (Eds.), *Proc. Dynamical Systems and Bifurcations*, Lecture Notes in Mathematics, Springer Verlag, Berlin, vol. 1125, pp. 99-106, 1985.

[93] J. C. Schouten, F. Takens, and C. M. van den Bleek, "Maximum likelihood estimation of the entropy of an attractor," *Phys. Rev. E*, vol. 49, pp. 126-129, 1994

[94] J. Theiler, S. Eubank, A. Longtin, B. Galdrikian, and J. D. Farmer, "Testing for nonlinearity in time series: the method of surrogate data," *Physica D: Nonlinear Phenomena*, vol. 58, pp. 77-94, 1992.

[95] M. B. Kennel and S. Isabelle, "Method to distinguish possible chaos from colored noise and to determine embedding parameters," *Phys. Rev. A*, vol. 46, pp. 3111-3118, 1992.

[96] F. Takens, "Detecting nonlinearities in stationary time series," *Int. J. Bifurcation Chaos*, vol. 3, pp. 241-256, 1993

[97] T. Schreiber and A. Schmitz, "Review Paper: Surrogate time series," *Physica D: Nonlinear Phenomena*, vol. 142, pp. 346-382, 2000.

[98] J. C. Schouten and C. M. van den Bleek, "Monitoring the quality of fluidization using the short-term predictability of pressure fluctuations," *AIChE J.*, vol. 44, pp. 48-60, 1998.

INDEX